Kister Distillation Symposium 2015 (Full Papers)

Topical Conference at the 2015 AIChE Spring Meeting and 11th Global Congress on Process Safety

Austin, Texas, USA
26-30 April 2015

ISBN: 978-1-5108-1241-3

Printed from e-media with permission by:

Curran Associates, Inc.
57 Morehouse Lane
Red Hook, NY 12571

Some format issues inherent in the e-media version may also appear in this print version.

Printed by Curran Associates, Inc. (2015)

For permission requests, please contact AIChE
at the address below.

AIChE
3 Park Avenue
New York, NY 10016-5991

Phone: (203) 702-7660
Fax: (203) 775-5177

www.aiche.org

Additional copies of this publication are available from:

Curran Associates, Inc.
57 Morehouse Lane
Red Hook, NY 12571 USA
Phone: 845-758-0400
Fax: 845-758-2634
Email: curran@proceedings.com
Web: www.proceedings.com

TABLE OF CONTENTS

DISTILLATION 2015

HAS DISTILLATION TECHNOLOGY REACHED THE GOLDEN AGE?

In his keynote address at this AIChE Spring Meeting, Dr. Schlanger posed the question whether this is the Golden Age of Chemical Engineering. A follow up for our specific interest is whether this is the Golden Age of distillation. Webster's dictionary defines "golden age" as "a period of great happiness, prosperity, and achievement". Some of us remember the days when we did not have simulations, no structured packings, no high capacity trays, no advanced controls, and no gamma scans. There is little doubt that if the golden age is not here yet, we are a lot closer to it than we were 50 years ago.

Yet not everything is golden. In a recent distillation class that I taught I was hit with questions like why do people dump random packings and what is the purpose of using trays in the towers. These did not come from passer-bys, but from graduate engineers who are proficient in running simulations. With the emphasis on new and exciting chemical engineering fields, attention is diverted away from the more traditional areas, like distillation, with distillation equipment taking the greatest hit. So our challenge is to help answer many of these questions raised by graduates entering our field. Yes, we have the tools to get us to and to sustain the golden age, but we must also rise up to the challenge of bringing our graduate engineers to effectively apply these tools. Excellence, innovation, and learning from past lessons have ushered in the golden age, and are the key for keeping us there.

In this 2015 Distillation Topical Conference we address some of the major challenges. We have many papers reporting improvements and innovations in debottlenecking, simulation, troubleshooting, control, and optimization that will help us sustain the golden age of distillation. We have papers describing new and exciting distillation processes, as well as cutting edge technology in exciting applications like energy utilization and process control and in fundamental modeling of equipment and processes. We will hear of wonderful developments in tray and packing technology, including the latest on the state of the art packings and the introduction of new fixed valve trays, and modeling work that will help us get better design, debottlenecks, and evaluation. We have papers addressing some topics that have given us grief but were infrequently addressed in the literature, such as tray vibrations, multicomponent batch distillation, adapting refinery towers to the new fracked feedstocks, as well as new insights into fouling in towers, coke formation mechanisms, tower motion, sensitivity to changes in operating parameters, chimney trays, as well as of invaluable applications of gamma scans, x-rays, and thermal imaging.

Our 2015 Distillation Topical Conference has great papers from industry and academia, by some of today's legends in the field as well as by some of the most promising youngsters that will turn into tomorrow's top professionals. We are fortunate to have authors from 15 different countries.

Our leading session is dedicated to the legacy of one of the greats of distillation technology, Elzear or Zaz Lemieux, who left our world just over a year ago. Zaz was a great contributor to distillation technology, FRI, and to the profession. Two of his major contribution areas were sieve tray entrainment and baffle trays. For about 50 years his data on entrainment from sieve trays at close tray spacings, and from sieve trays with large diameter holes, have been the only data in the published literature, more than 50 years later, and were building blocks in most entrainment and flood correlations. In the early 1980's, we followed up on his entrainment work at the University of NSW in Australia, we actually used the method he developed to measure entrainment, and were impressed both by his wonderful data and the valuable tool he provided us with to expand on his leading work.

There were many like Zaz that dedicated their lives to our field and profession and in doing so paved the way to today's great distillation technology. Let us always remember their legacy on the path to the golden age of distillation. Let us hope that if we have not yet reached the golden age of distillation, we will reach it soon, and that once it is ushered in, it will stay with us for a long time. The theme of our Distillation Topical Conference engulfs excellence, improvement, and problem-solving to overcome challenges. Joined together, they can provide the engineering solutions that will bring us to a green planet and preserve our valuable resources for our children, grandchildren, and future generations to come....to the golden age of distillation.

Henry Kister
Chair
Distillation 2015

Bibliography of Elzear Lemieux

By

Anand Vennavelli
S Chambers
Tony Cai
Ken McCarley

Fractionation Research Inc.

424 S. Squires Street, Suite 200

Stillwater, OK 74074

11 June 2015

**Presented at the Kister Distillation Symposium,
AIChE Spring Meeting, Austin, TX, April 27, 2014**

BIOGRAPHY AND CAREER HIGHLIGHTS

Mr. Elzear Lemieux had a B.A. degree from Assumption College and B.S. and M.S. degrees in Chemical Engineering from Worcester Polytechnic Institute.

Mr. Lemieux began his career at M.W. Kellogg in 1950. Throughout his career, he was primarily involved in the design of trays and packings, and vessel design, in general. He was a strong proponent for the use of sieve trays in the chemical processing industry, at a time when sieve trays were new uncharted territory.

He and his team performed extensive testing of sieve tray capacity at the M.W. Kellogg laboratory in Piscataway, New Jersey. These tests studied and quantified the effects of hole diameter, tray spacing, weir rate, and vapor rate on tray capacity and they also led to an organized set of correlations for sieve tray capacity and downcomer backup.

Figures 1-3 are photographs of Mr. Lemieux. Figure 4 is a photograph of the pilot plant at Piscataway, NJ where Mr. Lemieux led numerous experiments. Figure 5 is a picture of tray designed by Mr. Lemieux specifically for the O2-He column at the New Jersey Piscataway pilot plant.

Mr. Lemieux was a supporter of FRI since its inception. He served on the FRI Technical Advisory Committee (TAC) from 1960 through 1975 for the M.W. Kellogg Company. He was extremely interested in basic fundamental research on tray and packing capacities and efficiencies. He retired from M.W. Kellogg in 1982, but continued to support FRI and attended the FRI TAC meetings. He was Kellogg's screener of FRI Topical and Progress Reports, until he stopped consulting for Kellogg in 1999.

He and his wife of 65 years, Claire, were very active members in the community and especially in their church, St. John Vianney Catholic Church in Houston. Mr. Lemieux was an avid golfer, clock & antique collector and loved to travel with his wife.

His wife passed away in February 2013. Mr. Lemieux passed away in November 2013.

This paper describes some of Mr. Lemieux's contributions to separations engineering.

Figure 1: Dr. Lemieux at Graduation

Figure 3: Dr. Lemieux at the pilot plant in NJ

Figure 2: Dr. Lemieux at his desk in Houston

Figure 4: Pilot plant at Piscataway, NJ

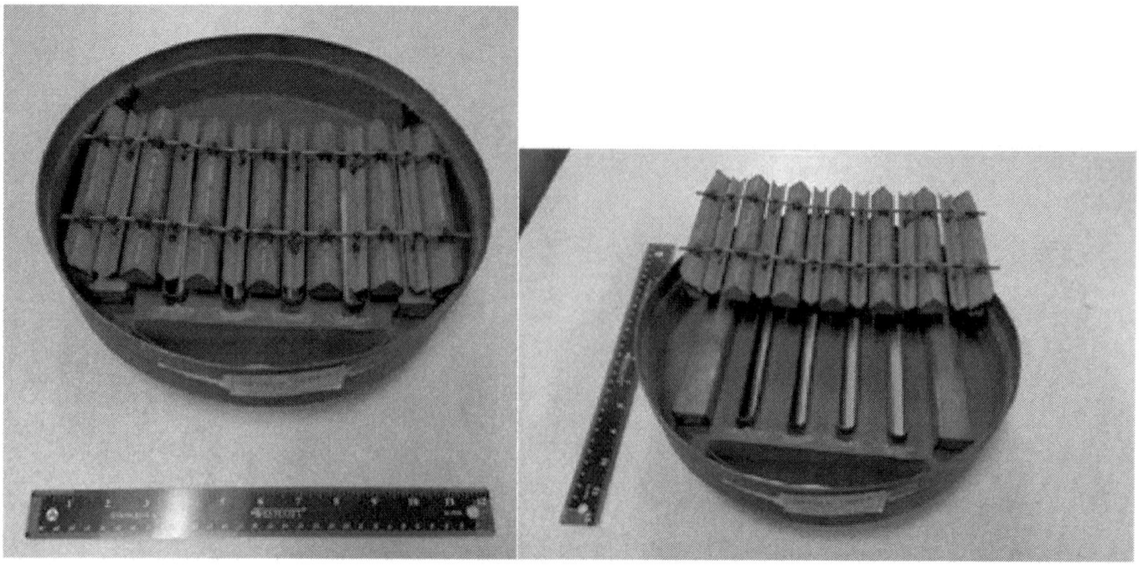

Figure 5:Photographs of the copper trays designed by Mr. Lemieux for the O2-He tower at the NJ pilot plant. The tray employed tunnel bubble caps and the device on the top was used as a de-entrainment device.

SELECTED PAPERS OF ELZEAR LEMIEUX

Mr. Lemieux had worked on diverse list of topics. The most published topics include entrainment, capacity, pilot plants, close tray spacings, bubble cap trays, dirty services, sieve trays, refinery fractionators, baffle trays, and downcomer backup. In the following pages, key elements of some of his papers are presented.

The Performance of Catalytic Cracking Unit Fractionating Towers - 1954 [1]

A schematic of the cracking-unit fractionation tower is shown in Figure 6. Data from 12 such operating units were obtained and used. An extensive research effort was expended to gather over 9 years of routine operating data and on occasional special tests. Furthermore, the data were collected in large towers that ranged 13 to 20 ft in diameter rendering the data extremely useful for industrial applications. New correlations for predicting ASTM 5%-95% point difference for the two sections were developed, as shown in Figure 7. Guidelines for troubleshooting baffle sections of the fractionating towers were also presented in the paper.

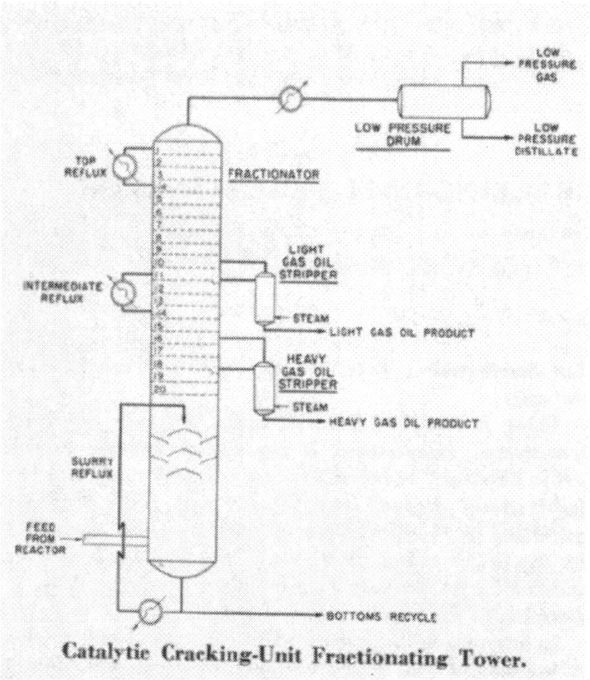

Figure 6: Schematic of a cracking-unit fractionation tower.

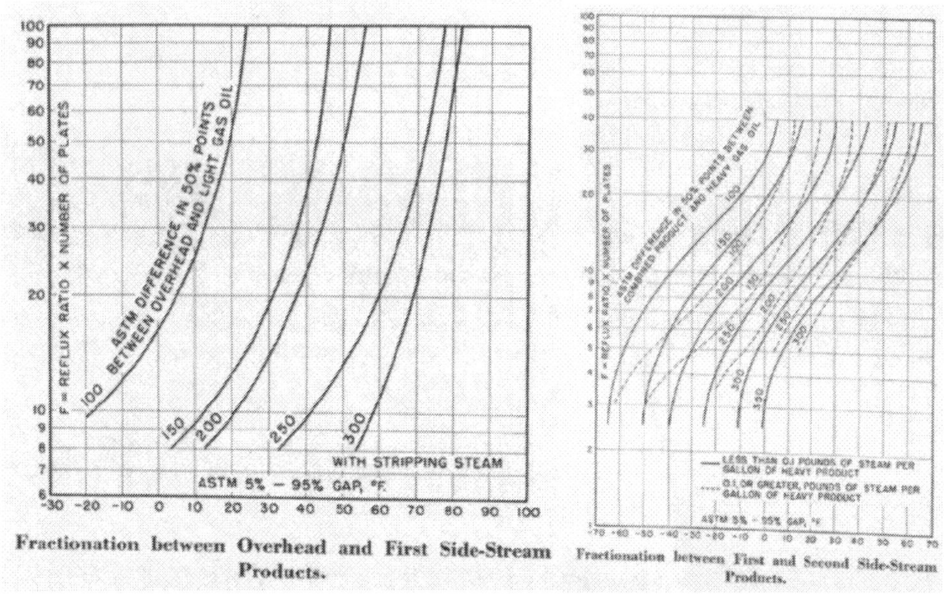

Figure 7: New correlations for predicting ASTM 5%-95% points difference for 2 sections

Entrainment from Bubble Cap Trays – 1956 [2]

This paper discusses the experiments on bubble cap trays in a rectangular column shown in Figure 8. The key results from the experiments showed that the entrainment from bubble cap trays (a) increases with decrease in the liquid-path length, (b) decreases with increasing cap-spacing, (c) does not change with decreasing the number of slots while maintaining the bubbling area constant. Furthermore, as shown in Figure 9, the entrainment curves showed a maximum or minimum when the liquid rate was varied at a given air-rate. Therefore, a surprising yet important finding was that entrainment could increase or decrease with increasing liquid rate.

Figure 8: The experimental setup for measuring entrainment from bubble cap trays. The figure shows a rectangular column equipped with bubble cap trays.

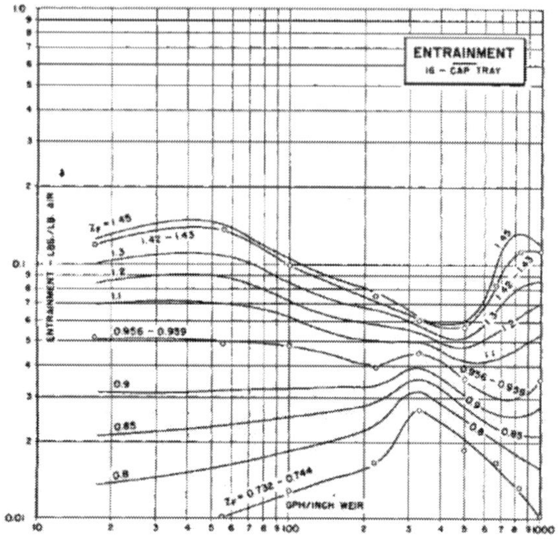

Figure 9: Effect of Liquid Rate on entrainment on bubble cap trays.

New Data on Entrainment From Perforated Trays at Close Spacings – 1960 [3]

Froth entrainment at close tray spacing becomes extremely important to characterize flood mechanisms. In this study, a number of key variables were studied at three different low tray spacings (6", 9", 12"). The variables include weir height (0", 1", 2"), hole diameter (9/16", ½"), slope (1 in/ft of tray length), plate thickness (1/8", 3/16").

Quantitative relationships between entrainment and tray design variables such as weir height, vapor rate, hole size, tray spacing and hole area were presented.

Results indicated that the dependence of entrainment on liquid rate was a function of the vapor rate, and entrainment could increase or decrease with increasing liquid rate depending on the vapor rate.

Entrainment was not affected by sloping the trays at low vapor rate; however, entrainment increased at high vapor rates. Tray thickness did not affect entrainment for the trays tested.

Performance of Perforated Trays in Dirty Service - 1964 [4]

The turndown capabilities of sieve trays, particularly in dirty services, were compared to bubble cap trays. Sieve tray data from commercial columns using 3/32-in and ½-in holes were studied. Sieve trays offered advantages of reduced cost and low pressure drop, and retained similar performance as bubble cap trays. Even with dirty service, trays remained clean even after catalyst carryover was detected.

Perforated Tray Performance - 1969 [5]

This paper provides entrainment data on sieve trays. A key finding was that entrainment for ½-inch holes (a) increased with decrease in hole area at low liquid rates, and (b) was independent of hole area at high liquid rates, as shown in Figure 11. Furthermore, entrainment measured on larger 1-in holes was greater than that on ½-inch holes. Tray froth heights observed on 1-inch hole trays were consistently higher than those observed on ½-inch hole trays.

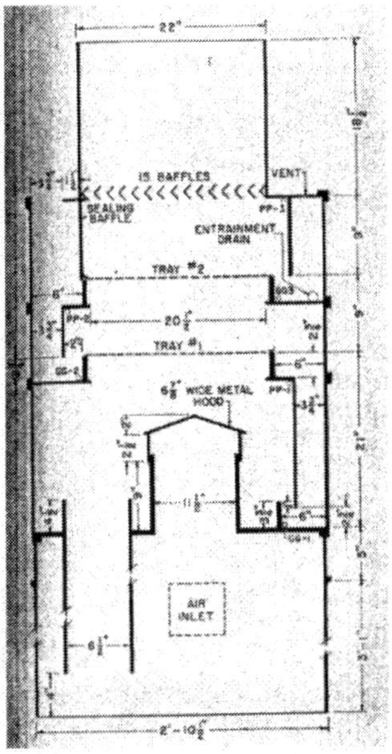

Figure 10: Experiemental setup used to obtain new data on entrainment from perforated trays at close tray spacings. [3]

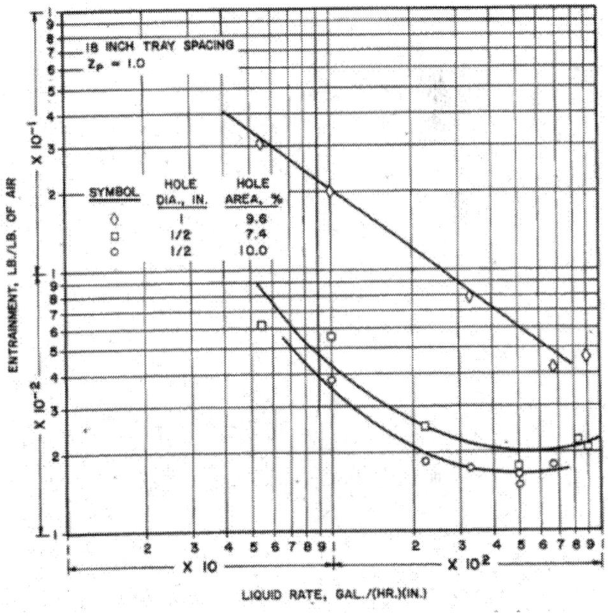

Figure 11: Entrainment data on perforated trays

Data for Tower Baffle Design - 1983 [6]

This paper includes flooding data on segmental and shower deck baffle trays using the air water system. The flood curves are shown in Figure 12 for segmental baffles and in Figure 13 for shower-deck baffles. A comparison with commercial data was also included in the study, as shown in Figure 14.

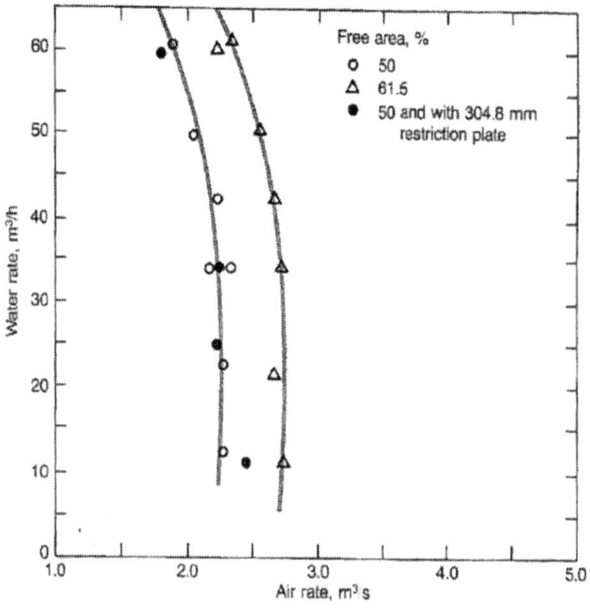

Figure 12: Flooding rates for segmental baffles

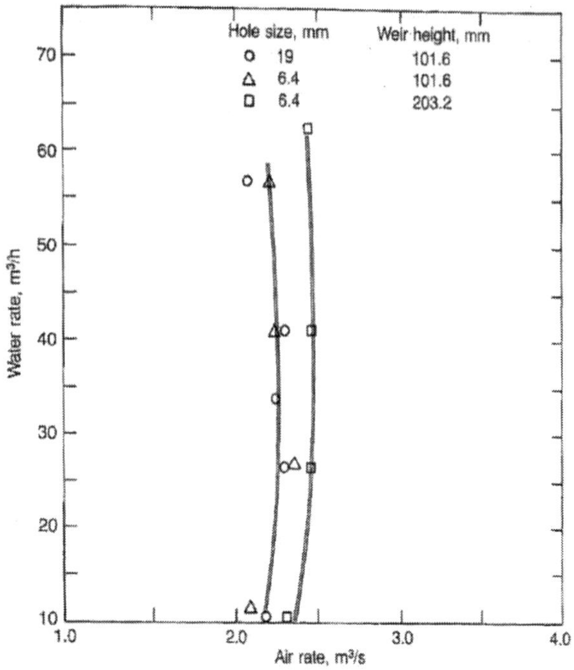

Figure 13: Flooding rates for shower-deck baffles

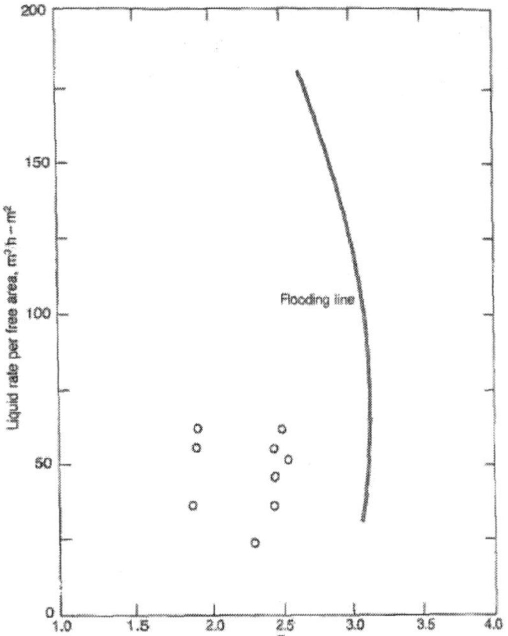

Fig. 13—Commercial data for segmental baffles in nine units compared with the air-water flooding line (50% free area).

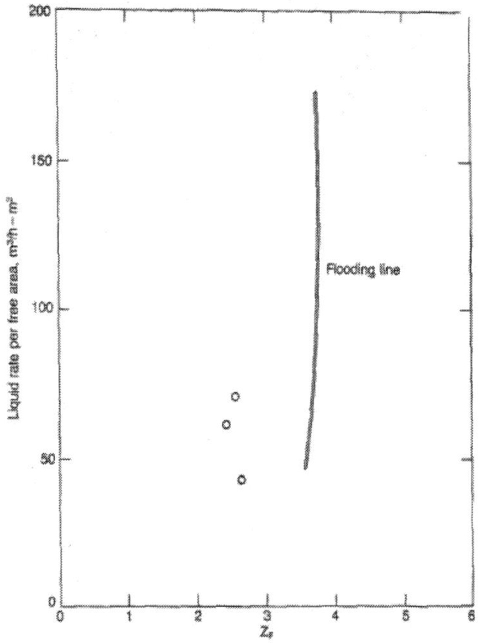

Fig. 14—Commercial data for shower-deck baffles also are compared with the air-water flooding line.

Figure 14: Comparison with commercial data

Acknowledgements

The authors would like to acknowledge Mr. Lemieux's son, Elzear J. Lemieux Jr, and Mr. Lemieux's friends and colleagues, particularly Ken Blanchard, Stann Stirling, Joe Parker, and Henry Kister for providing content, photographs, and references to this paper.

REFERENCES

[1] G.S. Houghland, E.J Lemieux, and W.C. Schreiner, "The Performance of Catalytic Cracking-Unit Fractionating Towers," in *19th Mid-Year Meeting of the American Petroleum Institute's Division of Refining*, May 1954.

[2] P.T. Atteridg, E.J. Lemieux, W.C. Scheiner, and R.A. Sundback, "Entrainment from Bubble-Cap Trays," *AIChE Journal*, vol. 2, no. 1, pp. 3-12, March 1956.

[3] L. Friend, E.J. Lemieux, and W.C. Schreiner, "New Data on Entrainment From Perforated Trays at Close Spacings," *Chemical Engineering*, vol. 67, no. 22, pp. 101-108, October 1960.

[4] R.S Eagle and E.J. Lemieux, "Performance of Perforated Trays in Dirty Service," *Chemical Engineering Progress*, vol. 66, no. 10, pp. 74-80, October 1964.

[5] E.J. Lemieux and L.J Scotti, "Perforated Tray Performance," *Chemical Engineering Progress*, vol. 65, no. 3, pp. 52-58, March 1969.

[6] E.J. Lemieux, "Data for tower baffle design," *Hydrocarbon Processing*, pp. 106-111, September 1983.

Flow-induced vibration of trays

Richard Darton[1], Wensong Zhang[1,2] and Paul Taylor[1]

1, Engineering Science, University of Oxford, UK
2, Centre for Offshore Foundation Systems, University of Western Australia, Australia

Presented at the 2015 Kister Distillation Symposium
Austin Texas, April 27th 2015. The Legacy of Dr Elzear Lemieux

1

Flow-induced vibration of trays

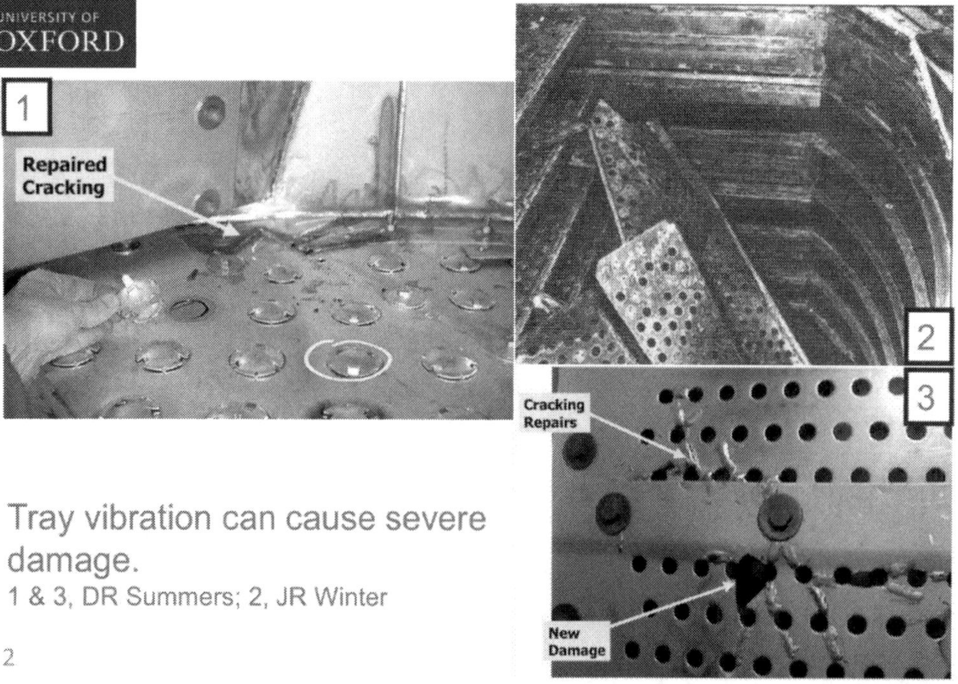

Tray vibration can cause severe damage.

1 & 3, DR Summers; 2, JR Winter

2

Flow-induced vibration of trays

UNIVERSITY OF OXFORD

3rd International Symposium on Distillation, London, 1979:

Brierley et al. (ICI) reported damage caused by flow-induced vibration at low vapour rates in various columns.

In a lab set-up Priestman et al. found the frequency *f*, increased with hole velocity *U* and with liquid head, H_L.

Hole separation (*s-d*) was crucial. *s* is hole pitch, *d* is hole diameter. See also Brown, 1958

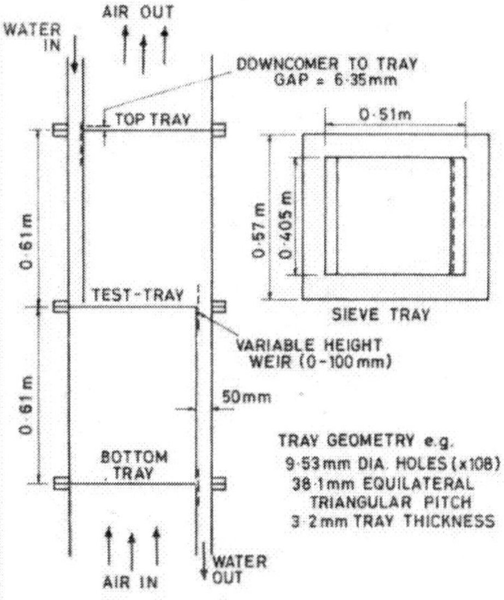

3

Flow-induced vibration of trays

UNIVERSITY OF OXFORD

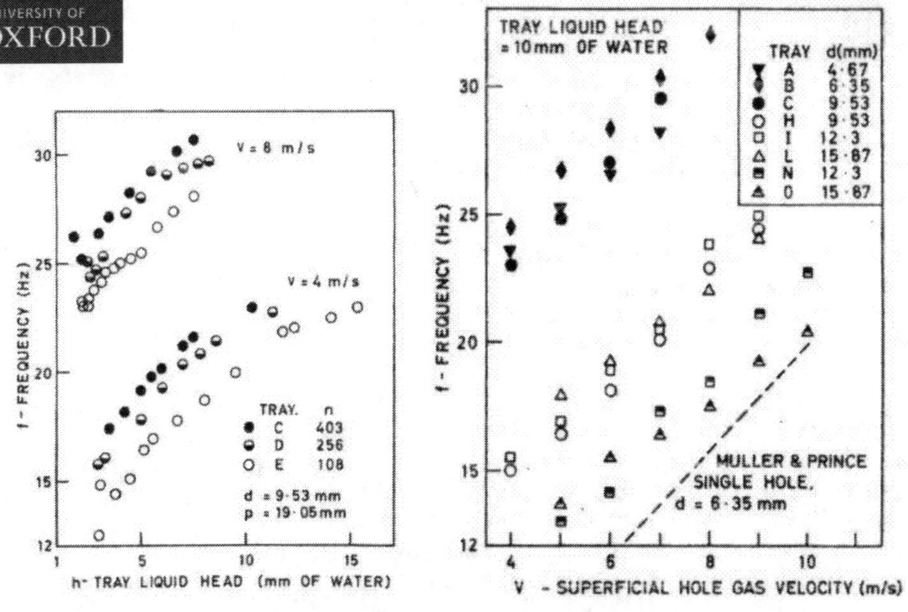

Data of Priestman et al., 1979

4

Flow-induced vibration of trays

Zhang et al., 10th International Symposium on Distillation and Absorption, Friedrichshafen, 2014. Correlates data of Priestman et al., r^2=0.974

$$\frac{fd}{U} = 0.005\left(1+\sqrt{\frac{\rho_L g d^2}{\gamma}}\right)+0.46\left(\frac{d}{s-d}\right)\sqrt{\frac{gH_L}{U^2}}$$

Strouhal number fd/U is ratio of time scales: gas flow through hole, to pulsation

Bond number $\rho_L g d^2/\gamma$ relates gravity to surface tension forces. Affects shape of waves, droplets, jets

Froude number $U/\sqrt{(gH_L)}$ relates hole velocity to speed of surface wave

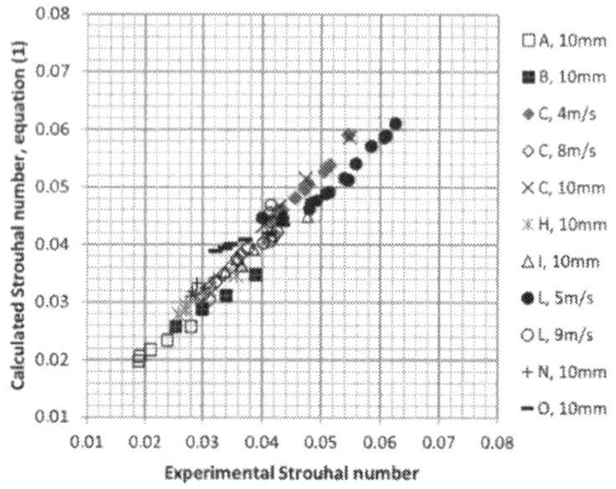

5

Flow-induced vibration of trays

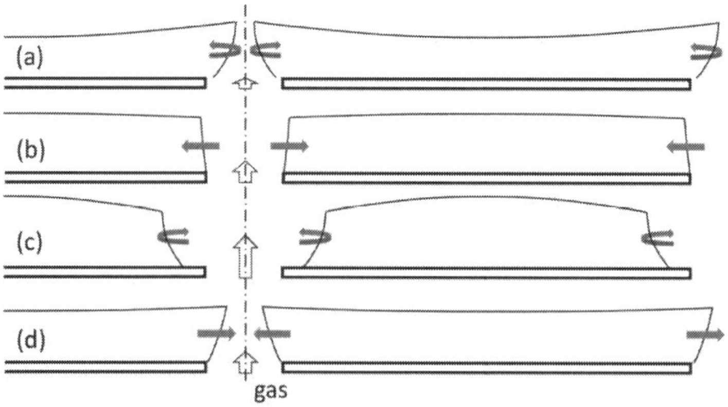

Fluid motion on a tray, as velocity through holes rises and falls. Flow oscillation on adjacent trays differs 180° in phase.

Helmholtz resonator: oscillating mass – fluid near hole; energy store – head on tray

6

Flow-induced vibration of trays

Darton et al., 2015. Correlates data of Priestman et al.,1979 with $r^2 = 0.991$

$$St = 5.537 \frac{d}{\lambda} \frac{c}{U}$$

Correlation scheme

$\lambda^{-1} = L^{-1} + 0.000784\, d^{-1} Bo^{0.5} Fr$

$L = 3.8\{s_m - 0.5d\,(1 - 3.1\, We^{-0.5})\}$

We is Weber number $\rho_G dU^2/\gamma$

$s_m = s\,(0.5 + 0.5\alpha)^{-0.5}$

s_m is mean hole pitch, s is hole pitch, α is fraction of plate uniformly perforated

c is wave speed from Lamb's result

$c^2 = \{g\lambda/2\pi + 2\pi\gamma/\rho_L\lambda\}\tanh(2\pi H_L/\lambda)$

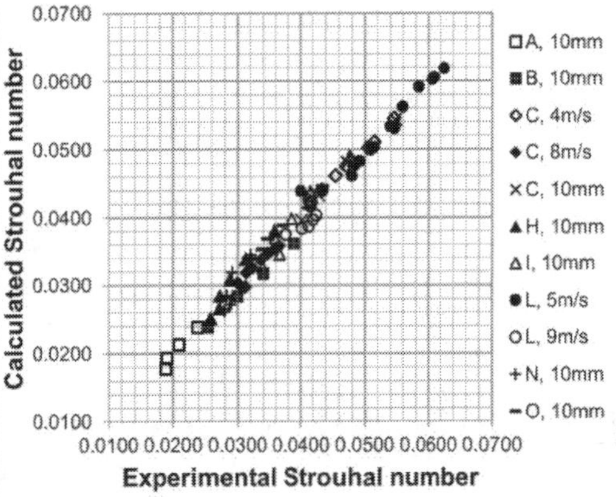

7

Flow-induced vibration of trays

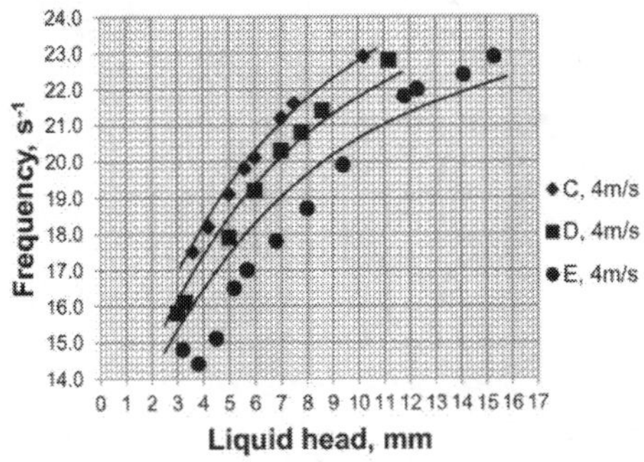

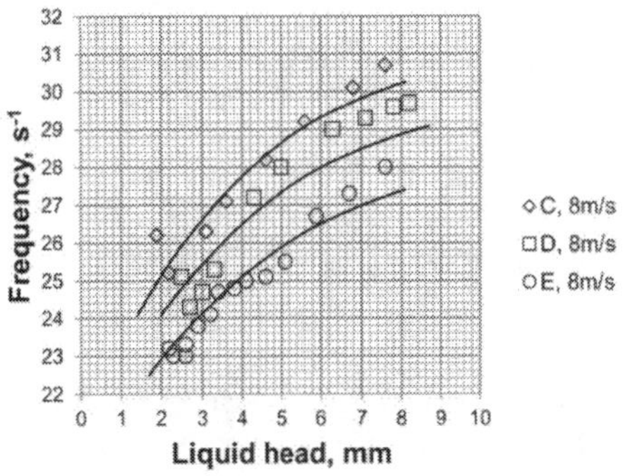

Pulsation frequency predicted by correlation scheme

Data of Priestman et al., 1979

Tray C	s =19.05 mm	d= 9.53mm	n=403	α = 0.763	s_m=20.29
Tray D	s =19.05 mm	d= 9.53mm	n=256	α = 0.485	s_m=22.11
Tray E	s =19.05 mm	d= 9.53mm	n=108	α = 0.204	s_m=24.55

8

Flow-induced vibration of trays

An alternative picture: Negative resistance theory

Waddington, 1973; Brierley et al., 1979

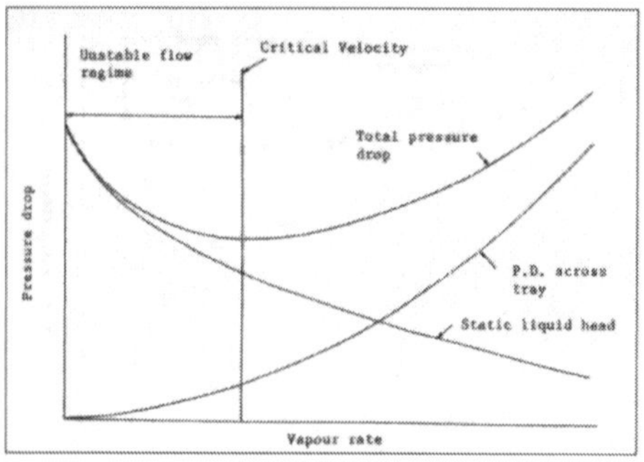

Near the *critical velocity*, two vapour rates give the same pressure drop. Might operation oscillate between these two conditions?

But: does a *critical velocity* exist? Pressure drop minimum is rarely seen

If the rate of oscillation depends on wave speed, the explanation starts to look like Helmholtz resonance. But this is independent of *critical velocity.*

9

Flow-induced vibration of trays

Mechanical vibration of tray structure

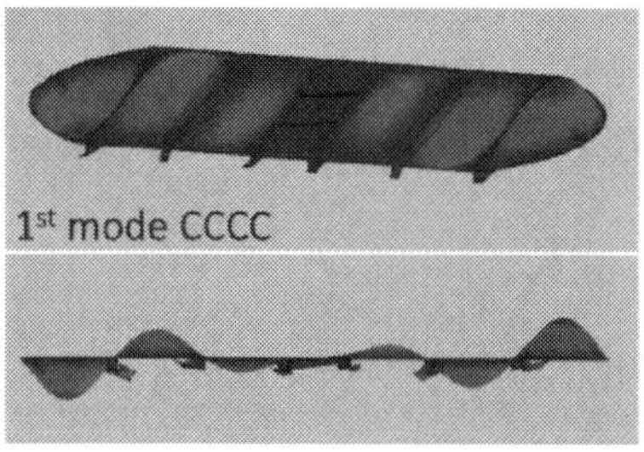

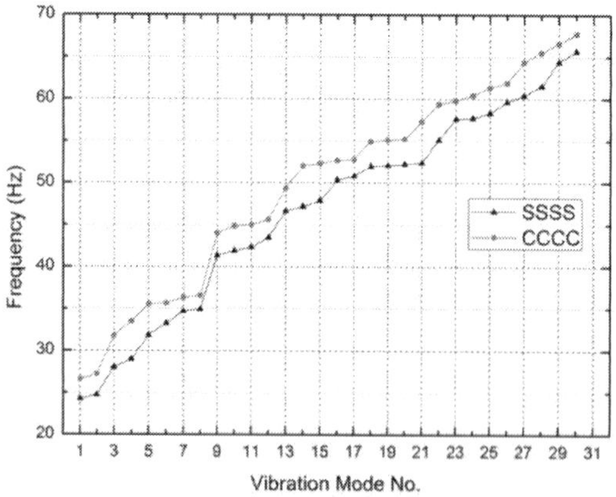

For odd modes of vibration, there will be net upward and downward motion of tray panels, reinforced by oscillating pressure drop at resonant frequencies. With low damping, damaging structural vibration can be expected.

SSSS = panel edges act as frictionless hinge

CCCC = panel edges are rigidly clamped

Zhang et al, 2014: vibration analysis with SOLIDWORKS

10

UNIVERSITY OF OXFORD

Flow-induced vibration of trays

Conclusions

1. Always specify to supplier if trays will operate much at low throughput
2. Risk factors: long flow path, low liquid head, low hole velocity, tray stacks with similar operations on adjacent trays – eg absorbing trace components, close boiling mixtures
3. Stiffen trays to increase mechanical vibration frequency
4. Check that first vibration mode exceeds fluid oscillation frequency
5. Stress concentrates at holes. Keep holes away from edges and each other!
6. Use irregular hole spacings to "de-tune" tray or tray stacks (?)
7. Always investigate, if vibration is audible outside column. Damage can be serious and very rapid. If noise stops when vapour rate is increased, you may have flow-induced vibration.
8. If plant throughput is reduced, check all trayed columns for flow-induced vibration. The trays may not have been designed for it.

11

UNIVERSITY OF OXFORD

Flow-induced vibration of trays

References

Brierley, R.J.P., Whyman, P.J.M. and Erskine, J.B., 1979, "Flow induced vibration of distillation and absorption column trays" in Distillation 1979, IChemE Symp. Ser No 56, 2.4/45

Brown, R.S., 1958, "Bubbling from perforated plates" PhD thesis Univ. California, Berkeley

Darton, R.C., Zhang, W. and Taylor, P.H. 2015 Chem Eng Res Des, http://dx.doi.org/10.1016/j.cherd.2015.03.023

Lamb, H., 1932, Hydrodynamics, 6th edition, Cambridge University Press, Cambridge UK

Priestman, G.H., Brown, D.J. and Kohler, H.K., 1979, Pressure pulsations in sieve-tray columns in Distillation 1979, IChemE Symp. Ser No 56, 2.4/45

Summers, D.R., 2003 AIChE Annual Meeting November 2003, paper 307g

Waddington, W., 1973 Vibration excitation of sieve tray columns by bubbling, MEng Thesis, Sheffield University

Zhang, W., Taylor, P.H. and Darton, R.C., 2014, The destructive vibration of column trays, in Distillation and Absorption 2014, Friedrichshafen, 14-17 September 2014, p. 588-593.

Acknowledgements

Anonymous referees of ChERD paper
DR Summers for publications otherwise difficult to obtain
John Brierley for encouraging RCD to look at this problem, in 1978

12

Paper 4c: Influence of tray geometry and hydraulic conditions on vibration induced damages of sieve and valve trays

Christian Geipel
RVT Process Equipment GmbH, Steinwiesen, Germany

Abstract

The present paper is a summary of observations on failures of trays in various industrial applications. The affected trays include fixed valve trays (type SVG and SRV), movable valve trays (V1-valves and A3-cage valves) as well as sieve trays.
Different incidents are investigated to derive general indications that promote vibrations and – as a consequence – failures of the trays. The failures are analyzed with regard to valve types, special design features as well as operating conditions that lead to the failure. The basis is operating data and pictures of the damaged trays.
Also the failure modes of the trays are investigated in detail to derive the destruction mechanism. Where possible video sequences through a sight glass were taken in operating columns to visualize gas and liquid flow as well as vibrations. These findings were compared to experimental results in a pilot plant with a transparent column.
General design features that promote vibrations in combination with special operating conditions like low liquid load could be identified.
The position of all operating points in the operating range has to be carefully investigated with respect to parameters that influence tray vibrations.
Based on these findings, guidelines for potentially critical trays were derived. These include mechanical design criteria as well as process and hydraulic design guidelines.

Keywords

Tray vibration, mechanical failure, design guidelines

1. Introduction

Vibration induced tray damage is a known phenomenon since long time. First approaches to predict when harmful vibrations will occur date back to the 1960s[1] or even earlier. However, reliable theoretical models that predict vibration are still scarce. Rigorous guidelines that could be used during design of trays are still not publically available.
The present paper gives an overview on some reasons for vibration induced damages.
A phenomenon that is not further discussed here is heavy pressure fluctuations in the column. These could be caused by high liquid level in the bottom of the column flooding the bottom trays, steam hammering, vacuum collapses or e.g. free water flashes. Often one single incident already leads to destruction of one or several trays.

Other vibration phenomena mentioned in literature are caused by the gas-liquid interaction on the tray. Different root causes for vibration induced tray damage are distinguished:

- One reason is so called "liquid swashing"[3,4]. Under certain circumstances a standing wave across the column diameter can built up. The wave is perpendicular to the liquid flow, pulsating at low frequency. Typical frequencies are below 5Hz.
- Another type of vibrations occurs when the pulsations created by the gas-liquid interaction excite the resonance frequency of the tray. The excitation is created by the forces resulting from bubble formation or jet flow pulses at the valves or orifices[2,5]. This phenomenon is the most investigated one among the listed effects, also because it often leads to tray damage within very short time. Several authors agree that resonance occurs at low liquid and gas loads. Therefore it can lead to tray damage already during turndown operation of the column at startup. Typical frequency of the tray vibration is in the range of approx. 15 – 40 (60) Hz.

Several authors have shown that only the gas load that was present during vibration did not lead to vibration on a dry tray.

2. Results and discussion

Subject of the present article is a summary of failures of operating trays in various industrial applications, including sieve, fixed valve and movable valve trays.
The following types of trays were affected by vibration cracking:

- Valve trays with big fixed valves: SVG and SRV
- Valve trays with movable valves: V1 and A3
- Failures for sieve and dual flow trays have also been observed

2.1 Characteristics of the affected trays

Mainly tray diameters in the range of 2000 mm or higher and with a one-pass-design were affected.
Less frequent are failures at tray diameters >3200 mm with two or more passes or at tray diameters below 1600 mm (1-pass).
All types of trays had integrated support beams (bent panel edges) and no lugs or support brackets below the downcomer.
Many of the affected trays had long weir lengths and huge enveloped downcomers. It was observed that often designs with long tray segments were used.
For movable valves mostly only one valve weight was used.
Often the trays were designed for low Δp of < 20mmWC (dry) respectively gas hole velocities of <11m/s at 100% load.
The failures happened at operating points with low liquid load combined with low gas load and therefore at unit references below 100%.
In the following, the effects occurring will be explained exemplarily for trays equipped with V1 valves.

2.2 Investigation of tray failures of V1-valves

Figure 1 and 2 show operating points that lead to tray damage in production columns.

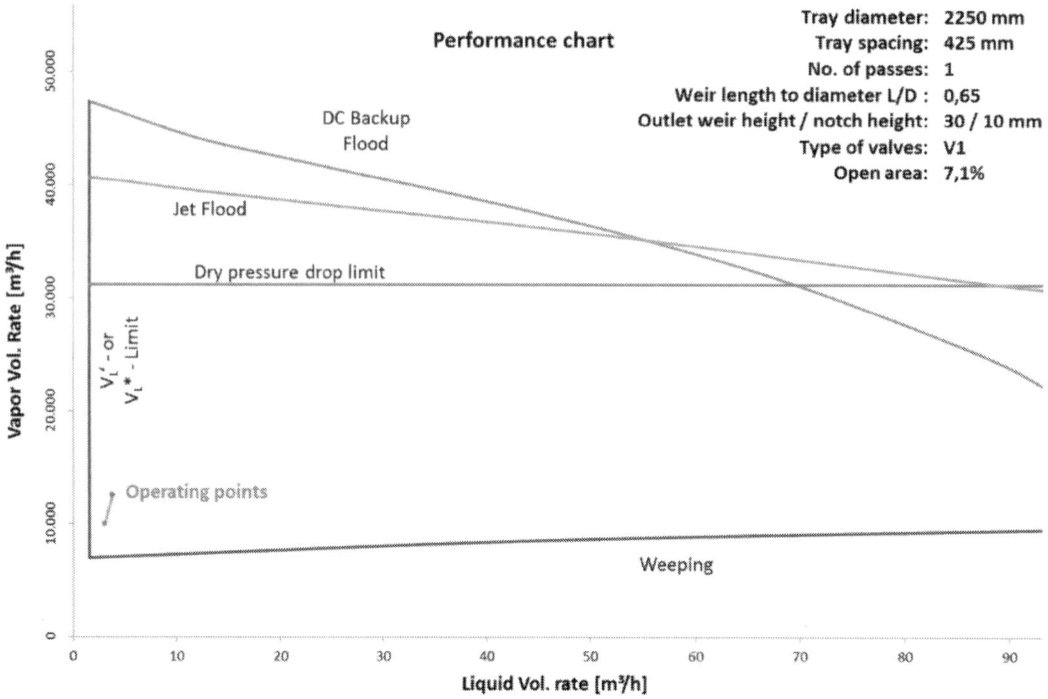

Figure 1: Operating points leading to damage, V1-valves, one pass

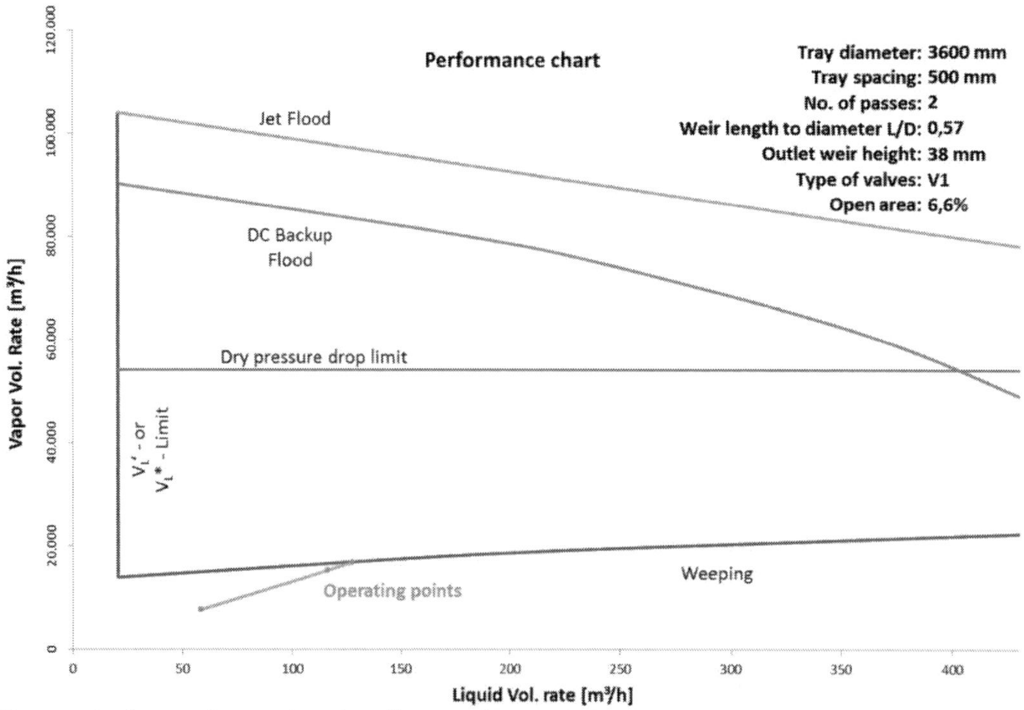

Figure 2: Operating points leading to damage, V1-valves, two passes

The failures in the two cases were quite similar to other cases examined: Vibration crackings in tray segments starting from the corner of the bent part through the next opening (screw or valve hole) and then in a zigzag pattern through the next valve or sieve holes (see Figure 3).

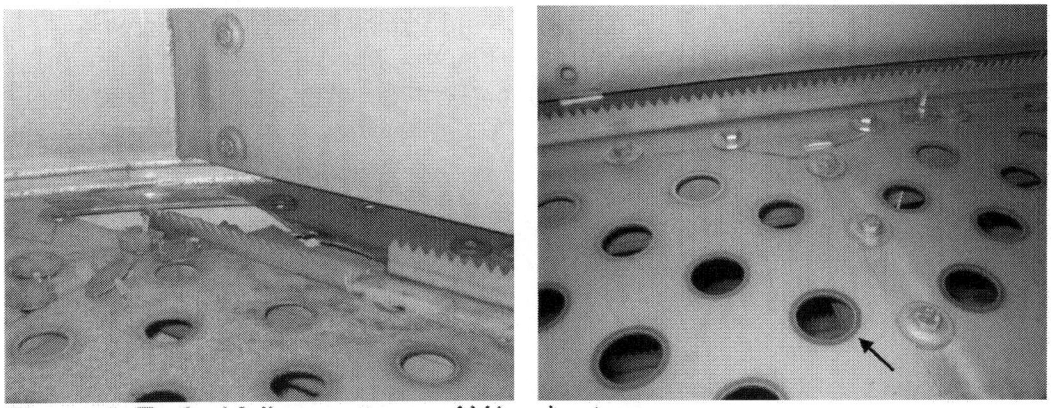

Figure 3: Typical failure patterns of V1-valve trays

Furthermore, whole tray segments were torn out of their bolting. The destruction often is more pronounced at the inlet side of the tray. It was observed that about 30-40% of the V1 valves were dislocated and traces of turning and scratching were observed on the top and bottom side of the tray (see Figure 3, right). This happened regardless of the valves being equipped with an anti-rotation device or not. From the valves that were found on the tray always one leg was broken off.
Cracks at the downcomer formed in parallel with the screw line at the small brackets fixing the downcomer wall. Also these brackets themselves often break (see Figure 4).

Figure 4: Broken downcomer bracket

For sieve and fixed valve trays, the failure patterns are quite similar: the cracks start at the bent corners of the integrated beam and continue in a random pattern from hole to hole. Often pieces up to the size of a hand palm break out of the tray.

The mechanisms leading to the failure were examined in several steps:

- For one alcohol distillation column a video could be taken through the sight glass. The column had a diameter of 2.3m and was equipped with 1-pass V1-valves. Even though the destruction had already happened, the trays were still vibrating at low frequency of < 5Hz with amplitude of approx. 20mm.
- In a second step pilot plant experiments in transparent 900 mm resp. 1600 mm columns were carried out. The gas factor was varied in ranges close to those in the real column. For dry operation it could be shown that in a certain range the valves started knocking synchronously.

- When the liquid load was added, heavy weeping was observed. Also, the height of the liquid level on the tray started pulsating. Consequently, also the flow over the weir pulsated heavily and there is an intermittent gas bypass flow through the downcomer.
- In the 1600 mm column equipped with fixed valves, these pulsations lead to heavy vibration of the trays. In the pilot plant column we reached a condition where the trays were ruptured out of their boltings within seconds. Due to the higher resonance frequency of the tray, no vibration occurred in the 900mm column.

With these experiments the failure mode became much clearer. Similar experiments and theoretical investigations were also done for other valve types. The following general observations could be derived.

2.4 General observations for different valve types

Observations in pilot plant and industrial size columns showed that:

- Generally low gas loads – often combined with low liquid loads – can cause vibration of tray segments und thus damage of the tray
- If V1-valves are not fully open, they tend to knock in the same frequency
- If A3-cage valves are not fully open, they can show strong vibration and beat or clash
- A11-cage valves can show very strong vibration before they are fully open
- Already deflection of the tray segments in the range of ≥2mm can lead to gas/liquid maldistribution. In critical operation regions this maldistribution can lead to vibration. Also inclination of the trays and long flow paths favor vibration.
- The visible amplitude of the integrated support beams is often >20mm (at variable frequency)

3. Conclusions

The analysis of the above mentioned phenomena leads to the following guidelines for tray design if the tray is operated in regions where vibration is considered is to be possible:

- Avoid integrated beams and use separate beams with gusset plates, lugs etc.
- Use more, narrow segments instead of bigger ones
- Use thicker tray decks, e.g. 3mm instead of 2mm. The effect on Δp respectively weeping has to be considered!
- Where possible, use more robust caged valves instead of V1-valves
- For trays that operate in a regime that could favor vibration inclinations or deflections have to be strictly avoided!
- Block weir length at low liquid load (no notches) and increase weir height (if possible)
- Use a combination of light and heavy valves if possible
- Use inlet weirs with liquid seals ≥15mm at low liquid loads
- Also the operating range of the column should be defined carefully. Especially huge operating ranges combined with strict pressure drop requirements favor vibration in low load regions.

The above investigations have pointed out indications of proneness to vibrations of trays. The desired load points have to be checked carefully concerning their location in the tray performance chart. All operating points including startup, turndown etc. have to be considered. If a risk of vibration is identified for a tray, the design criteria listed above should be considered.

Acknowledgements

We would like to thank all companies that were involved in our investigations on tray vibrations. The basis of the investigations was data from several plants and results could only be derived in valuable discussions with the plant and design engineers who shared their experience with us.

References

[1] Summers, D.R., "Bad vibrations", Hydrocarbon Engineering, March 2005, pp. 51-54
[2] Summers, D.R., "Harmonic vibrations cause tray damage", AIChE Annual Meeting 2003
[3] Winter, J. R., "Avoid vibration Damage to Distillation Trays", Chemical Engineering Progress, May 1993, pp. 42-47
[4] Priestman, G.H.; Brown, D.J., "Flow-induced Vibration and Damage in Sieve-tray Columns", Chem. Eng. Comm, 1988, vol 63, pp. 181-192
[5] Zeitsch, K., "Eine neue Theorie der Schwingungserregung an Kolonnenböden", Chem. Techn. Heft 10, Oct. 1984, pp. 415-418

Raschig High Capacity Mini Fixed Valves

FRI tested

Paper presented at 2015 AICHE Spring Meeting Austin, TX, April 2015

M. Schultes, A. Danninger, R. Altmann

For mass transfer trays standard sized round and rectangular valves have been proven their industrial advantages many times. For typical applications the deck hole diameter of a standard moveable or fixed round valve is approx. 38 mm. For fouling services even larger valves in rectangular shape have been designed. Over the past years a new trend is pushing into the market: mini moveable or fixed valves. They are getting more and more popular especially for mass transfer columns were operators are looking for high capacities.

Raschig has recently tested its Mini Fixed Valve in the Fractionation Research Inc. (FRI) test column to provide more information about capacity, pressure drop and efficiency to the industry. This article discusses the FRI test results and shows typical applications were Raschig's high capacity Mini Fixed Valve was successfully applied. An industrial example for a grass root design and another example for a revamp case will be presented.

Introduction

The mini valve technology is based on the principles that small deck openings on a tray results in smaller gas flow streams pushing the liquid to a smaller extend to the mass transfer tray above. As the total open area is comparable to standard large valves the deck is equipped with a large population of mini valves. Fig. 1 shows the Raschig Mini Fixed Valve deck layout that was tested at FRI. Fig. 2 shows the Raschig Mini Fixed Valve in comparison to the regular fixed V0-valve. Beside the number of valves the shape of the Mini Fixed Valve is advanced compared to standard designs, see Fig. 2. The umbrella shape of the mini valve hat pushes the gas flow into a more horizontal direction what reduces the liquid entrainment to the tray above. The umbrella shape is designed just smooth enough to allow a low pressure drop. The V-shape of Raschig's Mini Fixed Valve pushes the liquid over the deck towards the downcomer. In 2014 Raschig's Mini Fixed Valves had been tested at FRI.

Fig. 1: Raschig's high capacity Mini Fixed Valve as tested at FRI

FRI testing of Raschig's Mini Fixed Valves

The Mini Fixed Valves had been tested in the low pressure column at FRI. The tests were performed with the system cyclohexane/n-heptane at 1.62 bar (24 psia) and 0.31 bar (4.5 psia). The operation included various total reflux run conditions to verify the tray efficiency over a wide capacity range, as well as non-total reflux run conditions to verify the capacity limits for a wide loading range. In this paper the total reflux runs will be discussed. For these tests FRI has used a liquid collector tray below the bottom tray for the first time. With this collector tray the weepage could be measured at the various operation modes.

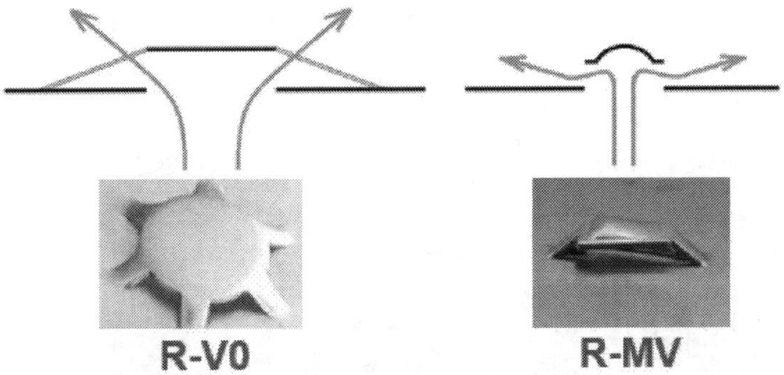

Fig. 2: Gas flow stream via a standard round fixed R-V0 valve and a high capacity R-MV Mini Fixed Valve by Raschig

The FRI column can be used for testing random packings, structured packings and mass transfer trays.This requires a very flexible column setup - therefore the column is not equipped with welded support rings. Instead, expansion rings get installed where required for the specific test. For the tray testing clips get welded to the column wall and expansion rings and support rings rest on the clips and get expanded towards the column shell. PTFE gasketing is applied to seal the expanded support ring as good as possible towards the column wall during tray installation. Using PTFE-sealed expansion rings is an industrial standard procedure if for example a tower revamp project is processed and new trays have to be positioned inside the tower without welding to the column shell.

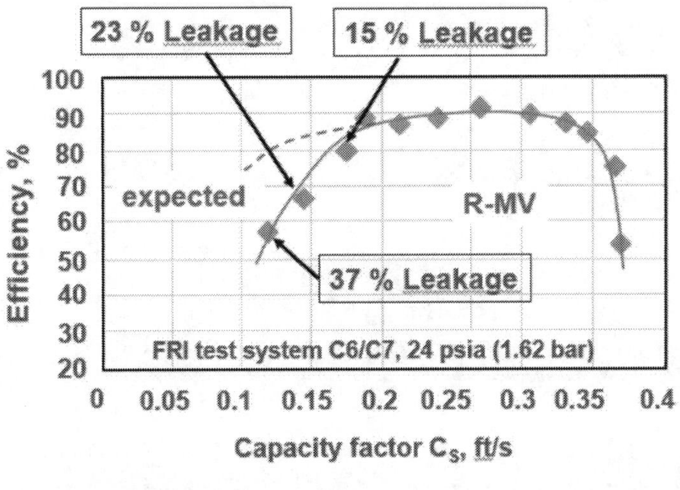

Fig. 3: Tray efficiency, capacity and tray support ring leaking fraction over column cross section area related Cs-factor of Raschig Mini Fixed Valves tested at FRI with cyclohexene/n-heptane at 1.62 bar (24 psia)

Six trays have been installed into the FRI low pressure column and an additional collector tray below. The tray spacing was 600 mm (24 inches) as a typical industrial standard. A standard sloped downcomer was used to enlarge the tray deck, providing a downcomer area of 6.7 % of the column cross section at the downcomer top and a 5 % downcomer area at the downcomer bottom. The tray active area was 87 % of the column cross section and the valve hole area was 12.4 % of the bubbling area. 549 Mini Fixed Valves were placed on each deck. A 51 mm (2 inches) outlet weir was incorporated in the tray design.

Fig. 3 shows the efficiency and capacity plot of Raschig's high capacity Mini Fixed Valve with the test system cyclohexane/n-heptane at 1.62 bar (24 psia). As described above the support ring expansion and PTFE sealing was controlled carefully. A liquid collector below the bottom tray was installed to read tray weepage.

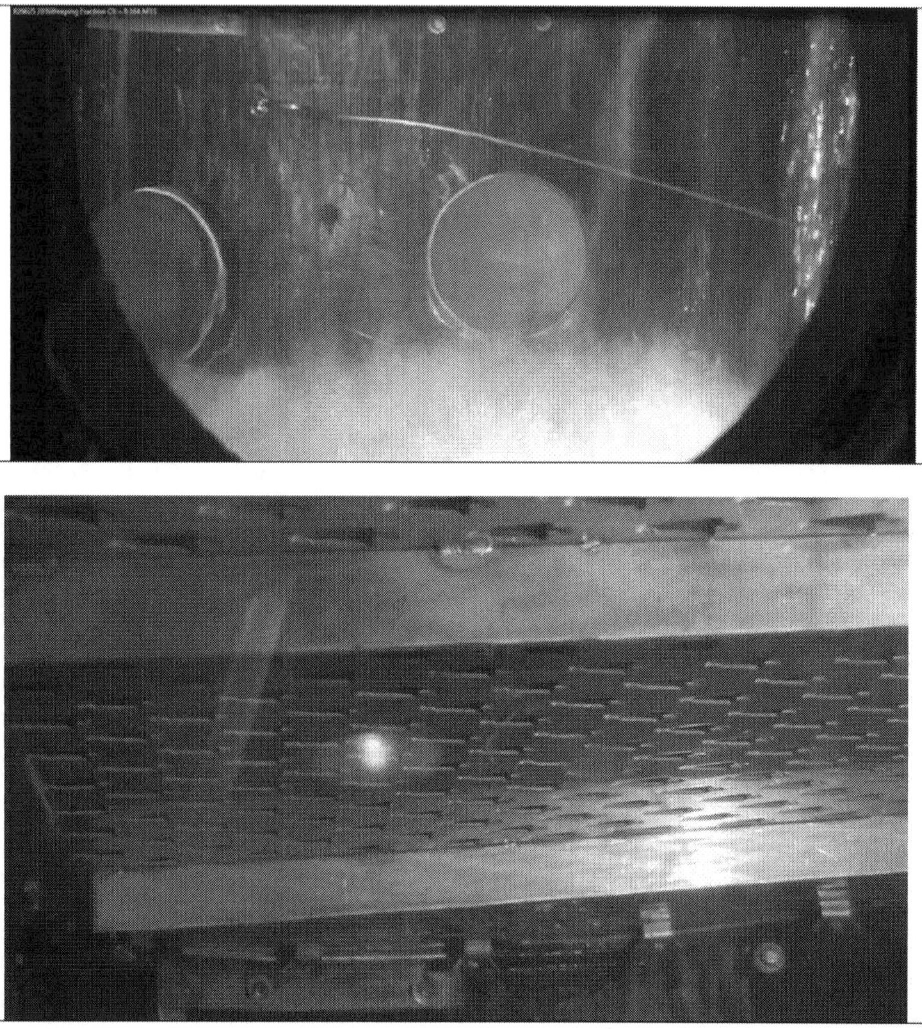

Fig. 4: Videotaped tray support ring leakage; top: column wall view between tray 3 and 4; bottom: view to tray 4 from below

Fig. 4 shows a snapshot during column operations between tray 3 and tray 4 during operation at a low column capacity. During operation of the trays noticeable support ring leakage was recognized at low loadings. The trays themselves were weepage free. Fig. 5 shows the pressure drop of the tray deck. One can see that the efficiency in Fig. 3 starts to drop when the ring leakage exceeded 15 % of the reflux rate. In the mid and upper capacity range the pressure drop of the trays were high enough to hold the liquid on the tray without leakage. In Fig. 3 the dotted line shows Raschig's expected efficiency curve assuming no support ring leakage.

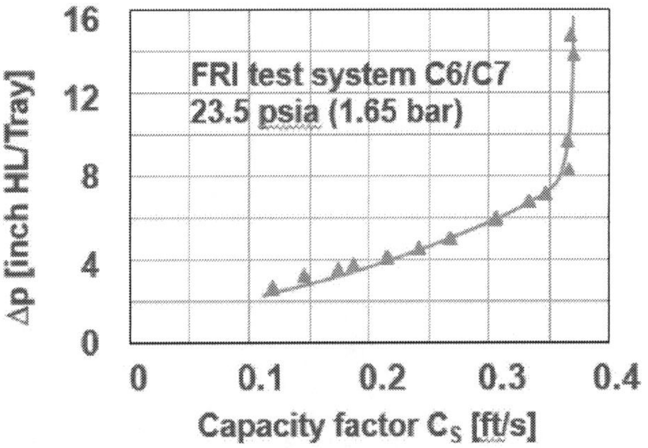

Fig. 5: Pressure drop and capacity over column cross section area related Cs-factor of Raschig Mini Fixed Valves tested at FRI with cyclohexene/n-heptane at 1.62 bar (24 psia)

Fig. 6 and Fig. 7 shows the equivalent test results for cyclohexane/n-heptane at 0.31 bar (4.5 psia) and provides corresponding information about support ring leakage and pressure drop. The dotted line shows again Raschig's expected efficiency curve without ring leakage. The turn-up and turn-down ratio for the Mini Fixed Valve is in the typical range for fixed valves. But Raschig's high capacity Mini Fixed Valve provides noticeable capacity and pressure drop advantages.

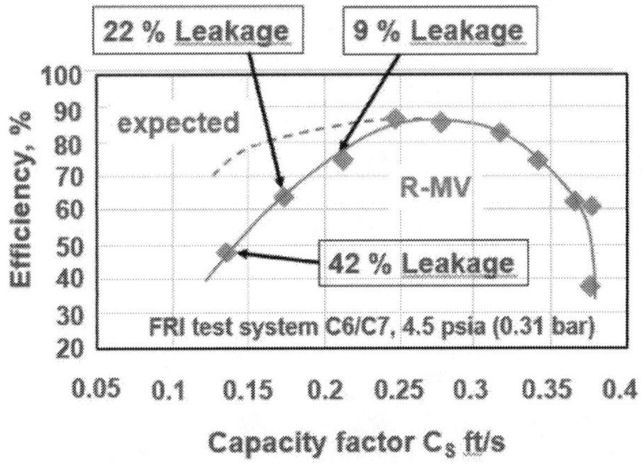

Fig. 6: Tray efficiency, capacity and tray support ring leaking fraction over column cross section area related Cs-factor of Raschig Mini Fixed Valves tested at FRI with cyclohexene/n-heptane at 0.31 bar (4.5 psia)

Fig. 8 shows the efficiency and capacity comparison to a standard movable V1 valve deck layout tested at FRI. Fig. 9 shows the comparison to a standard sieve tray deck layout and Fig. 10 shows the comparison to a standard fixed V0 valve deck layout, both also tested at FRI. These comparisons demonstrate the noticeable capacity and pressure drop advantage of Raschig's Mini Fixed Valve. A typical capacity advantage over standard fixed V0 valves is in the range of 10 % by using Raschig's high capacity Mini Fixed Valves.

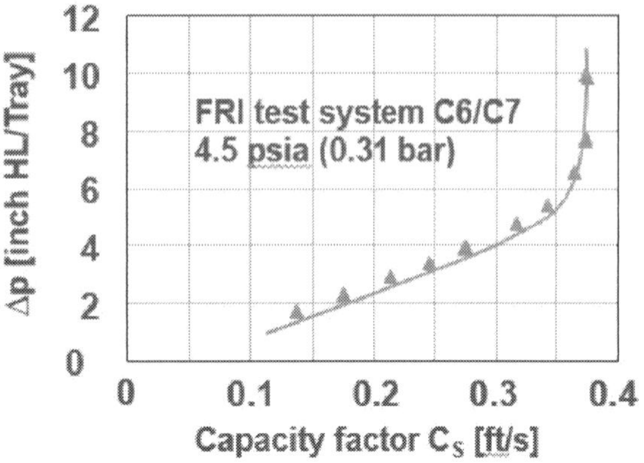

Fig. 7: Pressure drop and capacity over column cross section area related Cs-factor of Raschig Mini Fixed Valves tested at FRI with cyclohexene/n-heptane at 0.31 bar (4.5 psia)

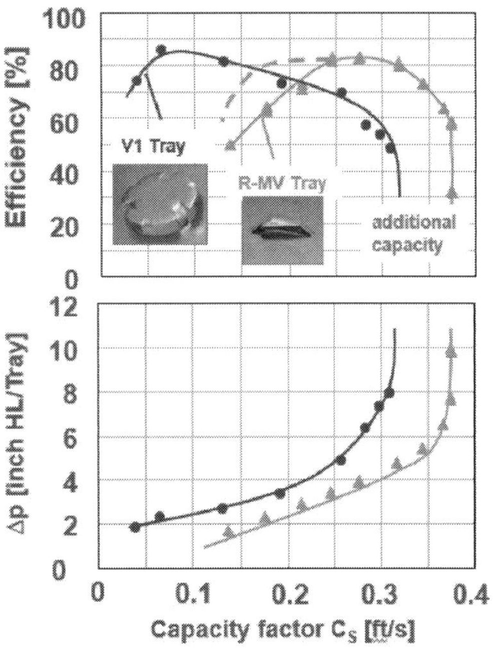

Fig. 8: Efficiency, pressure drop and capacity over column cross section area related Cs-factor of Raschig Mini Fixed Valves and movable V1 valves tested at FRI with cyclohexene/n-heptane at 0.31 bar (4.5 psia)

It has to be noticed that comparisons of tray efficiencies, capacities and pressure drop of different trays are very difficult to make as the whole tray design concept is determining its performance. For example the downcomer can be arranged straight, sloped, multi-chordal and truncated, the inlet and outlet weirs often have different heights, the orifice areas and bubbling areas can be different as well as the tray spacing - to name only a few parameters. Two, three and four pass designs have their own characteristics and are difficult to compare with one pass designs. But this is also typical for industrial practice. In a grass root design mini valves will have a different tray and downcomer layout than standard valve trays. In revamp cases for example the tower is performing with a given tray design at its capacity limit. Revamps often require an optimized downcomer, deck and valve design.

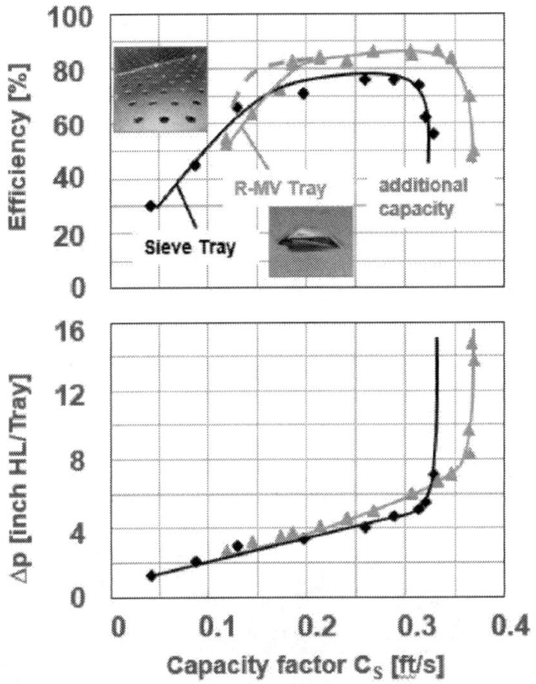

Fig. 9 Efficiency, pressure drop and capacity over column cross section area related Cs-factor of Raschig Mini Fixed Valves and sieve trays tested at FRI with cyclohexene/n-heptane at 1.62 bar (24 psia)

Application of Raschig's high capacity Mini Fixed Valve for a revamp case

An organic distillation column in Europe was equipped with movable round V1 valves. Tray spacing was 460 mm (18 inch) only and the column was operating at 2.2 bar (32 psia). The column was limited by its downcomer capacity. Consequently for increasing the capacity of the column the downcomer area had to be enlarged with the penalty that the active area had to be reduced, see Fig. 11. With the existing movable round V1 valve a jet flood value of 103 % was expected. No welding to the column shell was allowed in this revamp project.

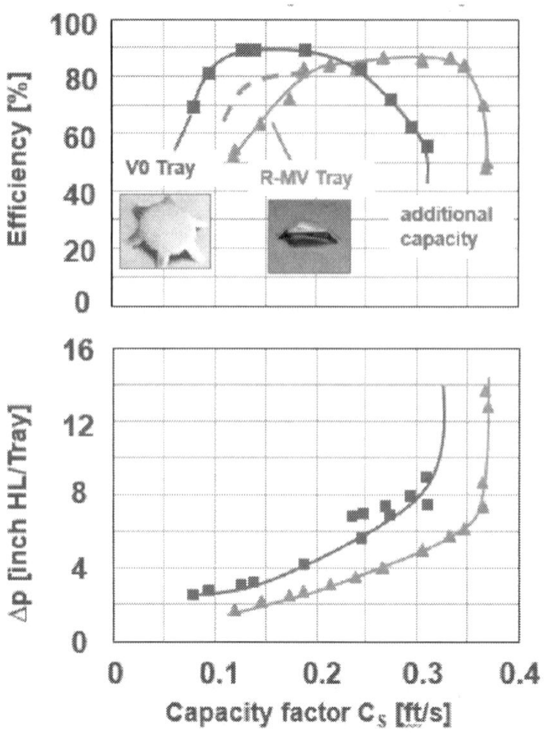

Fig. 10 Efficiency, pressure drop and capacity over column cross section area related Cs-factor of Raschig Mini Fixed Valves and fixed V0 valves tested at FRI with cyclohexene/n-heptane at 1.62 bar (24 psia)

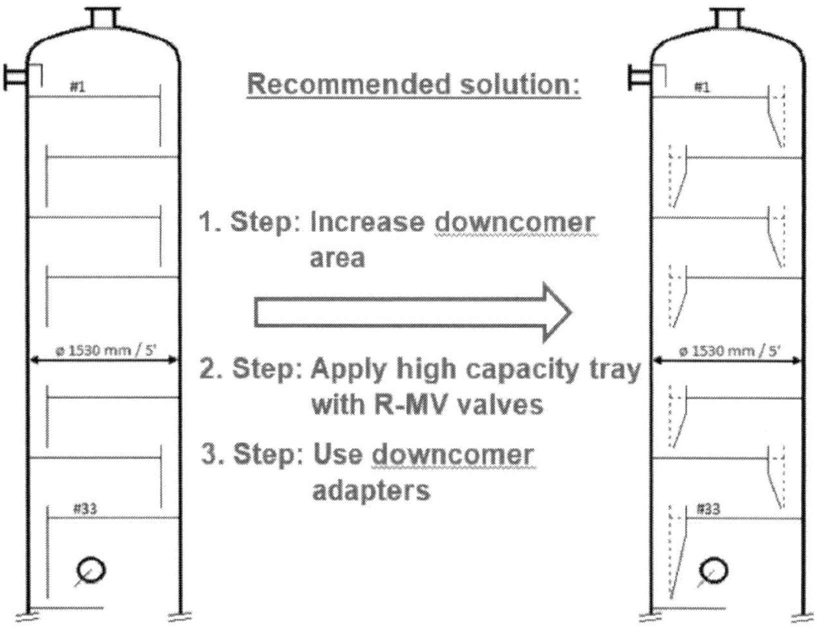

Fig. 11: Revamp of a distillation column from movable V1 valves to high capacity Raschig Mini Fixed Valves

The solution was as follows. First the downcomers were enlarged at their top end and received a sloped designed. Downcomer adapters were used, connected to the existing downcomer bars. Because of their high capacity Raschig's Mini Fixed Valves were applied on the tray deck. Fig. 12 shows the deck layout and the downcomer adapters. With this optimized design the capacity of the distillation tower was pushed by 13 % compared to its original design.

Fig 12: Tray deck layout applied for revamp project with Raschig Mini Fixed Valves and downcomer adapters

Application of Raschig's high capacity Mini Fixed Valve for a grass root design

A butane splitter had been designed to operate at 5.5 bar (80 psia), see Fig. 13. In the original design the column was specified with a 2-pass tray layout using fixed round V0 valves by applying a tray spacing of 400 mm (16 inches). During the detail engineering phase the customer has changed the column loadings to higher capacities. The specified V0 valves could not handle the new capacities as their capacity limit was 90 % of the new loadings. The limitation was jet flooding.

The solution was a change of valve type. Instead of using the standard V0 valves Raschig proposed its high capacity Mini Fixed Valves. This provided a capacity gain of 10 %. In addition Raschig proposed the application of NYE inserts at the downcomer outlets as shown in Fig. 14. The NYE inserts pushes the liquid out of the downcomer which helps the downcomer capacity. In addition the NYE inserts elevate the downcomer bottom area. Below the downcomer bottom area the NYE inserts are perforated. Also the front end area of the NYE inserts are perforated which provides more gas capacity. By applying the NYE insert technology a capacity gain of another 7 % was obtained.

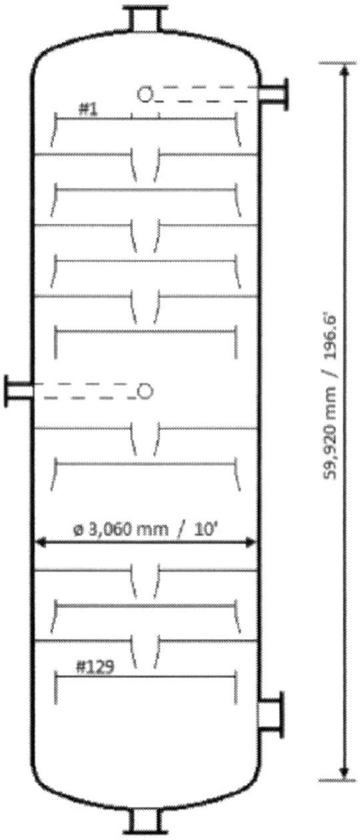

Fig. 13: Grass root design of a Butane Splitter with Raschig's to high capacity Mini Fixed Valves

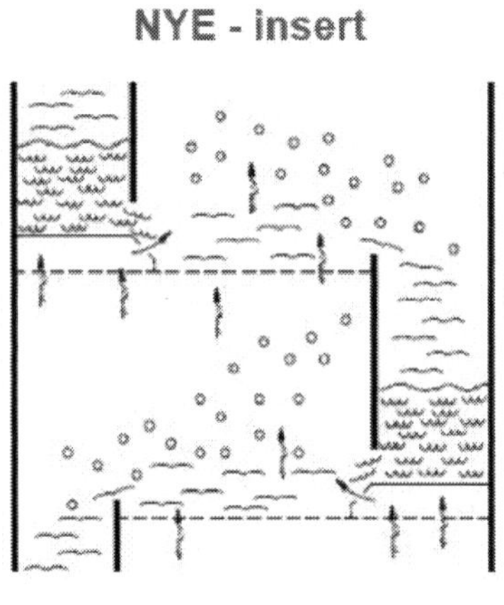

Fig. 14: Principle flow scheme of a NYE insert

Fig 15: Tray deck layout applied for a grass root butane splitter project with Raschig Mini Fixed Valves and NYE inserts

Conclusion

The high capacity Mini Fixed Valve technology designed by Raschig provides significant capacity and pressure drop advantages without any loss in efficiency compared to standard fixed valves. Typically a 10 % capacity gain can be expected changing from standard fixed valves to Raschig's high capacity Mini Fixed Valves. Modern downcomer designs can be applied to further bust capacities. The Mini Fixed Valve has widen Raschig's tray portfolio into high capacity devices. This technology helps our customers to increase their column capacities. Raschig has a long history in tray tower designs and has references available for several hundreds of trayed towers world-wide.

Nomenclature

Cs	ft/s, m/s	gas capacity factor: $u_V \sqrt{\frac{\rho_V}{\rho_L - \rho_V}}$
Δp	inchHL/tray	pressure drop in inch hot liquid per tray
u_V	ft/s, m/s	superficial gas velocity
ρ_V	kg/m^3	gas density

Literature

/1/ Fractionation Research, Inc.; February 1957 Progress Report Revision September 30, 1957

/2/ Fractionation Research, Inc.; Topical Report No. 41; December 15, 1967

/3/ Fractionation Research, Inc.; Topical Report No. 19; December 15, 1959

Troubleshooting and Solving a Gas Treater Downcomer Unsealing Problem

By

Henry Z. Kister

Fluor, Aliso Viejo, CA, USA

and

Nashaat M. Mohamed,

PMC for ADGAS IGD Project, Abu Dhabi,
WorleyParsons Engineering Pty Ltd

Paper presented at the Distillation Topical Conference, AIChE Spring Meeting, Austin, Texas, April, 2015

Troubleshooting and Solving a Gas Treater Downcomer Unsealing Problem

By Henry Z. Kister, Fluor, Aliso Viejo, CA, USA and Nashaat M. Mohamed, PMC for ADGAS IGD Project, Abu Dhabi, WorleyParsons Engineering Pty Ltd

Combining troubleshooting with a hydraulic analysis identified downcomer seal loss in moving valve trays as the root cause of premature flooding and instability in a gas treater.

ADGAS Integrated Gas Development (IGD) project transfers 900 MMSCFD of dry gas from the gas wells to the Habshan industrial site through a 30" pipeline. The project comprises a gas receiving station, three gas drying units (300 MMSCFD each), three gas turbine driven compressors, and a fuel gas treatment system with the capacity of 25 MMSCFD of sweet gas.

The objective of the Fuel Gas Treatment (FGT) is to supply sweet fuel gas for the turbines and also for other consumers (flare purge and pilots). The FGT system consists of an absorber and regenerator that use MDEA to absorb H_2S out of the gas to a spec of less than 20 ppm. The system was designed by an engineering contractor from Japan and the tower internals by a major tray supplier. The operation requirement of the gas turbine would dictate that the fuel gas treatment unit would supply the turbine at ramping rate of 4-5 MMSCFD per minute for the turbine to reach the minimum governor speed. The supply would slowly increase after that to about 6-7 MMSCFD as the compressor/turbine picks up the feed gas load.

The FGT system was pre commissioned and started up as per the instruction from the engineering contractor. When the plant was operated above 10 MMSCFD, the absorber column became very unstable and flooded, resulting in a sudden level rise in FG KO drum, tripping the whole system. Changes to the MDEA flow rates did not appear to have much influence. Any attempt to raise the gas load above 10 MMSCFD and keep stable column operation failed. An attempt to fix the problem by the contractor and vendor failed to provide a significant improvement.

A thorough troubleshooting investigation by ADGAS, together with hydraulic analysis by Fluor, yielded a correct diagnosis and a successful solution. This paper describes the troubleshooting investigation and accompanying hydraulic analysis.

Tray Details and Preliminary Hydraulic Evaluation

Fig. 1 shows the FGT absorber. Feed gas containing less than 2% H_2S enters via a vane distributor into the 1500 mm ID bottom section. Above the vane distributor the tower ID is diminished to 1200 mm. In the absorber, an MDEA solution contacts the gas over 20 trays to remove H_2S. The gas leaving the tower has a flow rate of 25 MMSCFD and contains less than 20 ppm of H_2S. The MDEA design rate is 38.5 M^3/h and it enters on the top tray. There are alternative feed points that permit introducing the MDEA solution above trays 16 or 12, but normally neither one of these is used.

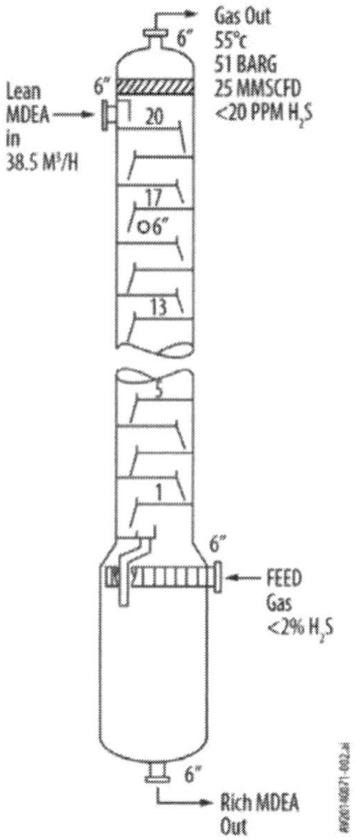

Figure 1 FGT Absorber

Fig. 2 shows the tray details. The trays were moving rectangular valve trays fabricated from 2 mm thick 316L stainless steel. Table 1 shows the hydraulic design flow rates and physical properties. Table 2 shows the results of our preliminary hydraulic evaluation.

Tables 1 and 2 do not reveal anything outstanding. The tray design does not look unusual. Table 2 shows that the design point does not appear to be close to any conceived capacity limits. Calculated values by both the tray supplier and Fluor were in excellent agreement.

Amine absorbers are known to be foaming systems. Perry's Handbook (1) recommends using a derating ("system") factor of between 0.73 and 0.8 for such systems. Indeed, the designers applied an even more conservative system factor of 0.7 in the design. The calculated percent flood values listed in Table 2 can be divided by the derating factor of 0.7 and still would maintain a comfortable margin from flood. The downcomer inlet velocity is also well below the value recommended (1) for a foaming system.

The only parameter that appears out of range for good design practices is the head loss under the downcomer. Most designers use a value of 15 to 30 mm of liquid. The number calculated here is about half of the lower value, about 8 mm. It results from a downcomer clearance larger than needed for handling the liquid flow rate. This head loss was not even shown on the supplier calculation list. Normally, this low value will not cause any problems, because the downcomer will be sealed by the outlet weir. In this case, too, the outlet weir of 75 mm was tall enough to give a static seal to the 50 mm downcomer clearance. However, in this case it turned into a major issue as will be discussed below.

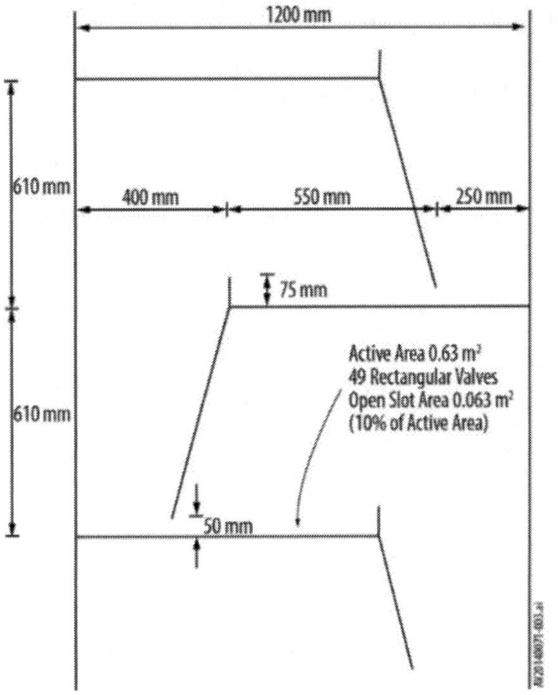

Figure 2 FGT Absorber Tray Details

Table 1 FGT Absorber Hydraulic Parameters Based on Bottom Tray

	Gas	Liquid
Flow rate, kg/h	28,580	42,133
Density, kg/m^3	52.7	1062
Volumetric flow rate, m^3/h	543.6	39.7
Liquid viscosity, cP		1.69
Liquid surface tension, mN/m		51

Table 2 Flood Hydraulic Evaluation[1] at Design Rates

	Tray Supplier	Fluor
Percent Jet Flood	54%	53%
Downcomer Backup, mm liquid	170	240
Percent froth in downcomer	48%	58%
Downcomer inlet velocity, m/s	0.033	0.033
Percent downcomer velocity flood	29%	24%
Weir load, M^3/h m	35	34
Pressure drop, mm liquid	106	122
mbars	11	13
Dry pressure drop, mm liquid	62	61
Liquid height at downcomer exit, mm		109
Head loss at downcomer exit, mm		8

Note 1: Values in Table 2 do not account for possible foaming or fouling

Initial Trials and Initial Fix Attempt

The tower was commissioned and operated at the MDEA solution design liquid rate of 38 M^3/hr and gas flow rate was raised to 10 MMSCFD. At about 10 MMSCFD, the column dP rose from 150 mbar to 450 mbar. Liquid carried over from the top of the absorber causing high level in the fuel gas knockout drum (FG KO Drum) downstream. This high level resulted in the trip of the system.

Several other trials were conducted over a period of two weeks without success. ADGAS engineers suspected a foaming problem or a major mechanical integrity incident that has happened during commissioning. The FGT absorber was shut down, then opened and thoroughly inspected, but neither mechanical damage nor fouling was found. The column was returned to operation.

To test the foaming theory, ADGAS circulated water instead of MDEA in the absorber. Water is non-foaming. Disappearance of the problem in the water test will verify foaming, while persistence of the problem in the water test will disprove foaming. Water circulation started at a rate of 10 M^3/hr, raised in steps to 20 M^3/hr, and gas rate raised from 5 to 10 MMSCFD. At 12 MMSCFD the column dP rose sharply to 245 mbar, with water carryover from the top of the absorber and high level in the FG KO Drum downstream. This trial proved that it is not a foaming problem.

Additional trials were conducted after replacing water with MDEA. Gas and MDEA flow rates were varied, and the MDEA feed point was changed between the 12th, 16th and 20th tray. The tower could not achieve stable operation at above a gas rate of 12 MMSCFD. The liquid rate did not appear to make much difference.

The tray supplier and engineering contractor conducted many additional trials and performed calculations. They concluded that the sudden increase of pressure drop is due to flooding of the trays. To fix the problem, they recommended the following changes in the trays:

a. Reduce the weir heights from 75 mm to 62 mm to decrease tray pressure drops and help open the valves (due to the shorter liquid head on the tray).
b. Weld-open all valve caps on Tray 1 and 2 (so they do not close) to prevent flood initiation at these trays.
c. Replace the 4" drain pipe from the bottom seal pan by a 6" pipe. A seal cup was installed at the bottom of this pipe, which was previously unsealed. Also, a vortex breaker was added where the new pipe exited the seal pan of tray 1.

Upon tower shutdown to perform the above modifications, it was thoroughly inspected. The tower was clean and there was no mechanical damage whatsoever. All tray and downcomer dimensions were checked and found to be according the drawings, denying any possible installation errors.

Upon return to service, there was only a minor improvement. The tower could stably operate only at less than 11-12 MMSCFD of sour gas and with maximum 20 M^3/hr MDEA circulation.

Troubleshooting

Much of the troubleshooting endeavors were already described above. Some of the tests were done before the initial fix attempt. Several theories were proposed by the tray supplier and engineering contractor, but none was supported by the trials. Some of the key theories were:

Fouling: This is a common issue in many amine systems, mostly MEA and to a lesser extent MDEA. However, the inspection during the two shutdowns found the trays shining clean. The tower was water-washed before the inspection, but even after a wash some deposits usually remain. No such deposits were found. This conclusively denied this theory.

Damage or Installation error: These theories were also denied when the tower inspection during the two shutdowns showed no damage and no installation error.

Foaming: This theory was conclusively denied when trials using water instead MDEA showed an identical tower limitation. This testing is novel. While amine systems may foam, a gas-water system is non-foaming. So testing a foaming system using clean water instead of the potentially-foaming process liquid can conclusively confirm or deny foaming.

In another test antifoam was injected at different concentrations and different rates. Antifoam injection gave no improvement to the system performance, again denying foaming.

In addition, hydraulic analysis by the tray supplier and us showed that even if foaming occurred, the tray design would handle the full design rates without flooding. In contrast, field observations showed that the column was flooding at about 40-50% of the design rates, both gas and liquid.

High Liquid Level in the Tower Base: This is a common cause of flooding (2). However, in this tower the bottom of the vane vapor distributor was a large margin, 1.7 meters, above the high-high liquid level. In some of the trials the liquid level was lowered, but this made no difference to the amount of gas that the tower could handle. This conclusively denied the high liquid level theory.

With all the above theories denied, there was no explanation to the premature flood.

Second Fix

ADGAS has another absorber in similar service that was working well. This tower contains bubble cap trays and showed no limitation with ramping up or down with gas turbine requirements. In the absence of a definition of the root cause of the problem, and in view of trays supplier recommendation that valve trays should not be ramped up more than 0.5 MMSCFD every 10 minutes, which does not fit with gas turbine fuel gas supply demand, ADGAS decided to scale up the design of the working absorber as the second fix to the absorber that was experiencing problems. It was decided to make the following changes:

1. Change the valve trays to Bubble cap trays.
2. Reduce weir heights further from 62 mm to 55 mm.
3. Reduce downcomer clearances from 50 mm to 40 mm.
4. Change the downcomer from sloped with top and bottom widths of 400 mm and 250 mm respectively to vertical 250 mm wide. This increased the tray active areas.
5. The open area was increased from 10.3% of the active area in the valve trays to 15.6% of the active area in the bubble caps.

In order to positively avoid another failure, and recognizing Fluor's distillation/absorption expertise, ADGAS requested Fluor to review the design of the second fix and evaluate whether Fluor believe it will solve the problem. Fluor's approach was to work with ADGAS on identifying the root cause before evaluating the solution. Our joint investigation that led to the root cause is described below.

Reanalysis of Pressure Drop Data in Search for the Root Cause

Fluor and ADGAS jointly went over the test and trial data that ADGAS obtained (Table 3). The erratic nature of the pressure drop per tray obtained from the measurements is astounding. At a gas rate of 5 MMSCFD, the pressure drop per tray varied from 2 to 8.5 mbar per tray. At and above 10 MMSCFD gas flow rates the pressure drop varied between 8 and 24 mbar per tray.

The design pressure drop for the trays in the absorber is of the order of 11 -13 mbar per tray (Table 3). Since the tower operated at lower vapor and liquid loads at the trials, a 7 to 9 mbar per tray pressure drop would be reasonable and would indicate good operation. High pressure drop of twice the design would indicate flooding as previously identified.

The measurements that caught our eyes were actually at the low end. At a gas flow rate of 5 MMSCFD and liquid flow rates of 10 M^3/h and less, the pressure drop was tiny, about 2 – 3 mbar. When the liquid load increased from 10 to 17 M^3/h, the pressure drop almost quadrupled to 8 mbar. This sudden change can only be explained by the downcomers losing their seal completely at the lower liquid flow rate. This, together with the earlier observation of a low head loss at the downcomer exit (Table 2), prompted us to explore a downcomer unsealing problem, which until then was not on the troubleshooting checklist.

Focus on Downcomer Sealing

The challenges of handling low liquid loads in tray towers are keeping the little liquid on the tray without drying up, and keeping the downcomers liquid-sealed (2). When a downcomer loses its liquid seal, gas rises up through it, and interferes with liquid descent. This interference has led to capacity bottlenecks, poor separation, instability, and total inability to start towers up.

On conventional trays, at liquid flow rates above about 20-30 M^3/h per meter of outlet weir length, the outlet weir generates a frothy pool (Fig.3a). As long as the downcomer clearance is lower than the liquid head in the pool, this liquid pool will seal the downcomer. In contrast, at low liquid flow rates (below about 20-30 M^3/h per meter of outlet weir), and especially at high gas velocities, the gas atomizes the tray liquid, forming a dispersion known as the spray regime (3, 4) or drop regime. In this regime, there is no frothy pool and the liquid resides as atomized drops in the inter-tray space (Fig. 3b). The continuous gas phase reaches the downcomer outlet. There is no static seal preventing gas rise up the downcomer. In the spray regime the downcomers are prone to seal loss.

Fig. 4 illustrates how downcomer unsealing can bottleneck a tower in the spray regime. This is a common scenario at startups at low liquid rates. Consider a tray with very little liquid on it. The gas, looking for the easiest path, ascends through both the trays and downcomers (Fig. 4a). Upon liquid introduction, some reaches the downcomer across (Fig. 4b). If the gas velocity through the downcomer is too high, it will not allow the liquid to descend into the downcomer (Fig.4c). Unable to descend, the liquid will accumulate on the tray above. The tray will flood, and liquid will carry over the top of the tower (Fig. 4d).

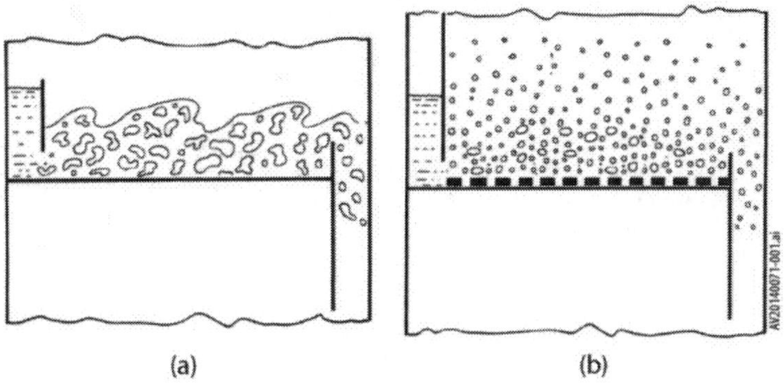

(a) (b)

Figure 3. Flow Patterns on trays (a) Froth regime (liquid phase is continuous) (b) Spray regime (gas phase is continuous). (Henry Z. Kister, excerpted by special permission from Chemical Engineering, September 8, 1980; copyright by McGraw-Hill, Inc., New York, NY, 10020).

Downcomer unsealing in the spray regime was modelled by Kister (5), with slightly revised equations published in reference 6. Successful applications of this model were reported (7, 8). No other models have been reported.

Previous applications of the Kister model were for sieve trays. There is a unique feature in valve trays that requires special consideration, as discussed below.

Downcomer Sealing in Valve Trays

In sieve and bubble cap trays, the dry pressure drop, or the pressure drop through the tray perforations, increases with the square of the gas velocity. Considering a tray with very little liquid on it, raising the gas rate will increase the hydraulic resistance to the ascending gas through both the tray perforations and the downcomer clearance (Fig. 4a) according to the equations:

$$\Delta P_T = K_T \frac{1}{\rho_V \rho_L}\left(\frac{V_T}{3600\, A_h}\right)^2 \qquad (1)$$

and

$$\Delta P_D = K_D \frac{1}{\rho_V \rho_L}\left(\frac{V_D}{3600\, A_{ud}}\right)^2 \qquad (2)$$

Where

ΔP= Dry pressure drop, mm liquid

V = Vapor (gas) flow rate, kg/h

K = Orifice constant, mm liquid $(s/m^2)^2$

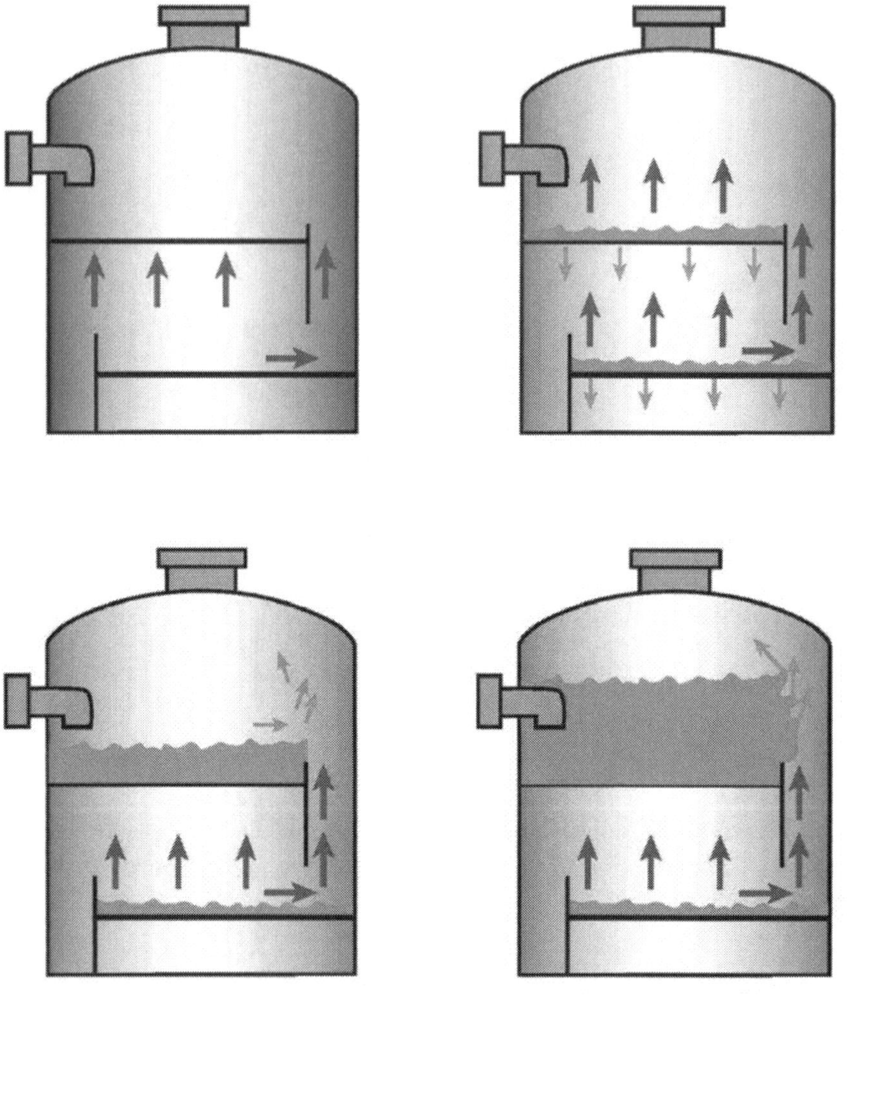

Figure 4. How Downcomer Unsealing Bottlenecks a Tower in the spray regime (a) Gas distributes between tray and downcomer; (b) Liquid introduced, reaches the downcomer, attempts to descend; (c) When gas velocity up the downcomer is too high, the gas does not permit liquid descent; (d) Unable to descend, liquid accumulates on the upper tray, generating flood.

Table 3 Pressure Drop Measurements at Different Gas and Liquid Loads

Gas Flowrate, MMSCFD	Liquid Flow rate, M^3/h	Inlet Tray for Lean Amine	Total Pressure Drop, mbar	Pressure drop mbar per tray	Liquid carryover?
5	10 (water)	20 (top)	39-44	2.0 – 2.2	No
5	10 (water)	12 (mid)	52	2.6	No
5	17 (water)	12 (mid)	162 - 170	8.1 – 8.5	No
5	20 (MDEA)	20 (top)	150	7.5	No
5	38 (MDEA)	20 (top)	170	8.5	No
10	17 (water)	12 (mid)	172	8.6	No
10	20 (MDEA)	20 (top)	160	8.0	No
10	38 (MDEA)	20 (top)	470	24	Yes
12	20 (water)	12 (mid)	245-295	12.3-14.7	Yes
15	20 (MDEA)	20 (top)	175	8.8	No
20	20 (MDEA)	20 (top)	230	11	Yes

Figure 5 Valve Tray Dry Pressure Drop Characteristic

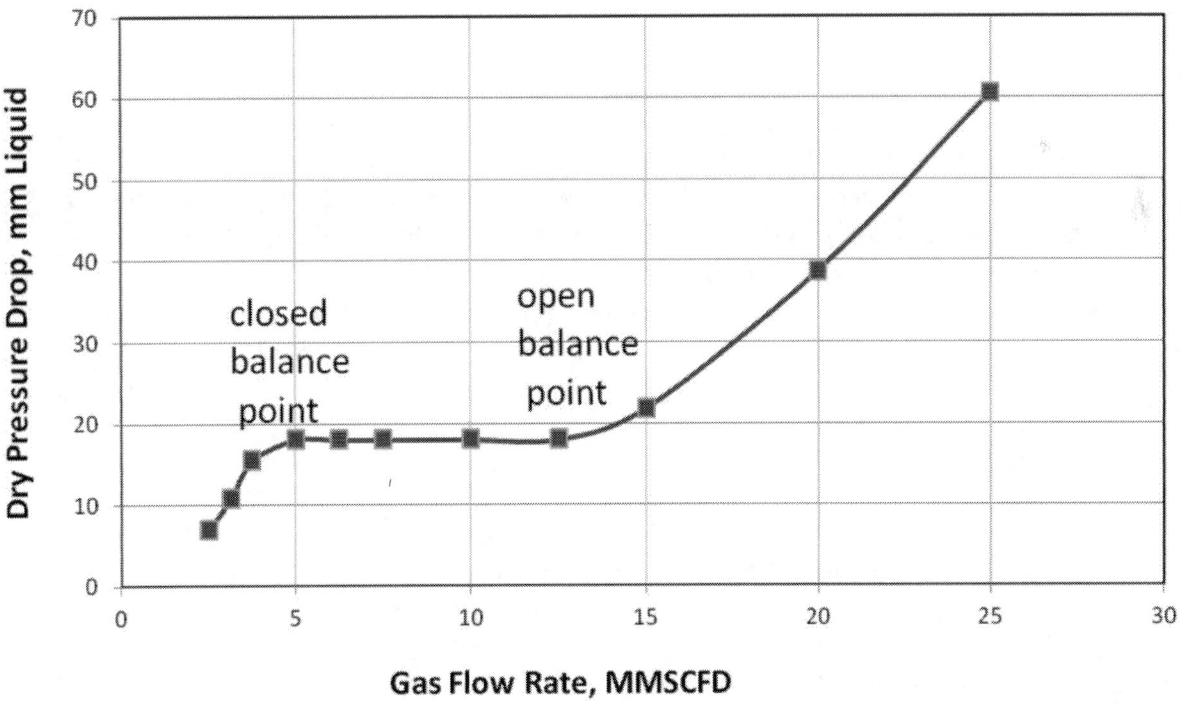

A = Area, m^2

ρ = Density, kg/m^3

Subscripts L indicate liquid, V vapor (or gas), T tray, D downcomer, h tray holes, ud clearance under the downcomer.

Since the gas splits between the tray and downcomer in a manner that minimizes hydraulic resistance, the pressure drops in Eq. 1 and 2 are equal, giving

$$V_D = \frac{V}{1+C^{-0.5}} \qquad (3)$$

and

$$V_T = \frac{V}{1+C^{0.5}} \qquad (4)$$

where

$$C = \left(\frac{K_T}{K_D}\right)\left(\frac{A_{ud}}{A_h}\right)^2 \qquad (5)$$

Eqs. 3 to 5 state that the split of gas between the tray holes and the downcomer depends only on the open areas and the orifice coefficients, but not on the gas flow rate. Eq. 3 also states that as the gas flow rate to the tower decreases, so does the gas flow rate through the downcomer.

With valve trays (Fig. 5), at low gas flow rates, the valves are closed. Eqs. 1 to 5 apply, with eq. 1 based on the area of the cracks under the valve floats left open when the valves are fully seated. Above a certain gas rate (termed "the closed balance point"), the valves begin to open, at which point the pressure drop becomes constant and no longer rises until all the valves open (termed "the open balance point"). At gas flow rates above the open balance point, the dry pressure drop again rises with the gas rate according to Eq. 1, this time with the hole area being the full open slot area. Between the closed and open balance points, Eq. 1 does not apply. and the pressure drop is given by

$$\Delta P_T = K_o \qquad (6)$$

Where K_o is a constant. For the trays in Fig. 5 the constant value is about 18 mm of liquid. Combining Eq. 2 and 6 means that while the valves are opening, the pressure drop across the downcomer remains constant, and so does the gas flow rate through the downcomer. The gas flow rate through the downcomer is given by

$$V_D = 3600\, A_{ud} \left(\frac{K_o \rho_V \rho_L}{K_D}\right)^{0.5} \qquad (7)$$

With sieve, bubble-cap, and valve trays above the open balance point, lowering the total gas flow rate reduces the gas flow rate through the downcomer, giving liquid a chance to descend and seal the downcomer (avoid the steps in Figure 4c and 4d). Not so with valve trays operating in the flat region in Fig. 5. In this region reduction of the total gas flow rate no longer reduces the gas flow rate through the downcomer until all valves close and the tray pressure drop again starts to decline.

The above model and discussion are only valid in the spray regime (Fig. 3b), i.e., when the liquid flow rate is 20-30 M^3/h per meter of outlet weir or less. At higher liquid loads there is a transition to the froth regime. In the fully-developed froth regime, the outlet weir generates a frothy pool (Fig. 3a) on the tray that keeps the downcomer liquid-sealed. During the transition, the frothy pool begins to form and cuts the effective downcomer escape area A_{ud}. This reduces the gas flow up the downcomer and helps seal the downcomer. The model does not account for this effect, and should not be applied for liquid loads above 20-30 M^3/h per meter of outlet weir.

Downcomer Unsealing in the FGT Absorber

Fig. 5 is the actual pressure drop characteristic for the trays in the FGT absorber. For gas flow rates between 5 MMSCFD and about 13 MMSCFD the valve trays operated in the flat region of Fig. 5.

Downcomer unsealing flood occurs when the gas velocity inside the downcomer is too high to permit all the liquid to descend. That gas velocity can be estimated by hydraulically modeling the unsealed downcomer as a counterflow column, through which gas ascends and liquid descends simultaneously, both fighting for the same opening. The counterflow column used in our past modeling work was a wetted-wall column (5). Here, considering the sloped downcomer and the availability of the newly-developed baffle tray flood correlation (8), we preferred to model the downcomer as a baffle tray with gas and liquid competing for the narrow opening at the bottom of the tapering downcomer. We also checked our calculations by comparing to the system limit correlation (1), and found that the gas velocities calculated from the baffle tray model were about 15% less than those from the system limit correlation.

Figs. 6 through 9 show our calculation results.

Fig. 6 shows that at the design liquid load of 38.5 M^3/h the gas velocity through the downcomers is too high to permit liquid descent whenever the gas flow rate exceeds about 6 MMSCFD. Above about 8 MMSCFD the gas flow rate through the downcomers is about 137% of flood. This figure is not accurate, and is likely to be high, because the design liquid load of Fig. 6, 34 M^3/h per meter of outlet weir length, is just outside the range of our model. As stated earlier, our model is valid when the liquid flow rate is 20-30 M^3/h per meter of outlet weir or less. While the 137% of flood is not an accurate value, we still expect the gas rate to exceed that of a downcomer seal flood.

The middle curve in Fig. 6 shows that the percent of downcomer seal flood declines to 102%, still above the flood point, when the liquid rate is lowered to 20 M^3/h. This liquid rate is well within the range of our model. At 20 M^3/h, the flood point is at and above 10 MMSCFD. The lower curve on Fig. 6 shows that at a liquid rate of 10 M^3/h, the percent flood at the downcomer is well below 100%, so that flooding would not be experienced. However, this liquid load was too low to permit good absorption and was only used at very low gas rates (Table 3).

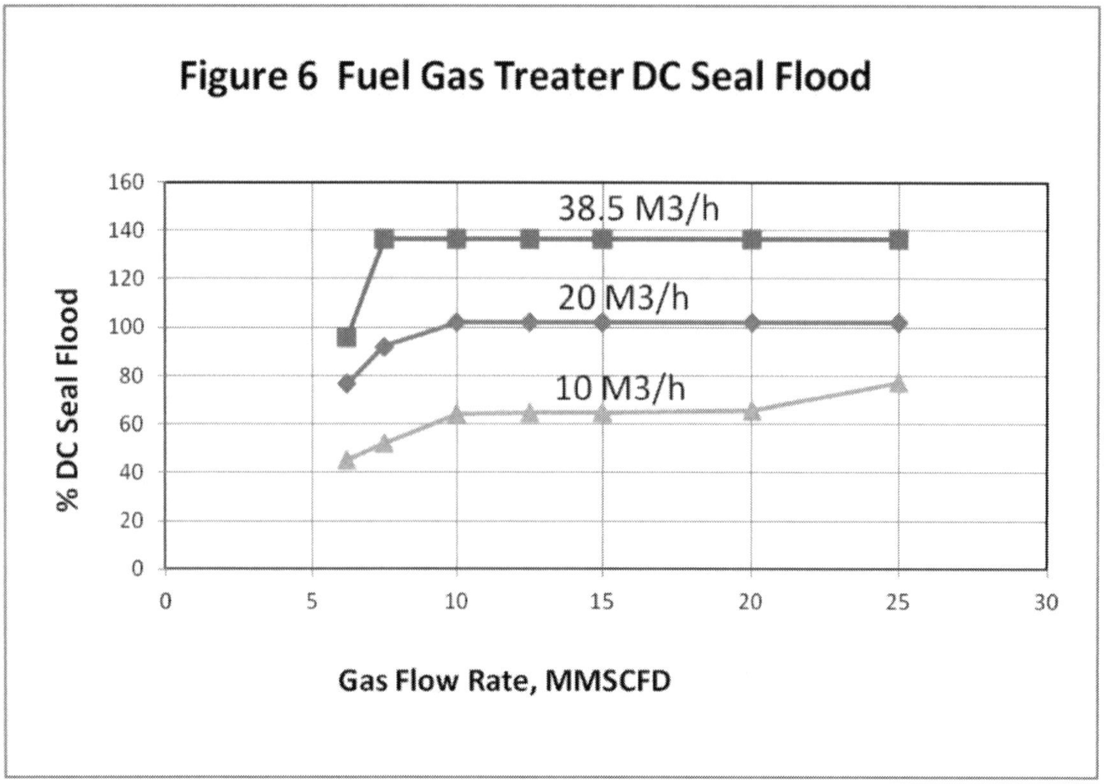

Figure 7 shows how the valve design could have been modified to circumvent the downcomer unsealing flood. The upper curve is for the original valve tray design with 50 mm clearance and sloped downcomers, and is the same as the middle curve in Fig. 6.

The lower curve Fig. 7 shows that a smaller clearance would have reduced the gas flow up the downcomer well below the flood point at all gas loads, allowing liquid to descend with no difficulty and seal the downcomers. The tower would have worked and there would have been no need for troubleshooting. The large impact of downcomer clearance on downcomer sealing was previously reported in sieve trays (5).

The middle curve in Fig. 7 shows the effect of using straight instead of the sloped downcomers in the original valve trays. The tray active areas were kept the same as in the original design (0.63 m^2 per Fig. 2), making the downcomer 328 mm wide. The downcomer clearance was kept at 50 mm per the original design. Compared to the original design (Fig. 1) the straight downcomer was narrower at the top but wider at the bottom.

The middle curve in Fig. 7 shows that the larger area near the bottom of the downcomer would have reduced the gas velocity through the downcomer well below the flood point at all gas loads, allowing liquid to descend and seal the downcomers. Had the downcomers been straight 328 mm wide, the tower would have worked well even with the 50 mm clearances, and there would have been no need for troubleshooting.

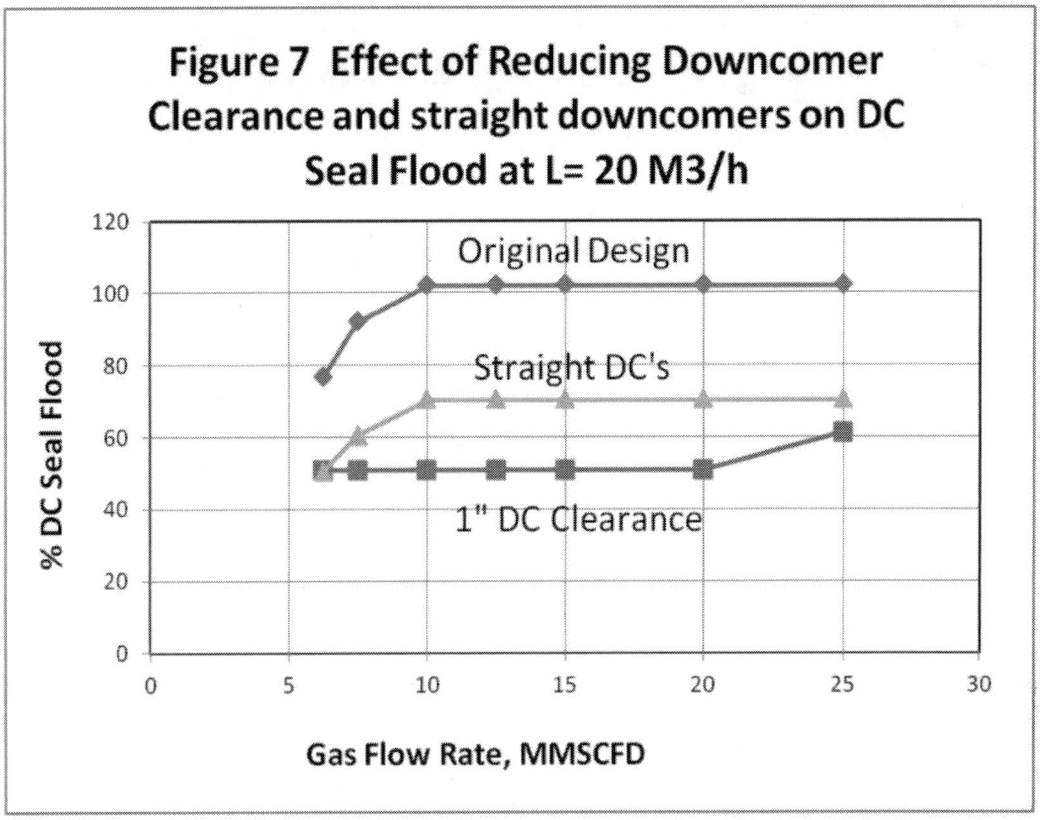

Fig. 8 compares the downcomer seal flood curve for the bubble-cap trays of the second fix (lower curve) to the downcomer seal flood curve for the original valve trays (upper curve, same as the middle curve in Fig. 6). It is important to appreciate that the bubble-cap tray design in the second fix had smaller downcomer clearances, larger active areas, and larger fractional hole areas than the original valve tray design, all of which will divert more gas through the trays and less through the downcomers, thereby reducing the percent downcomer seal flood.

Fig. 8 shows that the downcomers for the new bubble-cap trays would operate well below the downcomer seal flood rates at all gas rates. Liquid will be able to descend with no difficulty and form a seal at the bottom of the downcomer that would prevent further gas rise through it. Based on this finding, Fluor gave full endorsement for the second fix design.

The success of bubble-cap trays for solving downcomer sealing problems at low liquid rates is not unique to this tower. Lieberman (100 previously reported successful sealing experiences with bubble-cap trays at the extremely low liquid rates used in gas plant glycol dehydrators. Beside avoiding the constant dry pressure drop region of the valve trays (Fig. 5) which tends to divert a high gas flow through the downcomers, the use of bubble-caps allows incorporating more open area on the trays without fear of leaking the small amount of liquid that is held on the trays at these low liquid rates. A well-designed and well-sealed bubble-cap tray is highly leak-resistant because it has the gas entering about 70 mm above the tray floor.

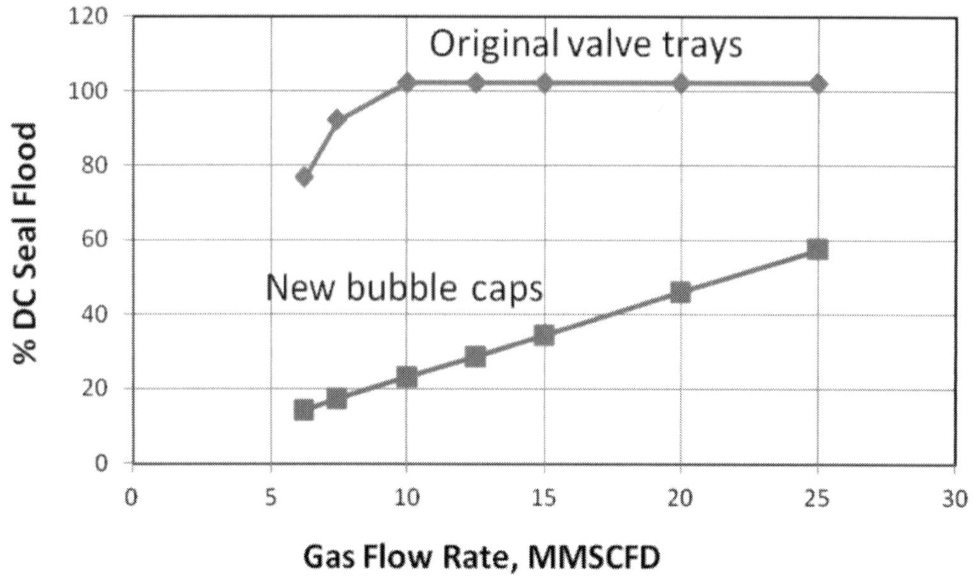

Figure 8 DC Seal Flood at L= 20 M3/h: New Bubble Caps vs. Original Valve Trays

Epilogue

The valve trays in the tower were replaced by the new bubble-cap trays described under the second fix in October 2013. The tower has been performing well ever since with no signs of flooding or instability, even well above the design rates.

Extensive tests after the second fix showed minimal pressure drop fluctuations with increases in the feed gas flow to the FGT absorber. Pressure drop did not exceed 7.5 mbars/tray even at 25 MMSCFD gas flow and 40 M^3/h MDEA flow. There have been no issues with the FGT absorber operation and its bubble-cap trays when feed gas was quickly ramped up. No sudden pressure drop increases were recorded and the tower did not take long to stabilize at the new gas flow rates. The H_2S content of the overhead gas ranged from 0 to 2 ppmv, well below the 20 ppmv max.

This experience sheds light on the unique experience of downcomer seal loss in moving valve trays. The constant dry pressure drop held by the valve tray as valves close off at reduced rates (the flat portion of the curve in Fig. 5) can make downcomer sealing difficult in moving valve trays, especially with large clearances. Large downcomer clearances should be avoided at low liquid rates, both with valve trays (this study) and other trays (5).

Finally, this study emphasizes the effectiveness of bubble-cap trays to deal with downcomer unsealing problems, as previously noted by Lieberman (10).

References

1. Kister, H. Z., P. M. Mathias, D. E. Steinmeyer, W. R. Penney, B. B. Crocker, and J. R. Fair, "Equipment for Distillation, Gas Absorption, Phase Dispersion and Phase Separation", in D. W. Green and R. H. Perry "Perry's Chemical Engineers' Handbook" 8th Ed., Section 14, McGraw-Hill, New York, 2008.

2. Kister, H. Z., "What Caused Tower Malfunctions in the Last 50 Years?", Trans. IChemE, Vol 81, Part A, p. 5, January 2003.

3. Lockett, M.J., "Distillation Tray Fundamentals," Cambridge Univ. Press, Cambridge, UK, 1986.

4. Kister, H.Z., "Distillation Design", McGraw-Hill, NY, 1992.

5. Kister, H. Z., "When Tower Startup Has problems", Hydrocarbon Processing, 58(2), p.89, 1979.

6. Kister, H. Z., "Distillation Operation", McGraw-Hill, NY, 1990.

7. Kister, H. Z., and M. Olsson, "Understanding Maldistribution in 3-Pass Trays", Chem. Eng. Res. and Dev., 89, p.1397 – 1404, Elsevier – IChemE, August 2011.

8. Bernard, A. "Distillation: Avoid Problems During Tower Startup", Chem. Eng., June 1, 2012. 1992.

9. Kister, H. Z., and M. Olsson, "Don't Let Baffle Tray Flood Baffle You" Chem. Eng. Progr., p. 22, July 2011

10. Lieberman, N. P. "Troubleshooting Natural Gas Processing", PennWell Books, Tulsa, OK, 1987.

OPTIMIZING PREFLASH OPTION FOR LIGHT TIGHT OIL PROCESSING

Soun Ho Lee
GTC Technology US LLC
Euless, TX

Presented for AIChE 2015 Spring Meeting
Austin, TX
April 26-30, 2015

"Kister Distillation Symposium"
SESSION 2 - Improved Design and Operation in Refinery Distillation
Paper 45a
April 27, 2015

INTRODUCTION

The basic function of a crude distillation unit is to provide the initial separation of crude oil feed into the desired fractions to feed downstream units. Now, newly-introduced light tight oil feed processing may limit existing crude distillation unit capacity. The addition of feed preflashing option may be one of feasible options to improve capacity and/or energy efficiency for light tight oil processing.

Two preflash options which are commonly implemented for crude distillation units are (1) preflash drum and (2) preflash columns. Each option has strengths and weaknesses.

This paper discusses improved semi-preflash column design that optimizes the traditional preflash drum and preflash column options. The case study, including revamp economics, will be demonstrated.

LIGHT TIGHT OIL

Light Tight Oil (LTO) is petroleum fractions produced from low permeability formations such as shale or tight sandstone.[1] Improved hydraulic fracturing technology with directional drilling has increased light tight oil production in North America, where processing light tight oil can be a game-changer for refiners. Light tight oil slates produced in North America have light and sweet natures typically. However, the characteristics of light tight oil significantly vary.

Adapting light tight oil as a new feed stock may require a significant retrofit in a refinery layout which was originally configured with traditional crude oil feed stock. Capacity imbalance of downstream conversion units and insufficient overhead train processing capacity are already well recognized issues with refiners. This imbalance issue is more exaggerated in refineries which are designed with heavy crude slates and a high conversion ratio. These mismatches between new light tight oil feed stock and current refinery configurations are initiated in the crude distillation unit which provides initial separation of feed stock for entire downstream refinery processing units.

There have been extensive researches to try and identify "magic juice", blending ratios of light tight oil and traditional heavy crude oil slates to fit existing refinery layouts. Meanwhile, it has been discovered that blending between light tight oil and traditional heavy crude oil can cause asphaltene destabilization due to the paraffinic nature of light tight oil. [2] This destabilization can cause unexpected fouling issues.

This paper focuses on debottlenecking and enhancing crude distillation unit capacity and energy consumption using various preflash options. Other retrofit strategies for light tight oil processing are not discussed in this paper.

CRUDE DISTILLATION UNIT PREFLASHING

Implementing crude feed preflashing has been one of the common options to debottleneck crude distillation units. Switching feed stock to lighter crude slate can cause undesired feed vaporization at the feed furnace pass control valve and limit capacities in the crude atmospheric column and overhead condensing circuits. Adding a preflash drum or column can reduce the charge rate to the feed furnace. In addition, transfer line vibration due to capacity limitation can be resolved through adapting feed preflashing.

Two preflash options that are commonly implemented to crude distillation unit are (1) preflash drum and (2) preflash column. Another preflash option, (3) GT-IPS^{SM} is a semi-preflash column design and may be optimized between traditional preflash drum and preflash column options.

PREFLASH DRUM OPTION

The preflash drum is the simplest option for crude preflashing. Complex equipment and instrumentation modification are not required for this option compared to preflash column, making it the most economical option with regards to CAPEX (capital expenditure).

One weakness of the preflash drum is that this option does not have capability of controlling preflashed vapor quality in foaming prone service. Therefore, the preflash drum is prone to have challenges with entrainment. Entrainment can be accelerated in foaming environment. Improper location of preflashed vapor feeding and/or undesirably entrained heavy oil boiling range materials can contaminate rundown product qualities.

Introducing preflashed vapor to a flash zone is the safest configuration to prevent product contamination. However, lower temperature of preflash vapor containing entrainment can quench the main feed vapor and can reduce feed lifting.[3]

Although recent technology such VTC (Vortex Tube Cluster) helps to discourage the chance of foaming, a preflash drum should be sized large enough to prevent foaming. [4]

PREFLASH COLUMN OPTION

The preflash column option produces preflashed products. Although it requires higher CAPEX than a preflash drum, this option is especially effective when the column top section and overhead condenser capacities are limited.

A rectifier is a common configuration for a preflash column. Crude feed is introduced at the preflash column bottom and preflashed vapor is rectified by reflux liquid. Overhead vapor can be condensed through a crude atmospheric column overhead condenser if the condenser has ample capacity. An exclusive preflash column overhead condenser needs to be installed if a crude atmospheric column overhead condenser is limited. Some deluxe preflash columns have bottom stripping section to control the front end of the crude atmospheric column top product.

High CAPEX is a major drawback of preflash column. Moreover, a large plot space is generally needed to add the equipment.

GT-IPS^{SM} SEMI PREFLASH COLUMN OPTION

GT-IPS^{SM} (Improved Preflash System) has a "semi-preflash column" arrangement compared to the traditional preflash column. Like the preflash drum option, the semi-preflash column does not produce any independent product. This column functions as a preflash column but the additional overhead condenser and receiver are not equipped for the preflash column.

Maintaining a desired preflash vapor quality is one of the most critical issues. To wash out the entrained components, some liquid is withdrawn from the crude atmospheric column and routed to the semi-preflash column as reflux. Then controlled preflashed vapor is safely fed to the middle of the column atmospheric column.

One of the benefits of the GT-IPS^{SM} concept is its flexibility. The preflash column reflux source and the preflashed vapor routing can be varied depending on process conditions, including crude slate and target performance, and refinery layout.

To compare three preflash options in more details, the revamping of a crude distillation unit for light tight oil processing is included as case study.

CASE STUDY: UNIT DESCRIPTION AND STUDY BASIS

Eagle Ford, a common light tight oil in the US, was selected as the feed stock. Eagle Ford tight oil has light and sweet characteristics with approximately 0.1 wt% sulfur content and is in a highly variable slate. Published gravity is varied between 40° and 62° API.

The case study's crude distillation unit receives 100% 60.7° API Eagle Ford; 50,000 BPD light tight oils are charged and heated through pre-heat trains and furnaces, and then introduced into the crude atmospheric column. This column separates the charged oils to intermediate products: (1) Unstabilized naphtha (2) Kerosene (3) Diesel and (4) Reduced Crude (R/C).

The crude atmospheric column equips three pumparound circuits: top pumparound, kerosene and diesel pumparound circuits. A top reflux stream which is recycled from the unstabilized naphtha (overhead distillate) is combined with the top pumparound stream before returning to the crude atmospheric column. The amount of the top reflux stream can be adjusted relative to the unstabilized naphtha boiling range of the processed crudes. The wash section is equipped with structured packing while the rest of the sections consist of trays. Two side strippers which belong to the crude atmospheric column are operated with stripping steam. Figure 1 illustrates the base case crude distillation unit configuration.

To evaluate the optimum preflashing option, three (3) case studies are conducted and compared to the base case. Light tight oil charge rate, quality, and product yields are maintained among four cases.

Since preflash equipment pressure is floated by crude atmospheric column operating pressure, each of the case's preflash system pressure is set through a pressure drop prediction between preflash and hydraulic end point in the crude atmospheric column. [5] In addition, no extra capacities are available in the crude atmospheric column overhead condenser and feed furnace. The base case crude atmospheric column overhead condenser and absorbed feed furnace duties are defined as maximum. Since preflashing options let down preheat train pressure, an additional booster pump is required to transport the remaining feed to the feed furnace and crude atmospheric column. Additional feed booster pump discharge pressure is set to prevent vaporization through the entire preheat train and the feed furnace pass control valve. Table 1 depicts process conditions of the crude distillation unit.

Figure 1 Crude Distillation Unit: Base Configuration

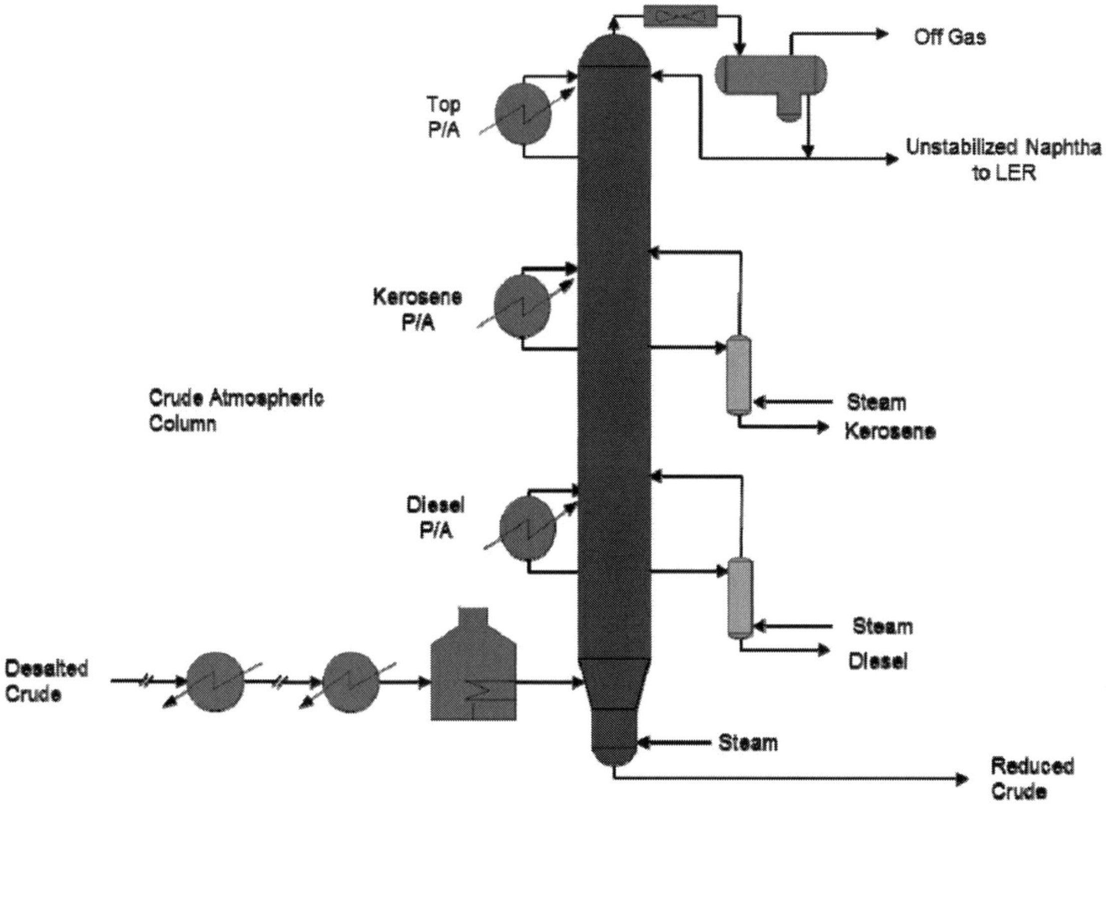

Table 1 Case Study Crude Distillation Unit

Case Parameter	Unit	BASE
Feed		
Crude slate	-	100% Eagle Ford
Crude charge	BPD	50,000
°API	-	60.7
- 300 °F Norminal Cut Fraction	LV %	58.2
300 – 450 °F Norminal Cut Fraction	LV %	14.7
450 – 650 °F Norminal Cut Fraction	LV %	13.4
+ 650 °F Norminal Cut Fraction	LV %	13.7
Operating parameter		
Furnace Coil Outlet Temperature	°F	674
Crude Atmospheric Column Top Pressure	psi (g)	24
Unit Bottom stripping steam rate	Lb / BBL	7.9
Unit Kerosene stripping steam rate	Lb / BBL	12.0
Unit Diesel stripping steam rate	Lb / BBL	6.2
Fractionation Performance		
Naphtha D86 (95 LV%)	°F	291
Kerosene Flash Point	°F	100.8
Kerosene Freezing Point	°F	- 58.0
Diesel D86 (90 LV%)	°F	649
Reduced crude D86 (5 LV%)	°F	617

CASE STUDY A: PREFLASH DRUM

Case study A illustrates how preflash drum implementation is conducted. A modified configuration is illustrated in Figure 2.

Considering potential rundown product contamination can occur by entrainment, preflashed vapor is introduced to the crude atmospheric column flash zone. Preflash drum operating pressure is determined per the crude atmospheric column flash zone pressure and the pressure drop through the preflash vapor line circuit.

Preflashing requires a higher furnace coil outlet temperature to maintain the current total distillate yields. Higher internal reflux ratios for fractionation sections are also required to achieve target distillate product yields / qualities. The preflash drum operating temperature is selected based on re-using atmospheric column overhead condenser duty. Higher preflash temperature can increase the amount of preflashing but it requires higher furnace coil outlet temperature and simulated overhead condenser duty that exceeds current value. Therefore, position of the preflash drum in this case study is determined upon overhead condenser duty limitation.

Figure 2 Crude Distillation Unit: Case Study A Preflash Drum Configuration

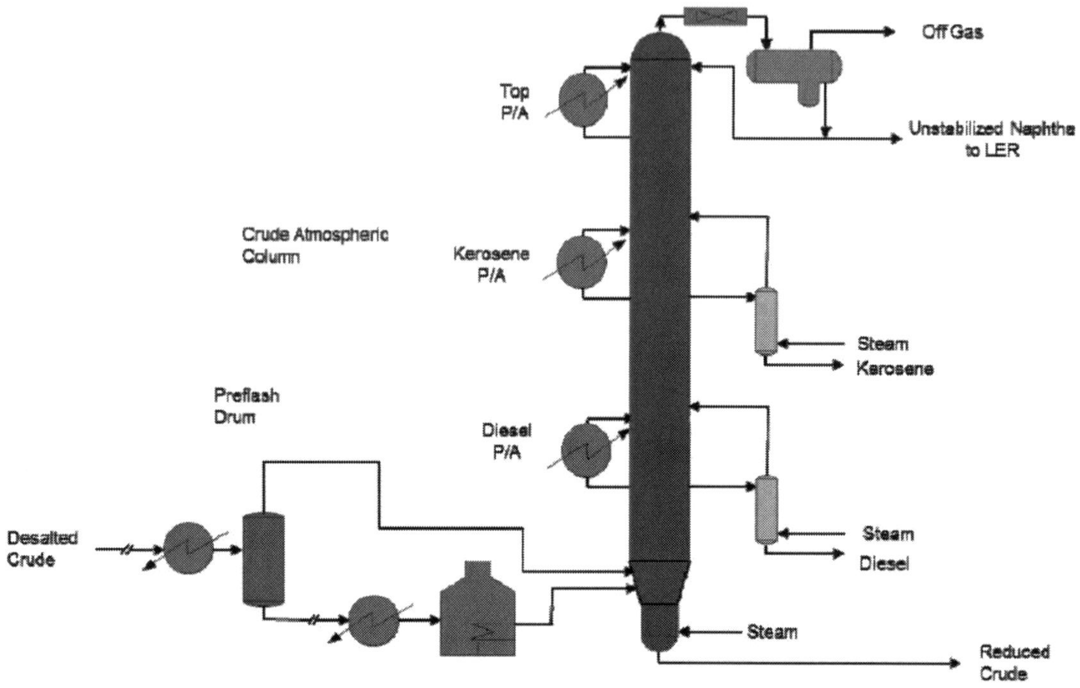

CASE STUDY B: PREFLASH COLUMN

Case Study B gives an overview of preflash column implementation. The preflash column is configured as a rectifier in this case study, and is equipped with an independent overhead receiver and overhead condenser. Preflash column is positioned in between the preheat train and furnace to maximize the amount of preflashing.

The preflash column overhead vapor streams are split into two streams. The amount of preflashed vapor stream charged to the crude atmospheric column overhead condenser is quantified by the crude atmospheric column overhead condenser limitation. The remaining preflashed vapor stream is condensed through an additional preflash column overhead condenser. The reflux rate for the preflash column is driven from the new preflash condenser and overhead receiver. Produced preflash overhead liquid distillate is combined with crude atmospheric overhead distillate and transported to a naphtha stabilizer for further separation.

A modified configuration for this case is illustrated in Figure 3.

Figure 3 Crude Distillation Unit: Case Study B Preflash Column Configuration

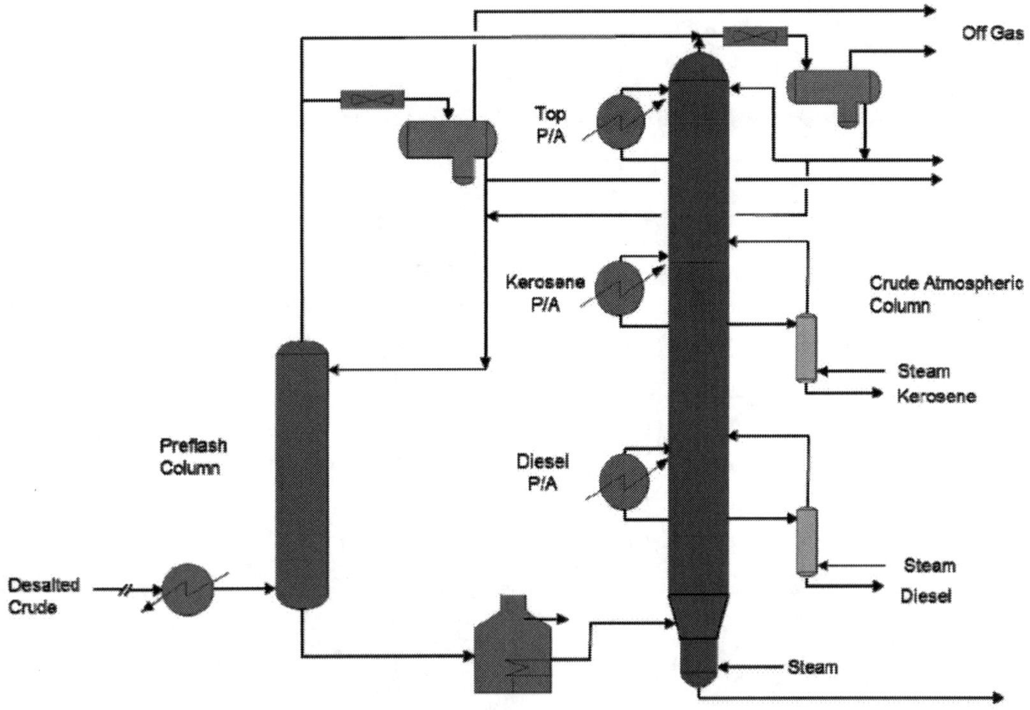

CASE STUDY C: GT-IPS^{SM} SEMI PREFLASH COLUMN OPTION

Case Study C illustrates the GT-IPS^{SM} semi-preflash column implementation scenario. A modified configuration for this case is illustrated in Figure 4.

Prior to being sent to the crude atmospheric column, produced preflashed vapor is rectified by the reflux stream recycled from the crude atmospheric column. The reflux source for the semi-preflash column is sourced from the crude atmospheric column overhead distillate in this particular case study. This reflux routing configuration is determined based on the amount of preflashed vapor and quality. Case study results show that the rectified preflashed vapor is suitable to mix with the unstabilized naphtha stream. Nevertheless, preflashed vapor is introduced underneath the naphtha/kerosene fractionation section taking floating semi-preflash column pressure under consideration.

The semi-preflash column operating pressure is determined by the crude atmospheric column operating pressure and the preflash vapor line pressure drop. The column operating temperature is selected based on re-using atmospheric column overhead condenser duty. Higher preflash temperature can increase the amount of preflashing but it requires higher furnace coil outlet temperature and simulated overhead condenser duty that exceeds current value.

Like the Case Study A preflash drum, the semi-preflash column is also located in the middle of the preheat train. However, a higher preflash percentage through the semi-preflash column allows the semi-preflash column to be located in a higher temperature zone of the preheat train compared to Case Study A.

Figure 4 Crude Distillation Unit: Case Study C GT-IPS Configuration

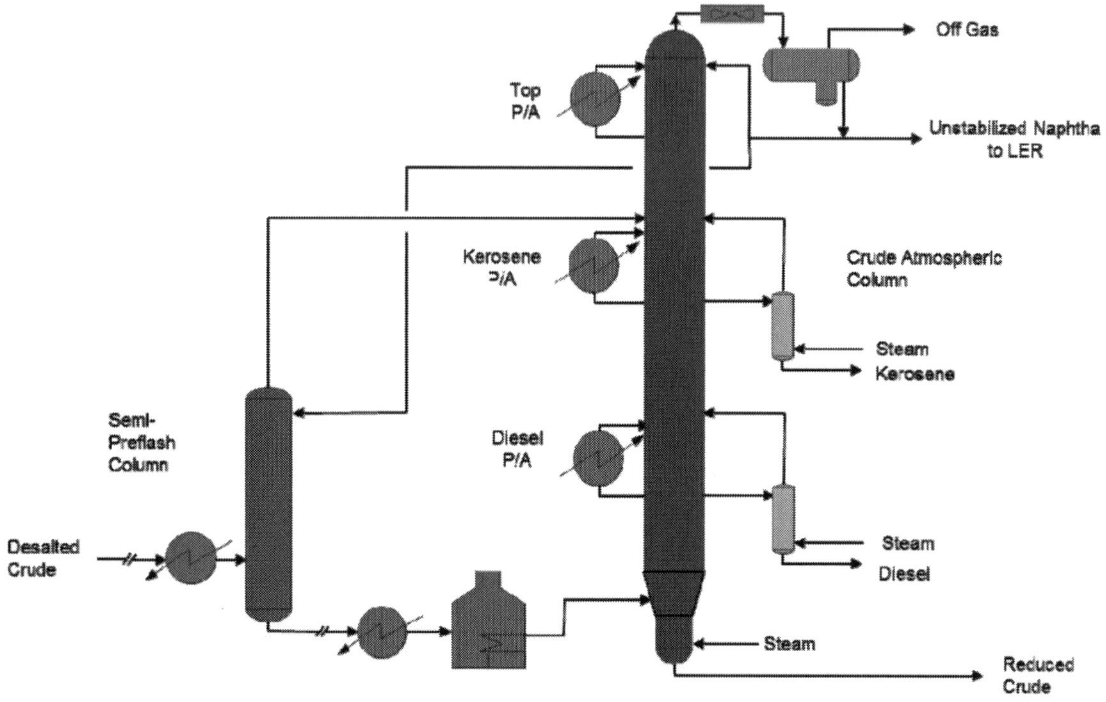

CASE STUDY RESULT EVALUATIONS

The three case study results are summarized and compared with base conditions in Table 2.

Since feed preflashinig results in a lower preflash drum or column bottom temperature, the temperature drop through each preflash option is quantified. The crude atmospheric column pumparound balance for each case is adjusted to maintain base case product yields and product qualities. These temperature drop values through preflashing and shifted pumparound balances are reflected in preheat train temperature profiles and furnace inlet temperature prediction. The cracking tendency identification of the chosen light tight oil slate can help optimize furnace coil outlet temperature further.

Crude atmospheric column distillation equipment capacities are calculated with simulated traffic. Calculated capacities are based on the packed wash section and fixed valve trays for rest of the section. Calculated capacity for each section is graphically plotted in Figure 5.

The Case study A preflash drum does not provide significant furnace duty savings. Moreover, calculated capacities of this preflash drum case are similar to those of the base case. Higher internal reflux ratios for fractionation sections erode distillation capacity gain. A lower preflash drum temperature requires higher discharge pressure of new booster pump installation compared to the other two cases: the preflash column and the semi-preflash column options.

Table 2 reveals that the case B preflash column option achieves the highest preflashing percentage among three cases. Meanwhile, the highest temperature drop is observed in the Case B preflash option, resulting in the lowest furnace inlet temperature. Nevertheless, the lowest net amount of furnace charge provides the minimum absorbed furnace duty. The preflash column also provides the highest distillation equipment capacity gain.

In Case C, the GT-IPS^{SM} semi-preflash column option shows lower furnace duty saving and distillation equipment capacity gain compared to the preflash column option. However, furnace duty saving and crude atmospheric column capacity gain are still substantial compared to the base case and the Case A preflash drum.

Table 2 Case Study Result Summary

Case / Parameter	Unit	BASE	Preflash Drum	Preflash Column	GT-IPS
Preflash Condition					
Preflash Inlet Temperature [note1]	°F	-	282	395	353
Preflash vapor/product %	Vol % of feed		16 %	41 %	32 %
Yield Structure					
Crude charge	BPD	BASE	0 %	0 %	0 %
Unstabilized Naphtha	BPD	BASE	0 %	0 %	0 %
Kerosene	BPD	BASE	0 %	0 %	0 %
Atmospheric Gas Oil	BPD	BASE	0 %	0 %	0 %
Reduced crude (R/C)	BPD	BASE	0 %	0 %	0 %
Operating parameter					
Furnace Inlet Temperature	°F	BASE	- Δ14 °F	- Δ68 °F	- Δ38 °F
Furnace Coil Outlet Temperature	°F	BASE	+ Δ31 °F	+ Δ45 °F	+ Δ42 °F
Absorbed Furnace Duty	MMBTU/hr	100 %	97 %	82 %	87 %
Preflash Condenser Duty	MMBTU/hr	100 %	-	Additional	-
Overhead Condenser Duty	MMBTU/hr	100 %	100 %	100 %	100 %
Top Pumparound Duty	MMBTU/hr	100 %	100 %	45 %	71 %
Kerosene Pumparound Duty	MMBTU/hr	100 %	86 %	100 %	77 %
Diesel Pumparound Duty	MMBTU/hr	100 %	76 %	100 %	91 %
Fractionation Performance					
Kerosene 5% -Naphtha 95% [note2]	°F	BASE	Δ 0 °F	+ Δ 2 °F	Δ 0 °F
LGO 5% -Kerosene 95% [note2]	°F	BASE	Δ 0 °F	Δ 0 °F	Δ 0 °F
Reduced crude 5% -Diesel 95% [note2]	°F	BASE	Δ 0 °F	Δ 0 °F	Δ 0 °F
Kerosene flash point	°F	BASE	Δ 0 °F	+ Δ1 °F	Δ 0 °F

Note

1. *Prior to preflashing*
2. *ASTM D86 (LV%)*

Figure 5 Crude Atmospheric Column Distillation Equipment Capacity

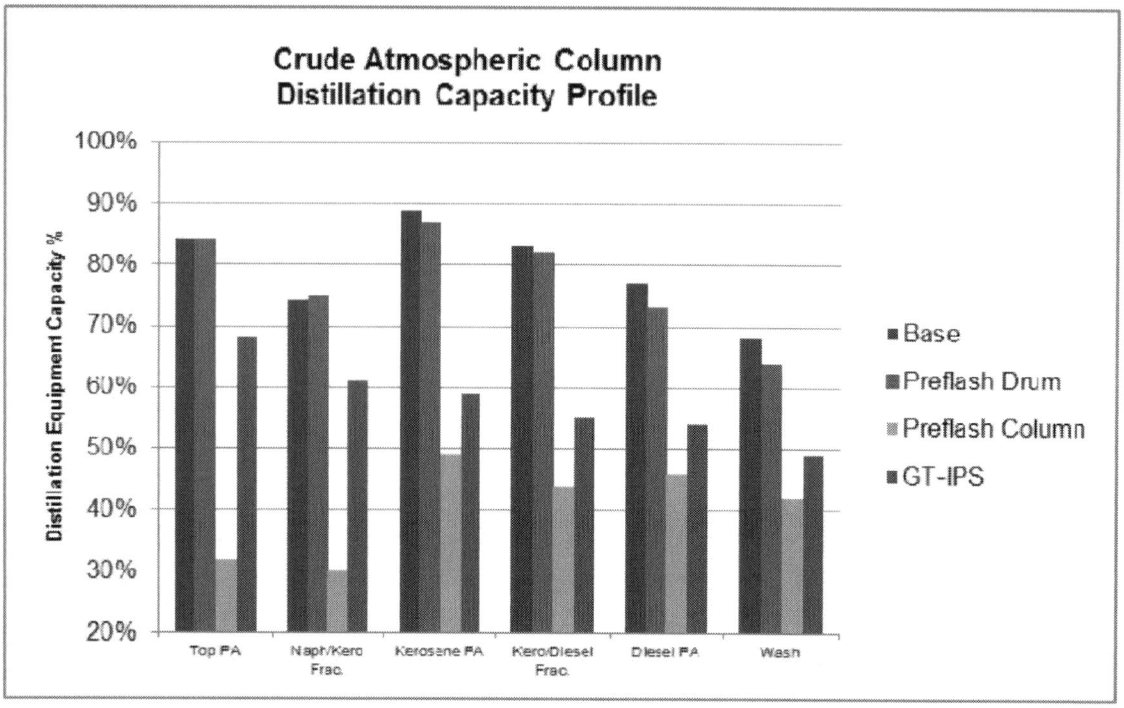

CASE STUDY: RETROFIT ECONOMIC EVALUATIONS

The Case A preflash drum option provides the simplest unit modification among the three cases. The required plot for the preflash drum implementation is also smaller than the other cases. However, the aforementioned higher discharge pressure of the new booster pump increases CAPEX substantially.

The Case B preflash column option provides the best performance improvement but it requires an additional overhead condenser and receiver for the preflash column and complex piping modification which influence retrofit cost. The highest CAPEX is identified for the Case B preflash column option. Larger plot requirement for the modification is another drawback in Case B. If space is limited in the existing unit, this case is not viable for implementation.

In Case C, GT-IPS^{SM} does not necessitate substantial equipment addition which is required in the Case Study B preflash column. Also, new booster pump differential pressure and capacity requirements in Case C are lower than those in Case A.

The pay-back period calculations are conducted to gauge retrofit economics. Crude atmospheric column capacity and/or furnace/transfer line capacity gains significantly improve retrofit profitability. However, profits through these factors are noticeably varied per refinery layout and equipment limitations. Therefore, these factors are not considered in the pay-back period calculations and only furnace heating medium saving is only used for economic evaluations in this paper.

Profitability indexes are expressed with regards to Pay-back Period. These indexes are charted in Figure 6. The modification costs are estimated based on the US Gulf region. Most US refiners utilize both refinery fuel gas and natural gas as furnace heating mediums. However, mixing ratio between two different utilities depends upon the particular refinery. Therefore, single EFO (Equivalent Fuel Oil) basis is selected. Three different fuel oil prices are selected through the last four year fuel gas and natural gas trend.[6] Typical furnace efficiency and EFO lower heating values are used for heating medium consumption predictions.

The chart in Figure 6 shows that Case C's GT-IPS^{SM} is identified to provide better retrofit economics compared to the other two cases. Adding extra profit through capacity gain can improve retrofit profitability further.

Figure 6: Profitability index – Payback Period

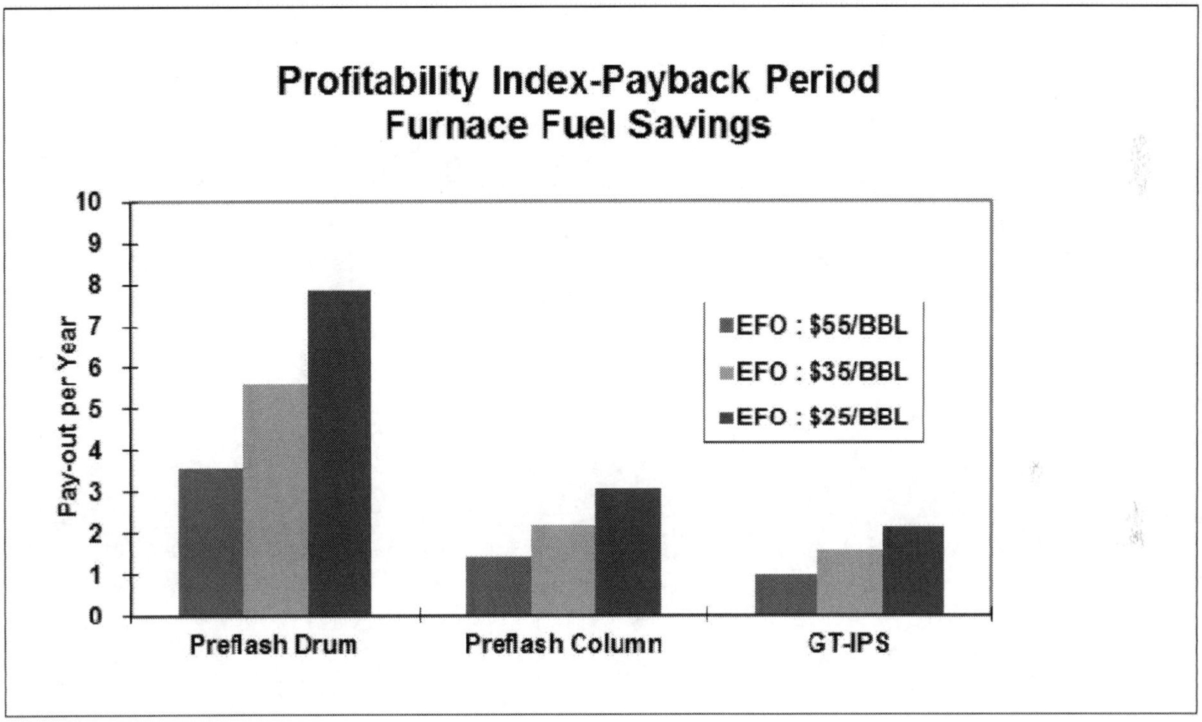

CONCLUSION

The technical results and economic evaluations from the three case studies show that the GT-IPS^{SM} semi-preflash column can deliver significant energy saving and capacity gain comparable to preflash column option. Meanwhile, the lower investment cost for GT-IPS^{SM} provides better retrofit economics compared to the preflash column case. These benefits can prove that GT-IPS^{SM} is one of viable options for efficient light tight oil processing Actual preflashing option and design can be customized after reviewing specific circumstances of crude distillation unit configuration and refinery layout. Retrofit economics can be precisely gauged after reflecting actual profit gains.

REFERENCES

1. Mills R, "The myth of the oil crisis: overcoming the challenges of depletion, geopolitics, and global warming", Greenwood Publishing Group, 2008

2. Sandu C, *et al*, "Innovative Solutions for Processing Shale Oils", Hydrocarbon Processing, July, 2013

3. Golden S W, "Prevent Preflash Drum Foaming" , Hydrocarbon Processing, May, 1997

4. Golden S W, "Crude Unit Preflash Drums and Columns" , Petroleum Technology Quarterly Revamp edition 2005.

5. Libermann N P, "Troubleshooting Process Operations" PennWell Books, 4th Edition.

6. US Energy Information Administration Web Site

Paper 45B

Atmospheric Preflash Towers

Andrew W. Sloley

CH2M HILL
21 Bellwether Way, Unit 111
Bellingham, WA 98225

Prepared for Presentation at the
American Institute of Chemical Engineers
2015 Spring Meeting
Austin, TX April 26-30, 2015

"UNPUBLISHED"

1. Abstract

Current trends in the crude petroleum supply are making large quantities of lighter crudes available. These crudes have API gravities of 40° or higher (SG of less than 0.8251). Over the last 20 years the refining industry invested heavily in capability for processing heavy crudes. Running light crudes stresses the capacity of the atmospheric crude unit in these plants. The most common method of debottlenecking light crude capacity adds a preflash drum or preflash tower to the crude unit. Crude preflash towers may be either independent towers or linked to an atmospheric distillation tower.

This presentation examines options for crude distillation units with independent preflash towers. Based on an author survey [1] approximately 20% of the crude distillation units in North America include an independent crude preflash tower. The overwhelming reason they are added to crude units is to increase unit capacity. However, when added they also change overall unit performance by modifying:

- Operating constraints
- Energy efficiency
- Yields
- Fractionation efficiency

The analysis here uses a constant capacity unit to examine how these factors change.

The four major topics covered are:

- Effects of removing light material from the atmospheric tower feed

1 Sloley 2014

- Differences between preflash tower configurations
- Importance of heat integration changes
- Effect of different crude types

Adding a preflash tower removes naphtha from the atmospheric tower feed. The atmospheric tower overhead duty drops. The atmospheric tower energy recovery also drops. Constraints around the atmospheric tower operation become more difficult. The tower moves closer to minimum internal liquid wetting rates. The atmospheric tower is typically constrained by approach to internal water condensation. Overall distillate yield may drop dramatically if the atmospheric tower heater outlet temperature is limited.

In general, a preflash tower with a fired heater for a reboiler will have the lowest energy use. However, in some plants with exceptionally low steam costs or in plants that desire to make a kerosene product from the preflash tower, a steam-stripped preflash tower may be an acceptable option.

Recovery of naphtha in the preflash tower shifts overall recovered duty to lower temperatures. Significant additions of heat exchanger surface may be required to maintain heat integration. One step that improves possible heat recovery is to vent light material from the preflash tower overhead drum.

Results from this analysis show that the reboiled tower had the largest advantage with the heaviest crude. As unit feed becomes lighter, the benefits of the reboiled tower drop. Nevertheless, even the lightest crude evaluated (49.7°API, 0.7809 SG) still favored the reboiled tower.

2. Preflash Applications

The most common crude distillation towers are atmospheric crude towers and vacuum crude towers. Atmospheric preflash towers and drums are the third most common type of crude distillation tower. In spite of the number of preflash installations, relatively little published discussion has focused on them.

Previous work [2] covered the heat and material balance implications of adding three basic preflash configurations to an atmospheric-vacuum crude unit:

- High-pressure preflash drum
- Low-pressure preflash drum
- Preflash tower with one product

These cases included examining two major constraints:

- Heater firing a constant duty (duty limit)
- Heater firing to a constant outlet temperature (temperature limit)

The conclusion of that analysis was that against a duty limit, adding a preflash improves unit yields. Against a temperature limit, adding a preflash reduces unit yields.

Other analysis [3] covered the differences between adding a preflash drum versus a preflash tower. The factors favoring a preflash tower include:

- Atmospheric tower has significant constraints.
- Atmospheric overhead system has significant constraints.

---

2 Sloley 2013a
3 Sloley 2014

With major constraints in the top of the atmospheric tower, more modifications are required to increase unit capacity. The preflash tower can take material completely out of the atmospheric tower. It also adds a new overhead system. However, capital requirements are higher, yield changes need to be carefully reviewed, and the preflash must be sized to deal with potential foaming.

Comparisons in previous work focused on preflash towers with all the energy provided by preheating the feed and having the bottom of the tower act as a flash drum. This work adds two major factors to the analysis of preflash towers:

- Preflash tower configuration
- Heat recovery and yield at economic optimum conditions

3. Evaluation Basis

The evaluation basis is a 100,000 bpd (15,900 m^3pd) crude unit without a preflash tower. Appendix A includes details on the base unit configuration. The evaluation covered crudes ranging from 23.3°API to 49.7°API (0.9141 to 0.7809 SG) The crudes studied were:

Table 1 Crudes Evaluated

°API Gravity	Crude	Source	Reference
23.3	Bachequero Field 24 (BCF 24)	Venezuela	Proprietary assay
31.6	Tia Juana Light (TJL)	Venezuela	Watkins 1973
42.1	Bakken	USA – North Dakota	Proprietary assay
49.7	Gippsland Blend (Bass)	Australia	ExxonMobil 2008

These crude cover the complete range likely to be seen in any single conventional crude unit. BCF 24 is a heavy crude with little light-ends. TJL is a medium crude with a density very close to the current average crude density being processed in North America. The Bakken assay is at the lighter end of material being currently produced from the North Dakota tight oil formations. The Gippsland Blend is very similar to stabilized Eagle Ford stream currently becoming available from West Texas.

The BCF 24, TJL, and Bakken assays were used as-is. The Gippsland Blend assay was provided without light-ends speciation. The C_4 minus was quoted as 2.00 volume percent. A light-ends speciation through C_7 was available from a different sample. The second sample had a C_4 minus content of 2.03 volume percent. C_3-C_6 light ends were defined based on the second sample. C_5-C_6 total content was adjusted to match the light naphtha distillation curve.

A constant standard volume was used for all crudes. This is different from a constant mass feed rate.

The results summarized in the tables and in Appendix B use the API gravity or specific gravity back calculated from the pseudocomponent blend used in the calculations for the crude assays. As such, they differ slightly from those shown in Table 1.

4. Preflash Configuration

Three types of preflash towers were evaluated:

- Steam stripped
- Reboiled
- Flash zone

These configurations cover the range of possible basic configurations for industrial units. The configurations are the same as discussed in the analysis of condensate splitters [4].

4.1. Steam Stripped

Steam stripped preflash towers use a configuration similar to conventional refinery atmospheric distillation. Figure 1 shows a general schematic of this unit. Feed is heated in the hot train [5] exchangers. After the hot train, the feed may go directly to the preflash tower or additional energy may be added by a fired preheater.

The preflash tower includes a conventional rectification section. Reflux for the rectification section comes from either a pumparound heat removal system or an overhead drum heat removal system. Figure 1 shows a unit with an overhead system. Fractionation differences between the pumparound option and the overhead condenser option are insignificant. The main difference is in the heat integration for energy recovery.

The stripping section uses inert gas stripping with steam to recover lighter material from the bottoms. Direct injection of the steam reduces the hydrocarbon partial pressures. With lower partial pressures, the lighter hydrocarbons in the bottoms vaporize and generate internal traffic in the tower to drive the desired separation.

One of the main benefits of the steam stripped condensate splitter is that many refineries have surplus low-pressure steam that has little other value. The steam pressure is too low to reboil a preflash tower, but stripping using partial pressure vaporization still works effectively.

One of the main disadvantages of having a rectifying section is potential operating problems from foaming of the condensate feed. Foam in the feed can overload the feed flash zone and lead to entrainment. Entrainment reduces unit capacity and degrades fractionation performance.

4.2. Reboiled

Reboiled preflash towers use a configuration similar to a classic distillation tower. Figure 2 shows a general schematic of this unit. The rectifying section configuration is the same as a steam-stripped preflash tower.

The stripping section uses a reboiler to generate the required vapor. One of the main benefits of the reboiled condensate splitter is that with sufficient duty, nearly any reflux ratio can be obtained. The reboiled condensate splitter achieves better separation or higher recovery for a fixed heat input.

As with the steam stripped preflash tower, foam in the flash zone leads to significant problems.

4.3. Flash Zone

Flash zone preflash towers use a rectification section only. The bottom of the tower acts as a flash drum. Figure 3 shows a general schematic of this unit. Feed is partially vaporized in a preheater and enters the bottom section of an atmospheric distillation tower. The rectifying section configuration is the same as a steam-stripped condensate splitter.

Many refineries use this configuration to avoid foam entrainment problems in the flash zone. As long as the tower boot has sufficient residence time and low enough vapor velocities, foam

---

4 Sloley 2015a

5 The glossary in Appendix C includes the definition of industry terms used.

problems can be avoided. Since the flash zone preflash tower has no stripping section, separation performance tends to be poor.

5. Operating Targets – Atmospheric Tower

5.1. Atmospheric Tower Constraints

Preflash towers change the feed composition to the atmospheric tower. The optimum operating point for the atmospheric tower changes with the new feed composition. For each cae study, the atmospheric tower was optimized against the constraints. Constraints on the atmospheric tower included both process objectives and equipment limitations.

The major equipment limitation was minimum tray wetting. This requirement might be relaxed with more elaborate equipment. One drawback of specialized configurations is reduced flexibility to handle different feeds. Refinery economics strongly benefit from flexibility in buying low-cost feeds. For this reason, truly pushed design criteria were not used.

Table 2 [6] identifies the limiting atmospheric tower constraints for each configuration. The overall objective was to:

- Match target product yield rates
- Optimize heat recovery and product yields

Process and equipment constraints were set at practical limits based on reliable operation, with reasonable flexibility, and for commercially available equipment. Every option was pushed to the limits of as many constraints as possible. Depending upon the configuration, the number of controlling constraints varied from three to five. The number of constraints that had flexibility to allow for optimization varied from one to three.

5.2. Heat Integration versus Separation

Heat recovery from the atmospheric crude tower provides a significant amount of energy for heating the crude oil. Both economic and equipment constraints limit heat recovery. Maximum heat recovery occurs when the most heat possible is removed at the highest temperature level. This discussion of heat recovery optimization applies to the atmospheric tower immediately downstream of the preflash tower.

Removing heat from the atmospheric tower using pumparounds reduces internal liquid traffic in the tower. The limitation selected for minimum liquid rate required to make the tower work defines the maximum heat recovery. Equipment characteristics set the minimum liquid rate. The minimum liquid rate rises with increasing tower diameter. For all cases the atmospheric crude tower was evaluated as a trayed tower with two liquid flow passes. The minimum liquid rate was set to 1.0 gpm/inch (1.49 lpm/cm). Maximum heat recovery has minimum fractionation between product streams.

The minimum heat recovery option rejects all heat to the overhead condenser. Minimum heat recovery maximizes internal liquid load in the tower. This results in maximum fractionation between products.

---

6 Large tables are collected in §10 at the end of the discussion.

Table 3 Heat Recovery versus Fractionation Effectiveness

Heat Recovery	Fractionation
Maximum	Minimum
Minimum	Maximum

When heat balances change in the atmospheric tower, product yields change as well as product separations. Optimum operation balances heat recovery versus product values. Other work [7] explores specifics of the optimization of heat recovery versus yield.

The key optimization criteria selected favor maximum middle distillate yield:

Table 4 Economic Basis for Heat Balance Optimization

Disposition	US$/gal or US$/Mbtu	US$/m³ or US$/GJ
Kerosene over naphtha	+0.0952	25.15
Diesel over ago	+0.1667	44.04
Kerosene versus diesel	0	0
Firing	4.00	3.80

All operations were optimized for maximum profit based on incremental operating revenue versus incremental investment cost. Table 5 summarizes the base operation without a preflash tower for the four crudes.

5.3. Atmospheric Tower Fired Heater

The atmospheric tower fired heater was kept at a constant heater outlet temperature.

6. Operating Targets and Constraints – Preflash Tower

Operating units must work within both process constraints and equipment constraints. Process constraints include meeting required operating conditions and stream compositions. Equipment constraints include expected mass-transfer efficiency, equipment sizes, and hydraulic performance of the equipment. This evaluation imposed both process constraints and equipment constraints. Appendix A summarizes the constraints imposed.

6.1. Yields

The base preflash operation was defined as the amount of naphtha recovery from the lightest crude (Gippsland) with a 10 lbs/bbl (28.5 kg/m^3) stripping steam rate. This gives a naphtha yield equal to a 336°F (169°C) cutpoint.

The 10 lbs/barrel (28.5 kg/m^3) rate creates enough volume on the bottom trays of the stripping section to give reasonable stripping tray effectiveness. Many units that have been built use lower steam rates. As steam rates drop, the performance of a steam stripped preflash column approaches the performance of a flash zone preflash column.

Comparisons for changes in preflash configuration and in feed type all kept the same preflash naphtha cutpoint. This is a simplifying assumption of downstream processing requirements.

---

7 Sloley 2015b

6.2. Operating Targets

The major process constraint on the preflash tower was the overhead product specification. A high-cutpoint naphtha operation was used as the basis. The full-range naphtha overhead product was drawn to a 350°F (177°C) D86 95 volume percent specification. Operating targets may shift depending upon customer requirements and values in the market. As of now (2015), many plants run with economics favoring naphtha overhead specification of approximately 350°F (177°C). The major driver behind this is the desire to maximize middle distillate production (kerosene plus diesel product).

6.3. Distillation Effectiveness - Rectification

All configurations used 14 theoretical stages (18 trays) in the preflash rectifying section plus one theoretical stage in the condenser. This gave a total of 15 theoretical stages for the naphtha product versus the feed.

The author's evaluations have always shown that optimization of investment for theoretical stages versus energy efficiency and product yields strongly drive units toward maximum theoretical stages. Incremental cost for an additional theoretical stage is low compared to the benefits in yields, energy, and reduced investment in other equipment (fired heaters, heat exchangers, pumps, and piping). Most existing units have too few trays in them. A sensitivity analysis using ±5 stages for the rectifying section showed that the 15 stage separation was comfortably inside the region where reflux strongly drives separation performance.

As a comparison, the ASTM D2892 method that defines the true boiling point (TBP) distillation curves uses 15 theoretical stages in the distillation. The refining industry settled on 15 stages because this was enough stages that minor equipment differences had no significant effect on the distillation results. While the D2892 method uses a batch distillation at high reflux ratios, the same logic applies to preflash towers.

6.4. Distillation Effectiveness - Stripping

As mentioned in §6.1, all the stripped preflash towers presented use 10 lbs of stripping steam per barrel of bottoms product (28.5 kg/m^3). This is a high rate compared to most units in service. One problem that causes low tray efficiency is very low vapor rates on the bottom stripping trays. The 10 lbs/bbl (28.5 kg/m^3) rate is high enough to make the lowest trays reasonably efficient. This allows the stripped towers presented to get three theoretical stages in the stripping section. A tray efficiency of 25% was used for the stripping units for a total of 12 actual trays. Plant testing on steam stripping sections supports a range of 25-33% efficiency for well designed stripping trays.

In comparison, the reboiled units used the same 12 actual trays. However, the tray efficiency used in the reboiled units was 75%. This gives 9 trays in the stripping section of the reboiled towers. Higher vapor rates and plant data from operating units justifies the 75% efficiency for the reboiled preflash columns.

6.5. Distillation Equipment - Hydraulics

Hydraulic limits and sizing constraints for the preflash towers were set using prudent design practices for trays. The rectifying section used conventional moving-valve trays. The stripping section used conventional fixed-valve trays. Only minor differences would be expected if sieve trays were used in the stripping section. When building a new tower the cheapest option is to normally use a larger diameter tower with trays instead of a smaller diameter tower with packing.

For the Gippsland feed case, a packed tower was used for the preflash tower. The Gippsland crude produced 51% of the preflash tower feed as overhead product. The tower diameter was extremely large compared to the internal liquid loads. For this case, the cost of packing is easily justified by minimizing the internal liquid rate. Lower internal liquid rates decrease the amount of energy shifted to a lower temperature level in the preflash condenser.

6.6. Overhead Configuration

All units included a partial condenser on the overhead stream. A small quantity of very light material may not show up in the feed analysis, but experience on operating units shows it is usually present. The small gas purge significantly increases the overhead drum temperature. Increased reflux temperature also reduces shock condensation in the preflash tower. Tower fouling rates are usually lower with higher reflux temperatures.

6.7. Flash Zone Entry

The flash zone is the part of the preflash tower where the two-phase feed enters. In the temperature range between 300°F and 500°F (149°C and 260°C) the two-phase mixture has a significant risk of forming stable foams. Foam accumulation is the balance between foam formation rates and foam decay rates. Both depend strongly upon system chemistry, liquid viscosity, and volume fraction vapor. Foam formation has been identified as an operating problem across a wide range of condensates and operating conditions. While the greatest risk is in the 300°F-500°F range (149°C-260°C), it can occur at both lower and higher temperatures.

Foam has normally been handled by providing ample residence time and decreasing vapor velocity. This can lead to very large towers. The basis here uses cyclonic entry devices to force foam decay. This allows the tower diameter to drop to the size required by the trays. The tower diameter at the flash zone is set as the larger of the sizes required by either the rectifying section above or the stripping section below.

6.8. Operating Conditions

All comparisons used a common basis of 30 psig (308 kPa) overhead drum pressure, 42.5 psig (394 kPa) overhead pressure and 45 psig (411 kPa) flash zone pressure. Sufficient pressure was used in the preheat train to prevent vaporization upstream of fired heaters. Pressure optimization was not included in this review. Higher pressure operation would make the preflash tower smaller and reduce exchanger area due to increased temperatures in the condenser. Higher pressure operations would increase required feed temperatures, make using low pressure steam in the stripping units more difficult, and make reboiler temperatures in the reboiled units higher.

6.9. Fired Heaters

For the steam stripped configurations, no fired heaters were required on the preflash feed. In fact, additional heat integration between the preflash bottoms and the available heat sources was possible.

For the reboiled configurations, all available preheat was put into the preflash tower feed. Since the reboiler heats up the preflash tower bottoms, the preflash tower feed is the only available heat sink for heat integration. This creates some thermodynamic losses in the preflash tower. However, some heat recovery is more cost effective than no heat recovery.

The units with a feed flash into the bottoms required additional feed preheat to match product rates. A fired heater provides this energy.

7. Results

Some preflash configurations do reduce energy costs. However, under current economics, the likely yield shifts from kerosene to naphtha and diesel to AGO would outweigh any energy savings. The major purpose of adding a preflash is to increase the capacity of an existing unit. This is especially true for units that have tight oils or condensates blended in the feed.

7.1. Yields

7.1.1. Total Distillate Yield

Appendix B includes tables comparing the results for the four feeds. As expected, against downstream temperature limits on the atmospheric tower heater outlet temperature, adding a preflash decreases overall distillate yield [8]. The preflash tower removes light material from the feed to the atmospheric tower.

- Against a constant atmospheric tower pressure, less light material increases the partial pressure of the heavier feed components.
- Therefore, at a constant temperature, vaporization in the atmospheric tower drops.
- The atmospheric tower bottoms rate increases, so the overall unit yield drops.

The cause of this is reduced vaporization in the atmospheric tower flash zone. The preflash tower removes light material from the atmospheric tower feed. Without the light material, the partial pressure of the heavier hydrocarbons in the atmospheric tower flash zone increases. This reduces overall vaporization. Yields shift diesel and AGO to atmospheric residue.

7.1.2. Distribution of Yields

Yields shift from kerosene to naphtha. The preflash tower has relatively little liquid because little kerosene is vaporized. The kerosene vaporized in the preflash tower feed provides the reflux for naphtha-kerosene separation.

The atmospheric tower top section liquid rate also drops because the bulk of the naphtha has been removed in the upstream preflash tower. Both of these changes create low L:V ratios in the towers. Low L:V ratios reduce fractionation effectiveness.

The operating target selected pushes for maximum kerosene yield against the kerosene flash point specification. The lower L:V ratios with the preflash units force more product into the naphtha. Kerosene yields drop.

7.2. Atmospheric Heater Limitations

For the lightest feed (Gippsland), the atmospheric heater outlet vaporization was limited to 90% for the operation without a preflash tower. This dropped the heater outlet temperature to 615°F (324°C). Adding the preflash reduced the amount of light material in the atmospheric heater feed.

---

8 Sloley 2013a

This allowed the atmospheric heater outlet temperature to rise to 690°F (366°C) without exceeding the 90% vaporization limit.

7.3. Atmospheric Tower Constraints

Adding a preflash tower forces the atmospheric tower against more of the unit constraints.

- Both the steam stripped preflash tower and the reboiled preflash tower selectively remove the lightest material from the atmospheric tower feed. This forces the operating temperature of the atmospheric tower overhead to rise to maintain a 30°F (16.7°C) margin above water dew point temperature in the tower.
- For the heavier crudes heat integration optimization is constrained by more equipment limits. The BCF24 and TJL crudes have flexibility in optimization for both the heavy naphtha pumparound and the kerosene pumparound in the base case. Adding either a stripped preflash or a reboiled preflash forced the heavy naphtha pumparound heat removal against the wetting limit inside the crude tower in both cases.
- The flash zone preflash tower forced the atmospheric tower against wetting limits for all pumparounds.

7.4. Atmospheric Tower Heat Recovery

Atmospheric tower heat recovery drops dramatically. In all cases, adding a preflash tower reduces the average temperature difference available to drive heat transfer. Table 6 shows the heat removal balance for the preflash tower and the atmospheric tower. All preflash options shift duty from all pumparounds and the atmospheric overhead condenser into the preflash condenser. The duty shift comes from two factors:

- Naphtha condensation duty shift
- Mass flow shift

The preflash tower shifts naphtha product from the atmospheric tower into the preflash tower. The condensation duty for the naphtha product moves from the atmospheric tower condenser to the preflash condenser.

The preflash tower removes mass flow from the atmospheric tower. Nearly all of the mass removed is naphtha product flow. Much of the cooling duty in the atmospheric tower comes from cooling the mass from the high temperature at the bottom of the tower to the low temperature at the top of the tower. When mass flow decreases, the available sensible duty decreases.

The consequences of the heat shift depend upon the unit configuration. The main variable is heat integration between the overhead condensers and the crude feed.

For new units, including a preflash tower forces some combination of three choices:

- Include extensive heat integration for the preflash tower overhead, the atmospheric tower overhead, or both.
- Add large surface areas to recover heat from product run-down.
- Accept low heat recovery and high fired heater duties.

The summaries shown in Appendix B assume overhead heat integration to maintain energy recovery.

For existing units, surveys by the author show that roughly 70% of existing crude units use the atmospheric tower condensers to preheat the crude. The other 30% do not. Adding a preflash tower to an existing unit requires careful analysis of the heat integration system. The most likely solutions all require adding extensive heat integration in the overhead systems. Improving heat integration in a unit that already has overhead-to-feed integration will be much more difficult and expensive than adding it to one that does not have already have heat integration with the overhead. This arises from the interplay of corrosion problems from water condensation and heat recovery [9].

7.5. Energy Intensity and Crude Heating

Energy intensity measures the energy input per unit product. In general, all distillate products have much higher values than the bottoms stream. Table 7 [10] shows energy intensity per unit volume of product made.

Table 7 Summary Energy Intensity

Crude	Topping	Stripped Preflash	Reboiled Preflash	Flash Zone Entry
BCF 24	1	3	2	4
Tia Juana Light	2	2	1	4
Bakken	3	2	1	3
Griffin	3	1	1	

1=lowest energy consumption, 4=highest energy consumption

- Heavier crudes favor the topping unit for maximum energy efficiency.
- Lighter crudes favor the reboiled preflash for maximum energy efficiency.
- Very light crudes are close to even for the reboiled preflash and the steam stripped preflash.

Many studies have stated that adding a preflash to a crude unit improves energy recovery. This is true for lighter crudes.

For heavier crudes energy recovery is only true in one situation – the unit has vaporization upstream of the fired heater. Vaporization upstream of the fired heater makes heat balance control in the fired heater unreliable. Coking and reduced heater run duration between shutdowns is the usual result. Other fired heater constraints may also create problems if the pressure in the unit gets too low. For this one situation, adding a preflash may enable other work to improve energy efficiency in an existing unit.

As a general rule, preflash towers should only be included in a new unit in rare cases. The purpose of preflash towers is to add capacity to an existing unit. It is cheaper with a new unit to build the atmospheric tower larger in the first place. Exceptions for a new unit would include:

- Very light feed to the unit, and
- Diesel as the heaviest product from the atmospheric tower, and
- No conventional vacuum tower in the unit, and
- Atmospheric tower bottoms directly feeds a vacuum preflash for diesel recovery

The preflash energy comparison in Table 8 do not include the effect of heat shifts from the atmospheric heater to the vacuum heater. The comparisons are for the preflash heater and

---

9 Sloley 2013b,c,d

10 The full Table 7 is included with the large tables are collected in §10 at the end of the discussion.

atmospheric heater only. If the unit includes a vacuum heater (for a vacuum tower) the vacuum heater duty will be much higher due to a higher feed rate. This makes the topping unit more attractive. Very light feeds such as Griffin may operate without a vacuum tower. However, for crudes of ~45°API (SG 0.8017) and heavier, adding a vacuum tower to the unit shifts the economics back toward a conventional atmospheric-vacuum unit.

7.6. Stripped versus Reboiled Preflash Towers

7.6.1. Overall Comparison

Table 8 includes an energy comparison for the stripped preflash tower versus the reboiled preflash tower. Stripped preflash towers use stripping steam to vaporize the naphtha. Reboiled preflash towers use fired heat to vaporize the naphtha. The value of heat in steam versus the value of heat in firing differs. In some plants it can differ greatly.

The energy for the vaporization in the stripped preflash tower comes from the bottoms liquid. As vaporization occurs in the stripping trays the bottoms stream cools. The stripped preflash bottoms temperature is lower than the feed temperature. The lower temperature increases the temperature difference between the heat recovery streams and the crude feed. Additional heat recovery into the stripped preflash tower bottoms is usually attractive.

Table 8 includes an estimate of the relative cost of the energy for steam versus firing that makes the energy cost of a stripped preflash tower equal to a reboiled preflash tower with the same yields. Two comparisons are made. The first shows the energy cost that balances the economics with improved heat recovery. The second shows the energy costs that balances the economics without improved heat recovery. Table 9 summarizes the comparison.

The heavier the crude, the cheaper steam must be compared to fuel for energy costs to favor the steam stripped unit. Improved energy recovery from the stripped preflash bottoms has significant economic benefits. Without the additional heat integration, the value of steam would have to be negative for energy costs to favor the stripped tower.

In general, the heavier the crude, the more attractive the reboiled preflash tower becomes.

Table 9 Stripped Preflash versus Reboiled Preflash Energy Equivalent*

	BCF24	TJL	Bakken	Gippsland
Additional Preheat	0.195	0.301	0.437	0.697
No additional Preheat	-0.915	-0.761	-0.388	-0.026

* Stripping steam-to-fuel cost to make a stream stripped preflash tower have the same energy cost as a reboiled preflash tower.

7.6.2. Why Are Stripped Preflash Towers Used? – Making Heavy Products

In spite of energy costs, stripped preflash towers are often used. One limitation on reboiled towers is bottoms temperature. The heavier the material going overhead, the higher the bottoms temperature. The full range naphtha recovered in these cases is already causing the bottoms temperature to approach maximum prudent temperatures. Too high a bottoms temperature increases thermal degradation and can cause product quality and operating problems.

For units that need to recover kerosene material in the preflash tower either a stripping configuration or very low-pressure operation of the preflash tower are needed.

7.6.3. Stripped Preflash Optimization

The yields of the stripping cases were adjusted by varying the feed preheat temperature. One option would be to increase the feed temperature and to reduce stripping steam rates. This could improve the relative energy efficiency of the stipping configuration.

However, at the limiting case of using 100% feed preheat and no stripping steam the stripping configuration becomes identical to the flash zone configuration. Table 8, Table 9, and Appendix B clearly show the flash zone configuration is the highest energy user. Optimization can make the stripped tower slightly more attractive, but the overall conclusion that reboiled towers are the best choice for most plants stands.

7.7. Flash Zone Preflash Towers

Table 8 shows that flash zone preflash towers have much higher energy consumption than either a stripped preflash tower or a reboiled preflash tower. In general, flash zone preflash towers should not be used.

Flash zone preflash towers are typically used when the plant is highly concerned about foam formation. A large bottoms section on the tower allows ample time for foam decay. However, the cost of not having a stripping section is high. Better solutions use either cyclonic entry devices, foam drums, or split towers [11]. All these options can be applied to either a stripped preflash tower or a reboiled preflash tower.

7.8. Heavy Naphtha Yields

The product specifications were set to maximize middle distillate yield (kerosene plus diesel). Enough stripping steam was used in the kerosene stripper for nearly all the heavy naphtha to be either forced into the overhead product or into the kerosene.

With the specifications used, operating experience shows that entrapment occasionally occurs with material boiling between the full range naphtha product and the kerosene product. This can lead to control instability in operating units.

More stable control can be achieved with either a higher endpoint specification on the overhead product or a small heavy naphtha slip-stream product. Higher middle distillate yields are possible with the slipstream product. For the BCF24, TJL, and Bakken crudes only a small slip-stream of 50 bpd (8.0 m^3/d) was required. For the Griffin crude the slip stream was larger at ~450 bpd (72 m^3/d). The most likely disposition of the heavy naphtha slip-stream would be to blend it into the full-range naphtha product taken overhead.

7.9. Other Factors

Appendix B shows systemic differences in fractionation effectiveness and water dew point limits. However, the differences between the different preflash options are minor. In general these trends are inversely correlated.

11 Sloley 2015b

Table 10 Minor Performance Trends

Configuration	Preflash Tower Fractionation	Atmospheric Tower Water Dew Point
Reboiled	Best	Worst
Stripped	Intermediate	Intermediate
Flashed	Worst	Best

The better the preflash tower fractionation, the sharper the separation of light versus heavy material in the preflash naphtha product. The sharper separation removes some light material that goes to the atmospheric tower overhead. At a constant atmospheric tower overhead pressure, removing the light hydrocarbons:

- Reduces hydrocarbon partial pressure
- Increases water partial pressure
- Increases the water dew point temperature

8. Revamps versus New Construction

8.1. Pressure Optimization

All comparisons presented here use a constant pressure in the atmospheric tower overhead. One possible benefit of using a preflash tower is to reduce pressure in the atmospheric tower system. Changes that support a lower operating pressure in the atmospheric tower might include:

- Less mass vaporizing in the atmospheric tower feed allows for lower pressure (lower vapor density) if atmospheric tower flooding was a constraint.
- Lower vapor rate to the atmospheric tower condenser reduces overhead system pressure drop.
- Lower duty in the atmospheric tower condenser allows for lower LMTD and hence lower pressure operation in the overhead.

Lower pressure operation would reduce the amount of yield loss in the atmospheric tower but would also make heat integration around the atmospheric tower more difficult.

8.2. Heat Integration

As mentioned in §7.4 adding a preflash tower may require extensive modifications to the heat integration system. The most likely addition is use of overhead condenser-to-crude heat recovery. This may be in either or both the preflash overhead and the atmospheric tower overhead.

9. Conclusions and Recommendations

For new units in general, independent preflash towers should be avoided. In nearly all cases making the crude tower and associated equipment larger requires less investment, has higher yields, and more effective heat recovery.

In the cases where a preflash may be beneficial in a new unit, the preferred option would be to use a preflash drum attached the atmospheric tower [12]. However, some unusual cases for processing light crudes with low levels of contaminants might justify a preflash tower.

---

12 Sloley 2014

For revamps, the factors that would favor independent preflash towers would include:

- Severe capacity constraints on the entire crude tower – a preflash tower removes volume from the entire atmospheric tower system (feed heater, tower, overhead system).
- Strict limits on the duty available in the existing fired heaters – a preflash tower adds extra steam stripping or an additional fired heater.
- Severe limits on the existing overhead system – a preflash tower would add a new overhead system.

Economics for preflash towers need to be examined carefully. It is conceivable that an entire parallel crude unit may be a more cost-effective expansion option.

Once the decision to build a preflash tower has been made, configuration selection depends upon perceived foam risk, products required, and energy costs. Figure 4 includes a flow chart for the selection criteria. No recommended outcome ends up with a flash zone on the bottom of the tower. If foaming is considered a serious risk, a preflash tower with a foam drum on the feed would be the preferred solution [13].

---

13 Sloley 2015b

10. Tables and Figures

Table 2 Constraints

Feed	BCF24	BCF24	BCF24	BCF24	TJL	TJL	TJL	TJL
Configuration	Topping	Stripped	Reboiled	Flashed	Topping	Stripped	Reboiled	Flashed
No Preflash	X				X			
Overhead duty set by overhead temp	X				X			X
Overhead duty set by water dew approach		X	X	X		X	X	
Naphtha PA duty set by optimization	X				X			
Naphtha PA to zero duty due to optimization		X	X	X		X	X	X
Naphtha PA duty set by internal liquid wetting								
Kero PA duty set by optimization	X	X	X	X	X	X	X	
Kero PA duty set by internal liquid wetting								X
Diese PA duty set to internal liquid wetting	X	X	X	X	X	X	X	X
Kero stripping steam set by flash spec	X				X			
Kero stripping steam set to minimum rate		X	X	X		X	X	X
Heavy naphtha product as drag stream	X	X	X	X	X	X	X	X
Fired heater before preflash				X				X
Heat integration after preflash		X				X		

Feed	Bakken	Bakken	Bakken	Bakken	Gipps	Gipps	Gipps	Gipps
Configuration	Topping	Stripped	Reboiled	Flashed	Topping	Stripped	Reboiled	Flashed
No Preflash	X				X			
Overhead duty set by overhead temp	X			X	X			X
Overhead duty set by water dew approach		X	X			X	X	
Naphtha PA duty set by optimization		X	X					
Naphtha PA to zero duty due to optimization				X		X	X	X
Naphtha PA duty set by internal liquid wetting	X				X			
Kero PA duty set by optimization	X				X	X	X	
Kero PA duty set by internal liquid wetting		X	X	X				X
Diese PA duty set to internal liquid wetting	X	X	X	X	X	X	X	X
Kero stripping steam set by flash spec	X	X	X	X	X	X	X	X
Kero stripping steam set to minimum rate								
Heavy naphtha product as drag stream	X	X	X	X	X	X	X	X
Fired heater before preflash				X				X
Heat integration after preflash		X				X		

(shaded) Equipment or process constraints did not limit optimization.

Table 5 Base Cases – No Preflash Tower, Atmospheric Distillation Results Only

Feed		BCF24	TJL	Bakken	Gipps		BCF24	TJL	Bakken	Gipps
Config		Topping	Topping	Topping	Topping		Topping	Topping	Topping	Topping
Feed API		23.2	31.5	42.0	51.1		0.9147	0.8681	0.8156	0.7749
Rates	BPD					m3/d				
Feed		99999	99999	100000	100011		15900	15900	15900	15902
Preflash Vapor										
Preflash Naphtha										
Atm Light Nap		9841	19532	33477	49916		1565	3106	5323	7937
Atm Heavy Nap		50	50	50	447		8	8	8	71
Kerosene		14398	20146	24530	22538		2289	3203	3900	3584
Diesel		9576	10663	12519	8748		1523	1695	1991	1391
AGO		7383	9672	12066	10984		1174	1538	1918	1746
ARC		58752	39936	17359	7377		9342	6350	2760	1173
Total Distillate		41248	60063	82642	92633		6558	9550	13140	14729
Preflash Cutpoint	F					C				
Atmos Cutpoint	F	714	747	797	777	C	379	397	425	414
Preflash Dew Margin	F					C				
Atmos Dew Margin	F	35.9	39.3	40.7	58.7	C	19.9	21.8	22.6	32.6
Atmos Ovhd Temp	F	265	265	265	265	C	129	129	129	129
Preflash Nap D86 95	F					C				
Atmos Heavy Nap D86 95	F	335	333	328	341	C	168	167	164	172
Kero D86 99	F	560	560	560	560	C	293	293	293	293
Diesel D86 95	F	640	640	640	640	C	338	338	338	338
Kero Flash	F	120	120	120	120	C	49	49	49	49
Preflash Cond Q	Mbtu/hr					GJ/hr				
Preflash Stripping	lb/bbl					Kg/m3				
Preflash Stripping	wt%					wt%				
Atm Cond Q	Mbtu/hr	100.1	100.3	153.5	179.1	GJ/hr	105.6	105.8	161.9	189.0
HN PA Q	Mbtu/hr	9.4	57.8	110.8	93.5	GJ/hr	9.9	61.0	116.9	98.6
Kero PA Q	Mbtu/hr	28.7	17.6	0.1	14.5	GJ/hr	30.3	18.6	0.1	15.3
Dies PA Q	Mbtu/hr	53.3	77.1	104.0	74.0	GJ/hr	56.2	81.3	109.7	78.1
LH Gap	F	-27.3	-53.3	-52.5	-45.6	C	-15.2	-29.6	-29.2	-25.3
HK Gap	F	20.9	25.7	24.9	24.2	C	11.6	14.3	13.8	13.4
KD Gap	F	-73.5	-94.1	-107.6	-124.6	C	-40.8	-52.3	-59.8	-69.2
DA Gap	F	-28.6	-22.0	-18.2	-46.3	C	-15.9	-12.2	-10.1	-25.7
AR Gap	F	-162.0	-153.7	-167.1	-200.6	C	-90.0	-85.4	-92.8	-111.4
AGO 650- D86	Vol %	10.5	9.1	8.8	12.4	Vol %	10.5	9.1	8.8	12.4
AGO 700- D86	Vol %	43.7	40.8	41.3	40.3	Vol %	43.7	40.8	41.3	40.3
Atm Stripping	lb/bbl	10.0	10.0	10.0	10.0	Kg/m3	28.5	28.5	28.5	28.5
Atm Stripping	wt%	3.0	3.0	3.0	3.3	wt%	3.0	3.0	3.0	3.3
Kero Stripping	lb/bbl	9.8	12.1	36.6	24.3	Kg/m3	28.0	34.5	104.4	69.3
Kero Stripping	wt%	3.2	4.2	12.8	8.4	wt%	3.2	4.2	12.8	8.4
Preflash heater	Mbtu/hr					GJ/hr				
Preflash reboiler	Mbtu/hr					GJ/hr				
Atm heater	Mbtu/hr	194.5	222.6	256.0	217.0	GJ/hr	205.2	234.8	270.1	228.9
Prelfash Stm	lb/hr					kg/hr				
Atm steam	lb/hr	24522	16655	7255	3077	kg/hr	11121	7553	3290	1395
Preflash Feed Temp	F					C				
Energy Balance										
Total Firing	Mbtu/hr	194.5	222.6	256.0	217.0	GJ/hr	205.2	234.8	270.1	228.9
Total Steam	lb/hr	24522	16655	7255	3077	kg/hr	11121	7553	3290	1395
Total Firing + PF Steam	Mbtu/hr	194.5	222.6	256.0	217.0	GJ/hr	205.2	234.8	270.1	228.9
PA Recovery	Mbtu/hr	91.4	152.5	214.9	182.0	GJ/hr	96.4	160.9	226.7	192.0
Intensity	btu/bbl distillate	4715	3706	3098	2343	kJ/m3	31288	24591	20554	15544
After-PF Integration	Mbtu/hr					GJ/hr				
Steam/Fired	For equal value									
Steam/Fired	No after PF-Integration									

Table 6 Heat Recovery Balance

Feed		BCF24	BCF24	BCF24	BCF24	TJL	TJL	TJL	TJL
Config		Topping	Stripped	Reboiled	Flashed	Topping	Stripped	Reboiled	Flashed
Feed API		23.2	23.2	23.2	23.2	31.5	31.5	31.5	31.5
Preflash Cond Q	Mbtu/hr		89.0	56.7	95.1		103.8	78.3	116.4
Preflash Stripping	lb/bbl		10.0				10.0		
Preflash Stripping	wt%		3.1				3.2		
Atm Cond Q	Mbtu/hr	100.1	63.8	57.2	54.3	100.3	57.8	54.1	44.4
HN PA Q	Mbtu/hr	9.4	0.1	0.1	0.1	57.8	0.1	0.1	0.1
Kero PA Q	Mbtu/hr	28.7	13.2	17.7	21.3	17.6	19.2	20.0	32.1
Dies PA Q	Mbtu/hr	53.3	30.1	28.0	29.3	77.1	39.3	38.3	39.8
PA Recovery	Mbtu/hr	91.4	43.4	45.8	50.7	152.5	58.6	58.4	72.0

Feed		Bakken	Bakken	Bakken	Bakken	Gipps	Gipps	Gipps	Gipps
Config		Topping	Stripped	Reboiled	Flashed	Topping	Stripped	Reboiled	Flashed
Feed API	BPD	42.0	42.0	42.0	42.0	51.1	51.1	51.1	51.1
Preflash Cond Q	Mbtu/hr		134.6	111.8	149.8		149.1	128.7	174.4
Preflash Stripping	lb/bbl		10.0				10.0		
Preflash Stripping	wt%		3.3				3.4		
Atm Cond Q	Mbtu/hr	153.5	31.3	27.8	37.1	179.1	45.2	43.7	32.7
HN PA Q	Mbtu/hr	110.8	10.0	12.0	0.1	93.5	0.1	0.1	0.1
Kero PA Q	Mbtu/hr	0.1	34.6	33.4	44.0	14.5	39.8	44.2	36.7
Dies PA Q	Mbtu/hr	104.0	46.1	44.3	45.9	74.0	45.1	43.6	44.7
PA Recovery	Mbtu/hr	214.9	90.7	89.7	90.0	182.0	85.0	87.9	81.5

Feed		BCF24	BCF24	BCF24	BCF24	TJL	TJL	TJL	TJL
Config		Topping	Stripped	Reboiled	Flashed	Topping	Stripped	Reboiled	Flashed
Feed SG		0.9147	0.9147	0.9147	0.9147	0.8681	0.8681	0.8681	0.8681
Preflash Cond Q	GJ/hr		93.9	59.8	100.3		109.5	82.6	122.8
Preflash Stripping	kg/m3		0.7				0.7		
Preflash Stripping	wt%		3.1				3.2		
Atm Cond Q	GJ/hr	105.6	67.3	60.3	57.3	105.8	61.0	57.1	46.8
HN PA Q	GJ/hr	9.9	0.1	0.1	0.1	61.0	0.1	0.1	0.1
Kero PA Q	GJ/hr	30.3	13.9	18.7	2.2	18.6	20.3	21.1	3.4
Dies PA Q	GJ/hr	56.2	31.8	29.5	30.9	81.3	41.5	40.4	42.0
PA Recovery	GJ/hr	96.4	45.8	48.3	53.5	160.9	61.8	61.6	76.0

Feed		Bakken	Bakken	Bakken	Bakken	Gipps	Gipps	Gipps	Gipps
Config		Topping	Stripped	Reboiled	Flashed	Topping	Stripped	Reboiled	Flashed
Feed SG		0.8156	0.8156	0.8156	0.8156	0.7749	0.7749	0.7749	0.7749
Preflash Cond Q	GJ/hr		142.0	117.9	158.0		157.3	135.8	184.0
Preflash Stripping	kg/m3		0.7				0.7		
Preflash Stripping	wt%		3.3				3.4		
Atm Cond Q	GJ/hr	161.9	33.0	29.3	39.1	189.0	47.7	46.1	34.5
HN PA Q	GJ/hr	116.9	10.6	12.7	0.1	98.6	0.1	0.1	0.1
Kero PA Q	GJ/hr	0.1	36.5	35.2	4.6	15.3	42.0	46.6	3.9
Dies PA Q	GJ/hr	109.7	48.6	46.7	48.4	78.1	47.6	46.0	47.2
PA Recovery	GJ/hr	226.7	95.7	94.6	95.0	192.0	89.7	92.7	86.0

Table 7 Energy Intensity

Feed		BCF24	BCF24	BCF24	BCF24	TJL	TJL	TJL	TJL
Config		Topping	Stripped	Reboiled	Flashed	Topping	Stripped	Reboiled	Flashed
Feed API		23.2	23.2	23.2	23.2	31.5	31.5	31.5	31.5
Energy Balance									
Total Firing	Mbtu/hr	194.5	162.4	171.4	211.6	222.6	169.3	183.1	222.4
Total Steam	lb/hr	24522	63688	27018	26869	16655	52369	19661	19522
Total Firing + PF Steam	Mbtu/hr	194.5	199.4	171.4	211.6	222.6	202.2	183.1	222.4
PA Recovery	Mbtu/hr	91.4	43.4	45.8	50.7	152.5	58.6	58.4	72.0
Intensity	btu/bbl distillate	4715	5559	4874	5957	3706	3790	3467	4185
After-PF Integration	Mbtu/hr		70.7				55.6		
Steam/Fired	For equal value		0.195				0.301		
Steam/Fired	No after PF-Integration		-0.915				-0.761		

Feed		Bakken	Bakken	Bakken	Bakken	Gipps	Gipps	Gipps	Gipps
Config		Topping	Stripped	Reboiled	Flashed	Topping	Stripped	Reboiled	Flashed
Feed API		42.0	42.0	42.0	42.0	51.1	51.1	51.1	51.1
Energy Balance									
Total Firing	Mbtu/hr	256.0	168.1	183.2	221.9	217.0	157.5	173.6	219.7
Total Steam	lb/hr	7255	37414	10908	10780	3077	24512	4256	4172
Total Firing + PF Steam	Mbtu/hr	256.0	194.8	183.2	221.9	217.0	178.0	173.6	219.7
PA Recovery	Mbtu/hr	214.9	90.7	89.7	90.0	182.0	85.0	87.9	81.5
Intensity	btu/bbl distillate	3098	2621	2482	2994	2343	1971	1934	2442
After-PF Integration	Mbtu/hr		30.9				17.7		
Steam/Fired	For equal value		0.437				0.697		
Steam/Fired	No after PF-Integration		-0.388				-0.026		

Feed		BCF24	BCF24	BCF24	BCF24	TJL	TJL	TJL	TJL
Config		Topping	Stripped	Reboiled	Flashed	Topping	Stripped	Reboiled	Flashed
Feed SG		0.9147	0.9147	0.9147	0.9147	0.8681	0.8681	0.8681	0.8681
Energy Balance									
Total Firing	GJ/hr	205.2	171.3	180.8	223.2	234.8	178.6	193.2	234.6
Total Steam	kg/hr	11121	28883	12253	12185	7553	23750	8917	8854
Total Firing + PF Steam	GJ/hr	205.2	210.3	180.8	223.2	234.8	213.4	193.2	234.6
PA Recovery	GJ/hr	96.4	45.8	48.3	53.5	160.9	61.8	61.6	76.0
Intensity	kJ/m3 distillate	31287.6	36884.9	32343.0	39526.3	24590.8	25146.0	23006.2	27766.9
After-PF Integration	GJ/hr		74.6				58.7		
Steam/Fired	For equal value		0.195				0.301		
Steam/Fired	No after PF-Integration		-0.915				-0.761		

Feed		Bakken	Bakken	Bakken	Bakken	Gipps	Gipps	Gipps	Gipps
Config		Topping	Stripped	Reboiled	Flashed	Topping	Stripped	Reboiled	Flashed
Feed SG		0.8156	0.8156	0.8156	0.8156	0.7749	0.7749	0.7749	0.7749
Energy Balance									
Total Firing	GJ/hr	270.1	177.3	193.3	234.1	228.9	166.2	183.1	231.8
Total Steam	kg/hr	3290	16968	4947	4889	1395	11117	1930	1892
Total Firing + PF Steam	GJ/hr	270.1	205.5	193.3	234.1	228.9	187.8	183.1	231.8
PA Recovery	GJ/hr	226.7	95.7	94.6	95.0	192.0	89.7	92.7	86.0
Intensity	kJ/m3 distillate	20553.9	17392.5	16468.3	19863.9	15543.5	13080.5	12831.1	16201.8
After-PF Integration	GJ/hr		32.6				18.7		
Steam/Fired	For equal value		0.437				0.697		
Steam/Fired	No after PF-Integration		-0.388				-0.026		

Table 8 Stripped Preflash Energy Tradeoffs

Feed		BCF24	TJL	Bakken	Gipps
Config		Stripped	Stripped	Stripped	Stripped
Feed API		23.2	31.5	42.0	51.1
Energy Balance					
Total Firing	Mbtu/hr	162.4	169.3	168.1	157.5
Total Steam	lb/hr	63688	52369	37414	24512
Total Firing + PF Steam	Mbtu/hr	199.4	202.2	194.8	178.0
PA Recovery	Mbtu/hr	43.4	58.6	90.7	85.0
Intensity	btu/bbl distillate	5559	3790	2621	1971
After-PF Integration	Mbtu/hr	70.7	55.6	30.9	17.7
Steam/Fired	For equal value	0.195	0.301	0.437	0.697
Steam/Fired	No after PF-Integration	-0.915	-0.761	-0.388	-0.026

Feed		BCF24	TJL	Bakken	Gipps
Config		Stripped	Stripped	Stripped	Stripped
Feed API		0.9147	0.8681	0.8156	0.7749
Energy Balance					
Total Firing	GJ/hr	171.3	178.6	177.3	166.2
Total Steam	kg/hr	28883	23750	16968	11117
Total Firing + PF Steam	GJ/hr	210.3	213.4	205.5	187.8
PA Recovery	GJ/hr	45.8	61.8	95.7	89.7
Intensity	kJ/m3 distillate	36884.9	25146.0	17392.5	13080.5
After-PF Integration	GJ/hr	74.6	58.7	32.6	18.7
Steam/Fired	For equal value	0.195	0.301	0.437	0.697
Steam/Fired	No after PF-Integration	-0.915	-0.761	-0.388	-0.026

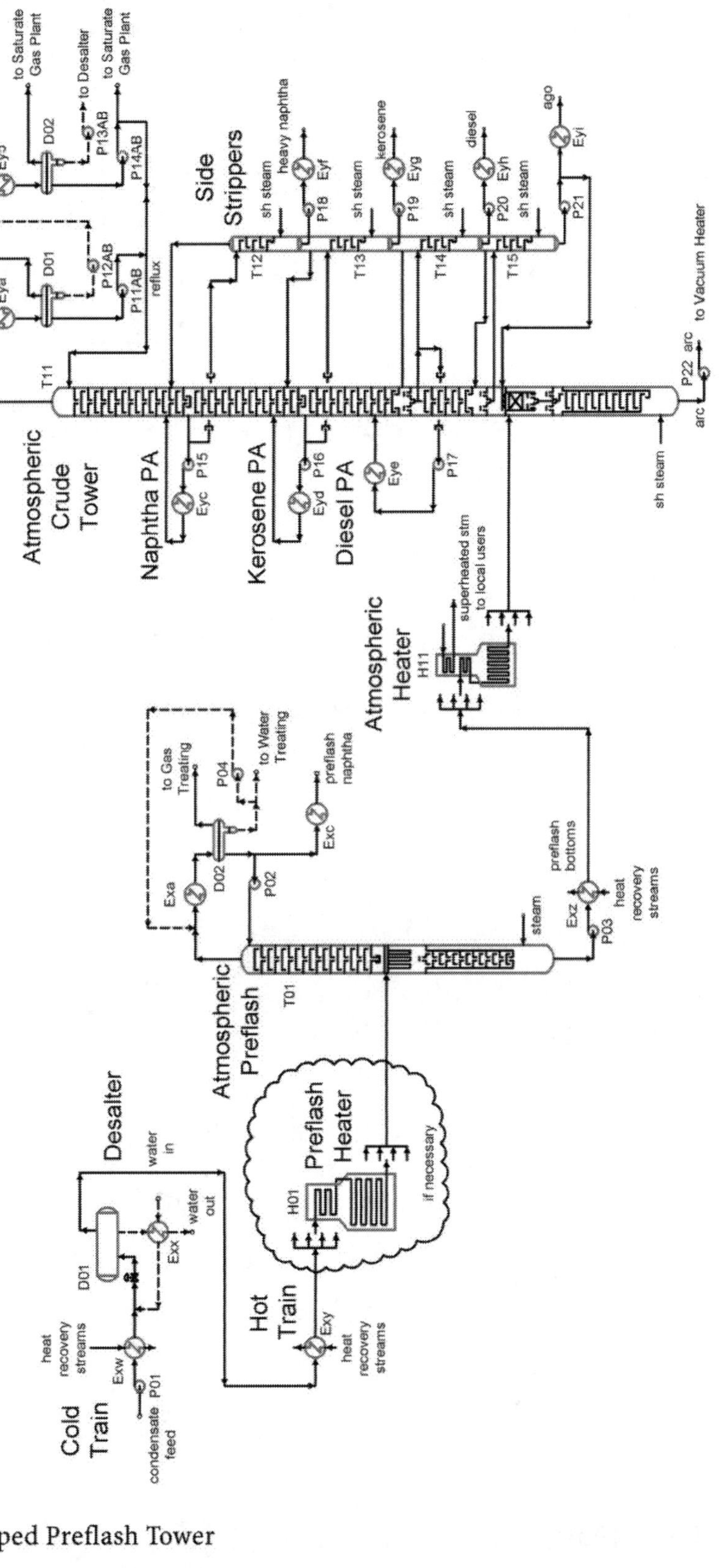

Figure 1 Steam-Stripped Preflash Tower

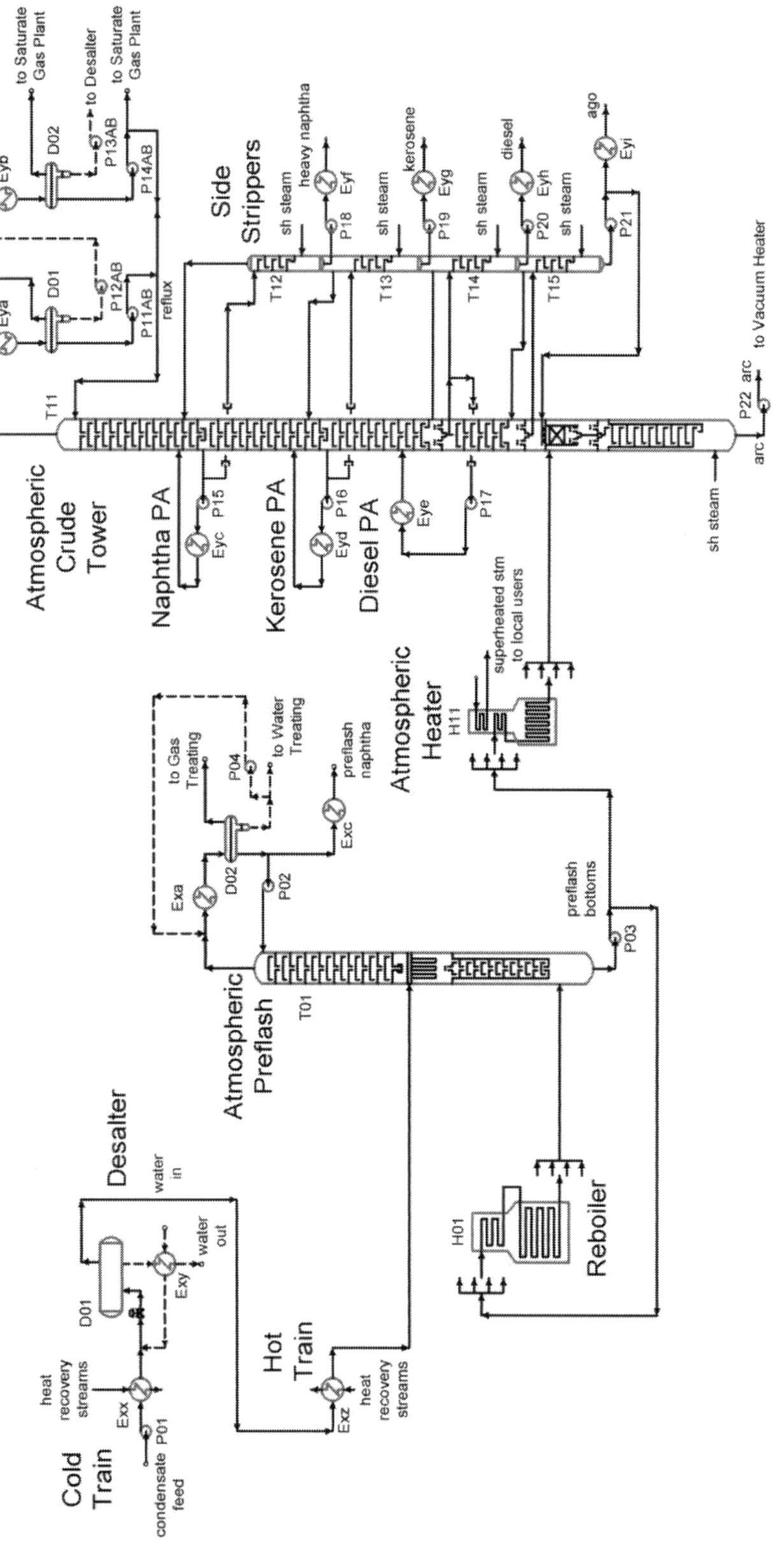

Figure 2 Reboiled Preflash Tower

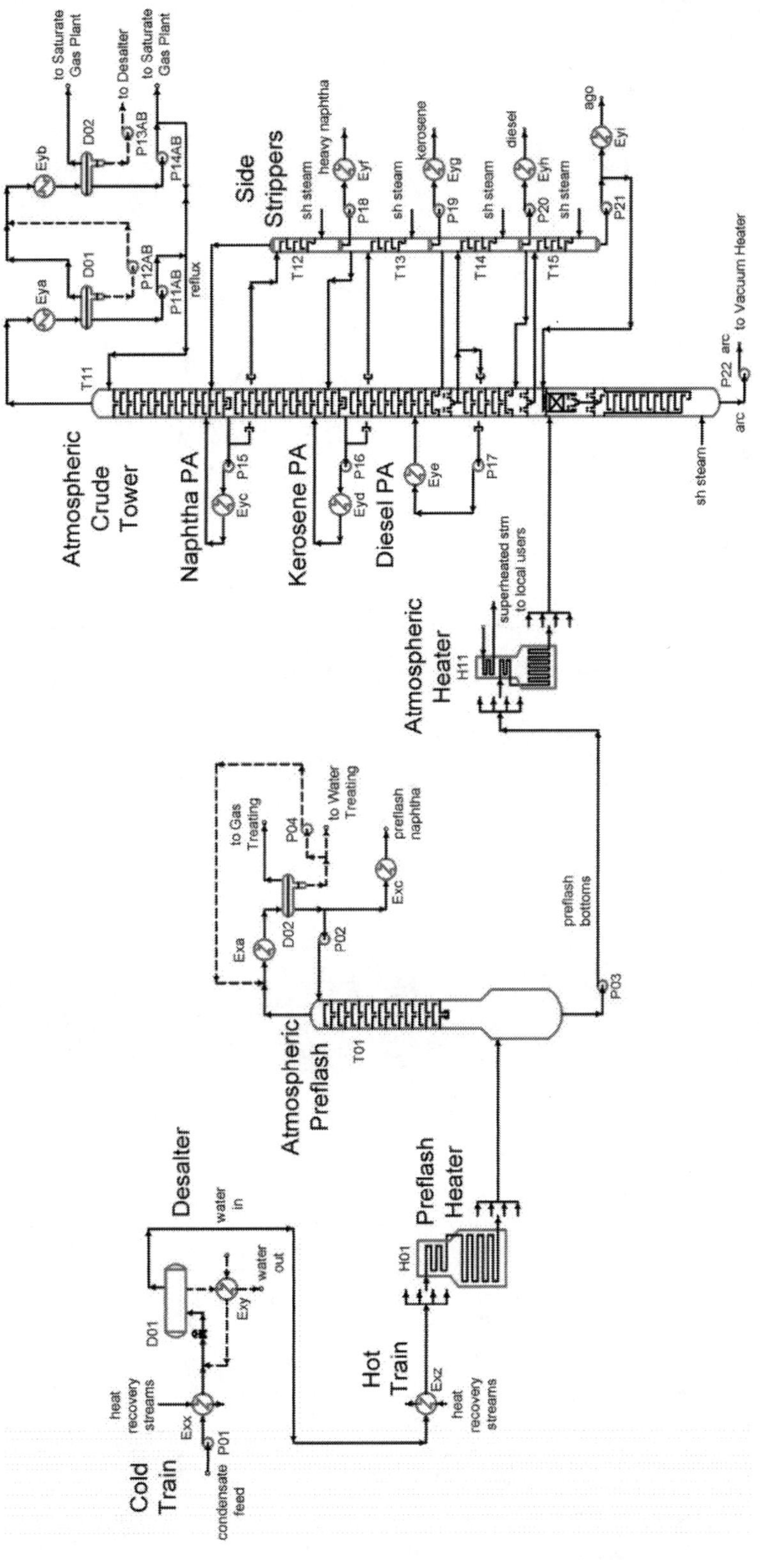

Figure 3 Flash Zone Preflash Tower

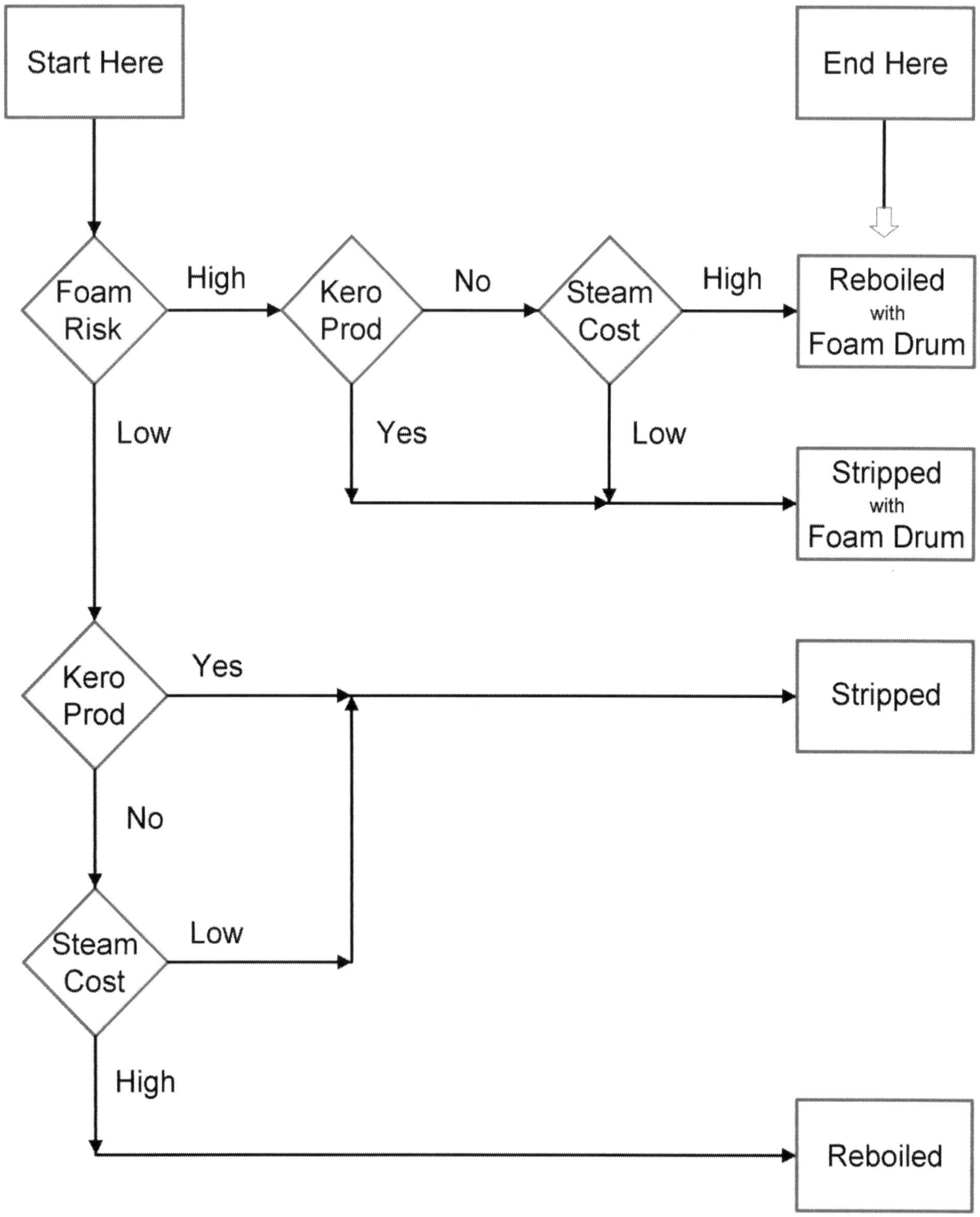

Figure 4 Selection Flowchart

11. References

EXXONMOBIL. About Gippsland Blend. Http://www.exxonmobil.com/crudeoil/download/GIPSB13X.pdf, 2008. (17 June 2008).

SLOLEY, A. W. Crude distillation sequences, AIChE National Meeting. San Francisco, 3-8 November, 2013.

SLOLEY, A. W. Mitigate fouling in crude unit overhead - part 1. *Hydrocarbon Processing* 92 (9): 73-81, Sept, 2013.

SLOLEY, A. W. Mitigate fouling in crude unit overhead - part 2. *Hydrocarbon Processing* 92 (11): 73-75, Nov, 2013.

SLOLEY, A. W. Mitigate fouling in crude unit overhead - part 2. *Hydrocarbon Processing* 92 (12): 75-79, Dec, 2013.

SLOLEY, A. W. Atmospheric preflash options, AIChE Southwest Process Technology Conference. Galveston, TX, 9-10 October, 2014.

SLOLEY, A. W. Crude unit heat recovery - Energy and product economics. In: *Improved Design and Operation in Refinery Distillation*, AIChE Spring Meeting. Austin, 26-30 April, 2015.

SLOLEY, A. W. Condensate splitting - what are the options? In: *Improved Design and Operation in Refinery Distillation*, AIChE Spring Meeting. Austin, 26-30 April, 2015.

WATKINS, R. N. *Petroleum Refinery Distillation*. Houston: Gulf Publishing Company, 1973.

Rev 01 2015-04-26

12. Appendix A Basis and Constraints

Parameter				
Streams				
Feed Rate	100,000	bpd	15,900	cmpd
Stripping steam rate	10	lbs/bbl residue	0.72	kg/m^3
Products				
Full range naphtha D86 95%	350	°F	177	°C
Heat Balance				
Feed temperature	70	°F	21	°C
Product temperatures	110	°F	43	°C
Desalter operating temperature		°F		°C
BCF 24	290		143	
TJL	275		135	
Bakken	250		121	
Gippsland	235		113	
Desalter water mixing temperature loss	15	°F	8.3	°C
Equipment Constraints and Performance				
Heat exchanger approach temperatures	40	°F	22.2	°C
Desalter free water at outlet		Wt %		
BCF 24	0.3			
TJL	0.2			
Bakken	0.1			
Gippsland	0.1			
Maximum heater outlet vaporization	90	Vol %		
Tray efficiency, rectifying	78	%		
Tray efficiency, steam stripping	25	%		
Tray efficiency, reboiled stripping	75	%		
Tray sizing, diameter increments	6	inches	15.2	cm
Tray sizing, maximum one-pass tray diameter	114	inches	2.90	m
Stripping vapor rate, minimum	0.058	ft^3/min-ft^2	0.018	m^3/min-m^2
Minimum liquid rate on weirs	1	gpm/in	1.49	lpm/cm

13. Appendix B Detailed Tables

Table B-1 BCF24 and Tia Juana Light

Feed		BCF24	BCF24	BCF24	BCF24	TJL	TJL	TJL	TJL
Config		Topping	Stripped	Reboiled	Flashed	Topping	Stripped	Reboiled	Flashed
Feed API		23.2	23.2	23.2	23.2	31.5	31.5	31.5	31.5
Rates	BPD								
Feed		99999	99999	99999	99999	99999	99999	99999	99999
Preflash Vapor			38	28	28		30	30	30
Preflash Naphtha			11285	11285	11285		20975	20975	20975
Atm Light Nap		9841	445	224	1487	19532	802	340	2743
Atm Heavy Nap		50	50	50	50	50	50	50	50
Kerosene		14398	11740	11936	10523	20146	16855	17239	14536
Diesel		9576	8174	7807	8191	10663	9540	9366	9878
AGO		7383	4171	3861	3985	9672	5144	4838	4963
ARC		58752	64104	64808	64450	39936	46598	47161	46826
Total Distillate		41248	35865	35163	35521	60063	53366	52808	53145
Preflash Cutpoint	F		335	337	337		335	336	336
Atmos Cutpoint	F	714	655	647	651	747	675	668	672
Preflash Dew Margin	F		56.5	117.3	136.2		74.1	128	139.8
Atmos Dew Margin	F	35.9	30	30	30	39.3	30	30	30.1
Atmos Ovhd Temp	F	265	273	276	273	265	269	271	271
Preflash Nap D86 95	F		350	350	350		350	350	350
Atmos Heavy Nap D86 95	F	335	299	298	355	333	298	280	354
Kero D86 99	F	560	560	560	560	560	560	560	560
Diesel D86 95	F	640	640	640	640	640	640	640	640
Kero Flash	F	120	134	132	142	120	125	121	127
Preflash Cond Q	Mbtu/hr		89.0	56.7	95.1		103.8	78.3	116.4
Preflash Stripping	lb/bbl		10.0				10.0		
Preflash Stripping	wt%		3.1				3.2		
Atm Cond Q	Mbtu/hr	100.1	63.8	57.2	54.3	100.3	57.8	54.1	44.4
HN PA Q	Mbtu/hr	9.4	0.1	0.1	0.1	57.8	0.1	0.1	0.1
Kero PA Q	Mbtu/hr	28.7	13.2	17.7	21.3	17.6	19.2	20.0	32.1
Dies PA Q	Mbtu/hr	53.3	30.1	28.0	29.3	77.1	39.3	38.3	39.8
LH Gap	F	-27.3	-3.4	-4.8	-7.6	-53.3	-8.6	-7.0	-21.7
HK Gap	F	20.9	69.4	65.4	36.9	25.7	57.2	67.6	1.9
KD Gap	F	-73.5	-56.4	-56.8	-56.1	-94.1	-67.9	-68.4	-68.8
DA Gap	F	-28.6	-48.7	-51.1	-49.6	-22.0	-46.8	-48.7	-47.2
AR Gap	F	-162.0	-184.6	-189.9	-187.3	-153.7	-182.0	-185.9	-183.7
AGO 650- D86	Vol %	10.5	18.1	18.9	18.4	9.1	17.4	18.2	17.6
AGO 700- D86	Vol %	43.7	54.4	55.7	54.6	40.8	55.3	56.2	55.0
Atm Stripping	lb/bbl	10.0	10.0	10.0	10.0	10.0	10.0	10.0	10.0
Atm Stripping	wt%	3.0	3.0	3.0	3.0	3.0	3.0	3.0	3.0
Kero Stripping	lb/bbl	9.8	2.0	2.0	2.3	12.1	1.4	1.4	1.6
Kero Stripping	wt%	3.2	0.7	0.6	0.7	4.2	0.5	0.5	0.6
Preflash heater	Mbtu/hr				129.4				107.6
Preflash reboiler	Mbtu/hr			141.4				113.7	
Atm heater	Mbtu/hr	194.5	162.4	30.0	82.2	222.6	169.3	69.4	114.8
Prelfash Stm	lb/hr		36972				32945		
Atm steam	lb/hr	24522	26716	27018	26869	16655	19424	19661	19522
Preflash Feed Temp	F		455	525	635		446	500	584
Energy Balance									
Total Firing	Mbtu/hr	194.5	162.4	171.4	211.6	222.6	169.3	183.1	222.4
Total Steam	lb/hr	24522	63688	27018	26869	16655	52369	19661	19522
Total Firing + PF Steam	Mbtu/hr	194.5	199.4	171.4	211.6	222.6	202.2	183.1	222.4
PA Recovery	Mbtu/hr	91.4	43.4	45.8	50.7	152.5	58.6	58.4	72.0
Intensity	btu/bbl distillate	4715	5559	4874	5957	3706	3790	3467	4185
After-PF Integration	Mbtu/hr		70.7				55.6		
Steam/Fired	For equal value		0.195				0.301		
Steam/Fired	No after PF-Integration		-0.915				-0.761		

Table B-2 Bakken and Gippsland

Feed		Bakken	Bakken	Bakken	Bakken	Gipps	Gipps	Gipps	Gipps
Config		Topping	Stripped	Reboiled	Flashed	Topping	Stripped	Reboiled	Flashed
Feed API		42.0	42.0	42.0	42.0	51.1	51.1	51.1	51.1
Rates	BPD								
Feed		100000	100000	100000	100000	100011	100011	100011	100011
Preflash Vapor			30	30	30		31	31	30
Preflash Naphtha			35947	35947	35947		50958	50958	50958
Atm Light Nap		33477	2129	1411	4000	49916	2617	2266	3908
Atm Heavy Nap		50	50	50	50	447	475	476	473
Kerosene		24530	18174	18950	15955	22538	17110	17572	15291
Diesel		12519	11949	11666	12264	8748	10335	10100	10823
AGO		12066	6075	5789	5906	10984	8780	8400	8522
ARC		17359	25645	26157	25850	7377	9704	10208	10007
Total Distillate		82642	74324	73813	74122	92633	90275	89772	89975
Preflash Cutpoint	F		335	336	336		335	336	336
Atmos Cutpoint	F	797	689	684	687	777	749	743	745
Preflash Dew Margin	F		104	150	154		119	150	131.8
Atmos Dew Margin	F	40.7	30	30	39.5	58.7	30	30	43.7
Atmos Ovhd Temp	F	265	276	277	277	265	266	272	265
Preflash Nap D86 95	F		350	350	350		350	347	350
Atmos Heavy Nap D86 95	F	328	317	314	350	341	303	307	334
Kero D86 99	F	560	560	560	560	560	560	560	560
Diesel D86 95	F	640	640	640	640	640	640	640	640
Kero Flash	F	120	120	120	120	120	120	120	120
Preflash Cond Q	Mbtu/hr		134.6	111.8	149.8		149.1	128.7	174.4
Preflash Stripping	lb/bbl		10.0				10.0		
Preflash Stripping	wt%		3.3				3.4		
Atm Cond Q	Mbtu/hr	153.5	31.3	27.8	37.1	179.1	45.2	43.7	32.7
HN PA Q	Mbtu/hr	110.8	10.0	12.0	0.1	93.5	0.1	0.1	0.1
Kero PA Q	Mbtu/hr	0.1	34.6	33.4	44.0	14.5	39.8	44.2	36.7
Dies PA Q	Mbtu/hr	104.0	46.1	44.3	45.9	74.0	45.1	43.6	44.7
LH Gap	F	-52.5	-20.4	-18.6	-32.6	-45.6	-15.2	18.0	-28.0
HK Gap	F	24.9	32.8	32.8	4.8	24.2	47.1	41.5	32.6
KD Gap	F	-107.6	-73.2	-73.3	-70.4	-124.6	-80.2	-81.1	-72.1
DA Gap	F	-18.2	-51.0	-52.7	-51.7	-46.3	-40.3	-42.6	-43.6
AR Gap	F	-167.1	-178.6	-180.0	-179.5	-200.6	-185.9	-186.0	-187.2
AGO 650- D86	Vol %	8.8	21.0	21.9	21.2	12.4	14.4	15.0	14.8
AGO 700- D86	Vol %	41.3	62.8	63.6	62.7	40.3	43.8	45.5	44.4
Atm Stripping	lb/bbl	10.0	10.0	10.0	10.0	10.0	10.0	10.0	10.0
Atm Stripping	wt%	3.0	3.0	3.0	3.0	3.3	3.3	3.3	3.3
Kero Stripping	lb/bbl	36.6	1.3	12.7	4.6	24.3	13.2	16.8	3.1
Kero Stripping	wt%	12.8	0.5	0.4	1.6	8.4	4.6	5.8	1.1
Preflash heater	Mbtu/hr				90.4				90.5
Preflash reboiler	Mbtu/hr			88.7				72.5	
Atm heater	Mbtu/hr	256.0	168.1	94.5	131.5	217.0	157.5	101.1	129.2
Prelfash Stm	lb/hr		26721				20466		
Atm steam	lb/hr	7255	10693	10908	10780	3077	4046	4256	4172
Preflash Feed Temp	F		455	475	528		450	450	480
Energy Balance									
Total Firing	Mbtu/hr	256.0	168.1	183.2	221.9	217.0	157.5	173.6	219.7
Total Steam	lb/hr	7255	37414	10908	10780	3077	24512	4256	4172
Total Firing + PF Steam	Mbtu/hr	256.0	194.8	183.2	221.9	217.0	178.0	173.6	219.7
PA Recovery	Mbtu/hr	214.9	90.7	89.7	90.0	182.0	85.0	87.9	81.5
Intensity	btu/bbl distillate	3098	2621	2482	2994	2343	1971	1934	2442
After-PF Integration	Mbtu/hr		30.9				17.7		
Steam/Fired	For equal value		0.437				0.697		
Steam/Fired	No after PF-Integration		-0.388				-0.026		

Table B-1M BCF24 and Tia Juana Light

Feed		BCF24	BCF24	BCF24	BCF24	TJL	TJL	TJL	TJL
Config		Topping	Stripped	Reboiled	Flashed	Topping	Stripped	Reboiled	Flashed
Feed SG		0.9147	0.9147	0.9147	0.9147	0.8681	0.8681	0.8681	0.8681
Rates	m3/d								
Feed		15900	15900	15900	15900	15900	15900	15900	15900
Preflash Vapor			6	4	4		5	5	5
Preflash Naphtha			1794	1794	1794		3335	3335	3335
Atm Light Nap		1565	71	36	236	3106	128	54	436
Atm Heavy Nap		8	8	8	8	8	8	8	8
Kerosene		2289	1867	1898	1673	3203	2680	2741	2311
Diesel		1523	1300	1241	1302	1695	1517	1489	1571
AGO		1174	663	614	634	1538	818	769	789
ARC		9342	10193	10304	10248	6350	7409	7499	7445
Total Distillate		6558	5703	5591	5648	9550	8485	8396	8450
Preflash Cutpoint	C		168	169	169		168	169	169
Atmos Cutpoint	C	379	346	342	344	397	357	353	356
Preflash Dew Margin	C		31.4	65.2	75.7		41.2	71.1	77.7
Atmos Dew Margin	C	19.9	16.7	16.7	16.7	21.8	16.7	16.7	16.7
Atmos Ovhd Temp	C	129	134	136	134	129	132	133	133
Preflash Nap D86 95	C		177	177	177		177	177	177
Atmos Heavy Nap D86 95	C	168	148	148	179	167	148	138	179
Kero D86 99	C	293	293	293	293	293	293	293	293
Diesel D86 95	C	338	338	338	338	338	338	338	338
Kero Flash	C	49	57	56	61	49	52	49	53
Preflash Cond Q	GJ/hr		93.9	59.8	100.3		109.5	82.6	122.8
Preflash Stripping	kg/m3		28.5				28.5		
Preflash Stripping	wt%		3.1				3.2		
Atm Cond Q	GJ/hr	105.6	67.3	60.3	57.3	105.8	61.0	57.1	46.8
HN PA Q	GJ/hr	9.9	0.1	0.1	0.1	61.0	0.1	0.1	0.1
Kero PA Q	GJ/hr	30.3	13.9	18.7	22.5	18.6	20.3	21.1	33.9
Dies PA Q	GJ/hr	56.2	31.8	29.5	30.9	81.3	41.5	40.4	42.0
LH Gap	C	-15.2	-1.9	-2.7	-4.2	-29.6	-4.8	-3.9	-12.1
HK Gap	C	11.6	38.6	36.3	20.5	14.3	31.8	37.6	1.1
KD Gap	C	-40.8	-31.3	-31.6	-31.2	-52.3	-37.7	-38.0	-38.2
DA Gap	C	-15.9	-27.1	-28.4	-27.6	-12.2	-26.0	-27.1	-26.2
AR Gap	C	-90.0	-102.6	-105.5	-104.1	-85.4	-101.1	-103.3	-102.1
AGO 343- D86	Vol %	10.5	18.1	18.9	18.4	9.1	17.4	18.2	17.6
AGO 371- D86	Vol %	43.7	54.4	55.7	54.6	40.8	55.3	56.2	55.0
Atm Stripping	kg/m3	28.5	28.5	28.5	28.5	28.5	28.5	28.5	28.5
Atm Stripping	wt%	3.0	3.0	3.0	3.0	3.0	3.0	3.0	3.0
Kero Stripping	kg/m3	28.0	5.7	5.7	6.5	34.5	4.0	4.0	4.6
Kero Stripping	wt%	3.2	0.7	0.6	0.7	4.2	0.5	0.5	0.6
Preflash heater	GJ/hr				136.5				113.5
Preflash reboiler	GJ/hr			149.2				120.0	
Atm heater	GJ/hr	205.2	171.3	31.7	86.7	234.8	178.6	73.2	121.1
Prelfash Stm	kg/hr		16767				14941		
Atm steam	kg/hr	11121	12116	12253	12185	7553	8809	8917	8854
Preflash Feed Temp	C		235	274	335		230	260	307
Energy Balance									
Total Firing	GJ/hr	205.2	171.3	180.8	223.2	234.8	178.6	193.2	234.6
Total Steam	kg/hr	11121	28883	12253	12185	7553	23750	8917	8854
Total Firing + PF Steam	GJ/hr	205.2	210.3	180.8	223.2	234.8	213.4	193.2	234.6
PA Recovery	GJ/hr	96.4	45.8	48.3	53.5	160.9	61.8	61.6	76.0
Intensity	kj/m3 distillate	13450	15856	13903	16991	10571	10810	9890	11936
After-PF Integration	GJ/hr		74.6				58.7		
Steam/Fired	For equal value		0.195				0.301		
Steam/Fired	No after PF-Integration		-0.915				-0.761		

Table B-2M Bakken and Gippsland

Feed		Bakken	Bakken	Bakken	Bakken	Gipps	Gipps	Gipps	Gipps
Config		Topping	Stripped	Reboiled	Flashed	Topping	Stripped	Reboiled	Flashed
Feed SG		0.8156	0.8156	0.8156	0.8156	0.7749	0.7749	0.7749	0.7749
Rates	m3/d								
Feed		15900	15900	15900	15900	15902	15902	15902	15902
Preflash Vapor			5	5	5		5	5	5
Preflash Naphtha			5716	5716	5716		8102	8102	8102
Atm Light Nap		5323	339	224	636	7937	416	360	621
Atm Heavy Nap		8	8	8	8	71	76	76	75
Kerosene		3900	2890	3013	2537	3584	2720	2794	2431
Diesel		1991	1900	1855	1950	1391	1643	1606	1721
AGO		1918	966	920	939	1746	1396	1336	1355
ARC		2760	4078	4159	4110	1173	1543	1623	1591
Total Distillate		13140	11818	11736	11785	14729	14354	14274	14306
Preflash Cutpoint	C		168	169	169		168	169	169
Atmos Cutpoint	C	425	365	362	364	414	398	395	396
Preflash Dew Margin	C		57.8	83.3	85.6		66.1	83.3	73.2
Atmos Dew Margin	C	22.6	16.7	16.7	21.9	32.6	16.7	16.7	24.3
Atmos Ovhd Temp	C	129	136	136	136	129	130	133	129
Preflash Nap D86 95	C		177	177	177		177	175	177
Atmos Heavy Nap D86 95	C	164	158	157	177	172	151	153	168
Kero D86 99	C	293	293	293	293	293	293	293	293
Diesel D86 95	C	338	338	338	338	338	338	338	338
Kero Flash	C	49	49	49	49	49	49	49	49
Preflash Cond Q	GJ/hr		142.0	117.9	158.0		157.3	135.8	184.0
Preflash Stripping	kg/m3		28.5				28.5		
Preflash Stripping	wt%		3.3				3.4		
Atm Cond Q	GJ/hr	161.9	33.0	29.3	39.1	189.0	47.7	46.1	34.5
HN PA Q	GJ/hr	116.9	10.6	12.7	0.1	98.6	0.1	0.1	0.1
Kero PA Q	GJ/hr	0.1	36.5	35.2	46.4	15.3	42.0	46.6	38.7
Dies PA Q	GJ/hr	109.7	48.6	46.7	48.4	78.1	47.6	46.0	47.2
LH Gap	C	-29.2	-11.3	-10.3	-18.1	-25.3	-8.4	10.0	-15.6
HK Gap	C	13.8	18.2	18.2	2.7	13.4	26.2	23.1	18.1
KD Gap	C	-59.8	-40.7	-40.7	-39.1	-69.2	-44.6	-45.1	-40.1
DA Gap	C	-10.1	-28.3	-29.3	-28.7	-25.7	-22.4	-23.7	-24.2
AR Gap	C	-92.8	-99.2	-100.0	-99.7	-111.4	-103.3	-103.3	-104.0
AGO 343- D86	Vol %	8.8	21.0	21.9	21.2	12.4	14.4	15.0	14.8
AGO 371- D86	Vol %	41.3	62.8	63.6	62.7	40.3	43.8	45.5	44.4
Atm Stripping	kg/m3	28.5	28.5	28.5	28.5	28.5	28.5	28.5	28.5
Atm Stripping	wt%	3.0	3.0	3.0	3.0	3.3	3.3	3.3	3.3
Kero Stripping	kg/m3	104.4	3.7	36.2	13.1	69.3	37.7	47.9	8.8
Kero Stripping	wt%	12.8	0.5	0.4	1.6	8.4	4.6	5.8	1.1
Preflash heater	GJ/hr				95.4				95.5
Preflash reboiler	GJ/hr			93.6				76.5	
Atm heater	GJ/hr	270.1	177.3	99.7	138.7	228.9	166.2	106.7	136.3
Prelfash Stm	kg/hr		12118				9282		
Atm steam	kg/hr	3290	4849	4947	4889	1395	1835	1930	1892
Preflash Feed Temp	C		235	246	276		232	232	249
Energy Balance									
Total Firing	GJ/hr	270.1	177.3	193.3	234.1	228.9	166.2	183.1	231.8
Total Steam	kg/hr	3290	16968	4947	4889	1395	11117	1930	1892
Total Firing + PF Steam	GJ/hr	270.1	205.5	193.3	234.1	228.9	187.8	183.1	231.8
PA Recovery	GJ/hr	226.7	95.7	94.6	95.0	192.0	89.7	92.7	86.0
Intensity	kj/m3 distillate	8836	7477	7079	8539	6682	5623	5516	6965
After-PF Integration	GJ/hr		32.6				18.7		
Steam/Fired	For equal value		0.437				0.697		
Steam/Fired	No after PF-Integration		-0.388				-0.026		

14. Appendix C – Glossary, Definitions, Abbreviations and Conventions

Abbreviations and terms in text and figures that may not have been defined or that may be non-standard usage.

These definitions are tailored to the most common industry use of the term in crude distillation and petroleum refining. Many plants use their own variations on these definitions and other terms in place of those listed. Other technical fields may have different meanings for these terms.

Term or abbreviation	Definition
AEBP	Atmospheric equivalent boiling point. The temperature at which a compound would have boiled at atmospheric pressure if it could be heated to that temperature without thermal decomposition.
AET	Atmospheric equivalent temperature. See AEBP.
ANS	Alaskan North Slope - crude oil type
API degrees, API°	See API Gravity
API Gravity	A measure of liquid density defined by the American Petroleum Institute. The scale was derived from the reading marks on a specific type of measurement instrument. $API^{O} = \frac{141.5}{SG} - 131.5$ $SG = \frac{141.5}{API^{O} + 131.5}$ Where SG is specific gravity at 60°F (15.6°C). Water has an API gravity of 10. An API gravity lower than 10 is denser than water, an API gravity higher than 10 is less dense than water.
ASTM Distillation	Old usage. ASTM Distillation quoted without a descriptive qualifier refers to an ASTM D86 distillation. This is a batch distillation without reflux.
Boiling range	See distillation range.
bpd	Barrels per day, standard volume barrels at 60°F (15.6°C)
Cold train	Heat exchangers upstream of the desalter in a crude unit.
Condensate	Low density, naturally occurring petroleum liquid mixture. By industry usage this has a API gravity of 49° or higher (SG 0.7839 or lower). Condensates may come from direct production from wells or as captured liquids from pipeline systems (often called retrograde condensates).
CP	Cutpoint
Cutpoint	The ASTM D2892 distillation temperature that would achieve the same split between overhead products and bottoms products that the unit achieves. The D2892 distillation basis is atmospheric equivalent

	temperature. Typical usage is volume percent.
Distillation range	Range of AEBP boiling temperatures for a stream. Typically expressed as an ASTM D-2892 equivalent (true boiling range) or an ASTM D-86 equivalent (ASTM boiling range) on a 5-95 volume percent basis. Other basis may be used.
FBP	Final boiling point
Flash zone	Feed entry point into a main fractionator.
Gap-overlap	A measure of fractionation performance between the distillation of a light product and the distillation of the next heavier product from a main fractionators
IBP	Initial boiling point.
hnap	Heavy naphtha
Hot train	Exchangers in a crude preheat sequence downstream of the desalter.
IBP	Initial boiling point.
IP	Institute of Petroleum, UK
ISO	International Standards Organization.
LSR, lsr	Light straight run naphtha, same as light naphtha
MB	Million btu/hr (abbreviated to fit in figures), mmbtu/hr
MBtu/hr	Million btu/hr, mmbtu/hr
mmbtu/hr	Million btu/hr, industry standard usage remains the old form using a lower-case m for thousand. 'mm' equals thousand X thousand to get million.
NBP	Normal boiling point, see AEBP
NWS	Northwest Shelf, crude (condensate) type.
Overlap	See Gap
Specific gravity	Density of a stream at 60°F (15.56°C) compared to the density of water at 60°F (15.56°C).
Stabilizer	Distillation tower with a main purpose to control the vapor pressure of the bottoms product.
Stove oil	Alternate name for kerosene
T_{XX}	Temperature at which XX volume (or weight) percent of liquid is distilled.
TBP	True boiling point. Distillation curve based on ASTM D2892 or an equivalent.
Transfer line	Pipe connecting the outlet of a fired heater to the downstream main

	frationator.
TVP	True vapor pressure.
Wild naphtha	Unstabilized naphtha. Naphtha that still has a high vapor pressure due to a high butane and lighter content (typically more than one volume percent).

Figure conventions;

- Lower case text and abbreviations in figures refer to a stream.
- Upper case text and abbreviations in figures refer to an equipment item.

Marathon Petroleum Company LP

Tracerco
Providing Insight Onsite

Coking Mechanism in Refinery Fractionator Wash Beds

Gregory A. Cantley - Separations Technologist
James F. Johnson - Crude & Vacuum Technologist

Lowell Pless - Business Development Manager

2015 A.I.Ch.E. Spring Meeting | Improved Design and Operation in Refinery Distillation | April 27, 2015

Tracerco
Providing Insight Onsite

Crude Vacuum Tower

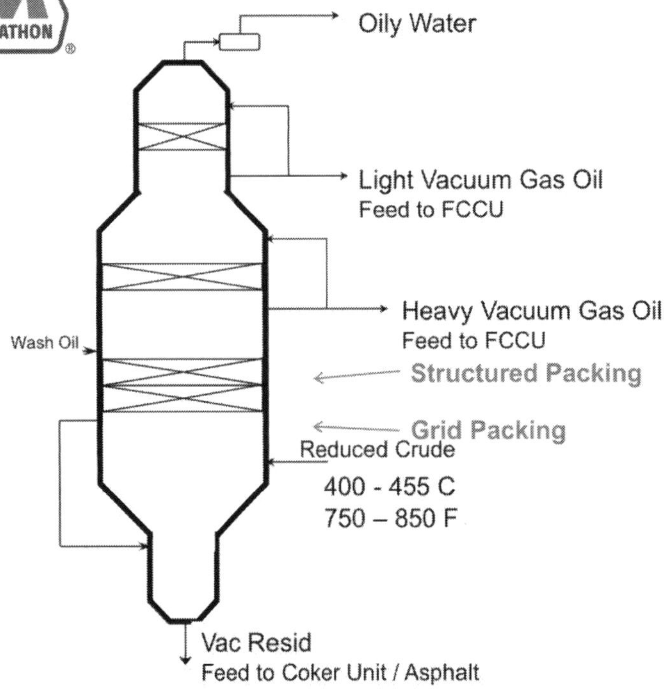

Vacuum Tower Wash Bed Design

- **Industry philosophy: Design for 0.2 gpm/ft^2 (1.4 E-04 m^3/s*m^2) of true Overflash from the bottom of the Wash Bed.**

- **Operating with this true Overflash rate typically avoids coking in the Wash Bed.**

- **Marathon Petroleum manages Vacuum Tower operating conditions to reach predetermined cycle length, BUT Maximize HVGO Yield by coking the Wash Bed.**

Tru-Grid™ vs. ThruVision™

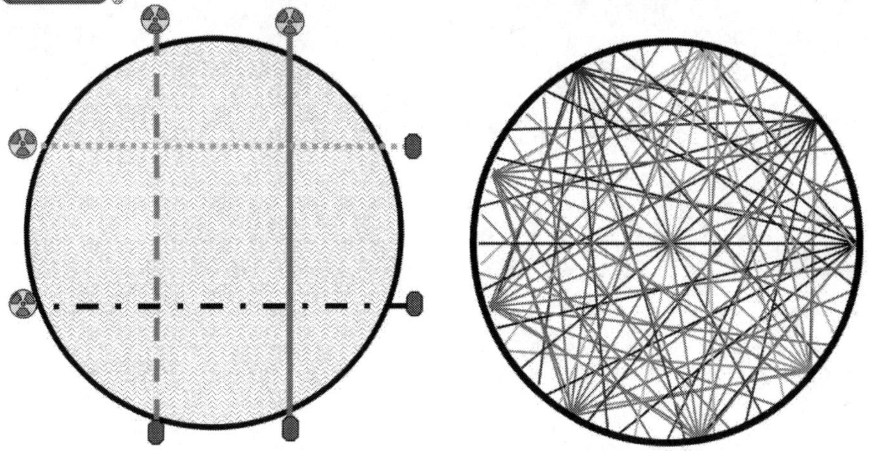

42 ft (12.8 m) Diameter:

~1.4% Area Coverage **~34%** Area Coverage

Tru-Grid™ - ThruVision™

Tracerco
Providing Insight Onsite

- Adequate for assessing damage
- Adequate for assessing gross liquid maldistribution
- Covers entire tower vertically
- Overall costs less
- Usually uses existing tower ladders & platform
- 4 data points (at each elevation)

- Detail information on liquid or density distribution
- Covers entire tower cross-sectional area
- Only done at one elevation
- Needs 360° access
- 81 (9x9) or 121 (11x11) data point grid

Tracerco
Providing Insight Onsite

Refinery A Vacuum Tower

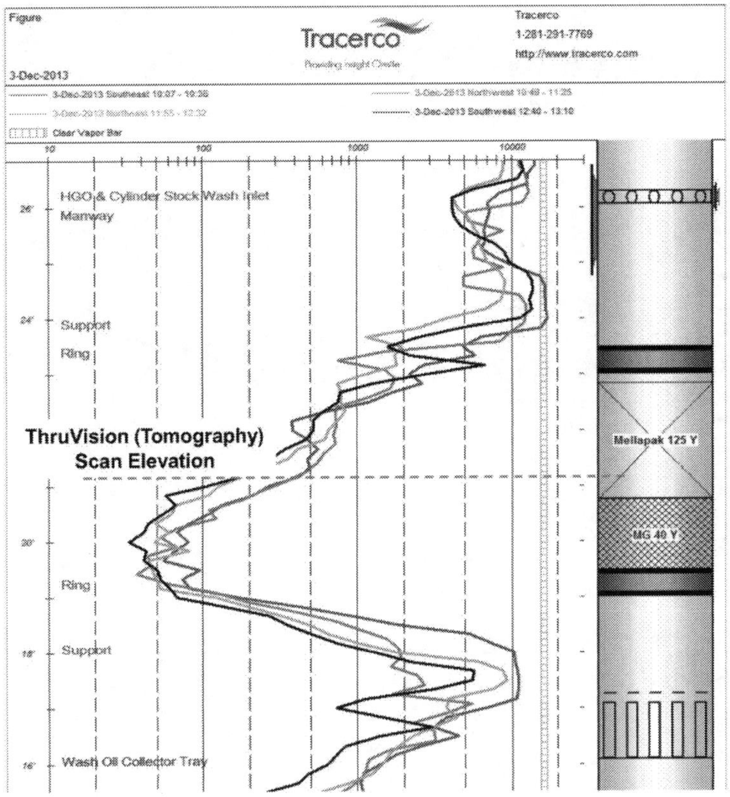

Refinery A Vacuum Tower

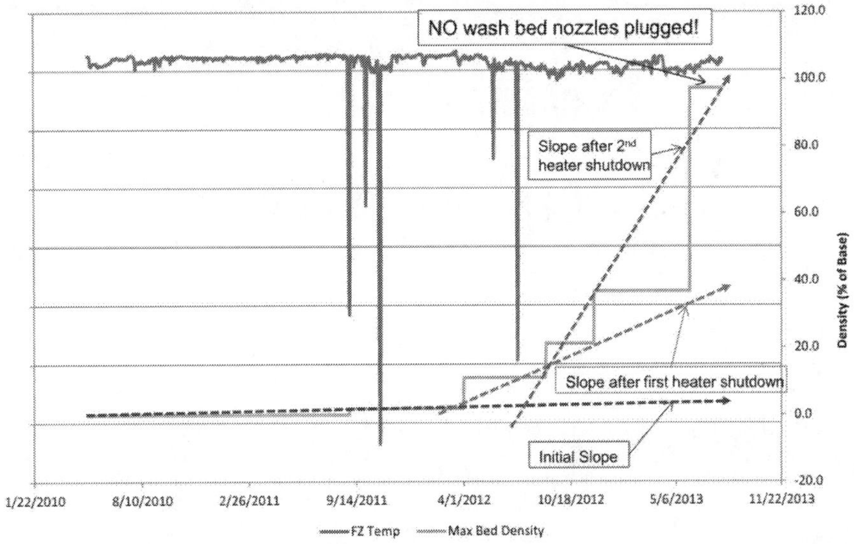

2015 A.I.Ch.E. Spring Meeting Improved Design and Operation in Refinery Distillation April 27, 2015 8

ThruVision™ Scan Results

Refinery A Vacuum Tower

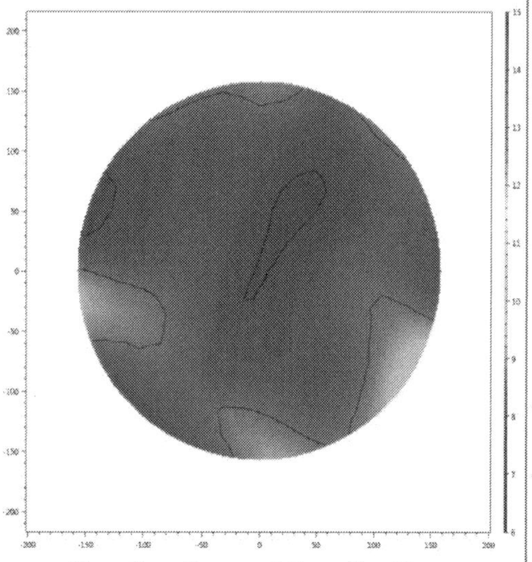

Baseline Scan w/ New Packing

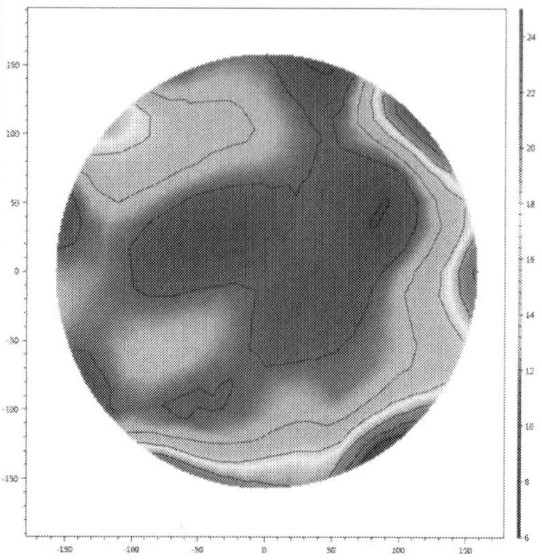

Scan ~ 3 years later

Refinery B Vacuum Tower

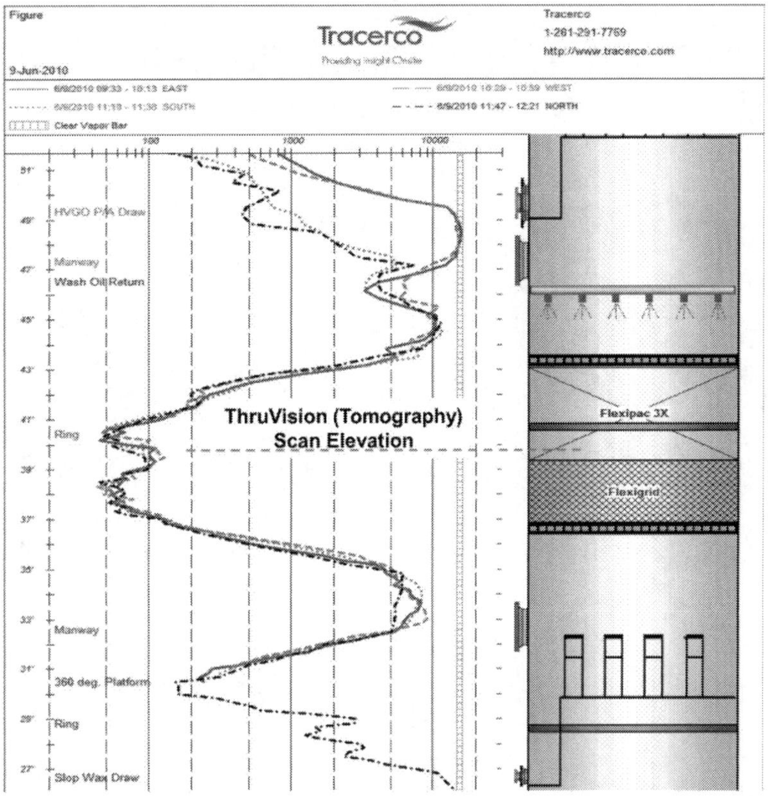

Refinery B Vacuum Tower

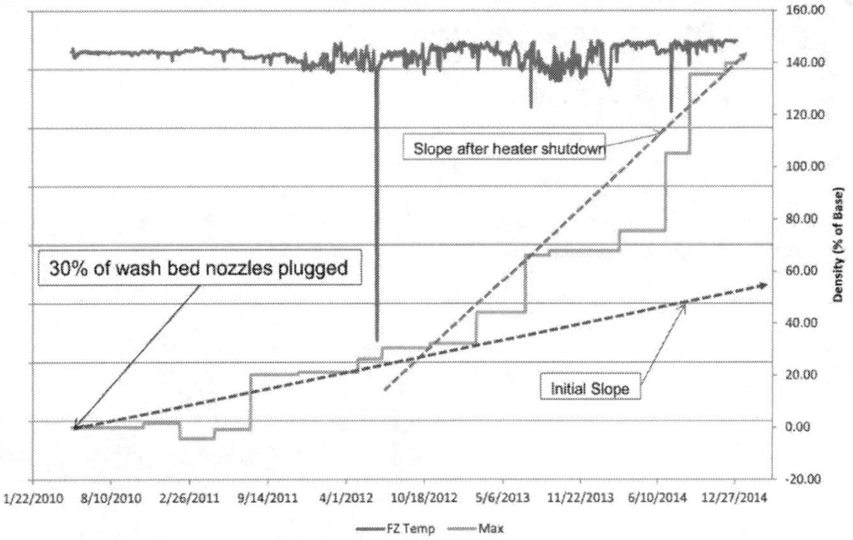

ThruVision™ Scan Results

Refinery B Vacuum Tower

Tracerco
Providing Insight Onsite

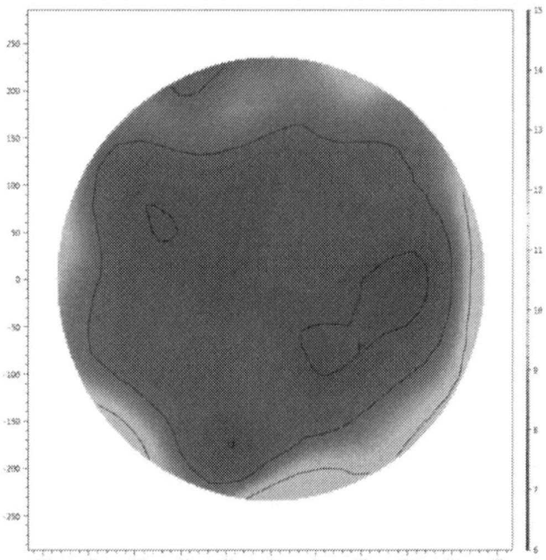

Baseline Scan w/ New Packing

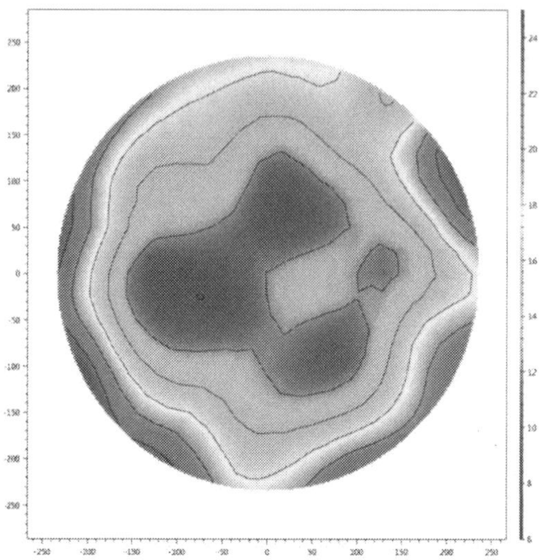

Scan 4.5 years later

Vacuum Tower Wash Bed Coking

Tracerco
Providing Insight Onsite

Thermosetting Polymer

- IUPAC* defines a thermosetting resin as a petrochemical in a soft-solid or viscous state that by curing changes irreversibly into an infusible, insoluble polymer network.
- Curing can be induced by the action of heat or suitable radiation (Not Tracerco's!), or both. A cured thermosetting resin is called a thermoset.
- Consequently, a thermoset cannot be melted after it is cured. Once the "hard candy" has setup in the packing, the packing must be discarded.

* International Union of Pure and Applied Chemistry. http://iupac.org/

Coking Mechanism in Wash Beds

Conventional Wisdom

- A total power outage leads to a loss of wash liquid causing the spray nozzles to plug.
- Plugged spray nozzles lead to dry portions (liquid maldistribution) of the wash bed.
- The dry portions of the wash bed then coke, causing plugging.

New Theory

- Coke precursor material is in the bed in a semi-fluid state, such that vapor/liquid can still move through with relatively little restriction.
- A power outage and/or loss of heat causes the bed to cool to a temperature where the coke precursor material solidifies.
- Once Operation returns to normal the thermoset restricts flow providing a site for continued coke growth.

Operating Conclusions

- Tomography Scans yield extensive cross-sectional coverage to monitor fouling/coking in packed beds.
- Tomography Scans can be used to monitor wash bed coking and to make decisions on operating conditions to target a run (cycle) length.
- In the event of a power failure or heater loss, the Wash Bed fouls by our new proposed mechanism and not by Spray Distributor nozzle plugage.

Thank You!

Questions?
Comments?

How Does a Collector Tray Leak Impact Column Operation?

Comparing the Impact of Two Leaking FCC Light Cycle Oil Collector Trays

Daryl W. Hanson, Valero, San Antonio, TX

Jill Brown Burns, Valero, San Antonio, TX

Mike Teders, Valero, San Antonio, TX

Presented at the AIChE Spring Meeting, Distillation Conference, Austin, Texas, April 27 - May 1, 2015

How Does a Collector Tray Leak Impact Tower Operation?

Daryl W. Hanson, Jill Brown Burns, and Mike Teders; Valero, San Antonio, TX

In the collective experience of the authors, one of the most troublesome pieces of refinery fractionation equipment is the collector tray[1,2,3]. Collector trays serve many different functions: liquid collection, liquid distributor to trays (possibly in a transition section), liquid re-distribution between packed sections, preliminary vapor distribution, solids collection device, and even as a de-entrainment device. Many times, the designer tries to accommodate many of the functions in the list above into a single tray design. Design complication often precedes tower failures.

Collector trays are most commonly used as liquid collection devices to provide positive material balance control in distillation columns. They are designed with utmost care; so that liquid can be collected, be drawn or flow to the section below without hydraulic restrictions and without leaking. Controlling liquid leakage from a collector tray is paramount; it ensures adequate column heat and material balance. Many heavy oil fractionators, including FCC Main Fractionators, use collectors for tower material balance control.

Engineers often envision that a collector tray leak results in a discreet, predictable stream of liquid from the collector tray, falling neatly onto the fractionation device below - a neat and convenient theory. Some even believe that leaks are as distributed as the liquid that exits a liquid distribution device (trough or pan distributors). Collector tray leaks can either be the ones accepted during a turnaround leak test or those developed during the run (corrosion, damage, or otherwise).

Leaks (can be called 'lost product') contribute additional reflux to the section below. Based on many discussions between distillation engineering experts, leaks have resulted in limited operational difficulty outside of lost revenue. Along with many in industry, the authors agreed with this theory before comparing the troubleshooting results of the two case studies discussed in this paper.

The authors want to point out that both of these case studies are from Fluid Catalytic Cracking (FCC) Main Fractionator towers. Therefore, their operation can be directly compared without having to address the complexities of different systems.

Case #1: Capacity of 100 MBD, trayed Main Fractionator, no instability

Case #2: Capacity of 30 MBD, packed Main Fractionator, severe instability

The product quality, tower layout, and operation of the towers are very similar. Yet, contrasting the two case studies shows that leakage (excess reflux) has a remarkable impact on one of the Main Fractionator's stability. The resultant instability caused the Main Fractionator to experience drastic operational swings and resulted in a large feed rate cut until a solution was identified and implemented. The other Main Fractionator only experienced the "lost in product" problem and no instability resulted.

Collector Tray - Basic Purpose

One of the simplest collector tray functions in a refinery fractionator is to collect liquid. Figure 1 below illustrates the simplest of refinery fractionator collector tray purposes.

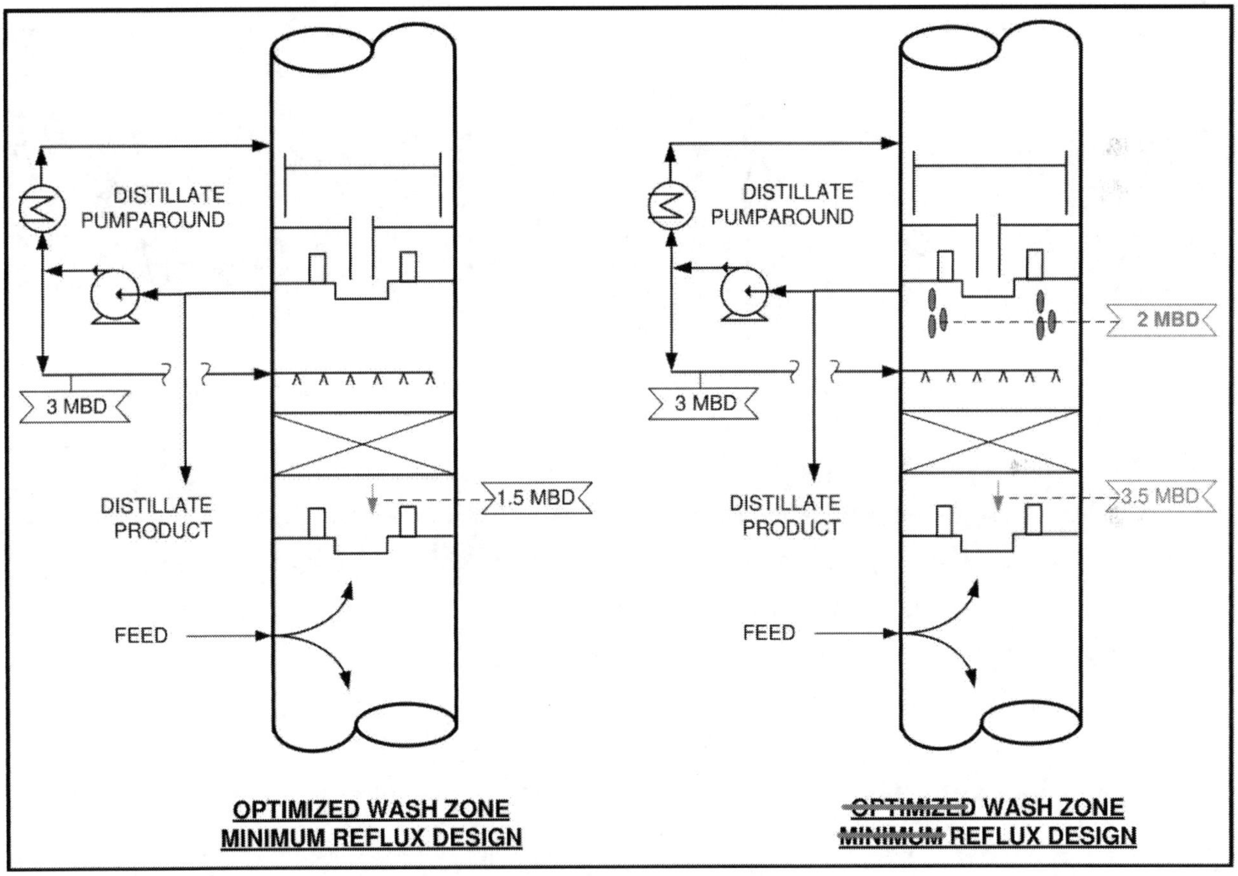

Figure 1: Collector Tray with Pumparound, Product and Leakage

Liquid from a trayed or packed section needs to be collected and withdrawn from the tower. The withdrawn liquid has three components: Product, Pumparound, and Excess Product (commonly called reflux).

Product is the liquid that is withdrawn from the tower and typically sent to a side stripper or product drum for further processing. Pumparound liquid flows through exchangers and is cooled before returning to the tower. Pumparounds are used to set the heat balance for the tower; and thus, the internal reflux in the section below (and vapor flow to the section above). The higher the pumparound duty (either rate or temperature differential), the more liquid is condensed in the section. The reflux is the condensed liquid that is allowed to flow (either pumped or gravity flowed, Figure 1 shows a pumped reflux) to the section below. Reflux provides the driving force for fractionation between the product and the next heaviest product lower in the tower.

Figure 2: Collector Tray Examples (rectangular riser, with overflow, round riser)

When an engineer wishes to control the material balance of a tower, a collector tray is used. Figure 2 shows some examples of collector trays. It is imperative for this function to be leak free. Most operating companies, including Valero, and many engineering companies have very tight leakage criteria for collector trays. Many times the specification will be stated in terms of a maximum liquid height loss on the tray deck in 30 minutes. For new collector trays, the criteria may be "less than ¼" in 30 minutes". Well designed "less seasoned" (ie. new) collector trays will easily meet the leakage criteria with good installation practices. Many times, "more seasoned" collector trays will barely meet the cold liquid leakage criteria. Typically, these trays need more attention as it has been shown that cold leakage rates are less than hot leakage rates (in operation, leakage will be more than witnessed in a cold test) due to thermal expansion.

Don't be fooled into thinking that "well-seasoned" collector trays can be made to be leak free! During a couple of recent jobs, the authors required the installation crews demonstrate that the trays would pass a leakage test. During the final inspection, there were several tubes of Silicone found in the tower sump section below the collector tray. Investigation found that the tray had been sealed with a water-proof Silicone caulk. In another instance, the tray was found to have black stripes along the seams of the tray. Evaluation of the material showed that it was a roofing rubber. While it was true that both of the trays passed the cold leak test, neither would have remained leak free through the tower heat-up, much less during actual operation. There is no substitute for good inspection skills to find these issues.

Case Study #1: "Large" Trayed FCC Main Column

One of the FCCs in Valero's circuit has a rated capacity of 100 MBD. Since its original construction, its capacity has been increased from 65 MBD to 100 MBD through many different revamps. The revamps included changes to the Main Column, gas plant, compressor, and reactor/regenerator sections to improve performance. Figure 2 shows the layout of the Main Column.

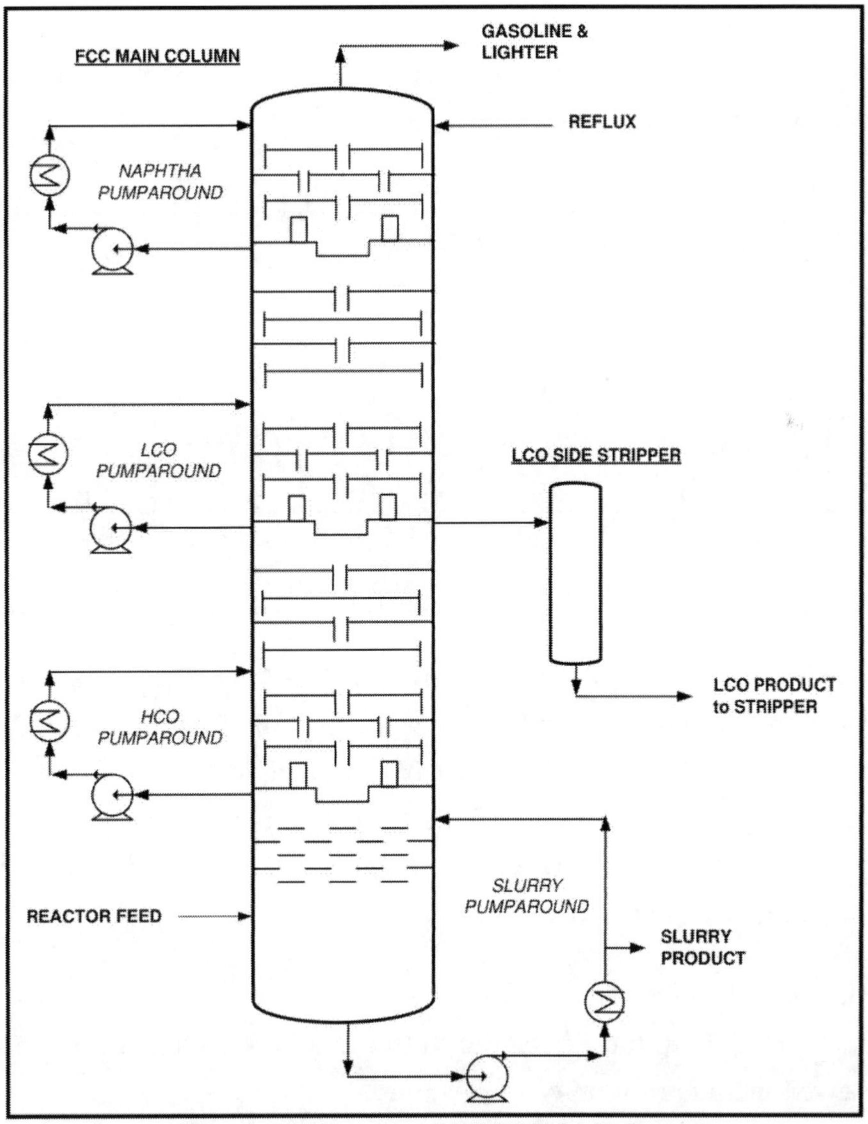

Figure 3: Diagram of FCC Main Column

In 2008, the unit had been operating near its rated capacity for several years. The unit operation was steady with no particular issues. Early in the morning, an operating event occurred and the unit had to be shutdown for a quick "squat" to fix an issue. When the unit started up a few days later, even at 75% of normal feed rates, LCO could not be drawn at the required rate to keep the Slurry yield low and on-specification (flash).

Due to the FCC unit operation, troubleshooting was put on the top of every priority list. The troubleshooting team was given 24 hours for problem identification and a path forward recommendation or the unit would be shutdown and column entry completed to resolve problems.

Field Troubleshooting

Arrival to the site revealed that the reactor/regenerator section was operating at normal conditions with normal feed slate composition; but at 75% of design rates. The summary of the observed products is shown below.

Table 1: Product Qualities

	LCO		Slurry	
	Normal	Present	Normal	Present
°API	14-16	21	(-2) to 0	4-6
Distillation, IBP	377	365	165	140
5%	435	425	500	420
50%	530	491	n/a	n/a
95%	687	570	n/a	n/a
EP	700	590	n/a	n/a

Quick review of the product qualities revealed that the Debutanized naphtha and lighter products were on-specification and within normal operating ranges. The LCO product was extremely light and appeared to be very well fractionated. The LCO product 50% thru EP distillation values were between 40 and 150 deg F lower and the API was about 5 °API numbers higher than typical. The LCO product color was the color of a light wheat ale beer, where its normal color should be close to that of an amber beer.

The tower temperature profiles mirrored the laboratory sample results. Figure 4 below shows the typical values (blue) with the values during the troubleshooting. Also, the unit material balance was skewed and it is also shown in the figure.

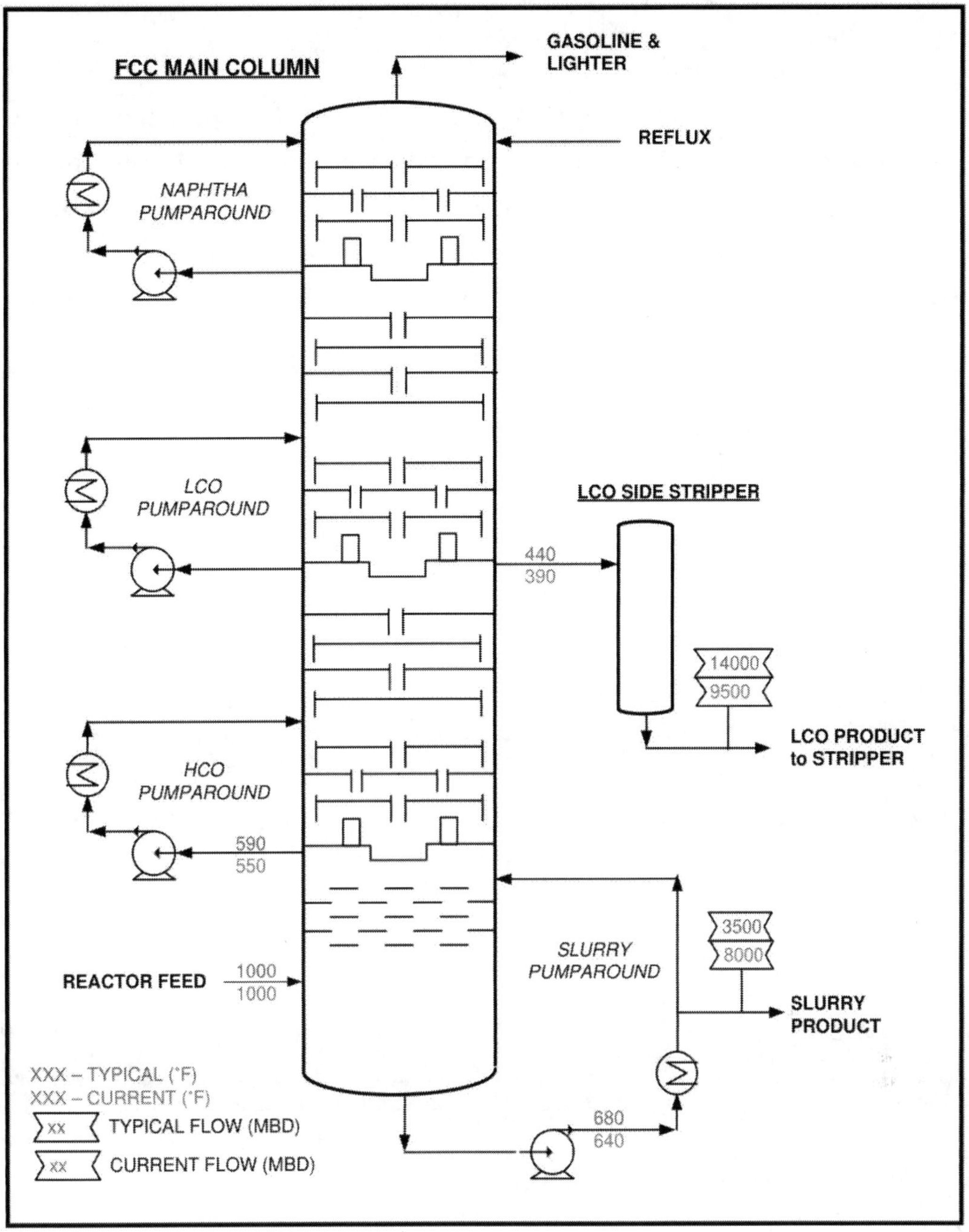

Figure 4: FCC Main Column Operating Differences

The unit was operating at 75% of design feed rates. At these feed rates, the Overhead, Naphtha, and LCO heat removal circuits were operating at maximum pumparound circulation rates. Operators were driving the return temperatures lower to maximize heat removal as much as possible, but the LCO product was still drying up and "disappearing" from the collector tray.

Whenever the LCO product flowrate was increased, the LCO stripper level would drop. This drop would signal the LCO stripper feed valve to open. But, this valve was operating at 50% open then would suddenly go 100% open. This signaled that there was no more LCO product that could be drawn off the draw tray. It seemed as though the LCO collector tray liquid level was independent from the LCO product yield and control valve position. Field pressure drop measurements around the valve indicated that the upstream valve pressure (ie. liquid level) was lost on the tower side. This meant that liquid was not getting into the LCO product draw nozzle. Barletta[4] analyzed a similar LCO draw capacity limitation that happened to reveal trapped vapor in the draw.

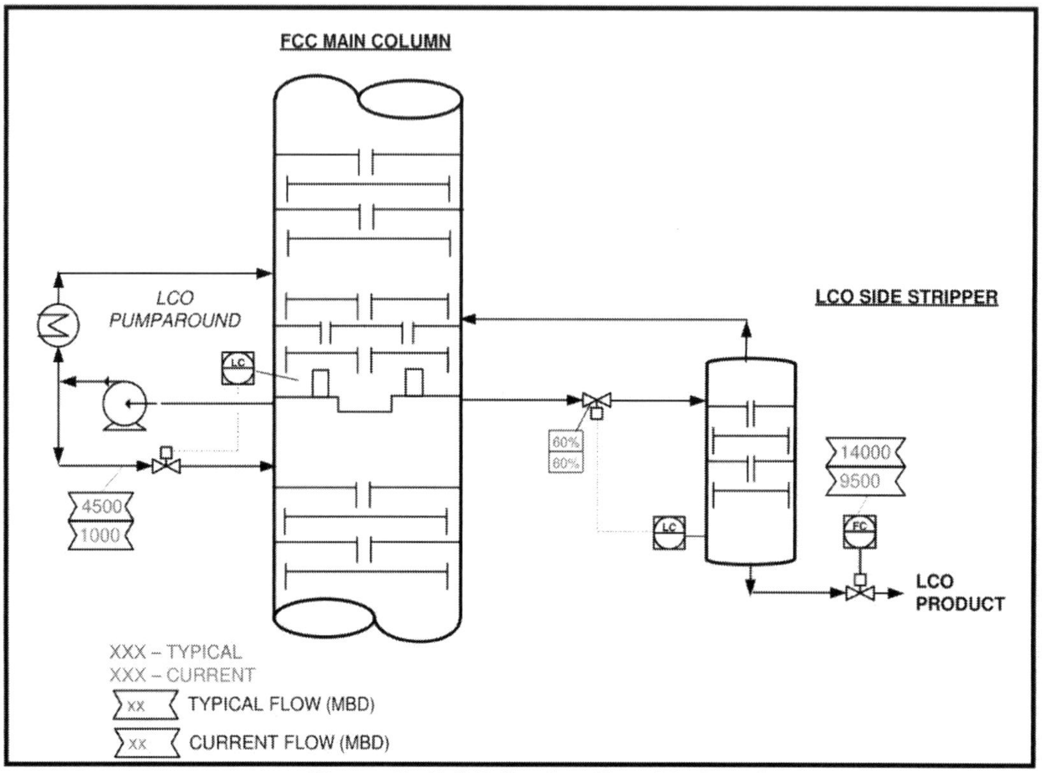

Figure 5: LCO Section Troubleshooting

Isotope Scans and Neutron Backscatter Results

As part of the routine troubleshooting steps, an isotope scan was commissioned for the tower. The scan revealed that all of the trays were intact and they all had a froth level indicating good tray activity and no damage. Liquid level on the LCO collector tray deck was confirmed and it agreed well with the level instrumentation.

There was a small anomaly between two scan chords on the LCO collector tray. A neutron backscatter was performed on the collector tray, its results did not agree with the liquid level instrumentation. Data from the neutron backscatter is shown below. This data was collected when the stripper feed valve was 100% open. Backscatter results around the periphery showed that an equal liquid level did not exist around the tray - liquid level above the pumparound nozzles showed a normal liquid level, but the liquid level was low above the LCO draw nozzle.

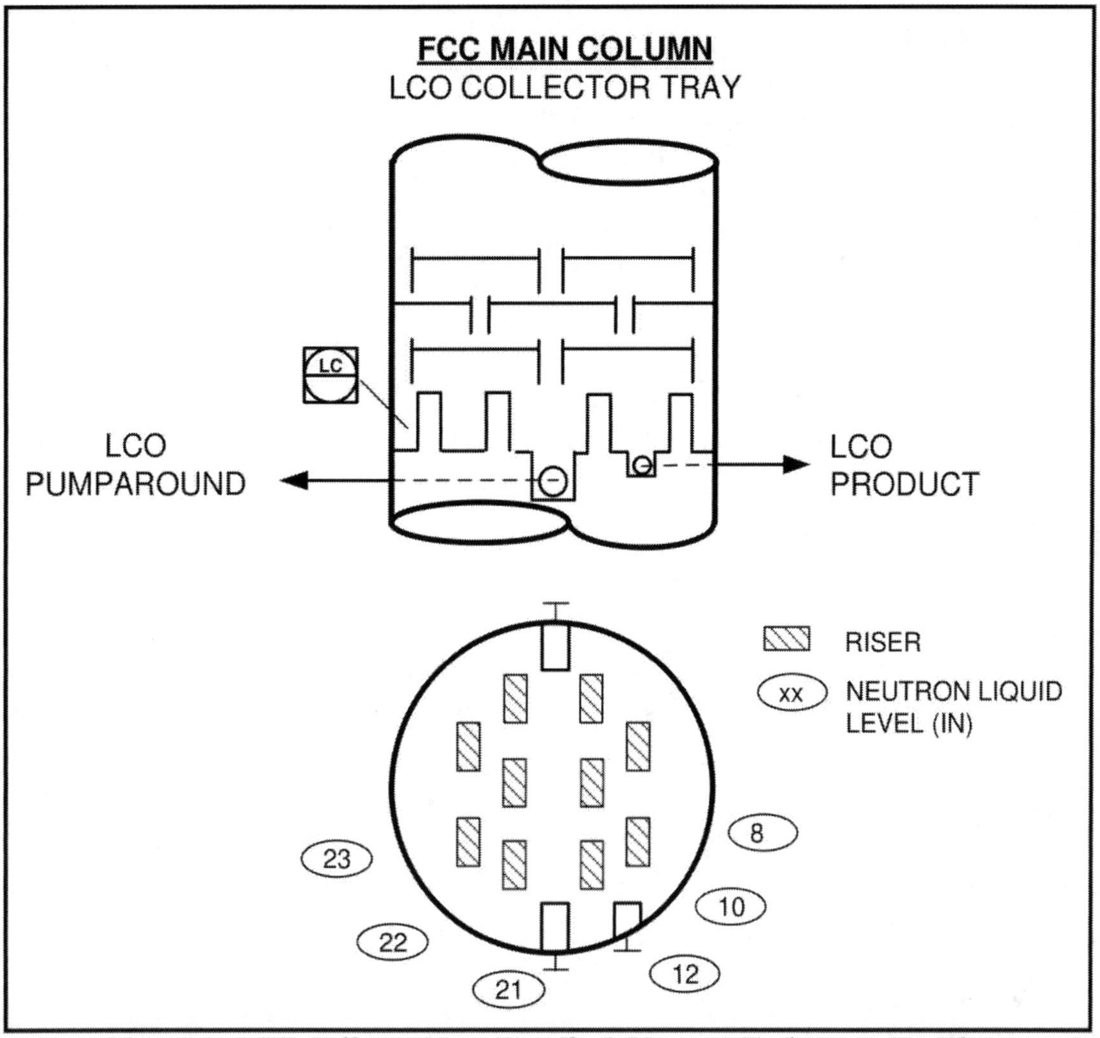

Figure 6: LCO Collector Tray Details & Neutron Backscatter Results

The data implied that liquid was only accumulating on half the tray. There was only a small liquid level on the LCO product draw side of the tower, while the other points showed an abundance of liquid level. Full access to the opposite side of the column was limited. Even though the results were confirmed with multiple measurements, they were considered odd at the time. It was not until additional steps were completed that the results made sense.

Importance of Operators

In the authors' experience, discussing any tower operational problem with the operators is one of the most important steps to identify root causes. Many times it yields the quickest and most direct way to problem identification. Even though they may not be able to explain the operation in complex terms, many operators have a keen sense of what has changed in the unit that can lead the troubleshooter quickly to the problem area. This was certainly the case in this troubleshooting study.

After gathering the pressure drop data around the LCO control valve, the operator was engaged in a discussion about the FCC tower operation. He explained through many stories that all of the symptoms boiled down to: "it does not matter how much heat we move up the column, we can't pull anymore LCO". As we discussed theories, each one of which could not be supported with evidence at the time, we walked up to the LCO Pumparound pump.

The LCO Pumparound pump was a very large pump that, along with its 100% spare, took up a lot of area around the Main Column. One author looked over at the operator and said, "I bet that pump takes a lot of effort to replace the seal. With the LCO tray running dry and the resulting cavitation, how many times have you had to retrofit the pump?" The expected answer was a description of a "bad actor" pump. But the operator responded with a puzzled look and said, "this pump has a 50 month mean-time between failure and we have not retrofit it in at least 3 years." Though surprising, that answer was the key to the troubleshooting solution.

Further Unit Troubleshooting

Troubleshooting now focused on the collector tray. We had to answer the question, "how could the LCO product circuit lose flow, while the LCO Pumparound circuit did not?" The answer to the question was provided by performing a time study on the circuits removing liquid from the collector tray.

Figure 7 summarizes the data from the time study. The study was conducted over several hours. The data proved that even though the two circuits withdrew liquid from the same collector tray, they were not linked hydraulically. Even though we lost flow to the Product circuit, the Pumparound circuit did not change. The Pumparound circuit was able to withdraw liquid independent of what happened in the Product circuit. We did several tests that confirmed that the LCO Pumparound rate could be varied and never did the pump lose suction pressure.

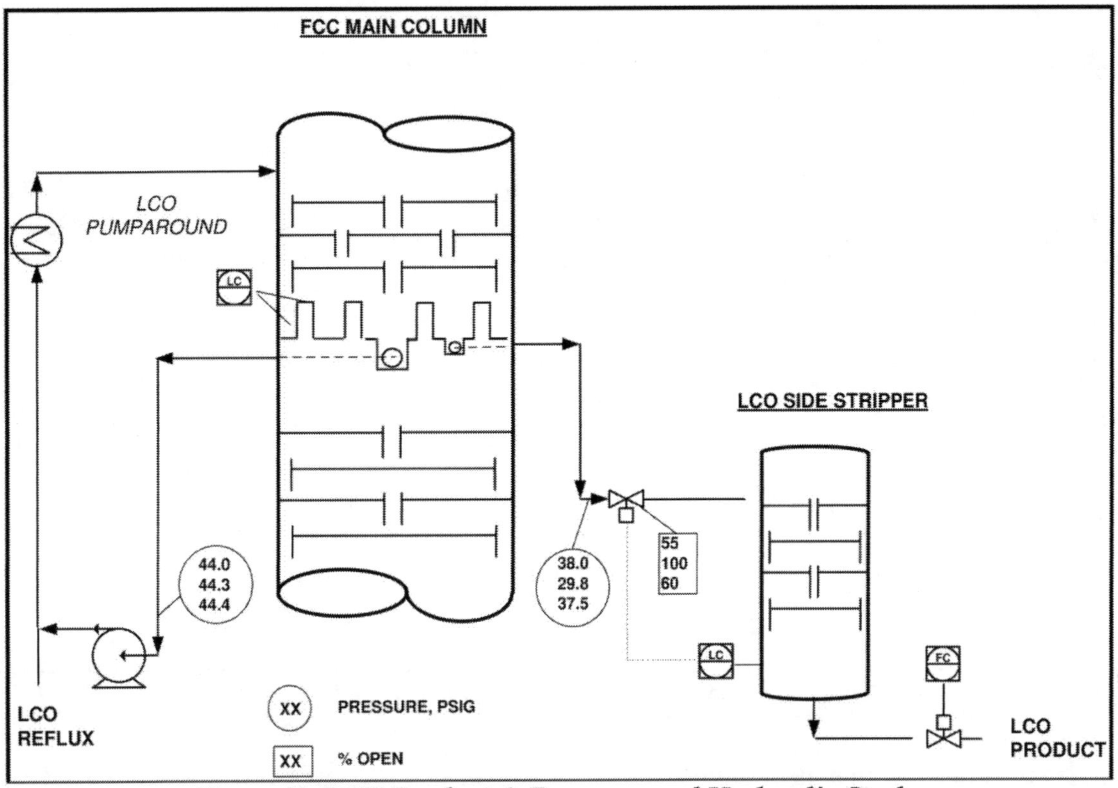

Figure 7: LCO Product & Pumparound Hydraulic Study

Many theories were provided to explain the problem. Nozzle blockage, mal-distribution across the tray deck due to the poor riser layout, tray damage, and nozzle limitations were provided. Many of these theories could not explain why it worked before but not now. Tray damage and nozzle blockage were the only theory that could not be completely dismissed - even though solids and tray parts dislodging could not be explained in this section of the tower.

Instituting a Fix and Problem Identification

Time was running out and a recommendation had to be made to management to either keep the FCC running or shut down. Two sets of data suggested that the liquid level above the LCO Pumparound draw nozzles was higher than the LCO product draw nozzles, so a recommendation was made to install a jump-over line to conduit LCO from the pumparound circuit to the product circuit. At the time, the authors did not know if this solution would enable the LCO product to increase by 1,000 bpd or 10,000 bpd. The question was could the liquid be drawn off the collector. Was the heat balance correct and we had an abundance of liquid on the tray deck that could not presently be drawn off? The goal was to get the excess liquid from the Pumparound circuit into the Product circuit.

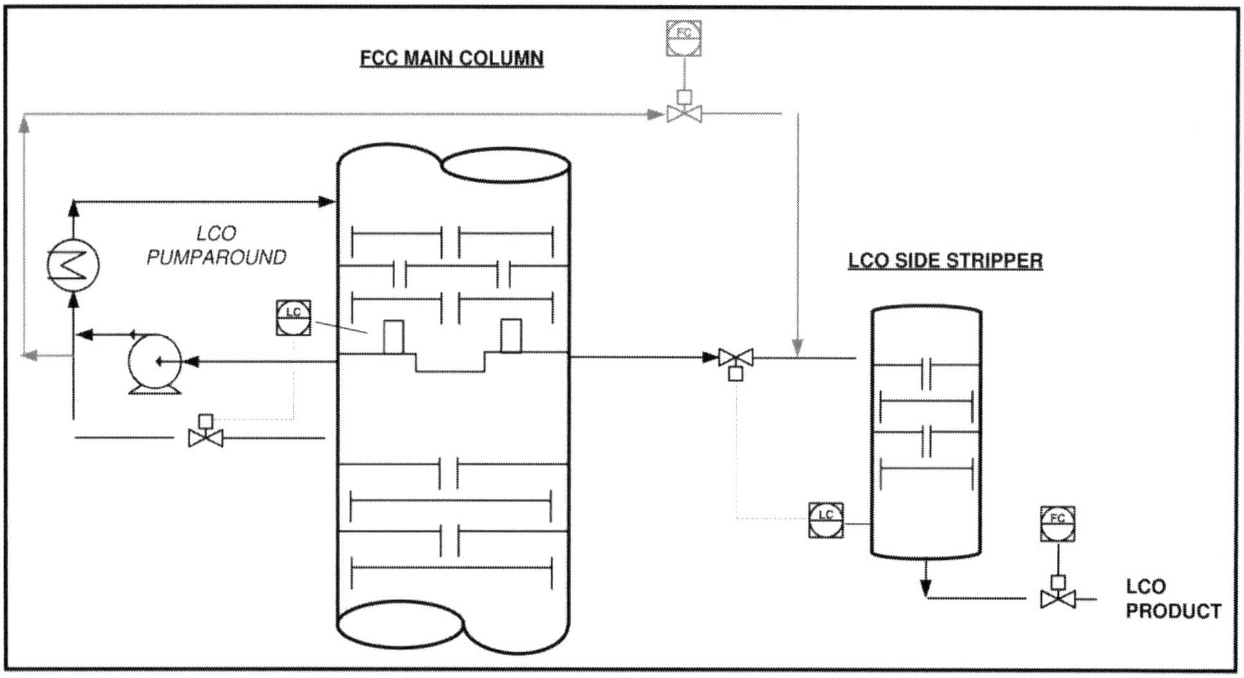

Figure 8: Excess Product Line to Side Stripper

A fast track project to install the line was initiated and is summarized in Figure 8 above. The line was installed in less than 3 days due to the dire nature of the Slurry product quality. After the line was approved for operation, the jump-over LCO rate was increased from no flow to 5,000 bpd without any operating issues. The LCO yield increased, product qualities returned to normal, and the Slurry yield returned to normal. Unit rates were increased to expectations.

During the next outage, there was keen interest to enter the tower and understand the circumstances that led to the unique operating issues with the LCO collector. Pictures from the initial inspection are included in the appendix. Unbeknownst, the collector tray was quite a bit more complicated than any drawings available. In a past installation, a "protection dam" was added in front of only the LCO product draw nozzle (to keep catalyst out of circuit). So, liquid level had to build up to 20" on the tray deck, before the product circuit could pull liquid. If the liquid level was below the dam height, the product circuit would not have liquid. This explains why the pumparound circuit always was flush with liquid, while the product circuit would run dry. The tray had sustained major weld damage as well as it had slipped off the tray ring. Both of these contributed to major tray leakage.

In this case, tray damage and a poorly designed LCO draw circuit contributed to tray leakage. Liquid level on the tray deck could not be increased enough to overcome the tray leakage. The collector leakage resulted in a loss of product and better fractionation due to the incremental reflux to the trays below, but no instability was observed.

Case Study #2: "Small" Packed FCC Main Column

The "small" FCC is a 30 MBD Fluid Catalytic Cracking unit with a 12-ft ID Main Fractionator. The column is fully packed with Slurry and Light Cycle Oil (LCO) pumparounds. The Slurry pumparound (SPA) is heat integrated with the Gas Con Debutanizer reboiler, heaters in the Alky unit, and steam generation. LCO is collected in a total draw tray and split to a product side stripper, a metered pumpdown to the LCO/Slurry fractionation bed below, and a pumparound that is also heat integrated with the Gas Con.

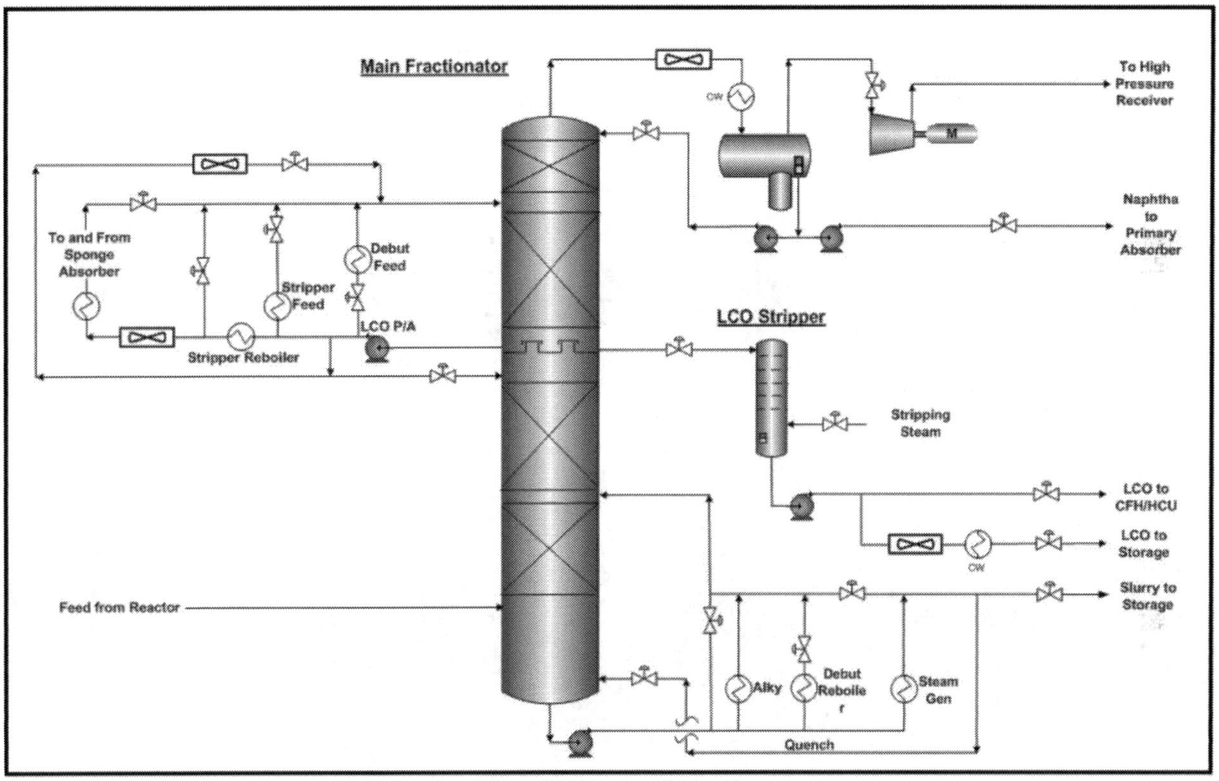

Figure 9: "Small" FCC Main Fractionator

During the most recent turnaround, the LCO/Slurry fractionation and SPA beds were replaced in kind. No significant coking was noted in the LCO/Slurry bed, while the SPA bed had some coking that was most pronounced on the side of the column opposite the reactor overhead vapor inlet nozzle. This type of Slurry bed coking is not uncommon in the industry, and in this case was not rate limiting to the unit during the previous cycle.

Halfway into its 5-year run cycle, the Slurry product gravity had a step change increase (got lighter) without a corresponding change in LCO distillation. The product quality shift was accompanied by a series of severe operational upsets where the Fractionator bottoms level was lost abruptly without any external change in tower heat balance. Until a solution was found for the upsets, the FCC feed rate was reduced to 80% of normal.

Slurry Product Quality Shift

For a given reactor yield set, Slurry gravity is most heavily influenced by tower heat balance and the amount of LCO dropping into the bottom of the tower. Generally, at higher heat removal rates in the bottom pumparounds, LCO endpoint is lowered as the heavier distillate components are dropped into the Slurry product. At the onset of the Fractionator's level control problems, the slurry product gravity increased by 5°API without a corresponding decrease in LCO distillation or increase in SPA duty (Figure 10). In fact, the LCO distillation shifted up slightly, indicating some loss of fractionation.

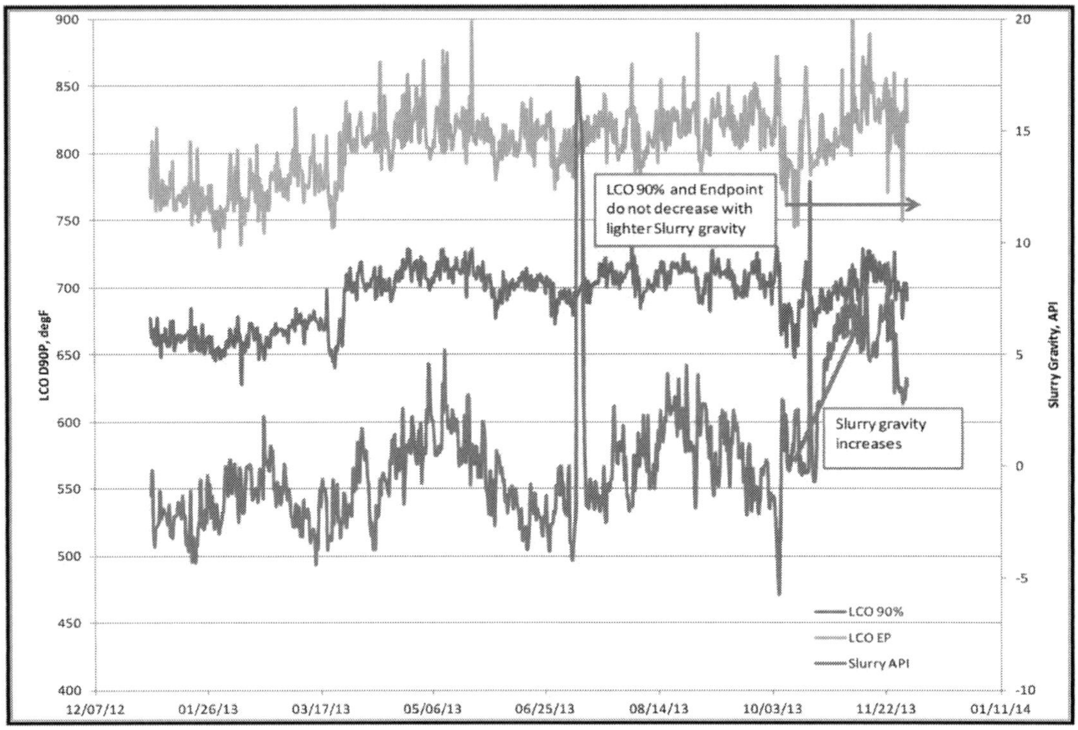

Figure 10: Trend of LCO and Slurry Product Qualities

External contamination of the slurry was initially suspected, but lab issues, high flush line flows, and exchanger leaks were all ruled out. The LCO pumpdown flow meter was investigated, calibrated, and found to be a reliable meter. There were no obvious indications that the Slurry gravity should have increased, but the timing connected it to the bottoms level control upsets.

Fractionator Bottoms Level Upsets

Figure 11 illustrates the instrumentation locations in the bottom of the Main Fractionator. The onset of the FCC Main Fractionator bottoms level upset occurred with a steep increase (approx +65degF) of the temperature recorded at the elevation of the reactor vapor inlet ("TI-A"). The vapor space temperatures ("TI-B" and "TI-C") at the SPA return also increased, but both the tower top temperature and reflux flow rate dropped rapidly (Figures 12 and 13). The response of the reflux and top temperature would be typical for an increase in heat removal lower in the column. Neither LCO Pumparound nor SPA flows or ΔT's changed immediately prior to an upset, while both charge rate and conversion were steady. The response from the SPA section did not match up with the reflux/top tower temperature response which signaled a heat balance issue.

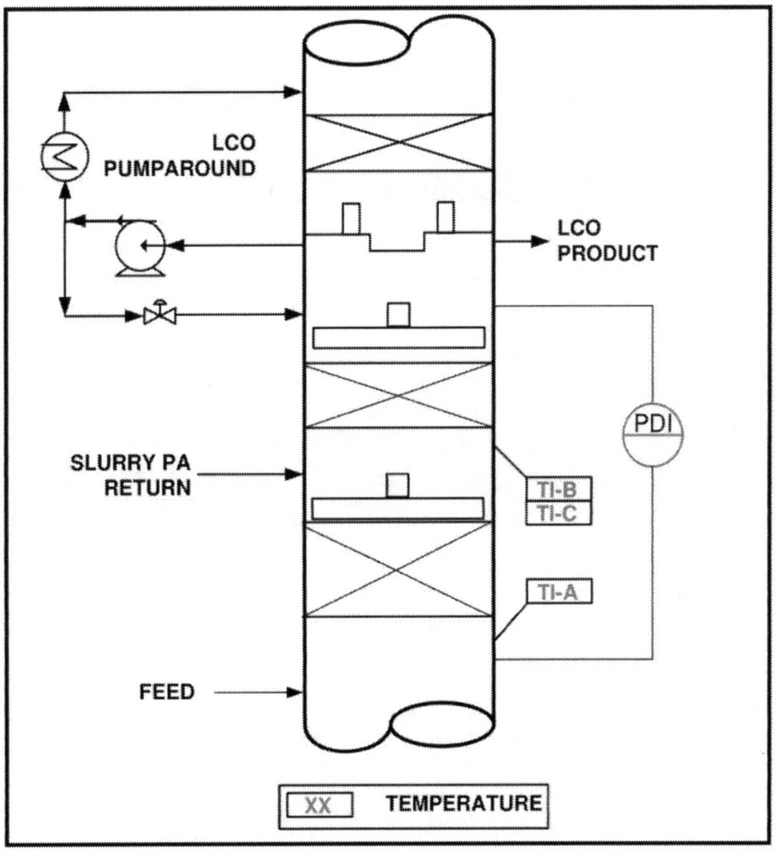

Figure 11: Main Fractionator Bottom Section Instrumentation

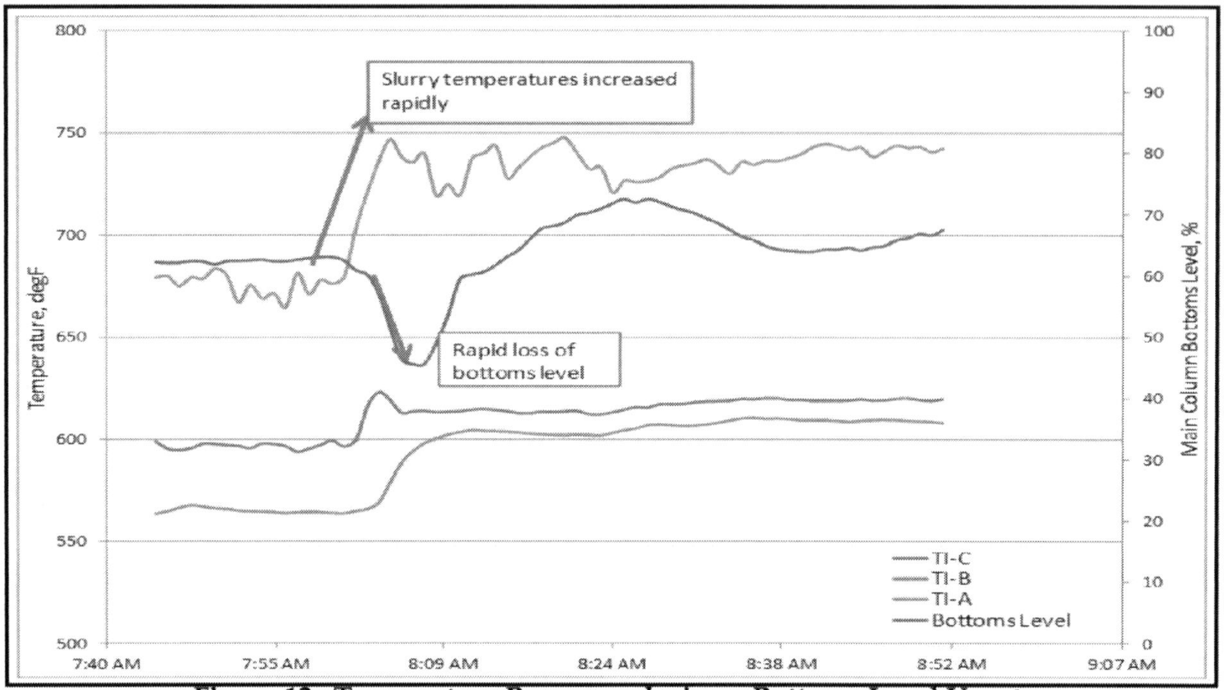

Figure 12: Temperature Response during a Bottoms Level Upset

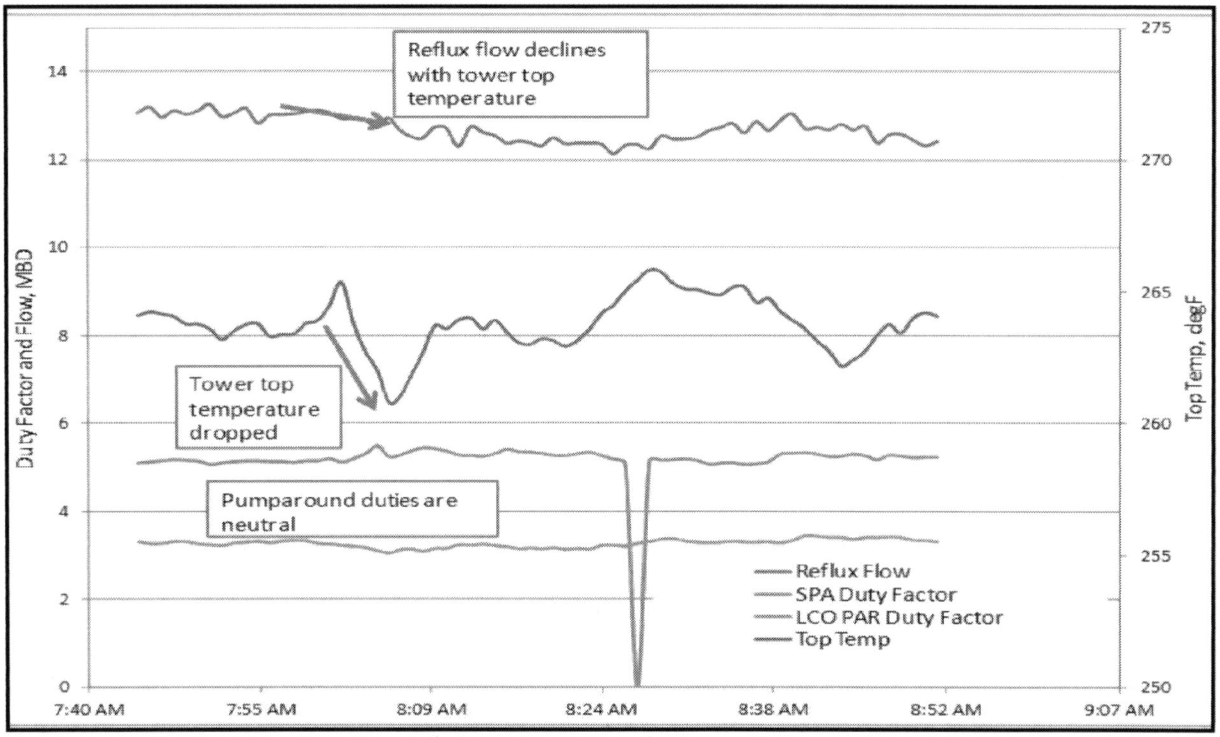

Figure 13: Heat Balance Response during a Bottoms Level Upset

Tower pressure drop instrumentation spanned from the Naphtha-LCO fractionation bed to the LCO-slurry fractionation bed (two fractionation beds and a Pumparound bed). A strong pressure drop response was noted during an upset (Figure 14), but due to the large span, the specific area of accumulation could not be identified.

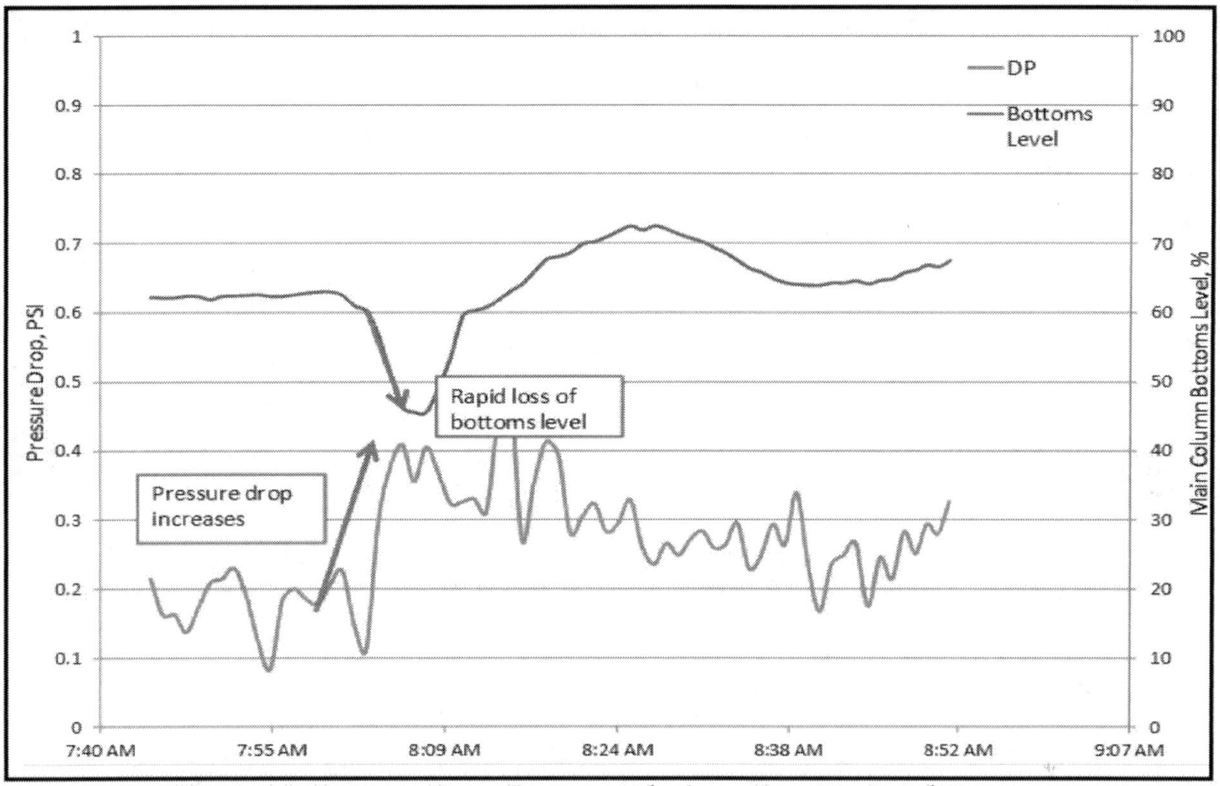

Figure 14: Pressure Drop Response during a Bottoms Level Upset

During an upset, the board operator took mitigation steps to recover the bottoms level, including increasing SPA flow rate (to increase duty) and increasing LCO pumpdown flow. These responses were targeted to hold traffic in the bottom of the tower, but were largely unsuccessful in controlling the bottoms level loss. Eventually the board operator discovered that as a last resort, lining up the cold feed start-up line to the SPA return helped recover bottoms level. It is generally industry-accepted knowledge that an un-cracked oil leak into the Main Fractionator generates coke, so this practice was not sustainable. If the bottoms level could not be recovered or the board operator was unable to act quickly enough with mitigating steps, the Slurry pumps would cavitate and the unit feed would be removed in order to have a controlled unit shut down. Upset identification and resolution became critical to keep the unit on line.

Isotope Scanning and Step Testing

A full elevation isotope scan of the column was conducted to identify possible maldistribution inside the column. An isotope grid scan had been conducted 4 months earlier in the year, which was used as a baseline to compare with the scan completed after the initial level upsets were observed. The updated isotope scan showed normal level on the LCO collector and fairly good distribution on all the packed beds, but with several anomalies at the elevation of the Slurry pumparound return distributor which were deviations from the baseline. Several regions of higher density (liquid) were observed in the vapor space above the SPA distributor, but with only four chords over the entire tower diameter and the complexity of equipment (distributor, feed pipe, beams) in the region, it is difficult to conclusively identify damage or maldistribution.

In order to better understand the liquid distribution at the SPA return elevation, twenty skin thermocouples were attached to the shell with continuous readings recorded by a local data logger. Surface temperatures[5] have been used to diagnose vapor and/or liquid maldistribution in fractionation towers. This device allowed for constant logging and retention of the surface temperatures. The temperature readings suggested a hotter zone on the west side of the tower. The temperatures located in the coldest areas near the SPA inlet nozzle tended to be the most reactive during an upset occurrence.

Operational step testing with stationary isotope monitoring of this region was inconclusive. During the first test, heat duty was shifted up the tower from the SPA to the LCO pumparound to induce more traffic through the LCO-slurry fractionation bed. During the second test, SPA return flow was increased at constant duty. There appeared to be no single threshold that consistently resulted in a bottoms level upset. For instance, at reduced SPA duty and high LCO pumpdown flow, the LCO section DP increased and the skin thermocouples and temperature instruments fluctuated, which was characteristic of an imminent level upset that never occurred. That operating point was not abnormal and often the unit operated there without experiencing an upset.

Importance of Operators

As discussed in Case Study #1, input from operators was important to understanding the issues in the main column. In this case, operators reported that one of the responses that helped to restore the main column bottoms level loss phenomena was to *reduce the LCO pumpdown rate* (pumped internal LCO reflux below the LCO draw/trap pan). Since this move was counter to the logical response for a loss of bottoms liquid level, the LCO draw/trap pan was suspected of a leak. The authors then began focusing on the LCO draw pan as the source of a leak that could upset the main column.

LCO Draw Tray

Combined with the distillation data (light Slurry gravity and no loss in LCO distillation), the tower responses during an upset (tower top temperature and reflux) seemed indicative of a tower heat sink; however, all external sources were already ruled out. A heat balance was completed around the LCO pumparound section. This calculation indicated that approximately 3,000 bpd additional distillate was being condensed and not accounted for in the product or pumpdown flow meters. This confirmed that a leak from the LCO collector was occurring.

In a parallel attempt to quantify a LCO collector leak, transient heat balance data was collected from the main column and a pattern emerged that indicated there was a difference in measured heat removal as the level in LCO collector tray changed.

Table 2: Transient Heat Balance Study with LCO Draw Tray Level

Time, hrs	LCO Collector Level %	OVHD Reflux, MBPD	LCO P/A Duty, MMBTU/hr	Slurry P/A Duty, MMBTU/hr	LCO + Slurry P/A Duty, MMBTU/hr
14:04	60.0	12.5	41.2	45.6	86.8
15:10	60.0	12.6	40.8	46.0	86.8
15:55	49.4	12.6	39.9	49.4	89.3
16:49	49.6	12.4	41.1	49.6	90.7
17:12	50.5	12.4	41.7	43.9	85.6
17:30	49.1	12.9	41.9	49.1	91.0
17:36	49.6	12.7	42.1	45.3	87.4
17:40	48.6	12.5	42.3	48.6	90.9
17:49	47.9	12.2	41.9	47.9	89.8
17:54	50.9	12.3	42.4	43.6	86.0

Average LCO + Slurry P/A duty with LCO collector tray level greater than 50% was 86.3 MMBTU/hr as compared to 89.9 MMBTU/hr when the LCO tray level was less than 50%. This difference in duty of 3.6 MMBTU/hr was the unidentified heat sink or leaking LCO from the collector tray. When the heat sink was converted into a LCO liquid flow rate, the leak rate amounted to 2,500 bpd. This independent calculation confirmed the approximately 3,000 bpd LCO draw tray leak when the draw tray level was greater than 50%.

To test this theory, the LCO collector tray liquid level was reduced. The setpoint selected would remove the liquid level from the tray floor (except for gradient) and contain the liquid pool in the draw sump. After moving the liquid level setpoint, the tower temperature distribution improved (Figure 15). The draw tray level stabilized, and bottoms level excursion and process upsets have been completely eliminated. Slurry product gravity made a corresponding improvement (Figure 16).

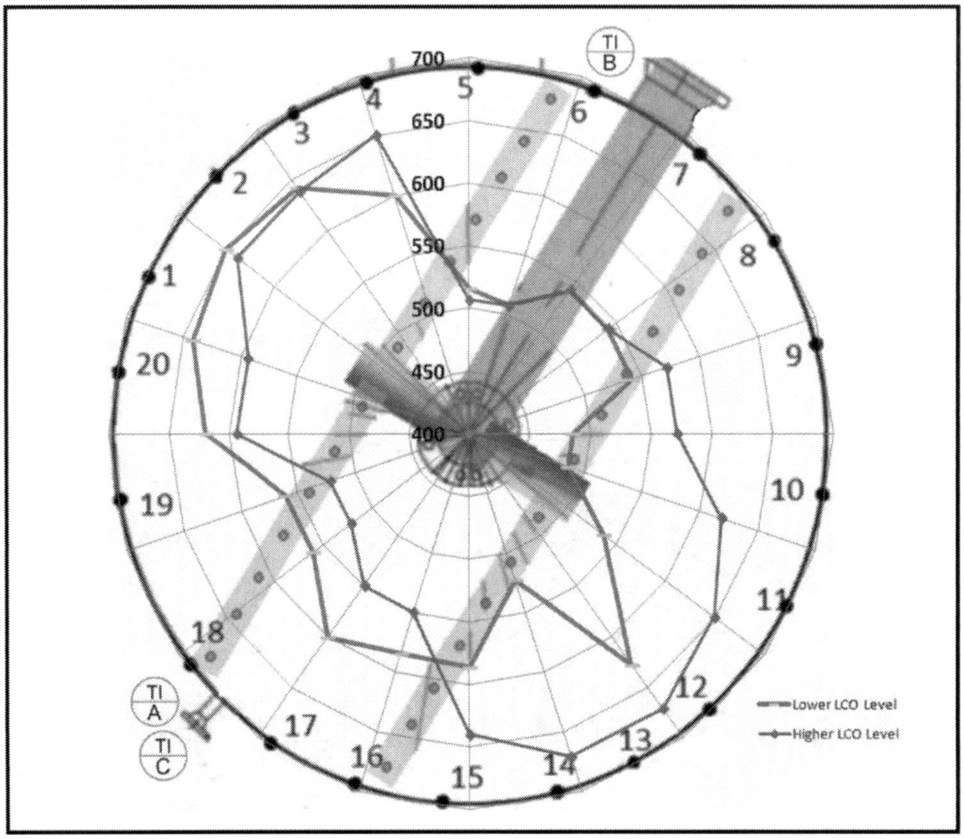

Figure 15: Spider Plot of Radial Skin Temperatures Before and After LCO Level Change

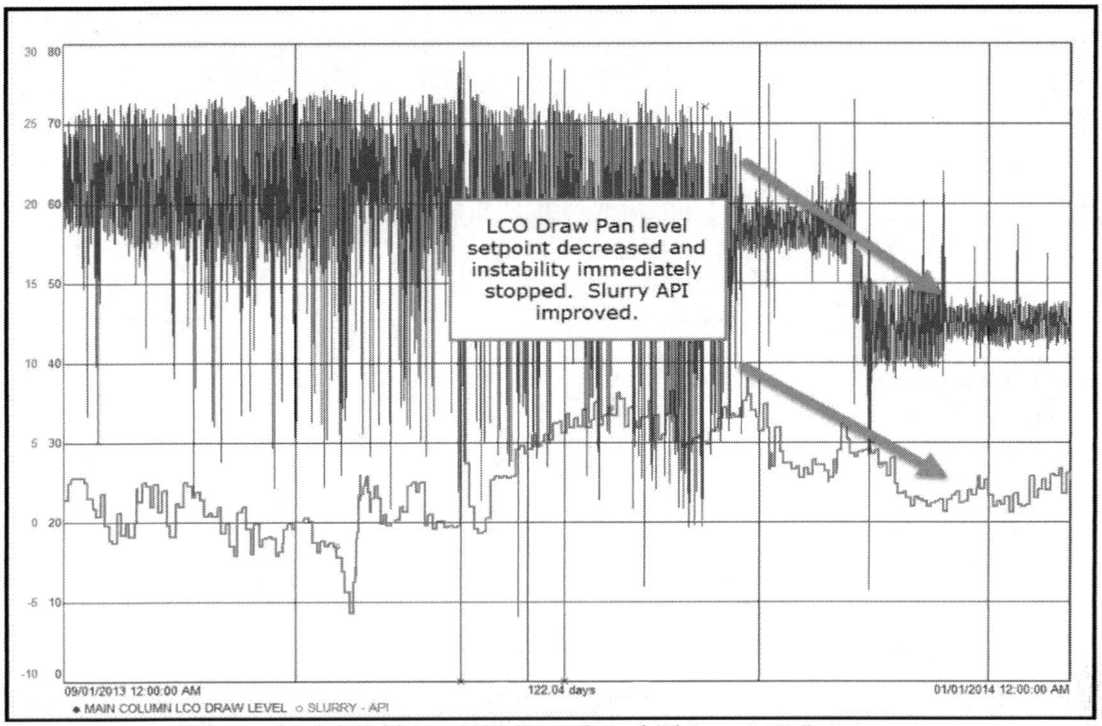

Figure 16: LCO Draw Pan Level and Slurry API Response

The authors believe that at higher tray liquid levels, liquid leaks from the collector. This behavior would be consistent with a weld seam failure that results in a concentrated liquid stream pouring into the LCO/Slurry fractionation bed. Because of its discreet nature, the stream is not redistributed and does not reach calculated equilibrium. It reaches the Slurry pumparound zone largely un-contacted by the up-flowing vapor (very limited heat or mass transfer occurring in the bed). The hot Slurry PAR section violently vaporizes cold liquid. This leakage and re-vaporization acts like a recycle heat sink and increases the hydraulic loading below the LCO PAR section. Ultimately, it results in higher liquid hold-up and a systematic loss of bottoms level. This similar phenomenon was cited by Bravo[6] in his testing with a hydrocarbon-water test system.

Conclusions

Both case studies had LCO tray leakage that resulted in loss of LCO production. The trayed Main Fractionator was able to operate in a controlled manner during the leakage. Leakage with the packed Main Fractionator resulted in unpredictable operation.

Table 3: Comparison of Collector Tray Leak Impacts

Section below Leak Location	Trayed Section	Packed Section
Operation Instability		
Material Balance	No	Yes
Heat Balance	No	Yes
Pressure Drop	No	Yes
Improved Fractionation (1)	Yes	No
LCO Quality	Improved	Not Impacted
Slurry Quality	Degraded	Degraded

(1) Based on product qualities and 5% / 95% gap (overlap) review.

In the trayed column, due to the natural re-distribution ability of the trays below, the reflux provided a process benefit of improved fractionation. The tower had a loss of LCO recovery (yield) and feed rate during this time, but the tower exhibited no instability. If it were not for the economic loss and Slurry product yield increase, the tower could have operated with LCO collector tray leakage until a scheduled outage. The leakage resulted in an improvement in LCO quality (lighter) with resultant lighter Slurry quality.

In contrast, the packed column had severe instabilities when the liquid level on the collector tray resulted in a large leakage rate. The instabilities were similar in nature to the performance swings noted by Bravo upon an introduction of free water into a packed bed. The free water migrated lower than expected into the bed and then an event of sudden and violent vaporization occurred. This is very synonymous with the cold, concentrated stream of LCO flowing un-vaporized through a bed of packing until reaching a point in the packing that contains sufficient heat to vaporize it. The sudden change in heat load in the tower required to vaporize the concentrated 3,000 bpd stream changes the tower heat balance and made it very hard to operate.

In both cases, the authors were able to remedy the leaking trays without a shutdown. The leaks were resolved with physical equipment additions in the Trayed case and with operational changes in the Packed case. Understanding fractionation and operational principles was key to solving the problems.

REFERENCES

1. Hanson, D. , "Recent Collector Tray Problems & Process Engineering Inspection", Unpublished, 2008.

2. Kister, H., et.al., "Troubleshoot Chimney Trays Effectively", Hydrocarbon Processing, April 2001, pp 101-107.

3. Sloley, A., "Don't Get Drawn Into Distillation Difficulties", Chemical Engineering Progress, June 1998, pp 63-78.

4. Barletta, T., "Pump Cavitation Caused by Entrained Gas", Hydrocarbon Processing, November 2003, pp 69-72.

5. Burns, J., et.al., "Targeted Liquid Injection Improves Shed Performance", Hydrocarbon Processing, June 2014, pp 75-81.

6. Bravo, J., et. al., "The Effects of Free Water on the Performance of Packed Towers in Vacuum Service", Ind. Eng. Chem. Res., 2001, *40*, pp 6181-6184.

Note:
The views expressed in this paper are that of the authors and do not necessarily reflect the views of Valero Energy Corp.

Pictures from Large FCC outage

Pictures show the "Protection Dam" that sheltered the LCO Product nozzle.

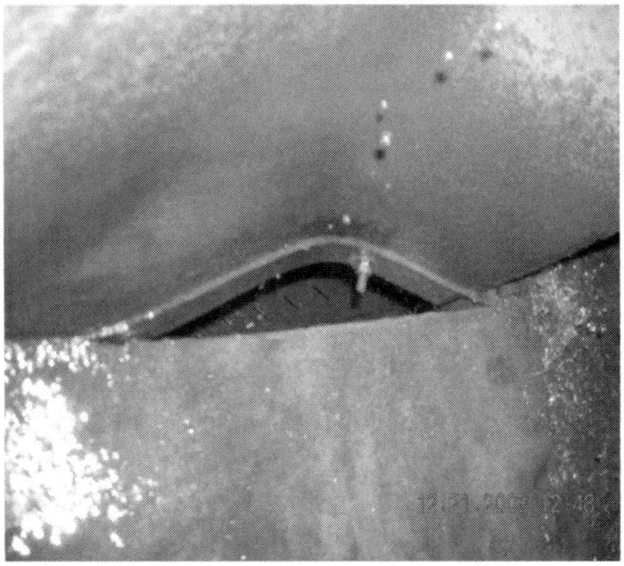

Pictures show damage to the LCO draw tray.
Weld failures and tray decks falling off rings created uncontrolled leakage.

Prevent Damage in FCC Main Column Slurry Pumparound Zones with Proper Design and Operation

Dennis Schmude & Oleg Karpilovskij
Koch-Glitsch

KOCH-GLITSCH

FCC Simplified Flow Diagram

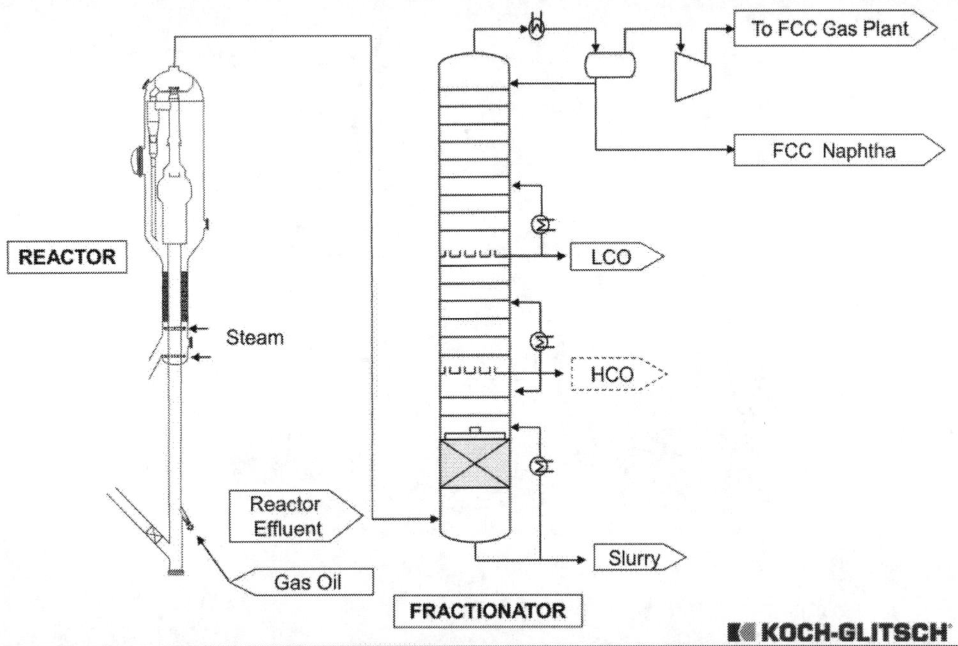

KOCH-GLITSCH

Slurry Pumparound Internals

- **Baffle trays**
  - Shed decks
  - Disc and donut trays
- **Grid packing**
  - GLITSCH GRID® severe service grid packing
  - FLEXIGRID® severe service grid packing

KOCH-GLITSCH

Coking in Slurry PA Bed

KOCH-GLITSCH

Grid Packing

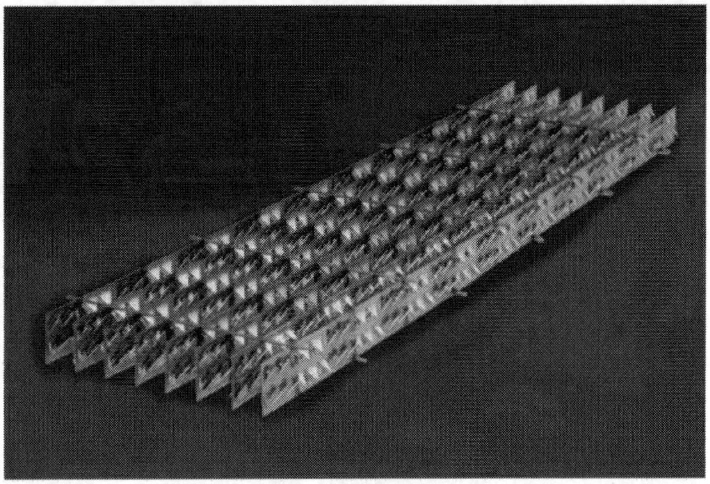

FLEXIGRID® Style 2 Grid Packing

KOCH-GLITSCH

Grid Packing in Slurry PA

- **Advantages include greater hydraulic capacity and improved heat removal**
- **Proven reliability**
  - Grid packing installation starting in 1970
  - Estimated 50% of FCC Units in USA currently contain grid type packing in slurry PA

KOCH-GLITSCH

FCC Main Column Feed Entry

- Superheated vapor 950-1000° F [510 – 540° C]
- Typical inlet velocity 90-180 ft/s [30 – 60 m/s]
- Contains catalyst fines
- Typically no inlet device

KOCH-GLITSCH

Grid Bed from Underside

KOCH-GLITSCH

Bottom Layer Attachment

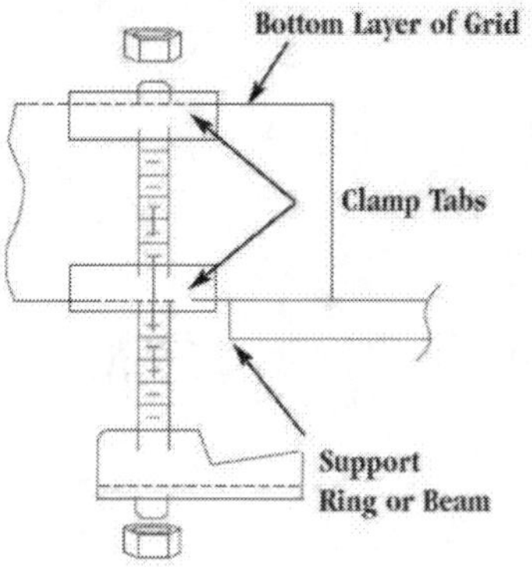

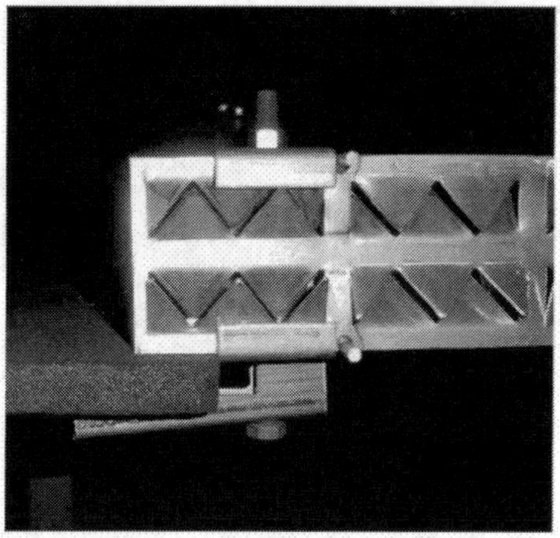

Top Layer Attachment

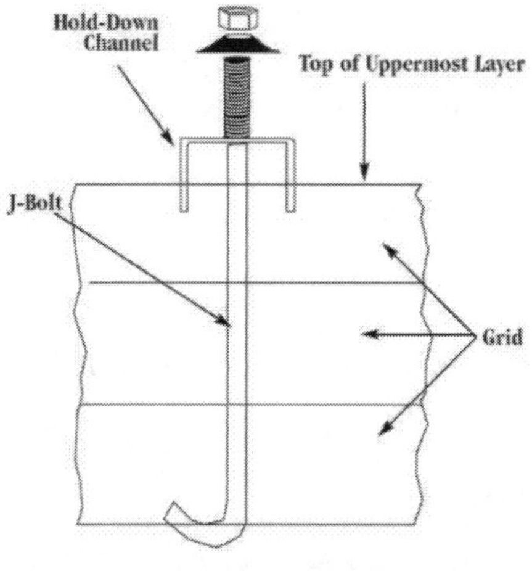

Damage at Bottom of Bed

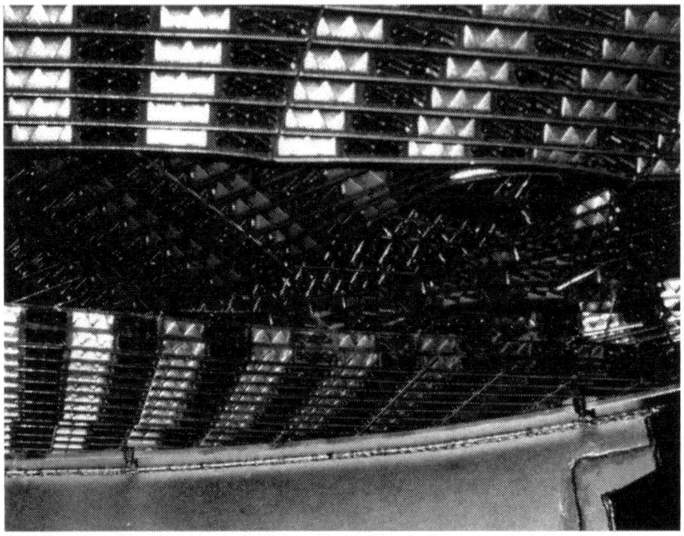

KOCH-GLITSCH

Recommended Mechanical Upgrades

- **Design for a minimum of 1 psi net uplift load (consider 2 psi depending on tower size & upset risk)**
  - Impact support grid and beam designs
  - May required beam clip tower attachments modifications
  - Heavy duty clamp assemblies
  - Coking considerations in support design
- **Separate packing support grid in lieu of grid packing layer**
- **Thru rod assemblies to tie bed together**

KOCH-GLITSCH

Net Uplift Loading

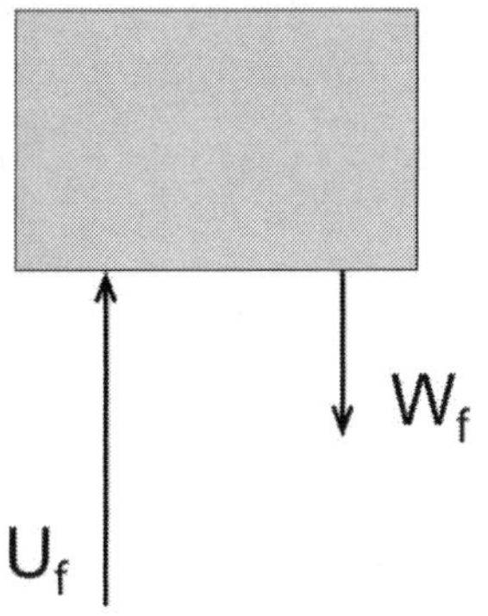

U_f = Total Upward Force

W_f = Weight of Packing
(typically in the range of 1 psi for FCC slurry pumparound beds depending on packing type & depth)

Net Uplift = $U_f - W_f$

KOCH-GLITSCH

Grid Packing with Support Grid

KOCH-GLITSCH

Thru Rod Assemblies

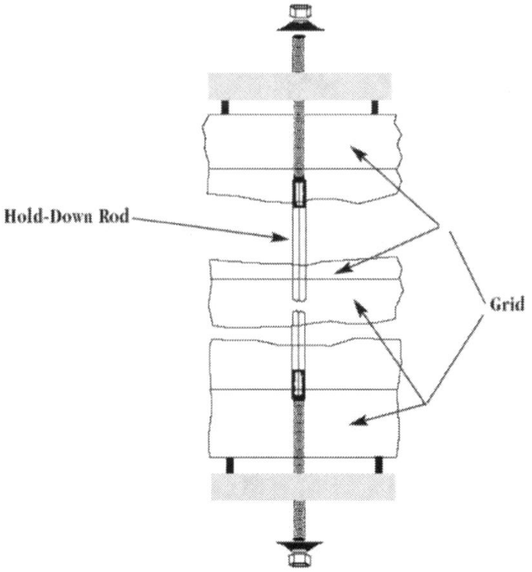

KOCH-GLITSCH

Slurry PA Grid Bed Installation

○ Position of thru rod assembly

KOCH-GLITSCH

Grid Bed Reliability

- Slurry PA bed damage is very rare with best practice mechanical design methods used
- Typical FCC run length 4 – 5 years (generally not limited by Fractionator internals)
- Rare cases of extreme bed damage

KOCH-GLITSCH

Severe Damage – Case 1

KOCH-GLITSCH

Severe Damage – Case 2

KOCH-GLITSCH

Severe Damage – Case 3

KOCH-GLITSCH

Consistent Observations

- Units with severe damage often have repeated incidents of severe damage
- Upward force originating from underside
- Location of the "hole" in the bed in the same region

Location of Bed Damage

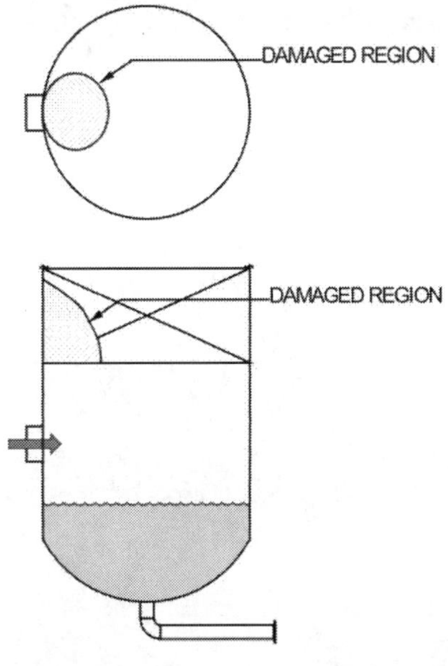

Typical Feed Inlet Profile

Particle Path Lines

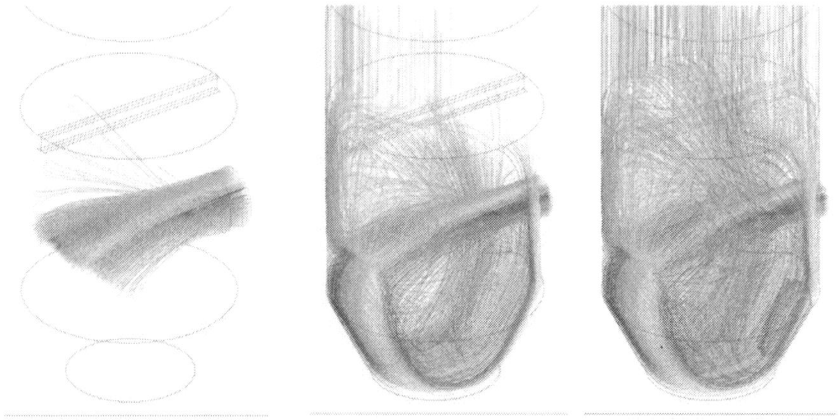

Typical Feed Inlet Profile

Normal Component of Velocity 6 in [.15 m] Below the Packing (m/sec)

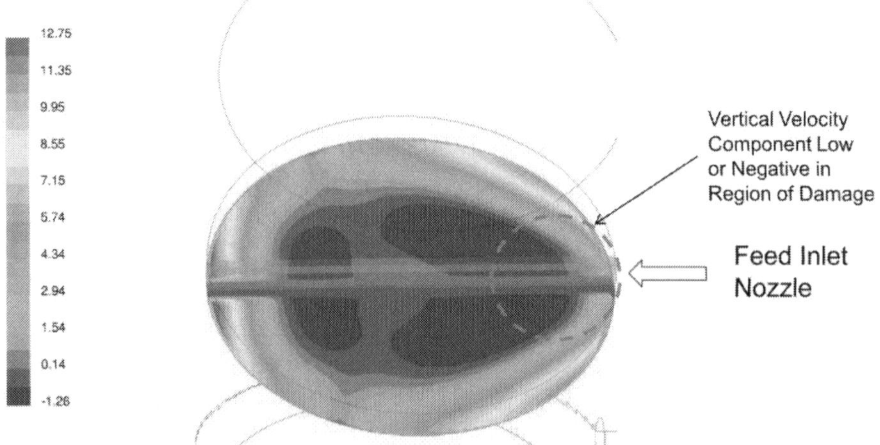

Severe Damage Case Study - Timeline

- **Approximate 10 year period with repeated damage with 1.5 – 2 year run lengths**
- **Koch-Glitsch engaged to troubleshoot**
  - Initially suspected excessive liquid level (however, not supported by level measurements)
- **Upgraded mechanical design installed**

KOCH-GLITSCH

Severe Damage Case Study - Timeline

- **Severe damage observed again after 2 year run**
- **Prompted further investigation of verification of level measurement / methodology**
  - Level measurement based on DP calculation
  - Incorrect density used in calculations to convert DP to level measurement
  - Corrected level calculations showed excessive liquid level – vapor feed nozzle submerged

KOCH-GLITSCH

Severe Damage Case Study - Timeline

- Level measurement methods corrected
- Cycle of repeated bed damage finally broken – no bed damage observed during the last shutdown

Conclusions

- Slurry PA bed damage is rare with proper equipment design
- Isolated cases of excessive damage have been resolved by proper tower operation / bottoms level control

KOCH

Thank you.

- **Questions?**

TRADEMARKS

The following trademarks are owned by Koch-Glitsch, LP in the United States and may be registered in other jurisdictions: AHOP® Automated Hardware Ordering Program, BALLAST® tray, BATCHFRAC® software, Bi-FRAC® tray, CASCADE MINI-RINGS® random packing, DEMISTER® mist eliminator, FLEXICHEVRON® mist eliminator, FLEXIFIBER® mist eliminator, FLEXIGRID® severe service grid packing, FLEXILOCK® tray construction, FLEXIMAX® random packing, FLEXIPAC® structured packing, FLEXIPAC® HC® structured packing, FLEXIRING® random packing, FLEXITRAY® valve tray, GEMPAK® structured packing, GLITSCH GRID® severe service grid packing, GOODLOE® mesh structured packing, HORIZON® technology, HY-PAK® random packing, IMTP® random packing, INTALOX® Packed Tower Systems technology, INTALOX PACKED TOWER SYSTEMS WITH PROCESS TOWER L® (logo), INTALOX® structured packing, INTALOX® SNOWFLAKE® random packing, INTALOX® ULTRA random packing, KARR® column, KATAMAX® reactive distillation structured packing, "K" KOCH-GLITSCH® (logo), KG-TOWER® software, KOCH-GLITSCH® technology, KOCH-OTTO YORK® separations technology, KY-FLEX® liquid-liquid coalescing media, MAX-FRAC® tray, MINIVALVE® tray, OMNI-FIT® technology, OTTO YORK and Design® (Football), PROFLUX® severe service grid, PROVALVE® tray, RATEFRAC® software, SCHEIBEL® column, Super INTALOX® Saddles random packing, SUPERFLUX® tray, SUPERFRAC® tray, THE SEPARATIONS COMPANY®, TRITON® tray, and ULTRA-FRAC® tray. The following registered trademarks are owned by Koch-Glitsch, LP in the CTM countries and may be registered in other jurisdictions: BETA RING® random packing and SATURN® technology. The following trademarks are owned by Koch-Glitsch, LP in the United States and may be registered in other jurisdictions: CMR random packing, COALEX liquid-liquid coalescing media, DEMISTER-PLUS mist eliminator, ETAPAK random packing, FLEXIPEAK mist eliminator, INTALOX® ULTRA random packing, KFBE fluidized bed structured packing, KITTEL tray, KY-MESH liquid-liquid coalescing media, LUBEMAX technology, PLUS technology, PURGE downcomer, ROSEMAX structured packing, SMV extraction packing, SMVP extraction packing, SPEED-WAY manway, TYPE "D" SAFETY SCRUBBER mist eliminator housing, VARIOFLEX tray, YORK-EVENFLOW vane inlet device, YORKMESH knitted wire mesh, and YOU CAN RELY ON US. All other trademarks, service marks, or registered trademarks that appear in this document are the trademarks or service marks of their respective owners.

PATENTS

The following technologies are protected by one or more patents in the USA; other foreign patents may be relevant: Column internals (US6889961, US5605654, US5632933, US6722639, US6889962, US7104529, US7114709, US7125004, US8317166, US8485504), FLEXILOCK® tray construction (US6592106), FLEXIPAC® HC® structured packing (US5632934, US6478290), FLEXIPEAK mist eliminator (US7905937), INTALOX® ULTRA random packing (US7722945), KFBE fluidized bed structured packing (US6224833), KITTEL® trays (US7028995), LUBEMAX technology (US6962661), MAX-FRAC® tray (US5632935), Mini Bubble Promoter (USD650466), OMNI-FIT® technology (US6736378, US7055810, US7125005), PLUS technology (US5762668), PROFLUX® severe service grid (US8298412), PROVALVE® tray (US5762834), SATURN® technology (US6736378, US7125005), SUPERFLUX® tray (US5895608), SUPERFRAC® tray (US5762668, US5895608, US5632935), TRITON® tray (US5618473, US5762834), ULTRA-FRAC® tray (US8500105, US8191870). Available exclusively from Koch-Glitsch, FLEXIPAC® HC® structured packing is protected by US Patent 5,632,934 and other patents worldwide assigned to Praxair Technology, Inc. Koch-Glitsch, LP is the exclusive worldwide licensee of Praxair Technology, Inc. for the manufacture and sale of this packing in all markets except for industrial gas separation.

LEGAL NOTICES

The Information in the presentation is not a guarantee of results to be achieved by any user and is not a statement of warranty, either express or implied. ALL EXPRESS AND IMPLIED WARRANTIES ARE EXPRESSLY DISCLAIMED, INCLUDING THOSE FOR MERCHANTABILITY AND FITNESS FOR A PARTICULAR PURPOSE. Individual results may vary and Koch-Glitsch cannot anticipate nor control conditions of application. IN NO EVENT SHALL KOCH-GLITSCH, ITS AFFILIATES, OR ANY PARTY INVOLVED IN CREATING, PRODUCING, OR DELIVERING THIS PRESENTATION BE LIABLE FOR ANY LOSS, DAMAGE, CLAIM, FINE, PENALTY OR ANY OTHER CLAIM, INCLUDING BUT NOT LIMITED TO CLAIMS FOR CONSEQUENTIAL, SPECIAL, GENERAL, INCIDENTAL, DIRECT, INDIRECT, PUNITIVE, PERSONAL INJURY OR PROPERTY DAMAGES, INCLUDING WITHOUT LIMIT LOSS OF PROFITS, REVENUES, OR OTHER ECONOMIC LOSSES, ARISING OUT OF YOUR ACCESS, USE, OR INABILITY TO USE THIS PRESENTATION OR ANY ERRORS OR OMISSIONS IN THE CONTENT THEREOF.

KOCH-GLITSCH

 CONFIDENTIAL

SULZER

Maximize Performance & Run length of a CDU MF with Advanced Mass Transfer Components

Katja Kolesnikov, Giuseppe Mosca* | AIChE, Spring Meeting, Austin, TX, 27 April 2015

* Speaker

SULZER

Outlook

- Typical CDU flow sheet
- CDU Main Fractionator revamp at MiRO Refinery
- Plant feed back data
- Conclusions

SULZER

CDU flow sheet

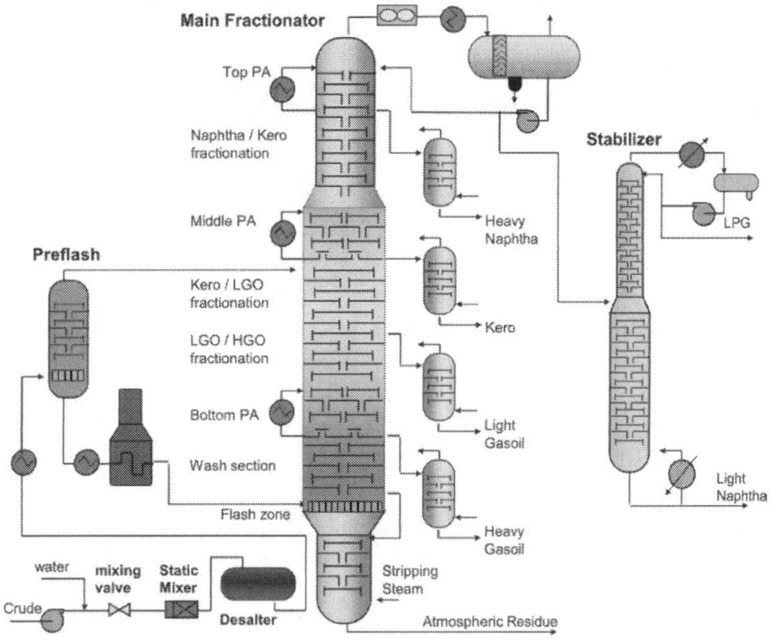

Maximize performance & Run length of a CDU MF with better utilization of active area / TBD / GM / June 2015 | slide 3

SULZER

MiRO Karlsrhue Refinery

- Capacity: 15 Mio t/a

Maximize performance & Run length of a CDU MF with better utilization of active area / TBD / GM / June 2015 | slide 4

SULZER

Concerns of existing Unit

- High corrosion at the top section of the Main Fractionator
- Compromised integrity of the top part of the Vessel
- Fouling at the top PA circuit
- Run length between turnarounds (2 -2.5 years) lower than desired:
  - excessive production loss
  - excessive maintenance cost
- Operating flexibility lower than desired

SULZER

Goals of the Revamp

- Minimize corrosion at top section of the Main Fractionator
- Increase the run length up to 5 years
- Maximize operation flexibility:
  - Process several different types of crudes
  - Operate the Unit at different modes: pro Diesel or pro Naphtha
- Reduce the Naphtha cut point :
  - 95% boiling point ASTM D86 from 150 to 140 °C
- Kero flash point >= 40 °C

SULZER

Major Constraints of the Revamp

- Keep the total height of the tower as per the existing one
- Minimize the external work around the tower
- Implement the work at site within the time frame of the turnaround:
  - 4 weeks
- Minimize the investment cost

SULZER

Major steps of the revamps

- Partial replacement of the vessel
- Upgrading the metallurgy of the top section of the column:
  - Alloy from overhead nozzle till top Pump Around draw off
- Revamp the top Pump-Around circuit:
  - Increase the duty
  - Increase the return temperature
- Upgrade the control system of the top section
- Upgrade the Mass Transfer Components

SULZER

Replaced sections of the Main Fractionator

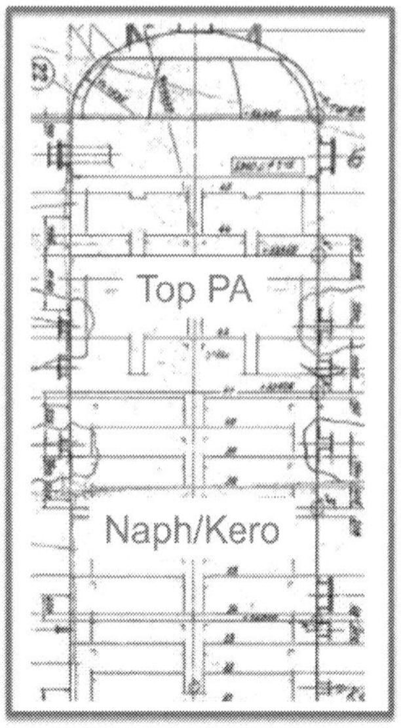

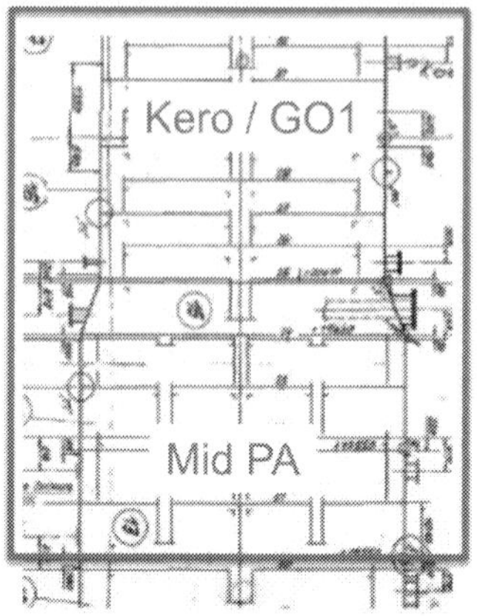

Maximize performance & Run length of a CDU MF with better utilization of active area / TBD / GM / June 2015 | slide 9

SULZER

Mass Transfer Components for the top PA

Before Revamp

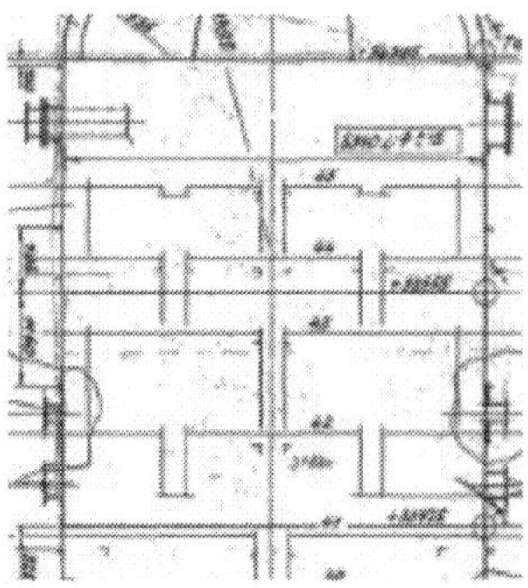

After Revamp

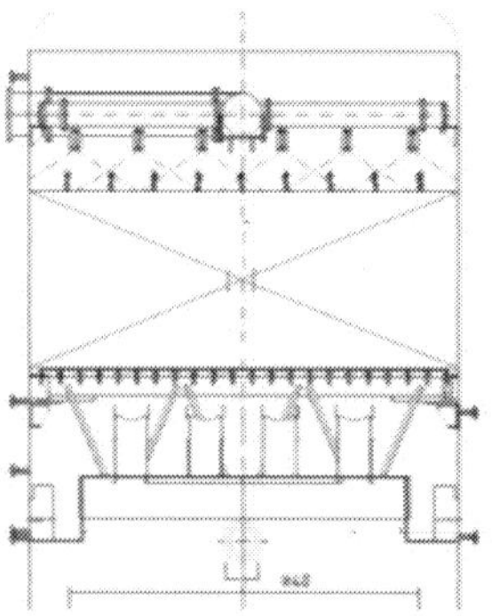

- New Spray Nozzle distributor
- New bed of Mellagrid
- Hold down & support grid for 1.2 psi uplift
- New total draw off Collector Tray
- Four new 20" draw off nozzles
- Material: Alloy 59

Maximize performance & Run length of a CDU MF with better utilization of active area / TBD / GM / June 2015 | slide 10

SULZER

Mellagrid installed at Top Pump Around

- Structured type of grid
- High heat transfer efficiency
- Smooth surface
- Great resistance to fouling
- Super mechanical stability
- Suitable for dismantling, cleaning, and reusing

SULZER

Mass Transfer Components at Naphtha / Kero fractionation section

Before Revamp

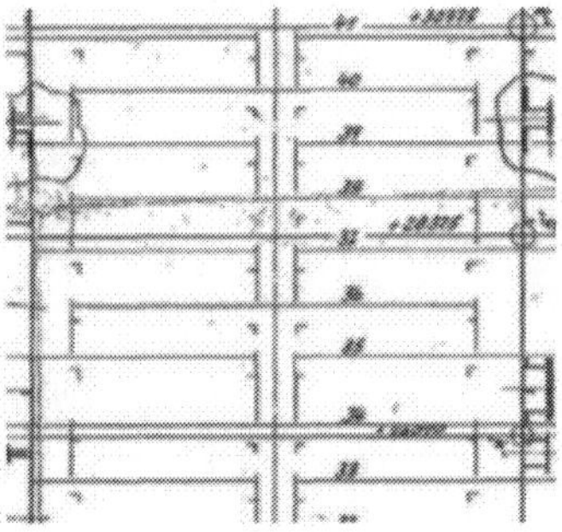

After Revamp

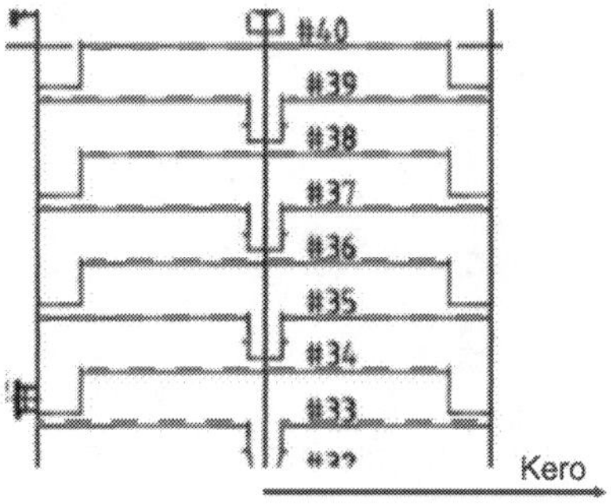

- The tray 41 was removed
- The existing 7 conventional trays 40-34 have been replaced with VGPlus™ High Performance trays
- The existing tray 33 was revamped with high performance active area and conventional central downcomer

SULZER

VGPlus trays

- It's a fractionation tray equipped with enhancement features:
  - Enhanced downcomer
  - Enhanced tray deck
  - Enhanced outlet weir
  - Redirecting devices

SULZER

Mass transfer components at Kero / Diesel fractionation section

Before Revamp

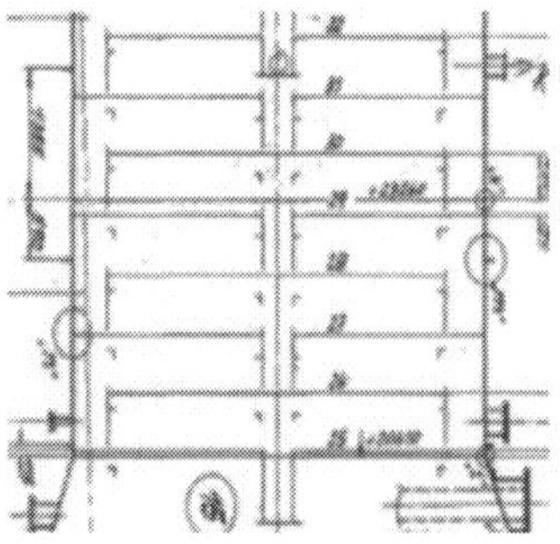

After Revamp

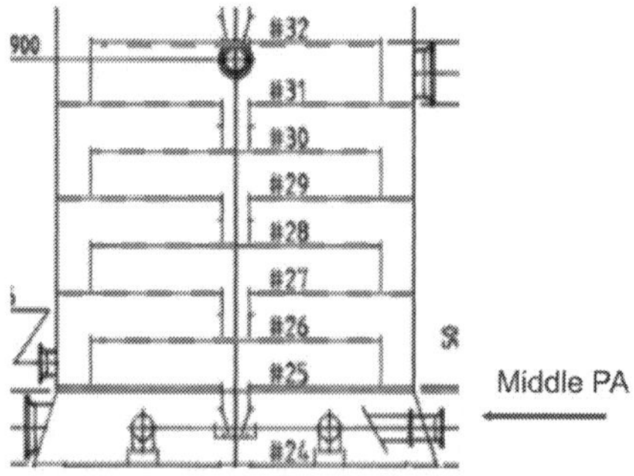

- The existing 8 conventional trays 32-25 have been replaced with MVG High Performance trays
- One transition from 2 to 4 pass tray

SULZER

Mass transfer components at Middle PA

Before Revamp | **After Revamp**

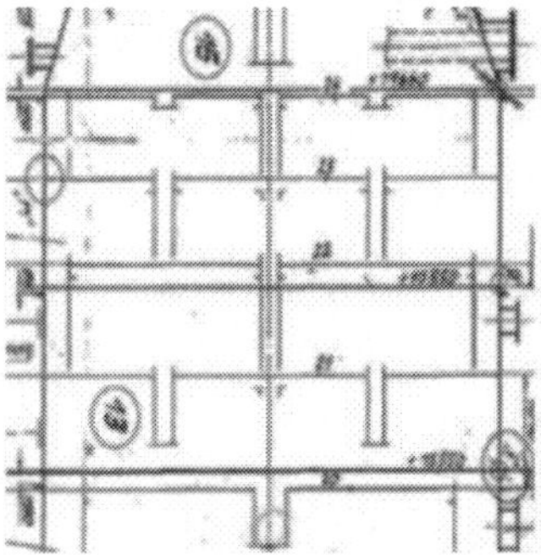

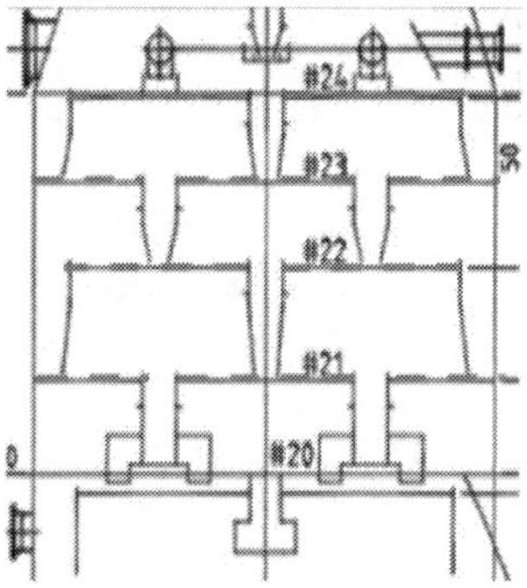

- The existing 4 conventional trays 24-21 have been replaced with MVG High Performance trays
- One transition from 4 to 2 pass tray
- Modified draw off sump below tray 20

SULZER

Mass transfer components at bottom PA

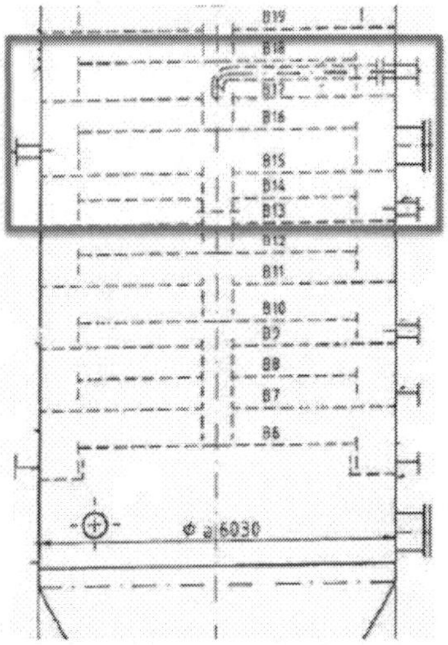

- Modified downcomer of tray 18
- The existing 3 conventional trays 17-15 have been replaced with MVG High Performance trays
- Modified draw off sump below tray 14
- Overflash section, Flash zone and Stripping section have been reused without any modifications

SULZER

Plant feedback data

- The Unit was commissioned on April 2010
- A test run was performed few months later to check the performances of the Unit at throughput close to revamp design

SULZER

Process Simulation of test run

- A process simulation was done to reproduce the plant data
- The feed to the tower was re-built by back mixing of the products
- The pressure profile was set as per plant data

Process Simulation Model of the Test Run

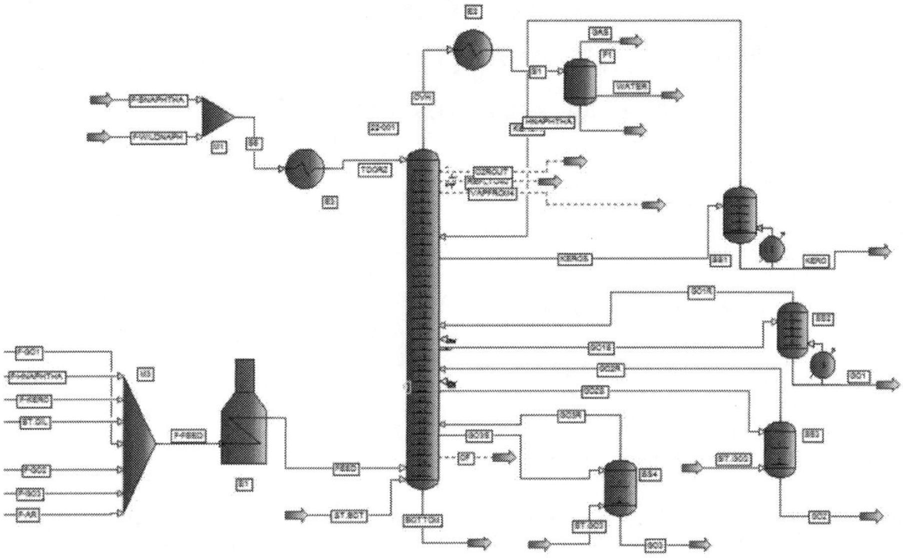

Simulation results VS plant data

- A good match between plant data and simulation results was achieved
- The temperature profile around the Main Fractionator was within 7°C max deviation: Flash zone, products draw off, PAs draw off and return, Overhead, and top receiver
- The heat balance at the PAs was within 6% max deviation
- The distillation curves of the main products were close to test run data with some small deviation in the front end

SULZER

Achievements

- The Top Pump Around provides the required performances:
  - Heat transfer capacity and hydraulic capacity
- The Naphtha cut point was reduced: the 95% ASTM D86 boiling point was even lower than 140 °C, i.e. 137 °C
- Kero Flash Point > 42 °C
- Tray efficiency higher than expected:
  - Over 85 % in the Naphtha / Kero section Vs 75 %
  - Over 78 % in the Kero / GO1 section VS 75%
  - Safe operation at Middle PA
  - Save operation at Bottom PA
- The Unit has run satisfactorily for 5 years

SULZER

Conclusions

- The structured type of grid "Mellagrid" is a great mean to debottleneck and/or upgrade the Top Pump Around of a CDU Main Fractionator

- It provides great hydraulic capacity, and high heat transfer capacity even with a low ΔT_{ML}

- A proper cooperation between the Engineering Company, End User, and Mass Transfer Components Vendor is mandatory while revamping Main Fractionators to achieve such challenging goals

SULZER

Thank you

Paper 45g

Condensate Splitting – What Are The Options?

Andrew W. Sloley

CH2M HILL
21 Bellwether Way, Unit 111
Bellingham, WA 98225

Prepared for Presentation at the
American Institute of Chemical Efngineers
2015 Spring Meeting
Austin, TX April 26 – 30, 2015

"UNPUBLISHED"

1. Summary

Tight oil production has dramatically increased the availability of 45°API to 60°API gravity condensates (0.8017 to 0.7389 specific gravity [1]). These light condensates produce too much vapor for many conventional atmospheric distillation units. Condensate splitting removes lighter material from the condensate. The objective is to produce a bottoms stream with a composition more closely matched to the feed expected in many atmospheric distillation units.

In past decades a few condensate splitting units have been built to handle these light streams. The trend towards heavier crude oil production between 1980 and 2005 made condensate splitters rare. The recent arrival of Bakken, Niobara, and Eagle Ford light crudes and condensates has rekindled interest in condensate splitting units. Various operating companies have released announcements for multiple new condensate splitting units. Many of the announced units, as well as others, are likely to be built.

To cover the range of likely feeds and operations a matrix of 30 combinations was evaluated for fractionation performance and energy consumption:

- Five feed compositions
    - Condensates covering the range from 50.6°API gravity to 63.6°API gravity (0.7770 to 0.7253 SG)
- Three process configurations

1 Specific gravity at 60°F (15.6°C).

    - Steam-stripped
    - Reboiled
    - Preflash
- Two operating modes
    - Low -cutpoint [2] (light naphtha production)
    - High-cutpoint (heavy naphtha production)

2. Recommendation

The recommended option is to use a reboiled condensate splitter:

- Best fractionation performance
- Lowest energy consumption

Capital investment for the reboiled and the steam stripped options is roughly the same. For sites with extremely low (or even negative) steam values a stripped condensate splitter might be usable.

The preflash configurations should only be used when high probability of foaming is considered to be extremely likely. However, addition of a foam drum to either a reboiled unit or a stripped unit allows those units to have equal foam handling capabilities. Therefore, the preflash configuration has little, if any, application.

3. Background

Recent production of tight oils has brought new low-density crudes and condensates onto the North American market. Due to logistic and processing limitations in the North American petroleum industry these streams have often sold at lower prices than what their assays would indicate as a historical value. Condensate splitting is one way of improving handling of these streams. Traditional crude production has removed light material by flashing or crude stabilization. Crude stabilization focuses on removing C_3 and lighter material and some C_4 material to control Reid Vapor Pressure (RVP) of the crude [3]. By contrast, condensate splitting focuses on removing naphtha and lighter material from the crude.

In the 1970s-1980s a few condensate splitting units were built to generate naphtha for use as a petrochemical feedstock. The major petrochemical use of this naphtha was steam cracking. In the early 1980s dramatically lower prices for ethane-propane for steam cracker feed eliminated the economic justification for most condensate splitting operations. After that point, few condensate splitters were built.

Recent trends, combined with government regulations, make condensate splitting of interest again.

- Increased light oil production from tight oil formations. This production contains large amounts of naphtha and lighter material.
- The new light oil naphtha tends to be very paraffinic:
    - High paraffin content may create asphaltene precipitation when mixed with heavier crudes.
    - High paraffin content naphtha has poor feed qualities for catalytic reforming.

---

2 The glossary in Appendix C includes the definition of cutpoint.

3 Maddox et al. 1979, Moins 1980a, Moins 1980b

- Refinery investment from 1980-2005 focused on improving capability to handle heavier crudes. Light oil processing often under-uses many refinery units.
- Finished product demand is shifting toward diesel and away from gasoline. Naphtha material goes into gasoline production.
- The areas with the growing gasoline demand include Africa, South America, and Asia. Gasoline demand in Europe and the USA varies from slightly rising to slightly decreasing.
- US law in the Energy Policy and Conservation Act (EPCA) stringently limits export of crude from the United States.

Condensate splitting benefits include:

- Removal of light, paraffinic material from the condensate:
    - Reduced asphaltene precipitation when mixing the remaining material with conventional crudes.
    - Reduced light-end handling requirements in a conventional refinery.
- Production of naphthas for shipment to growing gasoline markets.
- Production of naphthas that are allowed exports under the EPCA.

4. Condensates

Common industry usage has labeled condensates as produced streams with an API° gravity [4] range of 49° to 70°+ (0.7839 to 0.7022- SG). In 1980-2010 production of condensates often came from the North Sea and Australia/Indonesia. Current increase in North American condensate supply mostly comes from Eagle Ford production. Most Eagle Ford condensates vary from 50° to 60° API (0.7796 to 0.7389 SG).

Condensate may be either stabilized or unstabilized. Stabilization removes selected amounts of C_4 and lighter material for vapor pressure control. Vapor pressure control is often required to meet environmental regulation on volatile emissions from storage. In other cases, crude purchasers want to discount crudes containing large quantities of propane.

This paper examines condensate splitting options applied to five condensates. Table 1 lists the condensates and their properties. The five condensates were selected to illustrate specific differences due to the light end content of the crude. Figure 1 illustrates the logic for selecting the condensates examined.

Figure 1 characterizes the condensates by two criteria:

- API° gravity.
- Reid vapor pressure (RVP).

Both of these are routine measures used to set values on crude oils. RVP has also been used as a measure of crude oil volatility in safety and environmental regulations. Neither is a perfect way to measure crude oil value, environmental characteristics, handling safety, or processing characteristics [5]. Work is ongoing on tests to replace RVP. Nevertheless, both remain in current usage.

API gravity is a rough measure of hydrogen content. The higher the API gravity, the lower the density and the higher the hydrogen content.

---

4 Appendix C includes a glossary of standard terms and definitions.

5 Sloley 2014.

RVP is a rough measure of vapor pressure. The higher the light-end content (C_4 and lighter), the higher the RVP. RVP varies strongly with the exact light-end composition. Very small amounts of methane and ethane can dramatically increase RVP.

The condensates examined were selected to examine two trends:

- Higher density (low API gravity) versus lower density (high API gravity).
- Higher RVP (more light ends) versus lower RVP (less light ends).

In general, the expectation is that lighter crude (higher API gravity) will produce more light products. However, the high light-end content of some condensates may upset preconceptions about yield. The very light material in a heavier crude can lead to much higher than expected yields of light naphtha.

4.1. Comparison 1: Increasing API gravity at a high RVP

Åsgard, Sleipner Kårstø, and F3FB form a trend line for the upper range of RVP compared against API gravity. These condensates are unstabilized or have only relatively small amounts of C_3 and C_4 material removed. They produce relatively large quantities of light ends and large amounts of light naphtha product. These condensates have similar characteristics to much of the Eagle Ford production. The results shown for these crudes will cover much of the operating range expected on many of the new condensate splitters.

4.2. Comparison 2: Increasing API gravity at low RVP

Griffin and NWS form a trend line for the lower range of RVP compared against API gravity. These condensates are stabilized. They have significant amounts of C_3 and lighter removed. Stabilized condensates often have significant amounts of C_4 material removed as well.

As API gravity increases, even stabilized condensates have high RVPs. The low density required to get a high API gravity limits the amount of light ends removal possible. Figure 1 shows the low-RVP trend line converging with the high-RVP trend line as API gravity increases. However, selective reinjection of C_3-C_4 material can push RVP to high values even for relatively heavier crudes. Alaskan North Slope (ANS) is a commonly traded crude produced with selective reinjection of light material.

4.3. Comparison 3: Higher API Gravity and Lower RVP

The selected condensates show two comparisons for higher API gravity and lower RVP. These are:

- Åsgard versus Griffin.
- Sleipner Kårstø versus NWS

Each of these pairs shows an API increase with an RVP drop. The intent of these comparisons is to show how a high light-end content may confound yields and vaporization expected for a given API gravity.

4.4. Condensates Selected

Light-ends include propane and lighter material. Light-ends analysis of condensates comes from gas chromatography analysis. For accurate analysis of a condensate splitter system the chromatograph results for C_5 and lighter material are needed. The assays covered condensates with low RVPs through the highest values of RVP suitable for storage in floating roof tanks. All the condensates chosen have published assays with sufficient light-ends information to allow the RVP calculated from the composition to match the measured RVP reported in the assay. Small amounts of C_2 and C_1 were to fine tune the calculated RVP values.

Table 1 Condensates Evaluated

Condensate	Density		RVP		Origin	Assay Source
	API°	SG	psia	kPa		
Åsgard	50.6	0.7770	9.32	64.2	Norway	Total
Griffin	55.0	0.7587	5.40	37.2	Australia	BHP
Sleipner Kårstø	57.5	0.7487	10.3	70.7	Norway	Statoil
Northwest Shelf	59.5	0.7408	8.8	60.7	Australia	BHP
F3FB	63.6	0.7253	14.1	97.2	Netherlands	Total

4.4.1. Åsgard

Åsgard (50.6°API or 0.7770 SG) is a light North Sea Crude from Norway. It is close to the heavy end of the condensate range material. Åsgard contains a large amount of light ends for a heavier condensate. The relatively high RVP of 9.32 pisa (64.2 kPa) reflects the light ends in the condensate.

4.4.2. Griffin

Griffin (55.0°API or 0.7587 SG) is a mid-range condensate from Western Australia. Griffin has a higher API gravity but a lower RVP than Åsgard. Griffin is stabilized and has comparatively few light ends.

4.4.3. Sleipner Kårstø

Sleipner Kårstø (57.5°API or 0.7487 SG) is a light North Sea condensate from Norway. Sleipner Kårstø has high API gravity and relatively high vapor pressure. The condensate contains significant quantities of light ends.

4.4.4. Northwest Shelf (NWS)

NWS (59.5°API or 0.7408 SG) is a very light condensate from Western Australia. NWS has a high API gravity and a relatively low vapor pressure. The condensate contains relatively little light ends.

4.4.5. F3FB

F3FB (63.6°API or 0.7253 SG) is a very light condensate from the Netherlands North Sea. F3FB has a very high API gravity and a very high vapor pressure. The 14.1 psi (97.2 kPa) is at the technical limit for storage in floating roof tanks in a cool climate.

5. Condensate Splitter Types

Three types of condensate splitters have been evaluated:

- Steam stripped

- Reboiled
- Preflash

These configurations cover the range of possible basic configurations for industrial units.

5.1. Steam Stripped

Steam stripped condensate splitters use a configuration similar to conventional refinery atmospheric distillation. Figure 2 shows a general schematic of this unit. Feed is partially vaporized in a preheater and enters an atmospheric distillation tower. The tower includes a conventional rectification section. Reflux for the rectification section comes from either a pumparound heat removal system or an overhead drum heat removal system. Figure 2 shows a unit with an overhead system. Fractionation differences between the pumparound option and the overhead condenser option are insignificant. The main difference is in the heat integration for energy recovery.

The stripping section uses inert gas stripping with steam to recover lighter material from the bottoms. Direct injection of the steam reduces the hydrocarbon partial pressures. With lower partial pressures, the lighter hydrocarbons in the bottoms vaporize and generate internal traffic in the tower to drive the desired separation.

One of the main benefits of the steam stripped condensate splitter is that many refineries have surplus low-pressure steam that has little other value. The steam pressure is too low to reboil a condensate splitter, but stripping using partial pressure vaporization still works effectively.

One of the main disadvantages of having a rectifying section is potential operating problems from foaming of the condensate feed. Foam in the feed can overload the feed flash zone and lead to entrainment. Entrainment reduces unit capacity and degrades fractionation performance.

5.2. Reboiled

Reboiled condensate splitters use a configuration similar to a classic distillation tower. Figure 3 shows a general schematic of this unit. The rectifying section configuration is the same as a steam-stripped condensate splitter.

The stripping section uses a reboiler to generate the required vapor. One of the main benefits of the reboiled condensate splitter is that with sufficient duty, nearly any reflux ratio can be obtained. The reboiled condensate splitter achieves better separation or higher recovery for a fixed heat input.

As with the steam stripped condensate splitter, foam in the flash zone leads to significant problems.

5.3. Preflash

Preflash condensate splitters use a configuration similar to a conventional refinery atmospheric preflash tower. Figure 4 shows a general schematic of this unit. Feed is partially vaporized in a preheater and enters the bottom section of an atmospheric distillation tower. The rectifying section configuration is the same as a steam-stripped condensate splitter. The preflash condensate splitter has no stripping section.

Many refineries use the preflash configuration to avoid foam entrainment problems in the flash zone. As long as the tower boot has sufficient residence time and low enough vapor velocities, foam problems can be avoided. Since the preflash has no stripping section, separation performance tends to be poor.

6. Basis and Constraints

The comparison basis used 100,000 bpd (15,900 cmpd) of condensate for all cases. Operating units must work within both process constraints and equipment constraints. Process constraints include meeting required operating conditions and stream compositions. Equipment constraints include expected mass-transfer efficiency, equipment sizes, and hydraulic performance of the equipment. This evaluation imposed both process constraints and equipment constraints. Appendix A summarizes the constraints imposed.

Many of the tray design parameters imposed in this work can be relaxed with more elaborate equipment. One drawback of specialized configurations is reduced flexibility to handle different feeds. Refinery economics strongly benefit from buying whatever is today's cheapest feed. For this reason, truly pushed design criteria were not used.

6.1. Operating Targets

The major process constraint was the overhead product specification. Two distillation specifications were checked for each case; a high-cutpoint and a low-cutpoint operation.

- Low-cutpoint operation: production of light naphtha overhead product to a 230°F (110°C) D86 95 volume percent specification. The intent of low-cutpoint operation is to produce a light naphtha product to remove RVP from the refinery gasoline pool while leaving aromatics in aromatic precursors in the refinery feed for petrochemical production from catalytic naphtha reforming.
- High-cutpoint operation: production of heavy naphtha overhead product to a 350°F (177°C) D86 95 volume percent specification. The intent of high-cutpoint operation is to produce a heavy naphtha product that minimizes the size of the gasoline pool at the refinery while leaving as much material as possible that could be blended into kerosene for middle distillate (and diesel) production.

Operating targets may shift depending upon customer requirements and values in the market. As of now (2015), most economic operation would favor specifications around 350°F (177°C):

- Internal North American demand favors additional diesel production. High-cutpoint operation pushes the maximum amount of material into a naphtha export stream.
- All diesel range material still remains in the bottoms that can be sent to domestic refineries.

6.2. Yields

For each condensate, the yield on the steam-stripped configuration was defined as the base yield. As long as equipment constraints allowed, the yield for the reboiled configuration and the preflash configuration was matched to the base yield.

6.3. Distillation Effectiveness

All configurations used 14 theoretical stages (18 trays) in the rectifying section plus one theoretical stage in the condenser. This gave a total of 15 theoretical stages for the naphtha product versus the feed.

The author's evaluations have always shown that optimization of investment for theoretical stages versus energy efficiency and product yields strongly drive units toward maximum theoretical stages. Incremental cost for an additional theoretical stage is low compared to the benefits in yields, energy, and reduced investment in other equipment (fired heaters, heat exchangers, pumps, and piping). Most existing units have too few trays in them. A sensitivity analysis using ±5 stages for the rectifying section showed that the 15 stage separation was comfortably inside the region where reflux strongly drives separation performance.

As a comparison, the ASTM D2892 method that defines the true boiling point (TBP) distillation curves uses 15 theoretical stages in the distillation. The refining industry settled on 15 stages because this was enough stages that minor equipment differences had no significant effect on the distillation results. While the D2892 method uses a batch distillation at high reflux ratios, the same logic applies to condensate splitting units.

6.4. Distillation Equipment

Hydraulic limits and sizing constraints for the condensate splitter tower were set using prudent design practices for trays. The rectifying section used conventional moving-valve trays. The stripping section used conventional fixed-valve trays. Only minor differences would be expected if sieve trays were used in the stripping section. When building a new tower the cheapest option is to normally used a larger diameter tower with trays.

In the rectifying section packing would allow operation at lower liquid rates. In the stripping section, packing would cover a wider range of vapor rates. Based only on vapor and liquid loads, packing would be an attractive choice. Improved performance and lower operating costs would justify packing.

However, trays are more reliable. For this reason they were used in all configurations. Many of the operators producing the new condensates use complex additive programs to increase production. Often, the additives are volatile and end up vaporizing in the tower feeds. Multiple refineries running the new condensates have had severe fouling in the top sections of their crude towers. Fouling can result from direct deposition of solids from the vapor phase, shock condensation of water from cold reflux, localized heat loss, and water carryover in the reflux [6]. Trays are more resistant to fouling problems and easier to clean.

6.5. Overhead Configuration

All units included a partial condenser on the overhead stream. Unless aggressively stabilized, most condensates include very light material. The composition may include material as light as methane. The small quantity of very light material may not show up in the feed analysis, but experience on operating units shows it is usually present. All condensate splitters should have the capability of removing light material by making a gas purge product. This purge stream may go to recovery in a compression system or to fuel gas treating. The gas purge stream aids both heat integration and control.

The small gas purge significantly increases the overhead drum temperature. In the cases evaluated the minimum increase in temperature was ~20°F (11.1°C). The increased temperature in the overhead drum lowers the investment needed for heat recovery from the condenser to the feed. Condenser-to-feed heat integration provides the majority of heat recovery in most of the options.

Increased reflux temperature also reduces shock condensation in the crude tower. Tower fouling rates are usually lower with higher reflux temperatures.

6 Sloley 2013a,b,c

Unstabilized condensates, even those marketed under a single name, often have significant variations in light ends content. Variations may be seasonal or random. Seasonal variations come from limits on vapor pressure. In cooler seasons, more light material can be left in the crude without vapor pockets forming under the floating roof. In warmer seasons, more light material must be removed. Regulatory limits on emissions and technical limits on preventing tank roof collapse trend together.

Random variations come from shipping logistics limits. Increased condensate production often uses rail car and tanker truck shipment. Specific deliveries may have large variations in light-ends content. A continuous gas purge makes handling random spikes in very light material content much simpler.

6.6. Flash Zone Entry

The flash zone is the part of the condensate splitter where the two-phase feed from the heater enters. In the temperature range between 300°F and 500°F (149°C and 260°C) the two-phase mixture has a significant risk of forming stable foams. Foam accumulation is the balance between foam formation rates and foam decay rates. Both depend strongly upon system chemistry, liquid viscosity, and volume fraction vapor. Foam formation has been identified as an operating problem across a wide range of condensates and operating conditions. While the greatest risk is between in the 300°F-500°F range (149°C-260°C), it can occur at both lower and higher temperatures.

Foam has normally been handled by providing ample residence time and decreasing vapor velocity. This can lead to very large towers. The basis here uses cyclonic entry devices to force foam decay. This allows the tower diameter to drop to the size required by the trays. The tower diameter at the flash zone is set as the larger of the sizes required by either the rectifying section above or the stripping section below.

6.7. Operating Conditions

All comparisons used a common basis of 30 psig (308 kPa) overhead drum pressure, 42.5 psig (394 kPa) overhead pressure and 45 psig (411 kPa) flash zone pressure. Sufficient pressure was used in the preheat train to prevent vaporization upstream of the fired heater. Pressure optimization was not included in this review. Higher pressure operation would make the tower smaller and reduce exchanger area due to increased temperatures in the condenser. Higher pressure operations would increase required feed temperatures, make using low pressure steam in the stripping units more difficult, and make reboiler temperatures in the reboiled units higher. The most attractive cases for higher pressures would be the low-cutpoint operations.

6.8. Fired Heaters

For the low-cutpoint operations, preheat temperatures are low enough that a steam heater would be a feasible choice. Nevertheless, the energy balances used a fired heater on the tower feed in all cases. No velocity steam was injected into the fired heaters. In all cases, temperatures were relatively low for the fired heater outlet. Velocity was not needed to control heater outlet temperatures.

7. Constraints

Table 2 [7] identifies the limiting constraints for each configuration. The overall objective was to:

- Match target product yield rates
- Minimize heater firing

Process and equipment constraints were set at practical limits based on reliable operation, with reasonable flexibility, and for commercially available equipment. Every option was pushed to the limits of as many constraints as possible. Depending upon the configuration, the number of controlling constraints varied from 1 to 3.

7.1. Naphtha Yield

In most cases the naphtha yields were matched between the base case (stripping configuration) and the alternate units. The exceptions were;

- Griffin condensate, reboiled (both low-cutpoint and high-cutpoint operations).
- The lightest condensates (Sleipner Kårstø, NWS, and F3FB) on the preflash configurations.

For Griffin condensate the lack of light ends reduced the lift (vaporization) in the flash zone. This suppressed the yields for the steam stripping case. Resulting bottoms product rates were high. With a reboiler present, the large bottoms flow rate required a large tower diameter to handle the liquid rates. Creating enough reboiler vapor to make the bottoms trays work generated excess L/V ratios in the tower. At a constant naphtha specification extra overhead naphtha was produced.

The preflash configuration must have all the energy required to generate reflux enter with the tower feed. This increases the feed temperature and fraction vaporized. The light feeds ran into a 90% feed vaporization limit before generating sufficient reflux to match both product yield and product specification on the high-cutpoint operation. Less heavy naphtha was produced in these cases.

7.2. Rectifying (Top) Section Liquid Rates

Half the configurations (15 of 30) had constraints imposed by liquid rates in the top of the column. The condensates all have high naphtha yields. Little reflux is needed to meet the distillation specifications. The large vapor rates set a minimum tower diameter. The liquid rate is insufficient to effectively wet the trays.

A minimum liquid rate is required to make trays function effectively. The most common approach is to set a tray liquid rate based on flow rate per length of outlet weir. At the operating conditions present in condensate splitting units, low liquid rates create an operation where the liquid on the tray becomes the discontinuous phase. Tray efficiencies are low in this operation. The constraint limit was 0.72 gpm/inch (1.1 lpm/cm) of outlet weir length. This is just inside the region where efficiency starts to drop. However, reasonable equipment modifications plus some allowance for lower efficiency makes this liquid rate tolerable [8].

The choices at the constraint limit become:

- Impose minimum liquid rates; add more heat to get the liquid required.

---

7 Table 2 is collected in the large tables at the end of the discussion.

8 Summers & Sloley 2007

- Change operating point; increase pressure to reduce the size of tower.
- Accept low efficiency operation; add more trays to get the stages required.

For design this evaluation imposed minimum liquid rates. Feed or reboiler duty was increased to get the required liquid. The choice that gave the lowest duty requirement was usually to increase the feed duty (14 of 15).

7.3. Bottoms Vapor Rate

With the heaviest condensates (Åsgard and Griffin) the bottoms liquid rate set the tower bottoms section diameter. The reboiled splitters required extra bottoms duty to generate enough vapor for the trays to work properly.

7.4. Water Condensation Inside the Tower

One option was constrained by water condensation inside the tower. The Griffin condensate coupled with the low-cutpoint operation with steam stripping had a low overhead product rate and a high bottoms rate. These rates create a pincer forcing the unit toward condensing water in the tower.

Low overhead rates reduce the hydrocarbon vapor rate overhead. High bottoms rates increase the amount of steam required for stripping. The result is a high water partial pressure in the overhead system. This caused water condensation at high temperatures. To avoid water condensation in the tower, stripping steam rates were reduced.

7.5. Preheat Configuration

Six of the options had limitations on heat recovery imposed by preheat configuration. This was the least expensive of the constraints to deal with. The solution was to reject heat from the bottoms rundown around the pinch in the preheat train.

The heaviest condensates (Åsgard and Griffin) had easily recoverable heat available, but had no usable heat sink for the duty. The lighter condensates had very low bottoms rates. NSW, in particular, had some services with duties that would be so small that the heat exchangers could not be justified. F3FB was very close to the same limit.

7.6. Feed Vaporization

The lightest condensates (Sleipner Kårstø, NWS, and F3FB) coupled with high-cutpoint operation ran into a feed vaporization limit in the preflash configurations. The preflash configuration must have all the energy required to generate reflux enter with the tower feed. This increases the feed temperature and fraction vaporized. The light feeds ran into a 90% feed vaporization limit. The comparison kept the feed vaporization at 90% and allowed the naphtha yield to drop.

8. Analysis of Results

Appendix B includes the detailed results for each case evaluated. The results are shown in Table B-1 and Table B-2:

- Table B-1 Comparison by feed
- Table B-2 Comparison by operation and configuration

8.1. Steam Stripped versus Reboiled

The most important comparison is the steam-stripped configuration versus the reboiled configuration. Table B-1 and Table B-2 include two energy consumption totals;

- Total fired duty:
  - o Fired heater duty for feed preheat plus,
  - o Fired heater duty for the reboiler.
- Total fired duty plus steam fired duty:
  - o Fired heater duty for feed preheat plus,
  - o Fired heater duty for steam assuming 1,000 btu/lb (2.32 MJ/kg) for steam generation.

Table 4 [9] includes an analysis of energy consumption per volume of naphtha recovered and an estimated breakeven energy cost for steam-stripping versus reboiling. If the cost of steam is below this fraction of the cost of firing energy, steam stripping will usually have favorable economics. If the cost of steam is above this fraction, reboiling the condensate stripper will have favorable economics.

8.1.1. Low-cutpoint Operation

For low-cutpoint operation, if steam is relatively cheap compared to fuel, steam stripping may make economic sense. In general, the trend is that the lighter the condensate, the lower the steam value required for steam stripping to make sense.

8.1.2. High-cutpoint Operation

For high-cutpoint operation, all cases except the heaviest feed showed reboiled units as being more energy efficient. This efficiency comes from the higher heat recovery possible from the condensate splitter bottoms. The reboiled configurations all have hotter bottoms temperatures. The higher bottoms temperatures result from adding two effects:

- Steam stripping takes energy from the bottoms stream to vaporize light material. This cools the bottoms.
- Reboiling adds energy to the bottom stream to vaporize light material. By equilibrium this requires a hotter bottoms.

These add together to make the reboiled unit bottoms temperature much higher than the steam-stripped unit bottoms temperature.

The bottoms mass flow rate is lower than the feed flow rate. Heat integration between the bottoms and the feed pinches at the cold end of the heat exchanger train. The bottoms cools rapidly toward the feed temperature. The higher bottoms temperature from the reboiled cases significantly improves heat integration. The net result is lower total fired duty.

Increased heat recovery from the bottoms requires increased exchanger and pump investment. However, reduced investment in the fired heaters, steam generation, and wastewater handling offsets this. Unless steam has a negative value (costs money to not use), a reboiled condensate splitter is the better choice.

---

9 Table 4 is collected in the large tables at the end of the discussion.

8.1.3. Very High-cutpoint Operation

Bottoms temperature might limit some operations with condensate splitting. The bottoms temperature varied between 549°F and 629°F (287°C and 332°C). Forcing higher boiling point material overhead would require a hotter bottoms temperature. Increasing the overhead product cutpoint to 400°F (204°C) to make a heavy naphtha product overhead should be possible for most condensates. Taking kerosene boiling range material would push the overhead product cutpoint to 520°F (271°C) or higher. Trying to make a kerosene product overhead would require too high a bottoms temperature, exposing the bottoms to a risk of thermal cracking. The maximum allowable bottoms temperature will vary with the specific condensate.

An optimum unit for taking kerosene product overhead would use a combination of reboiling plus steam stripping. The best location for steam injection would be in the reboiler feed. The steam injection would increase the reboiler pressure drop, but would also decrease the reboiler outlet temperature required for vaporization.

8.2. Light-Ends and Feed Vaporization

Naphtha yields are strongly influenced by relatively small amounts of light ends. Modest changes in feed RVP can dramatically swing expected yields and unit operations. Condensate splitters must be built with sufficient equipment and control flexibility to handle swings in feed composition.

Stabilized condensates contain significantly less light naphtha. Normally, an increased API gravity increases light naphtha yields. However, a decrease in high RVP components can outweigh an increase in API gravity.

The low-cutpoint operation showed reduced naphtha yield for the lighter, but stabilized, condensates. The comparison looked at:

- Åsgard (partly unstabilized) compared to Griffin (stabilized)
- Sleipner Kårstø (partly unstabilized) compared to NWS (stabilized)

Each of these pairs shows:

- An API° increase and an RVP drop at the same time.
- Reduced light naphtha yield from the partly stabilized (or unstablized) condensate even though the API° increased.

Table 3 illustrates this by presenting naphtha yields for the reboiled preflash configuration. The other cases are similar.

Table 3 Naphtha Yields, Reboiled Condensate Splitter

Operation		Condensate				
		Åsgard	Griffin	Sleipner K	NWS	F3FB
Feed						
API° gravity		50.6	55.0	57.5	59.5	63.6
RVP, psia		9.32	5.40	10.3	8.8	14.1
SG		0.7770	0.7587	0.7487	0.7408	0.7253
RVP, kPa		64.2	37.2	70.7	60.7	97.2
Naphtha						
Low-cutpoint	bpd	20,706	9,238	44,803	41,223	43,759
Low-cutpoint	cmpd	3,292	1,469	7,124	6,554	6.958

8.3. Naphtha Recovery

The naphtha recovery in the high-cutpoint operation varied from 100% to 113%. This result can surprise. The result is a function of how distillation curves and yields are defined. Recovery has been defined as:

$$Recovery = 100 * \frac{naphtha\ product\ rate\ (volume)\ at\ D86\ spec}{naphtha\ product\ rate\ defined\ as\ possible\ by\ TBP\ curve\ (volume)}$$

The feed TBP is defined by ASTM D2892 method or an equivalent. This method defines stages and reflux, not a fixed separation effectiveness. When generating the TBP for the naphtha product estimate by the ASTM D2892 method, some naphtha remains un-recovered and some kerosene (the next heavier material) ends up in the naphtha product.

The results here show that the condensate splitter is performing better than an equivalent D2892 separation. The target of the condensate splitter configurations used here was to remove naphtha but to lose no kerosene range product into the naphtha. This choice is driven by favorable economics for blending as much material as possible into middle distillates (kerosene and diesel). All configurations succeeded in meeting this goal.

8.4. Fractionation Performance

Fractionation performance has been summarized by the D86 95-5 volume percent gap. Results with a positive number indicate a gap. Results with a negative number indicate an overlap. The more positive the number, the better the fractionation performance.

In general:

- The reboiled configuration has the best separation performance. Most of this is due to the more effective stripping section.
- The steam stripped configuration has reasonable performance. Except in cases with extremely high naphtha-kerosene values, the difference between the reboiled configuration and the stripped configuration is unlikely to be decisive.

- The preflash configuration had significantly poorer separation performance in all cases. The economic results of this will depend upon product values.

8.5. Investment Required

Investment requirements generated from the level of information available have an expected scatter of ±50%. The attempt to estimate investment for the steam-stripped versus reboiled configurations show cost differences much less than 50% for a fully independent unit (including utility supply). A sensitivity analysis showed that small differences in many of the assumptions required to generate an investment cost would significantly change the results. No justifiable conclusion could be drawn from the investment analysis. The preflash unit showed a consistent trend for slightly higher capital cost than the other unit types.

8.6. Feed Heat versus Bottoms Heat

The analysis showed that optimum energy use requires the ability to control both the feed enthalpy and the bottoms energy. The bottoms energy may be either stripping steam in the stripped configuration or reboiler duty in the reboiler configuration.

Sensitivity checks showed that, close to the optimum point, the effects of feed preheat and bottoms heat were:

- Roughly 80% of the feed duty change went into generating internal liquid in the column.
- Extra vaporization of the bottoms (either from steam stripping or reboiler) duty went roughly 80% into increased product yield until naphtha recovered was over 105%. At that point extra vaporization in the bottoms went nearly exclusively into generating internal liquid in the column.

One exception that did not require extra heating of the feed was found:

- Griffin condensate – a highly stabilized condensate
- Low-cutpoint operation
- Reboiled configuration

Sufficient heat was available in the rundown to reach the feed temperature with only heat-integration. However, using this configuration would severely restrict the flexibility of the unit.

The final conclusion is that all units require the ability to heat the feed to higher temperatures than feed preheat can achieve. The feed preheater could be either a fired heater or a steam heater. The fired heater would work for all cases. Units built specifically for low-cutpoint operation could use a steam heater.

8.7. Energy Efficiency and Vapor Rates

For the low-cutpoint operation, the light naphtha yield is relatively low for the Åsgard and Griffin condensates. Åsgard and Griffin were the heaviest condensates with the lowest API° gravity. Feed vaporization for both cases are low. This reduces the diameter enough for a 100,000 bpd (15,900 cmpd) unit that the top section can use one-pass trays. The one-pass trays have dramatically lower liquid rates to meet the liquid wetting constraint. Lower liquid rates require less feed vaporization. For this reason those cases look slightly more energy efficient. In general, smaller units that have low enough liquid rates to use one-pass trays will look more energy efficient.

8.8. Special Considerations for Low API° Gravity Condensates

The Åsgard and Griffin condensates deviate from the general trend shown by the lighter condensates. This results from the relatively large bottoms product rate. This leads to two results:

- The bottoms section of the condensate stabilizer has a relatively large diameter. The bottoms liquid rate, rather than the bottoms vapor rate, sets the minimum diameter required. Vapor rates must be increased to allow for operation of the larger diameter trays.
- For steam stripping units, the bottoms steam rate is high. In the case of Griffin, the steam rate must be reduced to prevent water condensation inside the condensate splitter.

8.9. Two-Step Condensate Splitting

If concern about foaming in the condensate splitter is the over-riding concern, the preflash condensate splitting option may be a desirable approach. Feed vaporization limits preflash splitting yields in high-cutpoint operation. One way to reach high-cutpoints with a preflash configuration would be to run the condensate splitting in two steps. The first step would be a stabilization-light naphtha recovery step. The second step would be a heavy naphtha recovery step.

As a proof-of-concept, using low-cutpoint operation on F3FB as the first step produces a bottoms stream that would be reasonably similar to Griffin. The second step would now be a smaller unit that would run a high-cutpoint operation. Table B-3 includes a simplified comparison based upon combining the F3FB low-cutpoint operation with a prorated Griffin high-cutpoint operation. This non-rigorous comparison predicts too high an energy consumption for the two-step splitter. Even taking this into account, the two-step processing is so inferior it is difficult to put together a reasonable set of assumptions that would justify a two-step splitter.

8.10. Alternatives to Two-Step Condensate Splitting and Foam Handling

Other alternates are available for increasing vaporization for preflash units and for handling foam:

- Keep the preflash configuration and add steam injection to the transfer line.
- Switching to a reboiled (or steam stripped) configuration and dividing the condensate splitter into two columns.
- Use a large foam drum upstream of the condensate splitter.

Figure 5 shows adding steam injection to the feed heater outlet. The steam injection reduces hydrocarbon partial pressure, allowing more vaporization before the feed enters the condensate splitter. This would allow the heater to operate at a lower vaporization but still vaporize more feed to reach a higher cutpoint. This preserves the preflash configuration for using a large tower boot to handle foam induced problems.

Figure 6 shows using a tower boot to handle foam problems. However, the overall reboiler configuration is still used. A separate stripping section is used with the reboiler. Essentially this layout mechanically divides the tower into two sections. The large boot acts as a separator vessel.

Figure 7 shows another alternative using a large upstream foam drum to handle any foam. Both vapor and liquid then go to the downstream condensate splitter. The downstream splitter may be either steam-stripped or reboiled. Figure 7 shows the reboiler configuration.

Other configurations are possible. All the alternatives have higher investment than the base stripped configuration or the base reboiled configuration. Nevertheless, they have better performance and lower investment than the two-stage option with preflashes as discussed in §8.9.

9. Conclusions and Selection Strategy

Figure 8 outlines the recommended selection strategy. The two major criteria are:

- Concern about foam handling
- Cost of steam versus cost of firing

Modern cyclonic separators are clearly capable of handling foam mixtures. However, their use is recent and many operators prefer to use a more traditional and proven technology. Alternatives would include a preflash configuration, split towers, or an upstream foam drum.

The preflash configuration comes with significant operating costs and limits on possible yields. The foam drum allows for either a stripped condensate splitter or a reboiled condensate splitter. If foam risk is a priority concern in selecting the condensate splitter configuration, the foam drum has the lower cost than a split tower and higher performance than a preflash tower.

If steam costs are low enough, a stripped condensate splitter may be viable. However, for a typical condensate with significant light ends present, the reboiled configuration benefits from:

- Lower energy consumption, and
- Improved separation performance.

For an independent facility that must include dedicated utility (steam) production, the reboiled configuration is nearly always the lower investment choice. The steam stripped configuration is only likely to be economic when added to an existing plant with surplus low-pressure steam that has a very low value.

10. Tables and Figures

Table 2 Constraints for Each Option

Feed	Asgard	Asgard	Asgard	Griffin	Griffin	Griffin	SleipK	SleipK	SleipK	NWS	NWS	NWS	F3FB	F3FB	F3FB
Operation	Low Cut	Low Cut	Low Cut	Low Cut	Low Cut	Low Cut	Low Cut	Low Cut	Low Cut	Low Cut	Low Cut	Low Cut	Low Cut	Low Cut	Low Cut
Configuration	Strip	Reboil	Preflash	Strip	Reboil	Preflash	Strip	Reboil	Preflash	Strip	Reboil	Preflash	Strip	Reboil	Preflash
Naph Yield	Base	X	X	Base		X	Base	X	X	Base	X	X	Base	X	X
Top Liquid - Feed Q	X			X			X	X		X	X		X	X	
Top Liquid - Rebl Q															
Bottom Vapor					X										
Water Condensation				X											
Preheat Config		X	X	X	X					X					
Feed Vaporization															

Feed	Asgard	Asgard	Asgard	Griffin	Griffin	Griffin	SleipK	SleipK	SleipK	NWS	NWS	NWS	F3FB	F3FB	F3FB
Operation	High Cut	High Cut	High Cut	High Cut	High Cut	High Cut	High Cut	High Cut	High Cut	High Cut	High Cut	High Cut	High Cut	High Cut	High Cut
Configuration	Strip	Reboil	Preflash	Strip	Reboil	Preflash	Strip	Reboil	Preflash	Strip	Reboil	Preflash	Strip	Reboil	Preflash
Naph Yield	Base	X	X	Base		X	Base	X		Base	X		Base	X	
Top Liquid - Feed Q	X			X			X			X	X		X	X	
Top Liquid - Rebl Q					X										
Bottom Vapor		X													
Water Condensation															
Preheat Config				X						X					
Feed Vaporization									X			X			X

Table 4 Energy Comparison

Configuration	Operation		Condensate				
Fired duty in feed heater plus reboiler							
	Low-cutpoint		Asgard	Griffin	SleipK	NWS	F3FB
Stripping		btu/bbl	2415	4891	2152	2144	2171
Reboiled		btu/bbl	3497	7101	2431	2421	2390
Preflash		btu/bbl	5623	6576	2940	3018	2916
	High-cutpoint						
Stripping		btu/bbl	2327	2131	2159	2041	2013
Reboiled		btu/bbl	2991	1914	2070	1971	1961
Preflash		btu/bbl	3730	3029	2464	2249	2391
Steam duty (@ 1,000 btu/lb)							
	Low-cutpoint		Asgard	Griffin	SleipK	NWS	F3FB
Stripping		btu/bbl	1884	1340	551	680	675
Reboiled		btu/bbl	0	0	0	0	0
Preflash		btu/bbl	0	0	0	0	0
	High-cutpoint						
Stripping		btu/bbl	185	288	89	93	107
Reboiled		btu/bbl	0	0	0	0	0
Preflash		btu/bbl	0	0	0	0	0
Fired duty in feed heater plus reboiler plus steam duty							
	Low-cutpoint		Asgard	Griffin	SleipK	NWS	F3FB
Stripping		btu/bbl	4300	6231	2703	2824	2846
Reboiled		btu/bbl	3497	7101	2431	2421	2390
Preflash		btu/bbl	5623	6576	2940	3018	2916
	High-cutpoint						
Stripping		btu/bbl	2512	2419	2248	2133	2120
Reboiled		btu/bbl	2991	1914	2070	1971	1961
Preflash		btu/bbl	3730	3029	2464	2249	2391
Energy comparison - Shifting from stripping to reboiling							
	Low-cutpoint						
Firing change		btu/bbl	1,081	2,210	279	277	219
Steam change		btu/bbl	-1,884	-1,340	-551	-680	-675
Breakeven steam:firing value		fraction	0.57	1.65	0.51	0.41	0.32
	High-cutpoint						
Firing change		btu/bbl	664	-217	-89	-70	-52
Steam change		btu/bbl	-185	-288	-89	-93	-107
Breakeven steam:firing value		fraction	3.60	-0.75	-1.00	-0.75	-0.48

Table 4 Energy Comparison

Configuration	Operation		Condensate				
Fired duty in feed heater plus reboiler							
	Low-cutpoint		Asgard	Griffin	SleipK	NWS	F3FB
Stripping		kj/m3	16026	32452	14276	14228	14405
Reboiled		kj/m3	23199	47115	16127	16063	15860
Preflash		kj/m3	37306	43632	19504	20022	19348
	High-cutpoint						
Stripping		kj/m3	15440	14138	14327	13540	13354
Reboiled		kj/m3	19848	12699	13735	13078	13012
Preflash		kj/m3	24751	20096	16348	14923	15866
Steam duty (@ 1,000 btu/lb)							
	Low-cutpoint		Asgard	Griffin	SleipK	NWS	F3FB
Stripping		kj/m3	12472	8866	3650	4501	4471
Reboiled		kj/m3	0	0	0	0	0
Preflash		kj/m3	0	0	0	0	0
	High-cutpoint						
Stripping		kj/m3	1223	1906	588	612	708
Reboiled		kj/m3	0	0	0	0	0
Preflash		kj/m3	0	0	0	0	0
Fired duty in feed heater plus reboiler plus steam duty							
	Low-cutpoint		Asgard	Griffin	SleipK	NWS	F3FB
Stripping		kj/m3	28527	41343	17935	18740	18885
Reboiled		kj/m3	23199	47115	16127	16063	15860
Preflash		kj/m3	37306	43632	19504	20022	19348
	High-cutpoint						
Stripping		kj/m3	16666	16048	14917	14154	14063
Reboiled		kj/m3	19848	12699	13735	13078	13012
Preflash		kj/m3	24751	20096	16348	14923	15866
Energy comparison - Shifting from stripping to reboiling							
	Low-cutpoint						
Firing change		kj/m3	7,173	14,663	1,851	1,835	1,455
Steam change		kj/m3	-12,472	-8,866	-3,650	-4,501	-4,471
Breakeven steam:firing value		fraction	0.58	1.65	0.51	0.41	0.33
	High-cutpoint						
Firing change		kj/m3	4,408	-1,438	-592	-462	-342
Steam change		kj/m3	-1,223	-1,906	-588	-612	-708
Breakeven steam:firing value		fraction	3.61	-0.75	-1.01	-0.76	-0.48

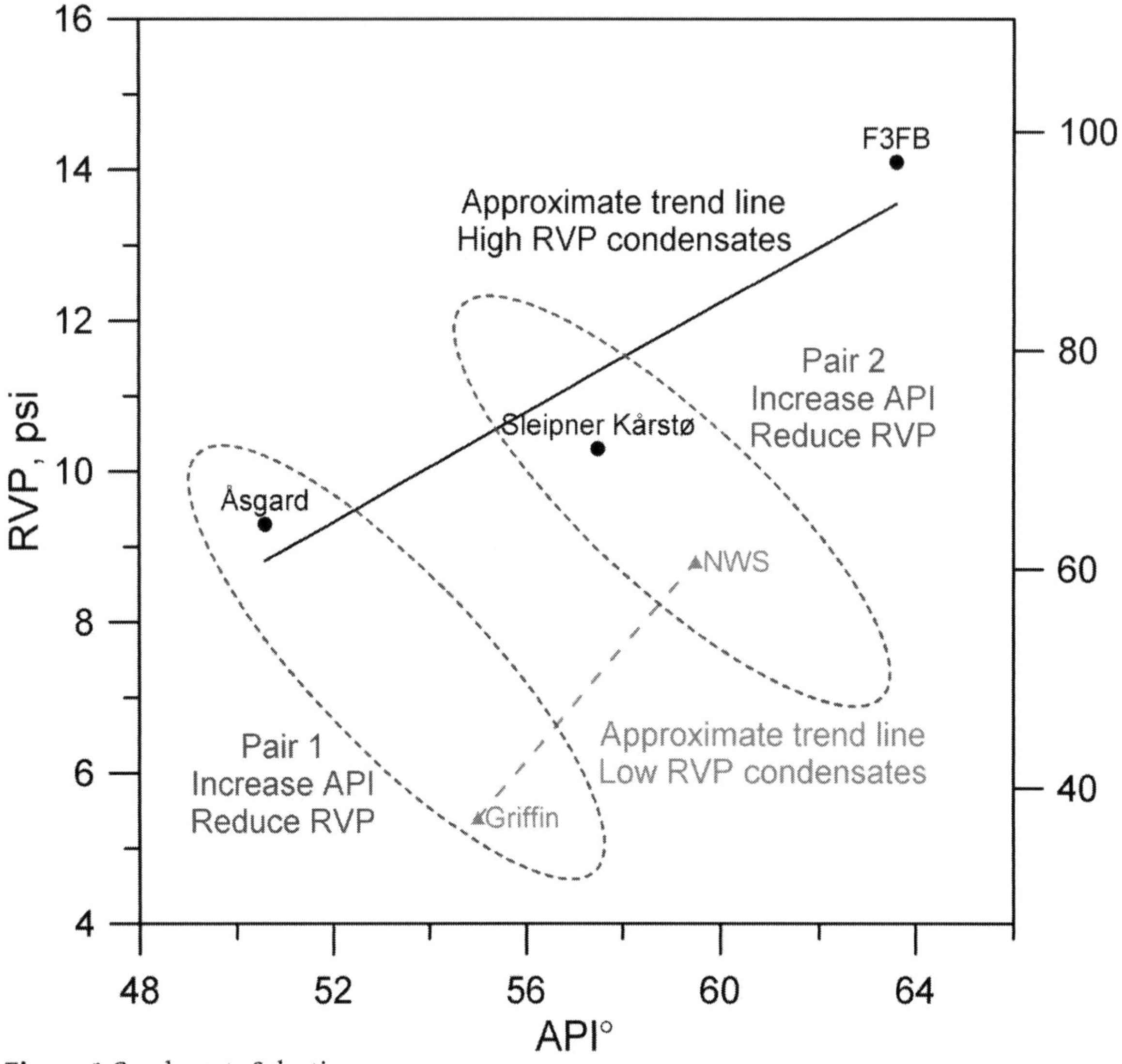

Figure 1 Condensate Selection

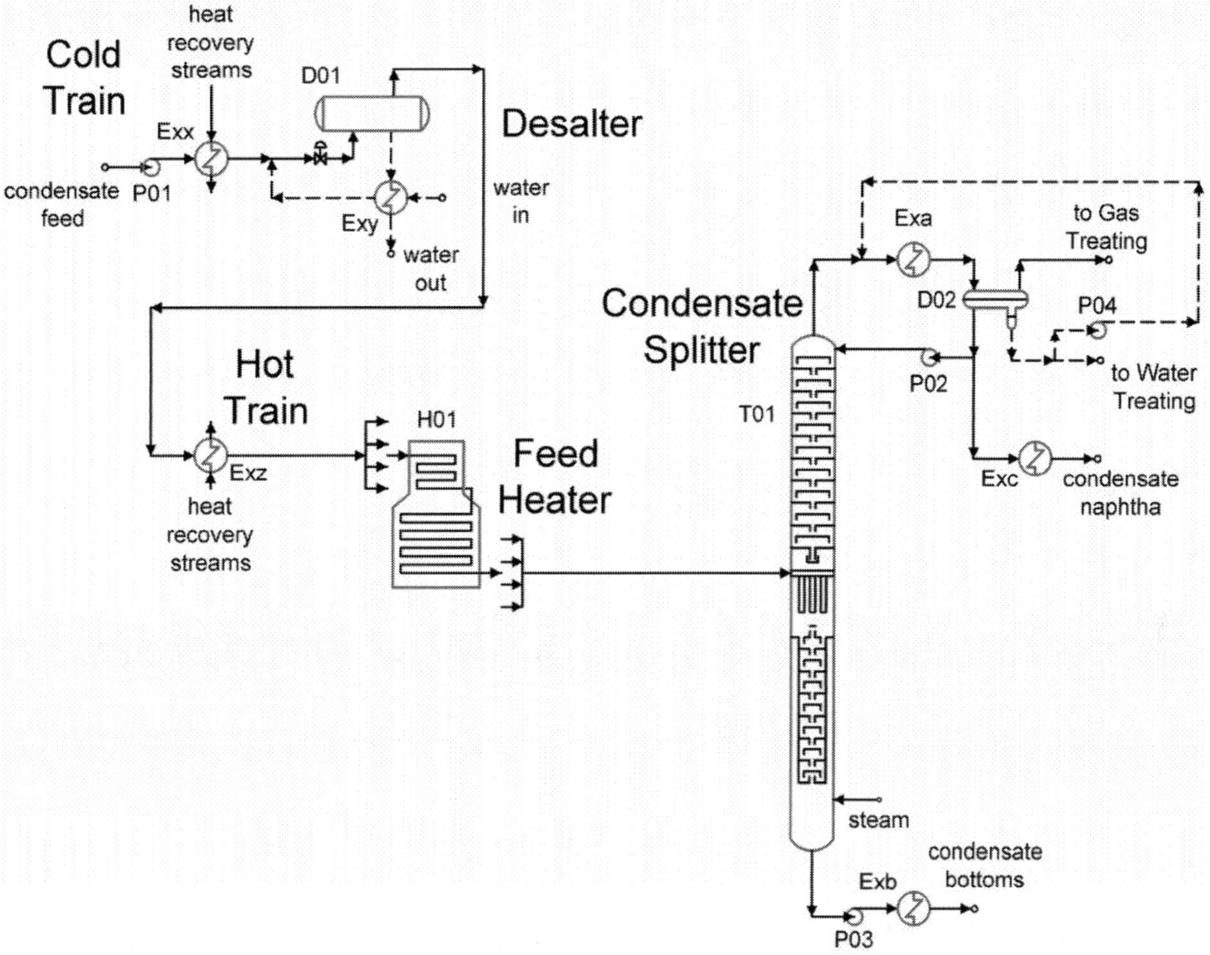

Figure 2 Steam-Stripped Condensate Splitter Schematic

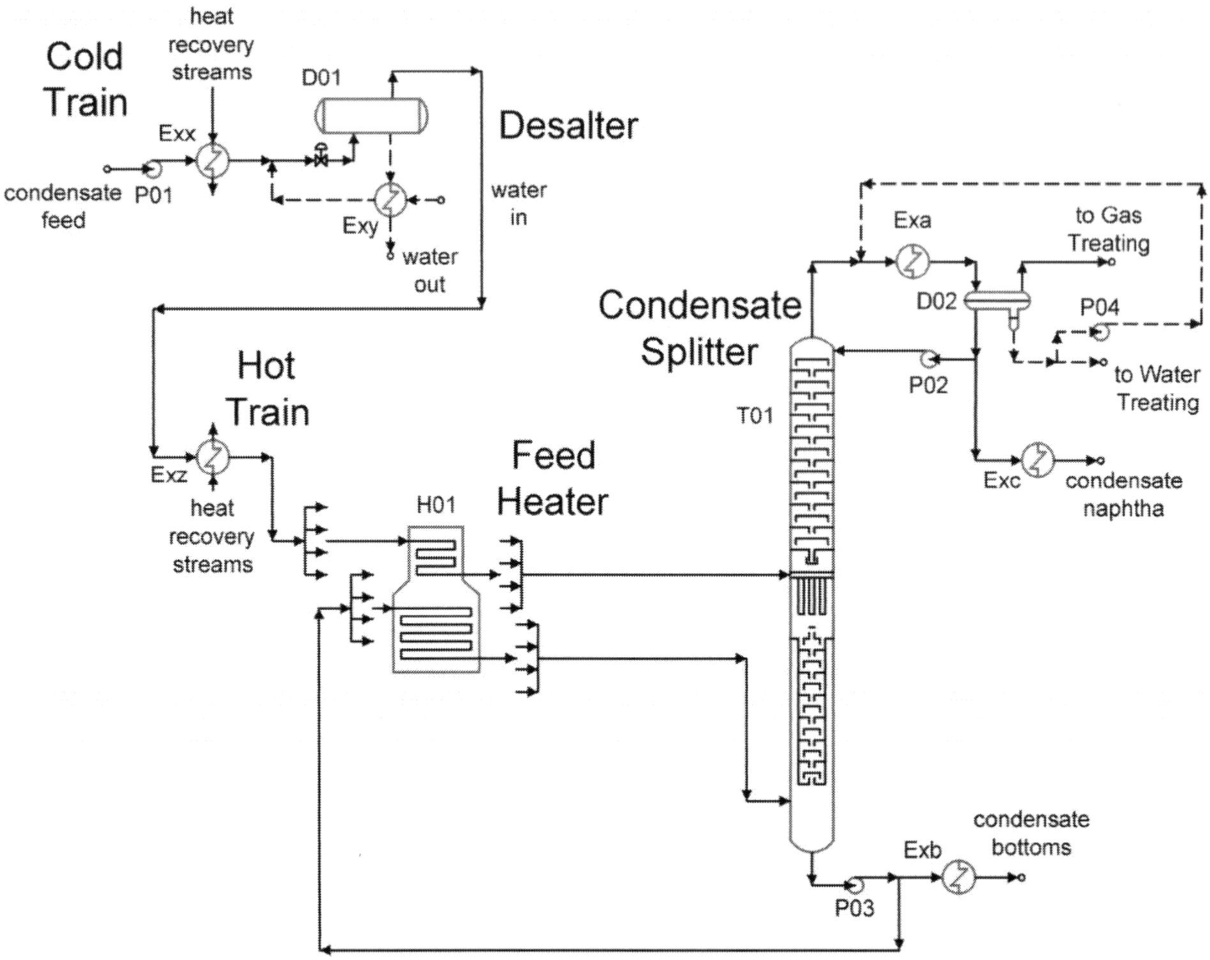

Figure 3 Reboiled Condensate Splitter Schematic

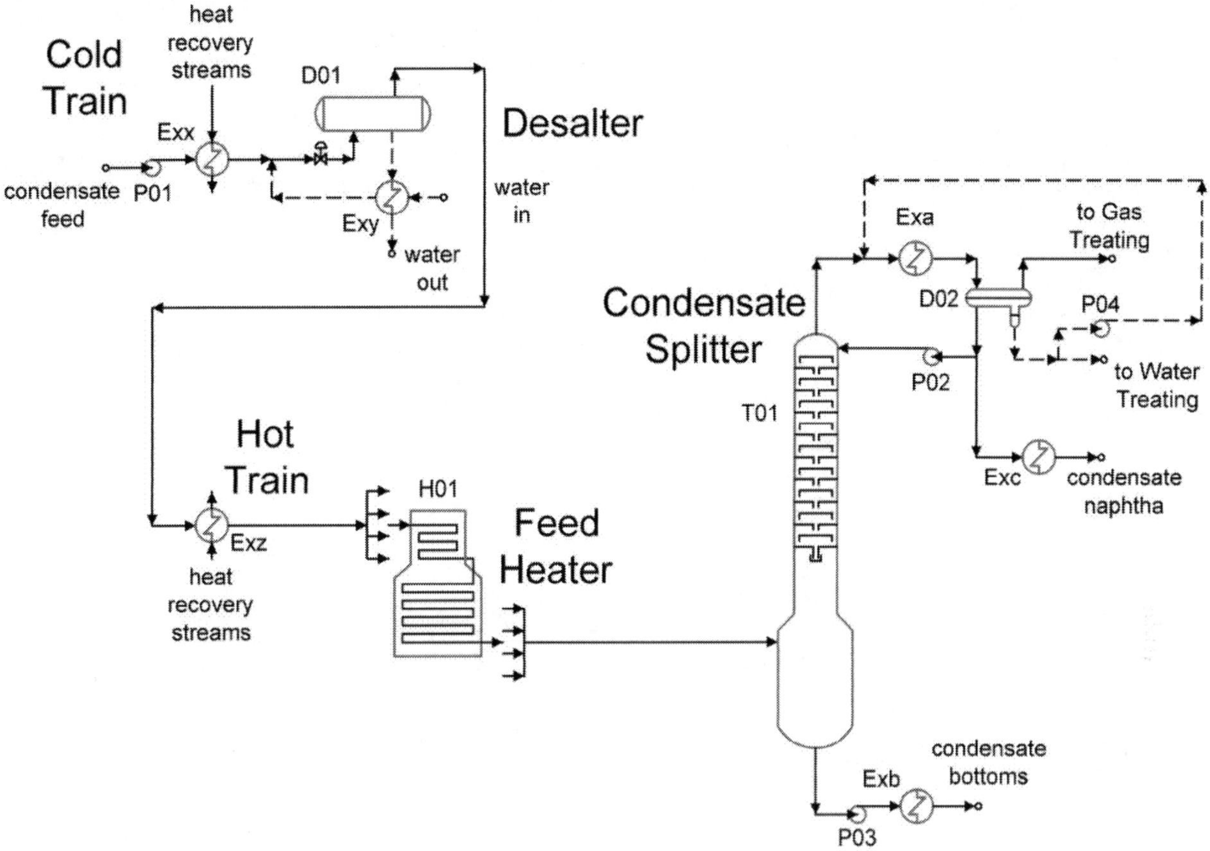

Figure 4 Preflash Condensate Splitter Schematic

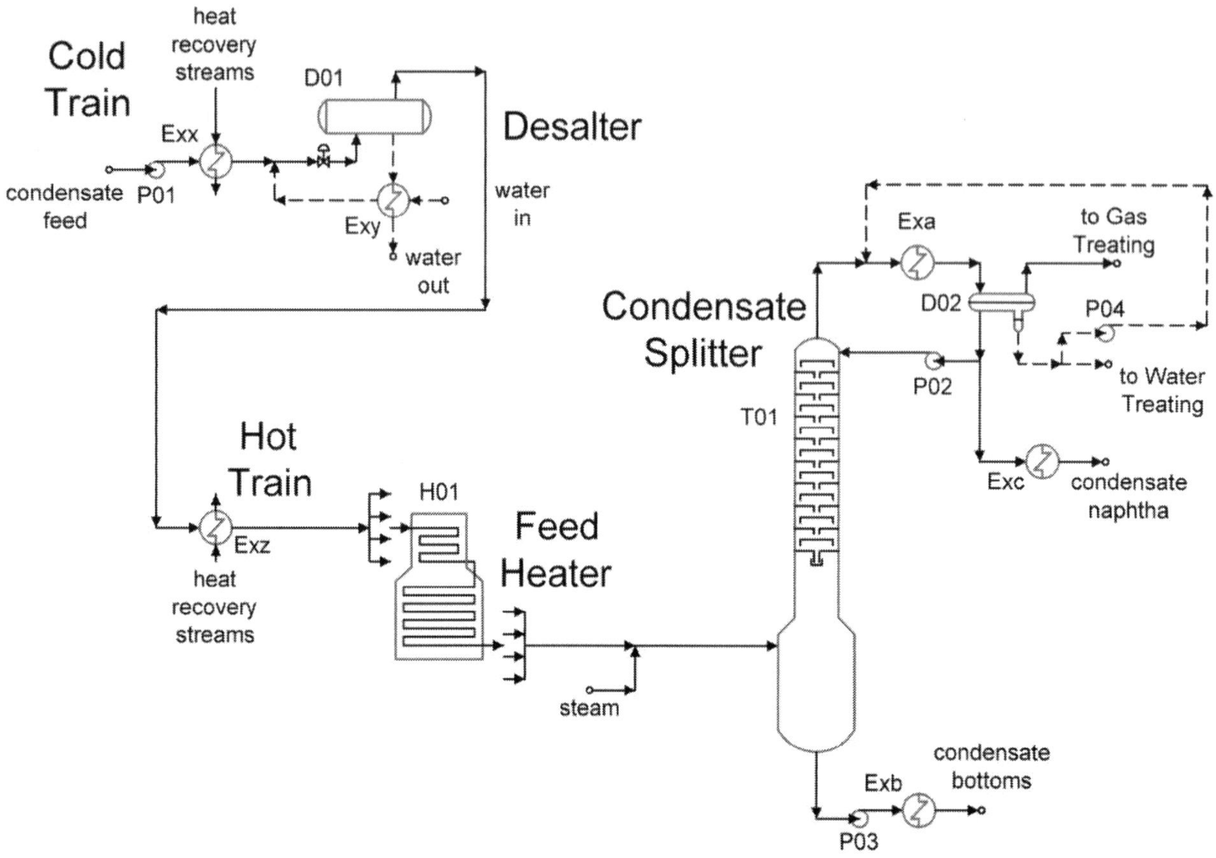

Figure 5 Preflash Condensate Splitter Schematic – Steam Injection Added

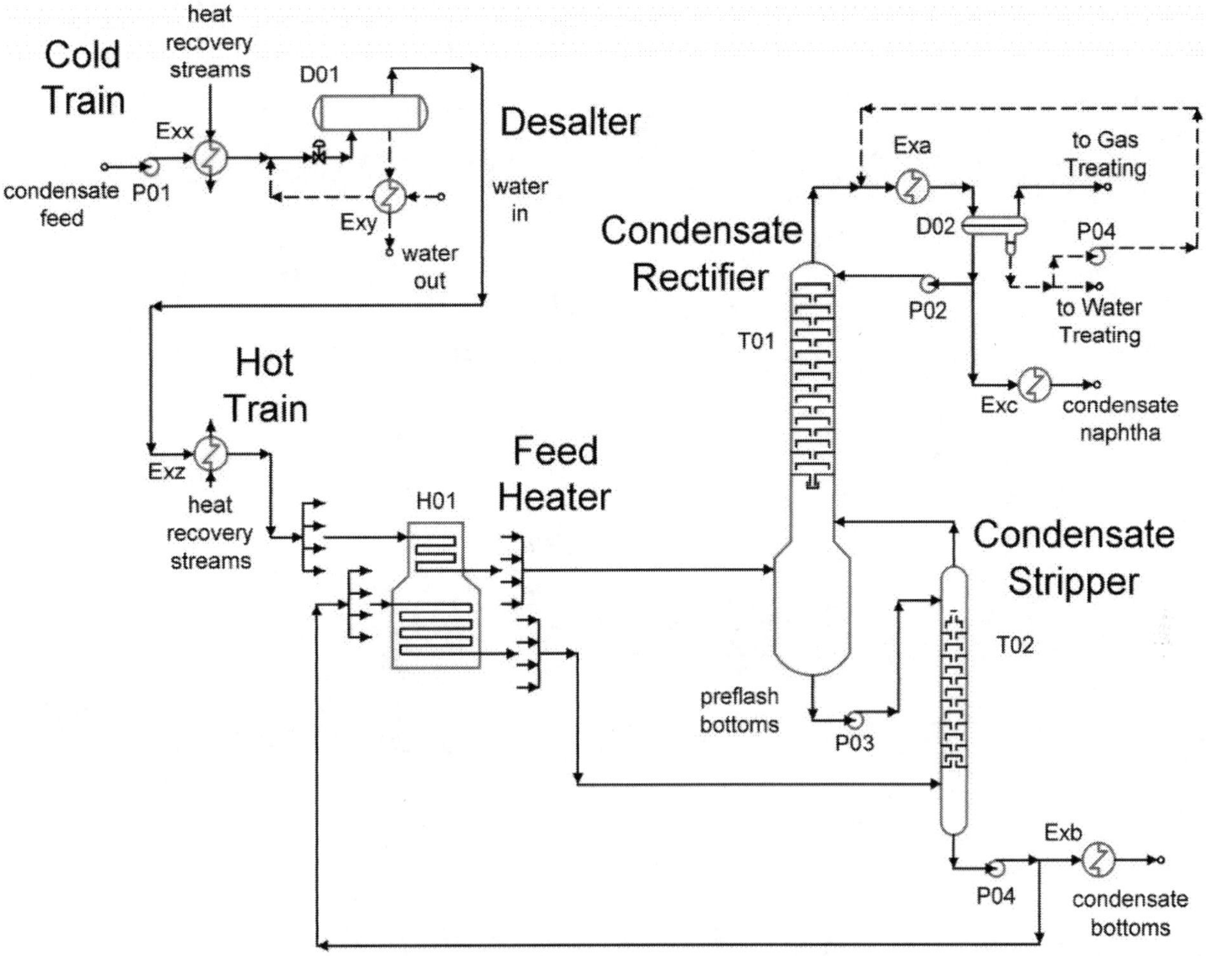

Figure 6 Reboiled Condensate Splitter Schematic – Split Tower

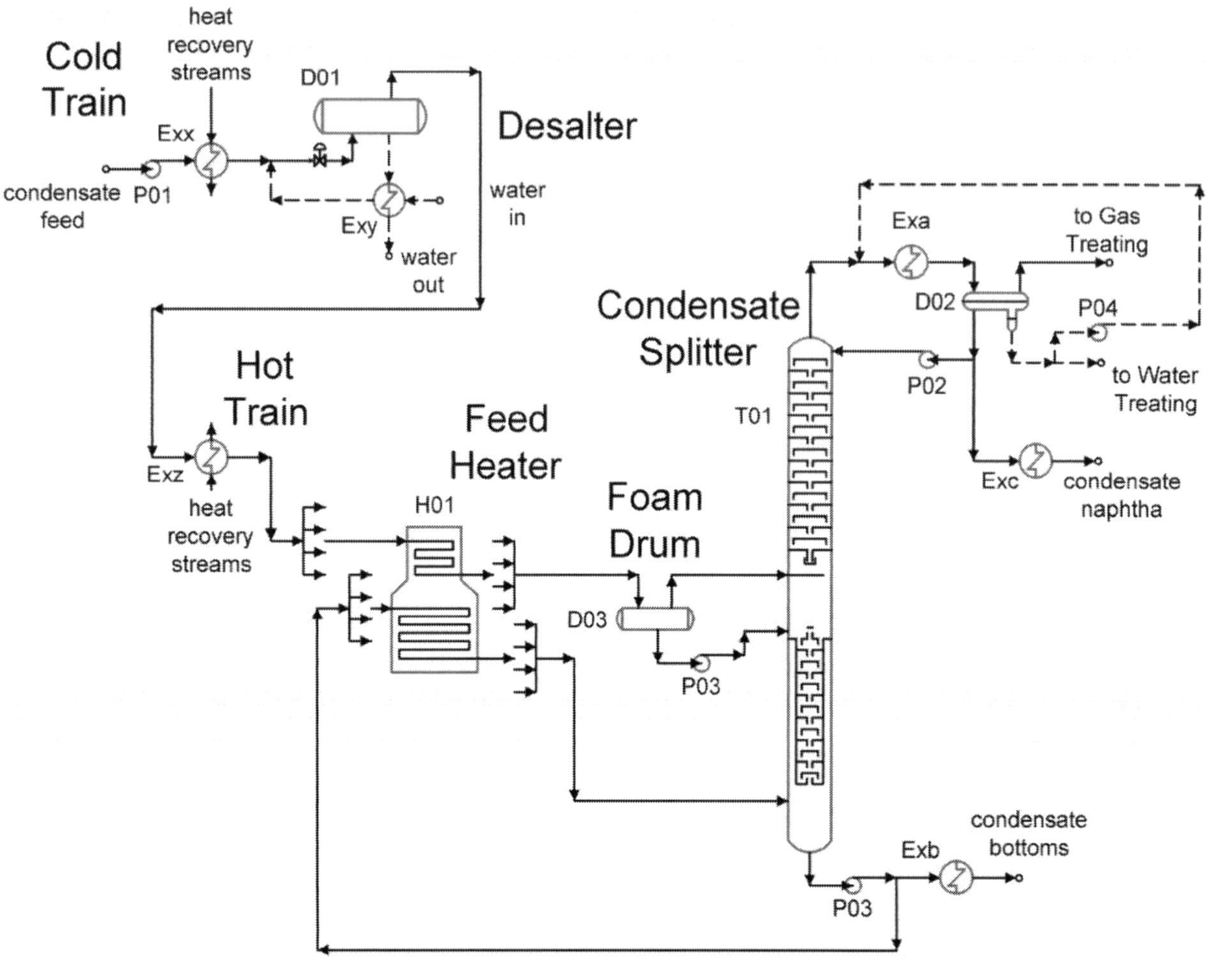

Figure 7 Reboiled Condensate Splitter Schematic – Foam Drum Added

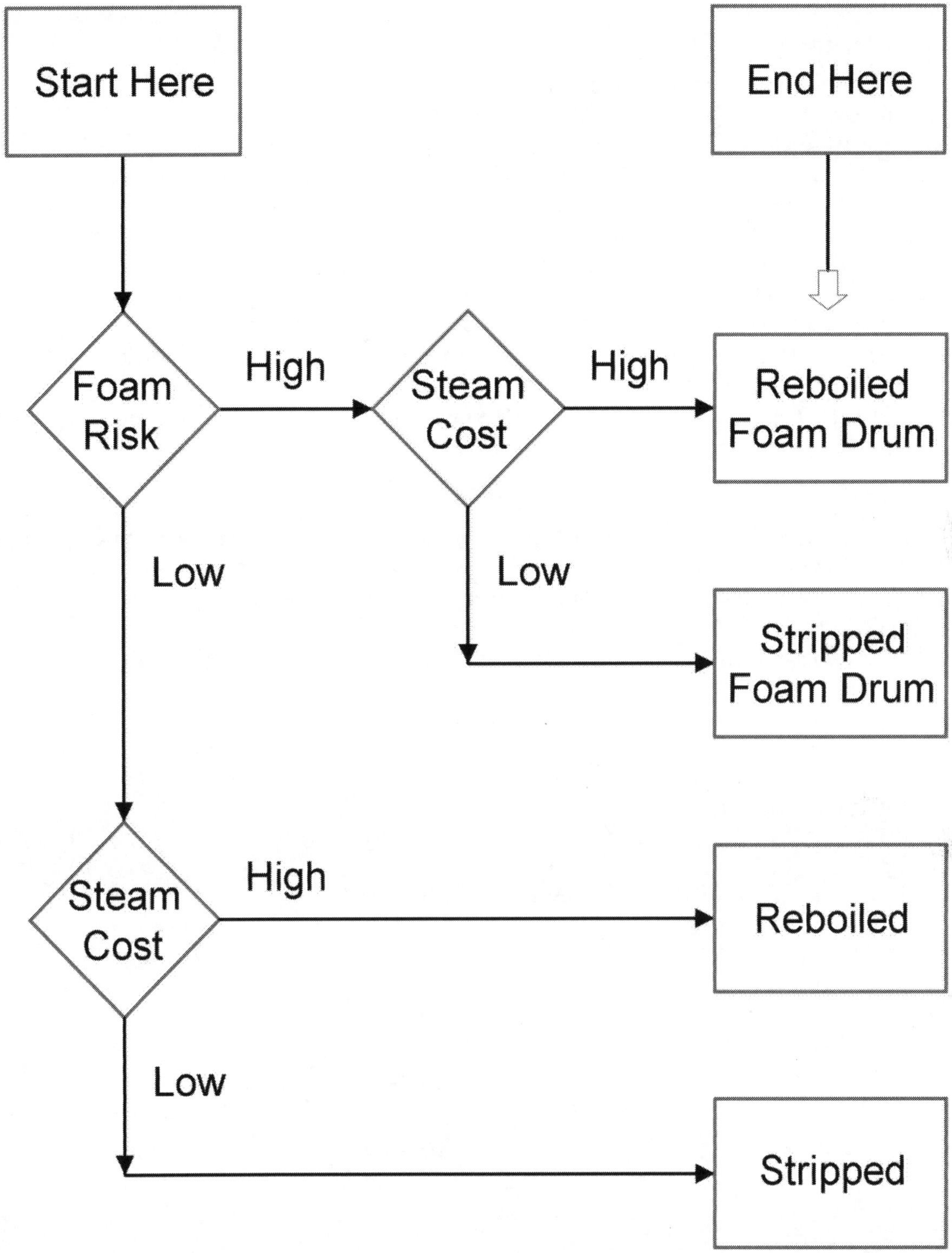

Figure 8 Selection strategy

11. References

<http://www.totsa.com/pub/crude/index2.php?expand=3&iback=3&rub=11&image=europe>, F3FB Condensate Assay. Total SA, 1999, p. CondF3FB0699.xls. (21-Jan-2015).

Asgard Norway. Total, 2007.

Griffin Crude Oil Assay (BHP Petroleum Laboratory). Mulgrave, Victoria, Australia: BHP Petroleum Trading and Marketing Pty Ltd, 1993. (16).

MADDOX, R. N., ERBAR, J. H., AND SHARIAT, A. Distillation crude stabilization boosts recovery. *Oil & Gas Journal* 77 (17): 70-76, 23-April, 1979.

MOINS, G. Crude stabilization updated - 1 - Optimization of crude stabilization units. *Oil & Gas Journal* 78 (3): 91-93, 21-Jan, 1980.

MOINS, G. Crude stabilization updated - 2 - Stabilization process comparision helps selection. *Oil & Gas Journal* 78 (4): 163-173, 28-Jan, 1980.

North West Shelf Condensate Assay (Inchcape Testing Services - Caleb Brett). Aberdeen, Scotland: BHP Petroleum Trading and Marketing Pty Ltd, 1995. (19).

Sleipner Karsto Crude Oil Assay (Product Technology and Customer Service Centre). Statoil, 1999. (17).

SLOLEY, A. W. Interpretting crude assays, AIChE Southwest Process Technology Conference. Galveston, TX, 9-10 October, 2014.

SLOLEY, A. W. Mitigate fouling in crude unit overhead - Part 1. *Hydrocarbon Processing* 94 (9): 73-81, Sept, 2013.

SLOLEY, A. W. Mitigate fouling in crude unit overhead - Part 2. *Hydrocarbon Processing* 94 (11): 73-75, Nov, 2013.

SLOLEY, A. W. Mitigate fouling in crude unit overhead - Part 3. *Hydrocarbon Processing* 94 (12): 75-79, Dec, 2013.

SUMMERS, D. R., AND SLOLEY, A. W. How to handle low-liquid loadings. *Hydrocarbon Processing* 86 (1): 67-78, Jan, 2007.

12. Appendix A Basis and Constraints

Parameter				
Streams				
Feed Rate	100,000	bpd	15,900	cmpd
Stripping steam rate;	10	lbs/bbl	0.72	kg/m^3
Reduced in increments of 1 lb/bbl of residue as required to avoid water condensation in main column (by 0.072 kg/m^3)		residue		
Products				
Light naphtha D86 95%	230	°F	110	°C
Heavy naphtha D86 95%	350	°F	177	°C
Heat Balance				
Feed temperature	70	°F	21	°C
Product temperatures	110	°F	43	°C
Desalter operating temperature	220	°F	104	°C
Desalter water mixing temperature loss	10	°F	5.6	°C
Equipment Constraints and Performance				
Heat exchanger approach temperatures	40	°F	22.2	°C
Desalter free water at outlet	0.1	Wt %		
Maximum heater outlet vaporization	90	Vol %		
Tray efficiency, rectifying	78	%		
Tray efficiency, steam stripping	25	%		
Tray efficiency, reboiled stripping	75	%		
Tray sizing, diameter increments	6	inches	15.2	cm
Tray sizing, maximum one-pass tray diameter	114	inches	2.90	m
Stripping vapor rate, minimum	0.058	$ft^3/min\text{-}ft^2$	0.018	$m^3/min\text{-}m^2$
Minimum liquid rate on weirs	0.72	gpm/in	1.07	lpm/cm
Maximum liquid rate on weirs	10	gpm/in	14.9	lpm/cm

13. Appendix B Detailed Tables

Table B-1 Condensate Comparison – Low-cutpoint, Stripping Configuration

Feed		Asgard	Griffin	SleipK	NWS	F3FB
Operation		Low Cut	Low Cut	Low Cut	Low Cut	Low Cut
Configuration		Strip	Strip	Strip	Strip	Strip
Streams						
Feed gravity	API	50.5	54.8	57.4	59.3	63.4
Bottoms gravity	API	43.9	52.3	41.8	47.2	52.2
Gas Rate	BBL/DAY	63	65	66	66	61
Nap Rate	BBL/DAY	20700	8546	44803	41223	43758
Btms Rate	BBL/DAY	79237	91389	55132	58707	56181
Steam Rate	#/HR	33402	11451	23107	24634	23549
Naphtha D86 95	F	230	230	230	230	230
Naphtha-Bottoms Gap	F	-22.4	-39.8	-24.0	-30.4	-19.6
Naphtha Cutpoint	F	175	131	212	198	205
Naphtha in Bottoms	vol %	17.2	23.4	20.9	24.0	21.0
Naphtha Recovery	%	64.4	29.4	87.3	80.8	85.9
Heat Balance						
Fired, Feed	Mbtu/hr	50.0	41.8	96.4	88.4	95.0
Fired, Reboiler	Mbtu/hr					
Cold Train Condenser	Mbtu/hr	71.4	31.8	81.6	82.1	83.1
Cold Train Rundown	Mbtu/hr	20.4	66.0	10.6	8.9	6.4
Cold Train Steam	Mbtu/hr	5.6		1.6	3.4	6.0
Cold Train Naphtha Rundown	Mbtu/hr					
Cold Train Duty Total	Mbtu/hr	97.4	97.8	93.8	94.4	95.5
Hot Train Condenser	Mbtu/hr					
Hot Train Rundown	Mbtu/hr					
Hot Train Duty Total	Mbtu/hr					
Condenser	Mbtu/hr	74.4	31.8	125.5	115.3	130.5
Condenser rejected	Mbtu/hr	3.0		43.9	33.2	47.4
Rundown rejected	Mbtu/hr	81.5	25.1	44.2	48.9	57.0
Light Naphtha Cooling	Mbtu/hr	0.6	0.4	8.2	9.2	1.4
Total Cooling Utility	Mbtu/hr	85.1	25.5	96.3	91.3	105.8
Above Pinch Lost	Mbtu/hr		3.7		1.9	
Total Fired		50.0	41.8	96.4	88.4	95.0
Total Fired + Steam Fired		89.0	53.3	121.1	116.4	124.5
Feed Temperature	F	276	277	299	295	298
Bottoms Temperature	F	266	266	268	265	267
Mechanical						
Stages Rectifying		14	14	14	14	14
Stages Stripping		3	3	3	3	3
Trays Rectifying		18	18	18	18	18
Trays Stripping		12	12	12	12	12
Diameter Rectifying	In	102	72	144	138	138
Diameter Stripping	In	120	108	102	108	108

Table B-1 Condensate Comparison – High-cutpoint, Stripping Configuration

Feed		Asgard	Griffin	SleipK	NWS	F3FB
Operation		High Cut	High Cut	High Cut	High Cut	High Cut
Configuration		Strip	Strip	Strip	Strip	Strip
Streams						
Feed gravity	API	50.5	54.8	57.4	59.3	63.4
Bottoms gravity	API	34.8	39.7	24.1	36.2	42.2
Gas Rate	BBL/DAY	58	56	62	61	60
Nap Rate	BBL/DAY	56294	59133	81786	81784	79547
Btms Rate	BBL/DAY	43645	40809	18151	18154	20394
Steam Rate	#/HR	10401	17030	7273	7567	8506
Naphtha D86 95	F	350	350	230	350	350
Naphtha-Bottoms Gap	F	7.5	-21.4	-3.8	-30.8	-13.3
Naphtha Cutpoint	F	364	349	406	389	382
Naphtha in Bottoms	vol %	8.0	15.0	8.3	12.7	11.1
Naphtha Recovery	%	103.2	99.6	107.4	106.4	106.0
Heat Balance						
Fired, Feed	Mbtu/hr	131.0	126.0	176.6	166.9	160.1
Fired, Reboiler	Mbtu/hr					
Cold Train Condenser	Mbtu/hr	97.4	97.8	93.7	94.4	95.5
Cold Train Rundown	Mbtu/hr					
Cold Train Steam	Mbtu/hr					
Cold Train Naphtha Rundown	Mbtu/hr					
Cold Train Duty Total	Mbtu/hr	97.4	97.8	93.7	94.4	95.5
Hot Train Condenser	Mbtu/hr	62.8	74.2	65.0	65.4	60.8
Hot Train Rundown	Mbtu/hr	15.7	4.8	7.0		4.5
Hot Train Duty Total	Mbtu/hr	78.5	79.0	72.0	65.4	65.3
Condenser	Mbtu/hr	190.0	181.3	256.8	239.4	245.0
Condenser rejected	Mbtu/hr	29.8	9.3	98.1	79.6	88.7
Rundown rejected	Mbtu/hr	72.6	72.3	29.7	33.4	33.6
Light Naphtha Cooling	Mbtu/hr	19.3	32.9	29.4	33.9	18.6
Total Cooling Utility	Mbtu/hr	121.7	114.5	157.2	146.9	140.9
Above Pinch Lost	Mbtu/hr		4.8		2.9	
Total Fired		131.0	126.0	176.6	166.9	160.1
Total Fired + Steam Fired		141.4	143.0	183.9	174.5	168.6
Feed Temperature	F	430	425	445	418	415
Bottoms Temperature	F	399	384	410	381	381
Mechanical						
Stages Rectifying		14	14	14	14	14
Stages Stripping		3	3	3	3	3
Trays Rectifying		18	18	18	18	18
Trays Stripping		12	12	12	12	12
Diameter Rectifying	In	162	162	192	186	186
Diameter Stripping	In	90	96	60	60	60

TableB-1 Condensate Comparison – Low-cutpoint, Reboiler Configuration

Feed		Asgard	Griffin	SleipK	NWS	F3FB
Operation		Low Cut	Low Cut	Low Cut	Low Cut	Low Cut
Configuration		Reboil	Reboil	Reboil	Reboil	Reboil
Streams						
Feed gravity	API	50.5	54.8	57.4	59.3	63.4
Bottoms gravity	API	44.2	52.2	41.7	47.2	52.4
Gas Rate	BBL/DAY	62	65	66	66	61
Nap Rate	BBL/DAY	20706	9238	44803	41223	43759
Btms Rate	BBL/DAY	79232	90696	55132	58711	56180
Steam Rate	#/HR					
Naphtha D86 95	F	230	230	230	230	230
Naphtha-Bottoms Gap	F	-14.0	-26.2	-12.6	-22.9	-10.8
Naphtha Cutpoint	F	175	139	212	198	205
Naphtha in Bottoms	vol %	17.0	22.9	20.7	23.9	20.9
Naphtha Recovery	%	64.4	31.8	87.3	80.8	85.9
Heat Balance						
Fired, Feed	Mbtu/hr	7.3		63.4	55.4	58.7
Fired, Reboiler	Mbtu/hr	65.1	65.6	45.5	44.4	45.9
Cold Train Condenser	Mbtu/hr	52.1	24.1	86.9	88.4	89.0
Cold Train Rundown	Mbtu/hr	45.3	73.7	6.8	6.0	6.5
Cold Train Steam	Mbtu/hr					
Cold Train Naphtha Rundown	Mbtu/hr					
Cold Train Duty Total	Mbtu/hr	97.4	97.8	93.7	94.4	95.5
Hot Train Condenser	Mbtu/hr					
Hot Train Rundown	Mbtu/hr	53.2	41.7	36.5	37.1	39.6
Hot Train Duty Total	Mbtu/hr	53.2	41.7	36.5	37.1	39.6
Condenser	Mbtu/hr	52.1	24.1	106.8	96.5	111.0
Condenser rejected	Mbtu/hr			19.9	8.1	22.0
Rundown rejected	Mbtu/hr	43.6	36.5	54.1	55.1	53.4
Light Naphtha Cooling	Mbtu/hr	0.6	0.2	8.0	8.9	1.3
Total Cooling Utility	Mbtu/hr	44.2	36.7	82.0	72.1	76.7
Above Pinch Lost	Mbtu/hr	7.4	13.6			
Total Fired		72.4	65.6	108.9	99.8	104.6
Total Fired + Steam Fired		72.4	65.6	108.9	99.8	104.6
Feed Temperature	F	287	277	302	298	299
Bottoms Temperature	F	376	362	380	366	376
Mechanical						
Stages Rectifying		14	14	14	14	14
Stages Stripping		9	9	9	9	9
Trays Rectifying		30	30	30	18	18
Trays Stripping		12	12	12	12	12
Diameter Rectifying	In	90	72	138	132	138
Diameter Stripping	In	150	150	126	132	132

Table B-1 Condensate Comparison – High-cutpoint, Reboiler Configuration

Feed Operation Configuration		Asgard High Cut Reboil	Griffin High Cut Reboil	SleipK High Cut Reboil	NWS High Cut Reboil	F3FB High Cut Reboil
Streams						
Feed gravity	API	50.5	54.8	57.4	59.3	63.4
Bottoms gravity	API	34.8	36.4	24.1	36.1	42.1
Gas Rate	BBL/DAY	58	54	62	61	60
Nap Rate	BBL/DAY	56293	66980	81785	81783	79547
Btms Rate	BBL/DAY	43648	32966	18153	18155	20394
Steam Rate	#/HR					
Naphtha D86 95	F	350	350	350	350	350
Naphtha-Bottoms Gap	F	-5.6	8.4	19.2	-6.8	9.1
Naphtha Cutpoint	F	364	395	406	389	385
Naphtha in Bottoms	vol %	8.1		8.1	12.5	10.9
Naphtha Recovery	%	103.2	112.8	107.4	106.3	106.0
Heat Balance						
Fired, Feed	Mbtu/hr	128.8		139.6	136.2	125.5
Fired, Reboiler	Mbtu/hr	39.6	128.2	29.7	25.0	30.5
Cold Train Condenser	Mbtu/hr	97.4	95.8	93.7	94.4	95.5
Cold Train Rundown	Mbtu/hr					
Cold Train Steam	Mbtu/hr					
Cold Train Naphtha Rundown	Mbtu/hr		2.0			
Cold Train Duty Total	Mbtu/hr	97.4	97.8	93.7	94.4	95.5
Hot Train Condenser	Mbtu/hr	61.2	84.7	67.6	67.7	63.2
Hot Train Rundown	Mbtu/hr	77.5	70.7	35.0	27.1	33.8
Hot Train Duty Total	Mbtu/hr	138.7	155.4	102.6	94.8	97.0
Condenser	Mbtu/hr	204.2	180.5	243.8	230.5	233.8
Condenser rejected	Mbtu/hr	45.6		82.5	68.4	75.1
Rundown rejected	Mbtu/hr	74.7	59.4	30.1	30.8	34.0
Light Naphtha Cooling	Mbtu/hr	19.4	37.5	29.4	33.9	18.6
Total Cooling Utility	Mbtu/hr	139.7	96.9	142.0	133.1	127.7
Above Pinch Lost	Mbtu/hr					
Total Fired		168.4	128.2	169.3	161.2	156.0
Total Fired + Steam Fired		168.4	128.2	169.3	161.2	156.0
Feed Temperature	F	490	383	438	416	412
Bottoms Temperature	F	575	629	600	549	561
Mechanical						
Stages Rectifying		14	14	14	14	14
Stages Stripping		9	9	9	9	9
Trays Rectifying		30	30	30	18	18
Trays Stripping		12	12	12	12	12
Diameter Rectifying	In	198	156	186	186	180
Diameter Stripping	In	90	210	84	90	96

Table B-1 Condensate Comparison – Low-cutpoint, Preflash Configuration

Feed Operation Configuration		Asgard Low Cut Preflash	Griffin Low Cut Preflash	SleipK Low Cut Preflash	NWS Low Cut Preflash	F3FB Low Cut Preflash
Streams						
Feed gravity	API	50.5	54.8	57.4	59.3	63.4
Bottoms gravity	API	44.7	52.6	42.9	48.4	53.1
Gas Rate	BBL/DAY	62	65	66	65	61
Nap Rate	BBL/DAY	20702	8546	44803	41223	43758
Btms Rate	BBL/DAY	79236	91390	55133	58711	56181
Steam Rate	#/HR					
Naphtha D86 95	F	230	230	230	230	230
Naphtha-Bottoms Gap	F	-58.1	-27.7	-72.9	-66.1	-51.2
Naphtha Cutpoint	F	176	131	212	198	205
Naphtha in Bottoms	vol %	17.6	23.5	21.7	24.4	21.5
Naphtha Recovery	%	64.4	29.4	87.3	80.8	85.9
Heat Balance						
Fired, Feed	Mbtu/hr	116.4	56.2	131.7	124.4	127.6
Fired, Reboiler	Mbtu/hr					
Cold Train Condenser	Mbtu/hr	63.9	26.6	90.0	91.2	91.0
Cold Train Rundown	Mbtu/hr	33.5	71.2	3.7	3.2	4.5
Cold Train Steam	Mbtu/hr					
Cold Train Naphtha Rundown	Mbtu/hr					
Cold Train Duty Total	Mbtu/hr	97.4	97.8	93.7	94.4	95.5
Hot Train Condenser	Mbtu/hr					
Hot Train Rundown	Mbtu/hr		31.7	29.8	25.5	25.4
Hot Train Duty Total	Mbtu/hr		31.7	29.8	25.5	25.4
Condenser	Mbtu/hr	63.9	26.7	132.7	120.8	134.2
Condenser rejected	Mbtu/hr			42.7	29.6	43.2
Rundown rejected	Mbtu/hr	86.4	26.1	48.9	56.8	54.2
Light Naphtha Cooling	Mbtu/hr	1.5	1.1	9.0	10.2	2.1
Total Cooling Utility	Mbtu/hr	87.9	27.2	100.6	96.6	99.5
Above Pinch Lost	Mbtu/hr	5.0				
Total Fired		116.4	56.2	131.7	124.4	127.6
Total Fired + Steam Fired		116.4	56.2	131.7	124.4	127.6
Feed Temperature	F	341	327	350	342	344
Bottoms Temperature	F					
Mechanical						
Stages Rectifying		14	14	14	14	14
Stages Stripping						
Trays Rectifying		18	18	18	18	18
Trays Stripping						
Diameter Rectifying	In	108	72	162	156	162
Diameter Stripping	In					

Table B-1 Condensate Comparison – High-cutpoint, Preflash Configuration

Feed		Asgard	Griffin	SleipK	NWS	F3FB
Operation		High Cut	High Cut	High Cut	High Cut	High Cut
Configuration		Preflash	Preflash	Preflash	Preflash	Preflash
Streams						
Feed gravity	API	50.5	54.8	57.4	59.3	63.4
Bottoms gravity	API	35.1	40.3	27.0	38.5	43.7
Gas Rate	BBL/DAY	58	55	62	61	60
Nap Rate	BBL/DAY	56293	59133	79913	79719	78077
Btms Rate	BBL/DAY	43648	40812	20026	20219	21863
Steam Rate	#/HR					
Naphtha D86 95	F	350	350	350	350	350
Naphtha-Bottoms Gap	F	-75.0	-77.2	-120.0	-126.7	-123.0
Naphtha Cutpoint	F	364	349	384	369	374
Naphtha in Bottoms	vol %	8.6	16.0	15.8	20.4	16.6
Naphtha Recovery	%	103.2	99.6	104.9	103.7	104.1
Heat Balance						
Fired, Feed	Mbtu/hr	210.0	179.1	196.9	179.3	186.7
Fired, Reboiler	Mbtu/hr					
Cold Train Condenser	Mbtu/hr	97.4	97.8	93.7	94.4	95.5
Cold Train Rundown	Mbtu/hr					
Cold Train Steam	Mbtu/hr					
Cold Train Naphtha Rundown	Mbtu/hr					
Cold Train Duty Total	Mbtu/hr	97.4	97.8	93.7	94.4	95.5
Hot Train Condenser	Mbtu/hr	62.1	74.8	65.2	65.8	61.5
Hot Train Rundown	Mbtu/hr	60.7	36.5	16.8	12.1	16.2
Hot Train Duty Total	Mbtu/hr	122.8	111.3	82.0	77.9	77.7
Condenser	Mbtu/hr	245.9	214.1	267.3	244.8	261.4
Condenser rejected	Mbtu/hr	86.4	41.5	108.4	84.6	104.4
Rundown rejected	Mbtu/hr	75.1	74.9	32.7	33.7	35.9
Light Naphtha Cooling	Mbtu/hr	19.6	33.4	28.5	32.8	17.9
Total Cooling Utility	Mbtu/hr	181.1	149.8	169.6	151.1	158.2
Above Pinch Lost	Mbtu/hr					
Total Fired		210.0	179.1	196.9	179.3	186.7
Total Fired + Steam Fired		210.0	179.1	196.9	179.3	186.7
Feed Temperature	F	561	506	482	445	461
Bottoms Temperature	F					
Mechanical						
Stages Rectifying		14	14	18	14	14
Stages Stripping						
Trays Rectifying		18	18	18	18	18
Trays Stripping						
Diameter Rectifying	In	222	192	204	192	204
Diameter Stripping	In					

Table B-2 Cutpoint and Configuration Comparison – Åsgard Condensate

Feed Operation Configuration		Asgard Low Cut Strip	Asgard Low Cut Reboil	Asgard Low Cut Preflash	Asgard High Cut Strip	Asgard High Cut Reboil	Asgard High Cut Preflash
Streams							
Feed gravity	API	50.5	50.5	50.5	50.5	50.5	50.5
Bottoms gravity	API	43.9	44.2	44.7	34.8	34.8	35.1
Gas Rate	BBL/DAY	63	62	62	58	58	58
Nap Rate	BBL/DAY	20700	20706	20702	56294	56293	56293
Btms Rate	BBL/DAY	79237	79232	79236	43645	43648	43648
Steam Rate	#/HR	33402			10401		
Naphtha D86 95	F	230	230	230	350	350	350
Naphtha-Bottoms Gap	F	-22.4	-14.0	-58.1	7.5	-5.6	-75.0
Naphtha Cutpoint	F	175	175	176	364	364	364
Naphtha in Bottoms	vol %	17.2	17.0	17.6	8.0	8.1	8.6
Naphtha Recovery	%	64.4	64.4	64.4	103.2	103.2	103.2
Heat Balance							
Fired, Feed	Mbtu/hr	50.0	7.3	116.4	131.0	128.8	210.0
Fired, Reboiler	Mbtu/hr		65.1			39.6	
Cold Train Condenser	Mbtu/hr	71.4	52.1	63.9	97.4	97.4	97.4
Cold Train Rundown	Mbtu/hr	20.4	45.3	33.5			
Cold Train Steam	Mbtu/hr	5.6					
Cold Train Naphtha Rundown	Mbtu/hr						
Cold Train Duty Total	Mbtu/hr	97.4	97.4	97.4	97.4	97.4	97.4
Hot Train Condenser	Mbtu/hr				62.8	61.2	62.1
Hot Train Rundown	Mbtu/hr		53.2		15.7	77.5	60.7
Hot Train Duty Total	Mbtu/hr		53.2		78.5	138.7	122.8
Condenser	Mbtu/hr	74.4	52.1	63.9	190.0	204.2	245.9
Condenser rejected	Mbtu/hr	3.0			29.8	45.6	86.4
Rundown rejected	Mbtu/hr	81.5	43.6	86.4	72.6	74.7	75.1
Light Naphtha Cooling	Mbtu/hr	0.6	0.6	1.5	19.3	19.4	19.6
Total Cooling Utility	Mbtu/hr	85.1	44.2	87.9	121.7	139.7	181.1
Above Pinch Lost	Mbtu/hr		7.4	5.0			
Total Fired		50.0	72.4	116.4	131.0	168.4	210.0
Total Fired + Steam Fired		89.0	72.4	116.4	141.4	168.4	210.0
Feed Temperature	F	276	287	341	430	490	561
Bottoms Temperature	F	266	376		399	575	
Mechanical							
Stages Rectifying		14	14	14	14	14	14
Stages Stripping		3	9		3	9	
Trays Rectifying		18	30	18	18	30	18
Trays Stripping		12	12		12	12	
Diameter Rectifying	In	102	90	108	162	198	222
Diameter Stripping	In	120	150		90	90	

Table B-2 Cutpoint and Configuration Comparison – Griffin Condensate

Feed		Griffin	Griffin	Griffin	Griffin	Griffin	Griffin
Operation		Low Cut	Low Cut	Low Cut	High Cut	High Cut	High Cut
Configuration		Strip	Reboil	Preflash	Strip	Reboil	Preflash
Streams							
Feed gravity	API	54.8	54.8	54.8	54.8	54.8	54.8
Bottoms gravity	API	52.3	52.2	52.6	39.7	36.4	40.3
Gas Rate	BBL/DAY	65	65	65	56	54	55
Nap Rate	BBL/DAY	8546	9238	8546	59133	66980	59133
Btms Rate	BBL/DAY	91389	90696	91390	40809	32966	40812
Steam Rate	#/HR	11451			17030		
Naphtha D86 95	F	230	230	230	350	350	350
Naphtha-Bottoms Gap	F	-39.8	-26.2	-27.7	-21.4	8.4	-77.2
Naphtha Cutpoint	F	131	139	131	349	395	349
Naphtha in Bottoms	vol %	23.4	22.9	23.5	15.0		16.0
Naphtha Recovery	%	29.4	31.8	29.4	99.6	112.8	99.6
Heat Balance							
Fired, Feed	Mbtu/hr	41.8		56.2	126.0		179.1
Fired, Reboiler	Mbtu/hr		65.6			128.2	
Cold Train Condenser	Mbtu/hr	31.8	24.1	26.6	97.8	95.8	97.8
Cold Train Rundown	Mbtu/hr	66.0	73.7	71.2			
Cold Train Steam	Mbtu/hr						
Cold Train Naphtha Rundown	Mbtu/hr					2.0	
Cold Train Duty Total	Mbtu/hr	97.8	97.8	97.8	97.8	97.8	97.8
Hot Train Condenser	Mbtu/hr				74.2	84.7	74.8
Hot Train Rundown	Mbtu/hr		41.7	31.7	4.8	70.7	36.5
Hot Train Duty Total	Mbtu/hr		41.7	31.7	79.0	155.4	111.3
Condenser	Mbtu/hr	31.8	24.1	26.7	181.3	180.5	214.1
Condenser rejected	Mbtu/hr				9.3		41.5
Rundown rejected	Mbtu/hr	25.1	36.5	26.1	72.3	59.4	74.9
Light Naphtha Cooling	Mbtu/hr	0.4	0.2	1.1	32.9	37.5	33.4
Total Cooling Utility	Mbtu/hr	25.5	36.7	27.2	114.5	96.9	149.8
Above Pinch Lost	Mbtu/hr	3.7	13.6		4.8		
Total Fired		41.8	65.6	56.2	126.0	128.2	179.1
Total Fired + Steam Fired		53.3	65.6	56.2	143.0	128.2	179.1
Feed Temperature	F	277	277	327	425	383	506
Bottoms Temperature	F	266	362		384	629	
Mechanical							
Stages Rectifying		14	14	14	14	14	14
Stages Stripping		3	9		3	9	
Trays Rectifying		18	30	18	18	30	18
Trays Stripping		12	12		12	12	
Diameter Rectifying	In	72	72	72	162	156	192
Diameter Stripping	In	108	150		96	210	

Table B-2 Cutpoint and Configuration Comparison – Sleipner Kårstø Condensate

Feed		SleipK	SleipK	SleipK	SleipK	SleipK	SleipK
Operation		Low Cut	Low Cut	Low Cut	High Cut	High Cut	High Cut
Configuration		Strip	Reboil	Preflash	Strip	Reboil	Preflash
Streams							
Feed gravity	API	57.4	57.4	57.4	57.4	57.4	57.4
Bottoms gravity	API	41.8	41.7	42.9	24.1	24.1	27.0
Gas Rate	BBL/DAY	66	66	66	62	62	62
Nap Rate	BBL/DAY	44803	44803	44803	81786	81785	79913
Btms Rate	BBL/DAY	55132	55132	55133	18151	18153	20026
Steam Rate	#/HR	23107			7273		
Naphtha D86 95	F	230	230	230	230	350	350
Naphtha-Bottoms Gap	F	-24.0	-12.6	-72.9	-3.8	19.2	-120.0
Naphtha Cutpoint	F	212	212	212	406	406	384
Naphtha in Bottoms	vol %	20.9	20.7	21.7	8.3	8.1	15.8
Naphtha Recovery	%	87.3	87.3	87.3	107.4	107.4	104.9
Heat Balance							
Fired, Feed	Mbtu/hr	96.4	63.4	131.7	176.6	139.6	196.9
Fired, Reboiler	Mbtu/hr		45.5			29.7	
Cold Train Condenser	Mbtu/hr	81.6	86.9	90.0	93.7	93.7	93.7
Cold Train Rundown	Mbtu/hr	10.6	6.8	3.7			
Cold Train Steam	Mbtu/hr	1.6					
Cold Train Naphtha Rundown	Mbtu/hr						
Cold Train Duty Total	Mbtu/hr	93.8	93.7	93.7	93.7	93.7	93.7
Hot Train Condenser	Mbtu/hr				65.0	67.6	65.2
Hot Train Rundown	Mbtu/hr		36.5	29.8	7.0	35.0	16.8
Hot Train Duty Total	Mbtu/hr		36.5	29.8	72.0	102.6	82.0
Condenser	Mbtu/hr	125.5	106.8	132.7	256.8	243.8	267.3
Condenser rejected	Mbtu/hr	43.9	19.9	42.7	98.1	82.5	108.4
Rundown rejected	Mbtu/hr	44.2	54.1	48.9	29.7	30.1	32.7
Light Naphtha Cooling	Mbtu/hr	8.2	8.0	9.0	29.4	29.4	28.5
Total Cooling Utility	Mbtu/hr	96.3	82.0	100.6	157.2	142.0	169.6
Above Pinch Lost	Mbtu/hr						
Total Fired		96.4	108.9	131.7	176.6	169.3	196.9
Total Fired + Steam Fired		121.1	108.9	131.7	183.9	169.3	196.9
Feed Temperature	F	299	302	350	445	438	482
Bottoms Temperature	F	268	380		410	600	
Mechanical							
Stages Rectifying		14	14	14	14	14	18
Stages Stripping		3	9		3	9	
Trays Rectifying		18	30	18	18	30	18
Trays Stripping		12	12		12	12	
Diameter Rectifying	In	144	138	162	192	186	204
Diameter Stripping	In	102	126		60	84	

Table B-2 Cutpoint and Configuration Comparison – NWS (Northwest Shelf) Condensate

Feed Operation Configuration		NWS Low Cut Strip	NWS Low Cut Reboil	NWS Low Cut Preflash	NWS High Cut Strip	NWS High Cut Reboil	NWS High Cut Preflash
Streams							
Feed gravity	API	59.3	59.3	59.3	59.3	59.3	59.3
Bottoms gravity	API	47.2	47.2	48.4	36.2	36.1	38.5
Gas Rate	BBL/DAY	66	66	65	61	61	61
Nap Rate	BBL/DAY	41223	41223	41223	81784	81783	79719
Btms Rate	BBL/DAY	58707	58711	58711	18154	18155	20219
Steam Rate	#/HR	24634			7567		
Naphtha D86 95	F	230	230	230	350	350	350
Naphtha-Bottoms Gap	F	-30.4	-22.9	-66.1	-30.8	-6.8	-126.7
Naphtha Cutpoint	F	198	198	198	389	389	369
Naphtha in Bottoms	vol %	24.0	23.9	24.4	12.7	12.5	20.4
Naphtha Recovery	%	80.8	80.8	80.8	106.4	106.3	103.7
Heat Balance							
Fired, Feed	Mbtu/hr	88.4	55.4	124.4	166.9	136.2	179.3
Fired, Reboiler	Mbtu/hr		44.4			25.0	
Cold Train Condenser	Mbtu/hr	82.1	88.4	91.2	94.4	94.4	94.4
Cold Train Rundown	Mbtu/hr	8.9	6.0	3.2			
Cold Train Steam	Mbtu/hr	3.4					
Cold Train Naphtha Rundown	Mbtu/hr						
Cold Train Duty Total	Mbtu/hr	94.4	94.4	94.4	94.4	94.4	94.4
Hot Train Condenser	Mbtu/hr				65.4	67.7	65.8
Hot Train Rundown	Mbtu/hr		37.1	25.5		27.1	12.1
Hot Train Duty Total	Mbtu/hr		37.1	25.5	65.4	94.8	77.9
Condenser	Mbtu/hr	115.3	96.5	120.8	239.4	230.5	244.8
Condenser rejected	Mbtu/hr	33.2	8.1	29.6	79.6	68.4	84.6
Rundown rejected	Mbtu/hr	48.9	55.1	56.8	33.4	30.8	33.7
Light Naphtha Cooling	Mbtu/hr	9.2	8.9	10.2	33.9	33.9	32.8
Total Cooling Utility	Mbtu/hr	91.3	72.1	96.6	146.9	133.1	151.1
Above Pinch Lost	Mbtu/hr	1.9			2.9		
Total Fired		88.4	99.8	124.4	166.9	161.2	179.3
Total Fired + Steam Fired		116.4	99.8	124.4	174.5	161.2	179.3
Feed Temperature	F	295	298	342	418	416	445
Bottoms Temperature	F	265	366		381	549	
Mechanical							
Stages Rectifying		14	14	14	14	14	14
Stages Stripping		3	9		3	9	
Trays Rectifying		18	18	18	18	18	18
Trays Stripping		12	12		12	12	
Diameter Rectifying	In	138	132	156	186	186	192
Diameter Stripping	In	108	132		60	90	

Table B-2 Cutpoint and Configuration Comparison – F3FB Condensate

Feed Operation Configuration		F3FB Low Cut Strip	F3FB Low Cut Reboil	F3FB Low Cut Preflash	F3FB High Cut Strip	F3FB High Cut Reboil	F3FB High Cut Preflash
Streams							
Feed gravity	API	63.4	63.4	63.4	63.4	63.4	63.4
Bottoms gravity	API	52.2	52.4	53.1	42.2	42.1	43.7
Gas Rate	BBL/DAY	61	61	61	60	60	60
Nap Rate	BBL/DAY	43758	43759	43758	79547	79547	78077
Btms Rate	BBL/DAY	56181	56180	56181	20394	20394	21863
Steam Rate	#/HR	23549			8506		
Naphtha D86 95	F	230	230	230	350	350	350
Naphtha-Bottoms Gap	F	-19.6	-10.8	-51.2	-13.3	9.1	-123.0
Naphtha Cutpoint	F	205	205	205	382	385	374
Naphtha in Bottoms	vol %	21.0	20.9	21.5	11.1	10.9	16.6
Naphtha Recovery	%	85.9	85.9	85.9	106.0	106.0	104.1
Heat Balance							
Fired, Feed	Mbtu/hr	95.0	58.7	127.6	160.1	125.5	186.7
Fired, Reboiler	Mbtu/hr		45.9			30.5	
Cold Train Condenser	Mbtu/hr	83.1	89.0	91.0	95.5	95.5	95.5
Cold Train Rundown	Mbtu/hr	6.4	6.5	4.5			
Cold Train Steam	Mbtu/hr	6.0					
Cold Train Naphtha Rundown	Mbtu/hr						
Cold Train Duty Total	Mbtu/hr	95.5	95.5	95.5	95.5	95.5	95.5
Hot Train Condenser	Mbtu/hr				60.8	63.2	61.5
Hot Train Rundown	Mbtu/hr		39.6	25.4	4.5	33.8	16.2
Hot Train Duty Total	Mbtu/hr		39.6	25.4	65.3	97.0	77.7
Condenser	Mbtu/hr	130.5	111.0	134.2	245.0	233.8	261.4
Condenser rejected	Mbtu/hr	47.4	22.0	43.2	88.7	75.1	104.4
Rundown rejected	Mbtu/hr	57.0	53.4	54.2	33.6	34.0	35.9
Light Naphtha Cooling	Mbtu/hr	1.4	1.3	2.1	18.6	18.6	17.9
Total Cooling Utility	Mbtu/hr	105.8	76.7	99.5	140.9	127.7	158.2
Above Pinch Lost	Mbtu/hr						
Total Fired		95.0	104.6	127.6	160.1	156.0	186.7
Total Fired + Steam Fired		124.5	104.6	127.6	168.6	156.0	186.7
Feed Temperature	F	298	299	344	415	412	461
Bottoms Temperature	F	267	376		381	561	
Mechanical							
Stages Rectifying		14	14	14	14	14	14
Stages Stripping		3	9		3	9	
Trays Rectifying		18	18	18	18	18	18
Trays Stripping		12	12		12	12	
Diameter Rectifying	In	138	138	162	186	180	204
Diameter Stripping	In	108	132		60	96	

Table B-3 Inspection Comparison for Two-step Preflash Splitting

Feed		F3FB	Griffin	Two-Step	F3FB	F3FB	F3FB
Operation		Low Cut	High Cut		High Cut	High Cut	High Cut
Configuration		Preflash	Preflash		Strip	Reboil	Preflash
Streams							
Feed gravi	API	63.4	54.8		63.4	63.4	63.4
Bottoms gr	API	53.1	40.3		42.2	42.1	43.7
Gas Rate	BBL/DAY	61	31		60	60	60
Nap Rate	BBL/DAY	43758	33233	76991	79547	79547	78077
Btms Rate	BBL/DAY	56181	22936	79117	20394	20394	21863
Steam Rate	#/HR				8506		
Naphtha D	F	230	350		350	350	350
Naphtha-B	F	-51.2	-77.2		-13.3	9.1	-123.0
Naphtha C	F	205	349		382	385	374
Naphtha ir	vol %	21.5	16.0		11.1	10.9	16.6
Naphtha R	%	85.9	99.6		106.0	106.0	104.1
Heat Balance							
Fired, Feec	Mbtu/hr	127.6	100.7	228	160.1	125.5	186.7
Fired, Rebc	Mbtu/hr					30.5	
Cold Train	Mbtu/hr	91.0	55.0	146	95.5	95.5	95.5
Cold Train	Mbtu/hr	4.5		5			
Cold Train	Mbtu/hr						
Cold Train	Mbtu/hr						
Cold Train	Mbtu/hr	95.5	55.0	150	95.5	95.5	95.5
Hot Train C	Mbtu/hr		42.0	42.0	60.8	63.2	61.5
Hot Train F	Mbtu/hr	25.4	20.5	46	4.5	33.8	16.2
Hot Train D	Mbtu/hr	25.4	62.6	88	65.3	97.0	77.7
Condenser	Mbtu/hr	134.2	120.3	255	245.0	233.8	261.4
Condenser	Mbtu/hr	43.2	23.3	67	88.7	75.1	104.4
Rundown r	Mbtu/hr	54.2	42.1	96	33.6	34.0	35.9
Light Naph	Mbtu/hr	2.1	18.8	21	18.6	18.6	17.9
Total Cooli	Mbtu/hr	99.5	84.2	184	140.9	127.7	158.2
Above Pinc	Mbtu/hr						
Total Fired		127.6	100.7	228	160.1	156.0	186.7
Total Fired		127.6	100.7	228	168.6	156.0	186.7
Feed Temp	F	344	506		415	412	461
Bottoms Te	F				381	561	
Mechanical							
Stages Rec			14		14	14	14
Stages Stri					3	9	
Trays Recti			18		18	18	18
Trays Strip					12	12	
Diameter F	In		192		186	180	204
Diameter S	In				60	96	

Table B-1 Condensate Comparison – Low-cutpoint, Stripping Configuration

Feed		Asgard	Griffin	SleipK	NWS	F3FB
Operation		Low Cut	Low Cut	Low Cut	Low Cut	Low Cut
Configuration		Strip	Strip	Strip	Strip	Strip
Streams						
Feed gravity	SG	0.7775	0.7595	0.7491	0.7416	0.7260
Bottoms gravity	SG	0.8067	0.7699	0.8165	0.7918	0.7703
Gas Rate	m3/day	10	10	10	10	10
Nap Rate	m3/day	3291	1359	7124	6555	6958
Btms Rate	m3/day	12599	14532	8766	9335	8933
Steam Rate	kg/hr	15148	5193	10479	11172	10680
Naphtha D86 95	C	110	110	110	110	110
Naphtha-Bottoms Gap	C	-12.4	-22.1	-13.3	-16.9	-10.9
Naphtha Cutpoint	C	79	55	100	92	96
Naphtha in Bottoms	vol %	17.2	23.4	20.9	24.0	21.0
Naphtha Recovery	%	64.4	29.4	87.3	80.8	85.9
Heat Balance						
Fired, Feed	GJ/hr	52.8	44.1	101.7	93.3	100.2
Fired, Reboiler	GJ/hr					
Cold Train Condenser	GJ/hr	75.3	33.5	86.1	86.6	87.7
Cold Train Rundown	GJ/hr	21.5	69.6	11.2	9.4	6.8
Cold Train Steam	GJ/hr	5.9		1.7	3.6	6.3
Cold Train Naphtha Rundown	GJ/hr					
Cold Train Duty Total	GJ/hr	102.8	103.2	99.0	99.6	100.8
Hot Train Condenser	GJ/hr					
Hot Train Rundown	GJ/hr					
Hot Train Duty Total	GJ/hr					
Condenser	GJ/hr	78.5	33.5	132.4	121.6	137.7
Condenser rejected	GJ/hr	3.2		46.3	35.0	50.0
Rundown rejected	GJ/hr	86.0	26.5	46.6	51.6	60.1
Light Naphtha Cooling	GJ/hr	0.6	0.4	8.7	9.7	1.5
Total Cooling Utility	GJ/hr	89.8	26.9	101.6	96.3	111.6
Above Pinch Lost	GJ/hr		3.9		2.0	
Total Fired	GJ/hr	291.2	292.2	315.4	311.2	314.4
Total Fired + Steam Fired	GJ/hr	280.6	280.6	282.7	279.6	281.7
Feed Temperature	C	136	136	148	146	148
Bottoms Temperature	C	130	130	131	129	131
Mechanical						
Stages Rectifying		14	14	14	14	14
Stages Stripping		3	3	3	3	3
Trays Rectifying		18	18	18	18	18
Trays Stripping		12	12	12	12	12
Diameter Rectifying	m	2.5908	1.8288	3.6576	3.5052	3.5052
Diameter Stripping	m	3.048	2.7432	2.5908	2.7432	2.7432

Table B-1 Condensate Comparison – High-cutpoint, Stripping Configuration

Feed		Asgard	Griffin	SleipK	NWS	F3FB
Operation		High Cut	High Cut	High Cut	High Cut	High Cut
Configuration		Strip	Strip	Strip	Strip	Strip
Streams						
Feed gravity	SG	0.7775	0.7595	0.7491	0.7416	0.7260
Bottoms gravity	SG	0.8509	0.8265	0.9094	0.8438	0.8146
Gas Rate	m3/day	9	9	10	10	10
Nap Rate	m3/day	8951	9403	13005	13004	12649
Btms Rate	m3/day	6940	6489	2886	2887	3243
Steam Rate	kg/hr	4717	7723	3298	3432	3858
Naphtha D86 95	C	177	177	110	177	177
Naphtha-Bottoms Gap	C	4.2	-11.9	-2.1	-17.1	-7.4
Naphtha Cutpoint	C	184	176	208	198	194
Naphtha in Bottoms	vol %	8.0	15.0	8.3	12.7	11.1
Naphtha Recovery	%	103.2	99.6	107.4	106.4	106.0
Heat Balance						
Fired, Feed	GJ/hr	138.2	132.9	186.3	176.1	168.9
Fired, Reboiler	GJ/hr					
Cold Train Condenser	GJ/hr	102.8	103.2	98.9	99.6	100.8
Cold Train Rundown	GJ/hr					
Cold Train Steam	GJ/hr					
Cold Train Naphtha Rundown	GJ/hr					
Cold Train Duty Total	GJ/hr	102.8	103.2	98.9	99.6	100.8
Hot Train Condenser	GJ/hr	66.3	78.3	68.6	69.0	64.1
Hot Train Rundown	GJ/hr	16.6	5.1	7.4		4.7
Hot Train Duty Total	GJ/hr	82.8	83.3	76.0	69.0	68.9
Condenser	GJ/hr	200.5	191.3	270.9	252.6	258.5
Condenser rejected	GJ/hr	31.4	9.8	103.5	84.0	93.6
Rundown rejected	GJ/hr	76.6	76.3	31.3	35.2	35.4
Light Naphtha Cooling	GJ/hr	20.4	34.7	31.0	35.8	19.6
Total Cooling Utility	GJ/hr	128.4	120.8	165.8	155.0	148.6
Above Pinch Lost	GJ/hr		5.1		3.1	
Total Fired	GJ/hr	453.7	448.4	469.5	441.0	437.8
Total Fired + Steam Fired	GJ/hr	420.9	405.1	432.6	402.0	402.0
Feed Temperature	C	221	218	229	214	213
Bottoms Temperature	C	204	196	210	194	194
Mechanical						
Stages Rectifying		14	14	14	14	14
Stages Stripping		3	3	3	3	3
Trays Rectifying		18	18	18	18	18
Trays Stripping		12	12	12	12	12
Diameter Rectifying	m	4.1148	4.1148	4.8768	4.7244	4.7244
Diameter Stripping	m	2.286	2.4384	1.524	1.524	1.524

TableB-1 Condensate Comparison – Low-cutpoint, Reboiler Configuration

Feed		Asgard	Griffin	SleipK	NWS	F3FB
Operation		Low Cut	Low Cut	Low Cut	Low Cut	Low Cut
Configuration		Reboil	Reboil	Reboil	Reboil	Reboil
Streams						
Feed gravity	SG	0.7775	0.7595	0.7491	0.7416	0.7260
Bottoms gravity	SG	0.8054	0.7703	0.8170	0.7918	0.7694
Gas Rate	m3/day	10	10	10	10	10
Nap Rate	m3/day	3292	1469	7124	6555	6958
Btms Rate	m3/day	12599	14421	8766	9336	8933
Steam Rate	kg/hr					
Naphtha D86 95	C	110	110	110	110	110
Naphtha-Bottoms Gap	C	-7.8	-14.6	-7.0	-12.7	-6.0
Naphtha Cutpoint	C	79	59	100	92	96
Naphtha in Bottoms	vol %	17.0	22.9	20.7	23.9	20.9
Naphtha Recovery	%	64.4	31.8	87.3	80.8	85.9
Heat Balance						
Fired, Feed	GJ/hr	7.7		66.9	58.4	61.9
Fired, Reboiler	GJ/hr	68.7	69.2	48.0	46.8	48.4
Cold Train Condenser	GJ/hr	55.0	25.4	91.7	93.3	93.9
Cold Train Rundown	GJ/hr	47.8	77.8	7.2	6.3	6.9
Cold Train Steam	GJ/hr					
Cold Train Naphtha Rundown	GJ/hr					
Cold Train Duty Total	GJ/hr	102.8	103.2	98.9	99.6	100.8
Hot Train Condenser	GJ/hr					
Hot Train Rundown	GJ/hr	56.1	44.0	38.5	39.1	41.8
Hot Train Duty Total	GJ/hr	56.1	44.0	38.5	39.1	41.8
Condenser	GJ/hr	55.0	25.4	112.7	101.8	117.1
Condenser rejected	GJ/hr			21.0	8.5	23.2
Rundown rejected	GJ/hr	46.0	38.5	57.1	58.1	56.3
Light Naphtha Cooling	GJ/hr	0.6	0.2	8.4	9.4	1.4
Total Cooling Utility	GJ/hr	46.6	38.7	86.5	76.1	80.9
Above Pinch Lost	GJ/hr	7.8	14.3			
Total Fired	GJ/hr	302.8	292.2	318.6	314.4	315.4
Total Fired + Steam Fired	GJ/hr	396.7	381.9	400.9	386.1	396.7
Feed Temperature	C	142	136	150	148	148
Bottoms Temperature	C	191	183	193	186	191
Mechanical						
Stages Rectifying		14	14	14	14	14
Stages Stripping		9	9	9	9	9
Trays Rectifying		30	30	30	18	18
Trays Stripping		12	12	12	12	12
Diameter Rectifying	m	2.286	1.8288	3.5052	3.3528	3.5052
Diameter Stripping	m	3.81	3.81	3.2004	3.3528	3.3528

Table B-1 Condensate Comparison – High-cutpoint, Reboiler Configuration

Feed Operation Configuration		Asgard High Cut Reboil	Griffin High Cut Reboil	SleipK High Cut Reboil	NWS High Cut Reboil	F3FB High Cut Reboil
Streams						
Feed gravity	SG	0.7775	0.7595	0.7491	0.7416	0.7260
Bottoms gravity	SG	0.8509	0.8428	0.9094	0.8443	0.8151
Gas Rate	m3/day	9	9	10	10	10
Nap Rate	m3/day	8951	10650	13004	13004	12649
Btms Rate	m3/day	6940	5242	2886	2887	3243
Steam Rate	kg/hr					
Naphtha D86 95	C	177	177	177	177	177
Naphtha-Bottoms Gap	C	-3.1	4.7	10.7	-3.8	5.1
Naphtha Cutpoint	C	184	202	208	198	196
Naphtha in Bottoms	vol %	8.1		8.1	12.5	10.9
Naphtha Recovery	%	103.2	112.8	107.4	106.3	106.0
Heat Balance						
Fired, Feed	GJ/hr	135.9		147.3	143.7	132.4
Fired, Reboiler	GJ/hr	41.8	135.3	31.3	26.4	32.2
Cold Train Condenser	GJ/hr	102.8	101.1	98.9	99.6	100.8
Cold Train Rundown	GJ/hr					
Cold Train Steam	GJ/hr					
Cold Train Naphtha Rundown	GJ/hr		2.1			
Cold Train Duty Total	GJ/hr	102.8	103.2	98.9	99.6	100.8
Hot Train Condenser	GJ/hr	64.6	89.4	71.3	71.4	66.7
Hot Train Rundown	GJ/hr	81.8	74.6	36.9	28.6	35.7
Hot Train Duty Total	GJ/hr	146.3	163.9	108.2	100.0	102.3
Condenser	GJ/hr	215.4	190.4	257.2	243.2	246.7
Condenser rejected	GJ/hr	48.1		87.0	72.2	79.2
Rundown rejected	GJ/hr	78.8	62.7	31.8	32.5	35.9
Light Naphtha Cooling	GJ/hr	20.5	39.6	31.0	35.8	19.6
Total Cooling Utility	GJ/hr	147.4	102.2	149.8	140.4	134.7
Above Pinch Lost	GJ/hr					
Total Fired	GJ/hr	517.0	404.1	462.1	438.9	434.7
Total Fired + Steam Fired	GJ/hr	606.6	663.6	633.0	579.2	591.9
Feed Temperature	C	254	195	226	213	211
Bottoms Temperature	C	302	332	316	287	294
Mechanical						
Stages Rectifying		14	14	14	14	14
Stages Stripping		9	9	9	9	9
Trays Rectifying		30	30	30	18	18
Trays Stripping		12	12	12	12	12
Diameter Rectifying	m	5.0292	3.9624	4.7244	4.7244	4.572
Diameter Stripping	m	2.286	5.334	2.1336	2.286	2.4384

Table B-1 Condensate Comparison – Low-cutpoint, Preflash Configuration

Feed Operation Configuration		Asgard Low Cut Preflash	Griffin Low Cut Preflash	SleipK Low Cut Preflash	NWS Low Cut Preflash	F3FB Low Cut Preflash
Streams						
Feed gravity	SG	0.7775	0.7595	0.7491	0.7416	0.7260
Bottoms gravity	SG	0.8031	0.7686	0.8114	0.7865	0.7665
Gas Rate	m3/day	10	10	10	10	10
Nap Rate	m3/day	3292	1359	7124	6555	6958
Btms Rate	m3/day	12599	14532	8767	9336	8933
Steam Rate	kg/hr					
Naphtha D86 95	C	110	110	110	110	110
Naphtha-Bottoms Gap	C	-32.3	-15.4	-40.5	-36.7	-28.4
Naphtha Cutpoint	C	80	55	100	92	96
Naphtha in Bottoms	vol %	17.6	23.5	21.7	24.4	21.5
Naphtha Recovery	%	64.4	29.4	87.3	80.8	85.9
Heat Balance						
Fired, Feed	GJ/hr	122.8	59.3	138.9	131.2	134.6
Fired, Reboiler	GJ/hr					
Cold Train Condenser	GJ/hr	67.4	28.1	95.0	96.2	96.0
Cold Train Rundown	GJ/hr	35.3	75.1	3.9	3.4	4.7
Cold Train Steam	GJ/hr					
Cold Train Naphtha Rundown	GJ/hr					
Cold Train Duty Total	GJ/hr	102.8	103.2	98.9	99.6	100.8
Hot Train Condenser	GJ/hr					
Hot Train Rundown	GJ/hr		33.4	31.4	26.9	26.8
Hot Train Duty Total	GJ/hr		33.4	31.4	26.9	26.8
Condenser	GJ/hr	67.4	28.2	140.0	127.4	141.6
Condenser rejected	GJ/hr			45.0	31.2	45.6
Rundown rejected	GJ/hr	91.2	27.5	51.6	59.9	57.2
Light Naphtha Cooling	GJ/hr	1.6	1.2	9.5	10.8	2.2
Total Cooling Utility	GJ/hr	92.7	28.7	106.1	101.9	105.0
Above Pinch Lost	GJ/hr	5.3				
Total Fired	GJ/hr	359.8	345.0	369.3	360.8	362.9
Total Fired + Steam Fired	GJ/hr					
Feed Temperature	C	172	164	177	172	173
Bottoms Temperature	C					
Mechanical						
Stages Rectifying		14	14	14	14	14
Stages Stripping						
Trays Rectifying		18	18	18	18	18
Trays Stripping						
Diameter Rectifying	m	2.7432	1.8288	4.1148	3.9624	4.1148
Diameter Stripping	m					

Table B-1 Condensate Comparison – High-cutpoint, Preflash Configuration

Feed		Asgard	Griffin	SleipK	NWS	F3FB
Operation		High Cut	High Cut	High Cut	High Cut	High Cut
Configuration		Preflash	Preflash	Preflash	Preflash	Preflash
Streams						
Feed gravity	SG	0.7775	0.7595	0.7491	0.7416	0.7260
Bottoms gravity	SG	0.8493	0.8236	0.8927	0.8324	0.8076
Gas Rate	m3/day	9	9	10	10	10
Nap Rate	m3/day	8951	9403	12707	12676	12415
Btms Rate	m3/day	6940	6489	3184	3215	3476
Steam Rate	kg/hr					
Naphtha D86 95	C	177	177	177	177	177
Naphtha-Bottoms Gap	C	-41.7	-42.9	-66.7	-70.4	-68.3
Naphtha Cutpoint	C	184	176	196	187	190
Naphtha in Bottoms	vol %	8.6	16.0	15.8	20.4	16.6
Naphtha Recovery	%	103.2	99.6	104.9	103.7	104.1
Heat Balance						
Fired, Feed	GJ/hr	221.6	189.0	207.7	189.2	197.0
Fired, Reboiler	GJ/hr					
Cold Train Condenser	GJ/hr	102.8	103.2	98.9	99.6	100.8
Cold Train Rundown	GJ/hr					
Cold Train Steam	GJ/hr					
Cold Train Naphtha Rundown	GJ/hr					
Cold Train Duty Total	GJ/hr	102.8	103.2	98.9	99.6	100.8
Hot Train Condenser	GJ/hr	65.5	78.9	68.8	69.4	64.9
Hot Train Rundown	GJ/hr	64.0	38.5	17.7	12.8	17.1
Hot Train Duty Total	GJ/hr	129.6	117.4	86.5	82.2	82.0
Condenser	GJ/hr	259.4	225.9	282.0	258.3	275.8
Condenser rejected	GJ/hr	91.2	43.8	114.4	89.3	110.1
Rundown rejected	GJ/hr	79.2	79.0	34.5	35.6	37.9
Light Naphtha Cooling	GJ/hr	20.7	35.2	30.1	34.6	18.9
Total Cooling Utility	GJ/hr	191.1	158.0	178.9	159.4	166.9
Above Pinch Lost	GJ/hr					
Total Fired	GJ/hr	591.9	533.8	508.5	469.5	486.4
Total Fired + Steam Fired	GJ/hr					
Feed Temperature	C	294	263	250	229	238
Bottoms Temperature	C					
Mechanical						
Stages Rectifying		14	14	18	14	14
Stages Stripping						
Trays Rectifying		18	18	18	18	18
Trays Stripping						
Diameter Rectifying	m	5.6388	4.8768	5.1816	4.8768	5.1816
Diameter Stripping	m					

Table B-2 Cutpoint and Configuration Comparison – Åsgard Condensate

Feed Operation Configuration		Asgard Low Cut Strip	Asgard Low Cut Reboil	Asgard Low Cut Preflash	Asgard High Cut Strip	Asgard High Cut Reboil	Asgard High Cut Preflash
Streams							
Feed gravity	SG	0.7775	0.7775	0.7775	0.7775	0.7775	0.7775
Bottoms gravity	SG	0.8067	0.8054	0.8031	0.8509	0.8509	0.8493
Gas Rate	m3/day	10	10	10	9	9	9
Nap Rate	m3/day	3291	3292	3292	8951	8951	8951
Btms Rate	m3/day	12599	12599	12599	6940	6940	6940
Steam Rate	kg/hr	15148			4717		
Naphtha D86 95	C	110	110	110	177	177	177
Naphtha-Bottoms Gap	C	-12.4	-7.8	-32.3	4.2	-3.1	-41.7
Naphtha Cutpoint	C	79	79	80	184	184	184
Naphtha in Bottoms	vol %	17.2	17.0	17.6	8.0	8.1	8.6
Naphtha Recovery	%	64.4	64.4	64.4	103.2	103.2	103.2
Heat Balance							
Fired, Feed	GJ/hr	52.8	7.7	122.8	138.2	135.9	221.6
Fired, Reboiler	GJ/hr		68.7			41.8	
Cold Train Condenser	GJ/hr	75.3	55.0	67.4	102.8	102.8	102.8
Cold Train Rundown	GJ/hr	21.5	47.8	35.3			
Cold Train Steam	GJ/hr	5.9					
Cold Train Naphtha Rundown	GJ/hr						
Cold Train Duty Total	GJ/hr	102.8	102.8	102.8	102.8	102.8	102.8
Hot Train Condenser	GJ/hr				66.3	64.6	65.5
Hot Train Rundown	GJ/hr		56.1		16.6	81.8	64.0
Hot Train Duty Total	GJ/hr		56.1		82.8	146.3	129.6
Condenser	GJ/hr	78.5	55.0	67.4	200.5	215.4	259.4
Condenser rejected	GJ/hr	3.2			31.4	48.1	91.2
Rundown rejected	GJ/hr	86.0	46.0	91.2	76.6	78.8	79.2
Light Naphtha Cooling	GJ/hr	0.6	0.6	1.6	20.4	20.5	20.7
Total Cooling Utility	GJ/hr	89.8	46.6	92.7	128.4	147.4	191.1
Above Pinch Lost	GJ/hr		7.8	5.3			
Total Fired	GJ/hr	291.2	302.8	359.8	453.7	517.0	591.9
Total Fired + Steam Fired	GJ/hr	280.6	396.7		420.9	606.6	
Feed Temperature	C	136	142	172	221	254	294
Bottoms Temperature	C	130	191		204	302	
Mechanical							
Stages Rectifying		14	14	14	14	14	14
Stages Stripping		3	9		3	9	
Trays Rectifying		18	30	18	18	30	18
Trays Stripping		12	12		12	12	
Diameter Rectifying	m	2.5908	2.286	2.7432	4.1148	5.0292	5.6388
Diameter Stripping	m	3.048	3.81		2.286	2.286	

Table B-2 Cutpoint and Configuration Comparison – Griffin Condensate

Feed Operation Configuration		Griffin Low Cut Strip	Griffin Low Cut Reboil	Griffin Low Cut Preflash	Griffin High Cut Strip	Griffin High Cut Reboil	Griffin High Cut Preflash
Streams							
Feed gravity	SG	0.7595	0.7595	0.7595	0.7595	0.7595	0.7595
Bottoms gravity	SG	0.7699	0.7703	0.7686	0.8265	0.8428	0.8236
Gas Rate	m3/day	10	10	10	9	9	9
Nap Rate	m3/day	1359	1469	1359	9403	10650	9403
Btms Rate	m3/day	14532	14421	14532	6489	5242	6489
Steam Rate	kg/hr	5193			7723		
Naphtha D86 95	C	110	110	110	177	177	177
Naphtha-Bottoms Gap	C	-22.1	-14.6	-15.4	-11.9	4.7	-42.9
Naphtha Cutpoint	C	55	59	55	176	202	176
Naphtha in Bottoms	vol %	23.4	22.9	23.5	15.0		16.0
Naphtha Recovery	%	29.4	31.8	29.4	99.6	112.8	99.6
Heat Balance							
Fired, Feed	GJ/hr	44.1		59.3	132.9		189.0
Fired, Reboiler	GJ/hr		69.2			135.3	
Cold Train Condenser	GJ/hr	33.5	25.4	28.1	103.2	101.1	103.2
Cold Train Rundown	GJ/hr	69.6	77.8	75.1			
Cold Train Steam	GJ/hr						
Cold Train Naphtha Rundown	GJ/hr					2.1	
Cold Train Duty Total	GJ/hr	103.2	103.2	103.2	103.2	103.2	103.2
Hot Train Condenser	GJ/hr				78.3	89.4	78.9
Hot Train Rundown	GJ/hr		44.0	33.4	5.1	74.6	38.5
Hot Train Duty Total	GJ/hr		44.0	33.4	83.3	163.9	117.4
Condenser	GJ/hr	33.5	25.4	28.2	191.3	190.4	225.9
Condenser rejected	GJ/hr				9.8		43.8
Rundown rejected	GJ/hr	26.5	38.5	27.5	76.3	62.7	79.0
Light Naphtha Cooling	GJ/hr	0.4	0.2	1.2	34.7	39.6	35.2
Total Cooling Utility	GJ/hr	26.9	38.7	28.7	120.8	102.2	158.0
Above Pinch Lost	GJ/hr	3.9	14.3		5.1		
Total Fired	GJ/hr	292.2	292.2	345.0	448.4	404.1	533.8
Total Fired + Steam Fired	GJ/hr	280.6	381.9		405.1	663.6	
Feed Temperature	C	136	136	164	218	195	263
Bottoms Temperature	C	130	183		196	332	
Mechanical							
Stages Rectifying		14	14	14	14	14	14
Stages Stripping		3	9		3	9	
Trays Rectifying		18	30	18	18	30	18
Trays Stripping		12	12		12	12	
Diameter Rectifying	m	1.8288	1.8288	1.8288	4.1148	3.9624	4.8768
Diameter Stripping	m	2.7432	3.81		2.4384	5.334	

Table B-2 Cutpoint and Configuration Comparison – Sleipner Kårstø Condensate

Feed Operation Configuration		SleipK Low Cut Strip	SleipK Low Cut Reboil	SleipK Low Cut Preflash	SleipK High Cut Strip	SleipK High Cut Reboil	SleipK High Cut Preflash
Streams							
Feed gravity	SG	0.7491	0.7491	0.7491	0.7491	0.7491	0.7491
Bottoms gravity	SG	0.8165	0.8170	0.8114	0.9094	0.9094	0.8927
Gas Rate	m3/day	10	10	10	10	10	10
Nap Rate	m3/day	7124	7124	7124	13005	13004	12707
Btms Rate	m3/day	8766	8766	8767	2886	2886	3184
Steam Rate	kg/hr	10479			3298		
Naphtha D86 95	C	110	110	110	110	177	177
Naphtha-Bottoms Gap	C	-13.3	-7.0	-40.5	-2.1	10.7	-66.7
Naphtha Cutpoint	C	100	100	100	208	208	196
Naphtha in Bottoms	vol %	20.9	20.7	21.7	8.3	8.1	15.8
Naphtha Recovery	%	87.3	87.3	87.3	107.4	107.4	104.9
Heat Balance							
Fired, Feed	GJ/hr	101.7	66.9	138.9	186.3	147.3	207.7
Fired, Reboiler	GJ/hr		48.0			31.3	
Cold Train Condenser	GJ/hr	86.1	91.7	95.0	98.9	98.9	98.9
Cold Train Rundown	GJ/hr	11.2	7.2	3.9			
Cold Train Steam	GJ/hr	1.7					
Cold Train Naphtha Rundown	GJ/hr						
Cold Train Duty Total	GJ/hr	99.0	98.9	98.9	98.9	98.9	98.9
Hot Train Condenser	GJ/hr				68.6	71.3	68.8
Hot Train Rundown	GJ/hr		38.5	31.4	7.4	36.9	17.7
Hot Train Duty Total	GJ/hr		38.5	31.4	76.0	108.2	86.5
Condenser	GJ/hr	132.4	112.7	140.0	270.9	257.2	282.0
Condenser rejected	GJ/hr	46.3	21.0	45.0	103.5	87.0	114.4
Rundown rejected	GJ/hr	46.6	57.1	51.6	31.3	31.8	34.5
Light Naphtha Cooling	GJ/hr	8.7	8.4	9.5	31.0	31.0	30.1
Total Cooling Utility	GJ/hr	101.6	86.5	106.1	165.8	149.8	178.9
Above Pinch Lost	GJ/hr						
Total Fired	GJ/hr	315.4	318.6	369.3	469.5	462.1	508.5
Total Fired + Steam Fired	GJ/hr	282.7	400.9		432.6	633.0	
Feed Temperature	C	148	150	177	229	226	250
Bottoms Temperature	C	131	193		210	316	
Mechanical							
Stages Rectifying		14	14	14	14	14	18
Stages Stripping		3	9		3	9	
Trays Rectifying		18	30	18	18	30	18
Trays Stripping		12	12		12	12	
Diameter Rectifying	m	3.6576	3.5052	4.1148	4.8768	4.7244	5.1816
Diameter Stripping	m	2.5908	3.2004		1.524	2.1336	

Table B-2 Cutpoint and Configuration Comparison – NWS (Northwest Shelf) Condensate

Feed Operation Configuration		NWS Low Cut Strip	NWS Low Cut Reboil	NWS Low Cut Preflash	NWS High Cut Strip	NWS High Cut Reboil	NWS High Cut Preflash
Streams							
Feed gravity	SG	0.7416	0.7416	0.7416	0.7416	0.7416	0.7416
Bottoms gravity	SG	0.7918	0.7918	0.7865	0.8438	0.8443	0.8324
Gas Rate	m3/day	10	10	10	10	10	10
Nap Rate	m3/day	6555	6555	6555	13004	13004	12676
Btms Rate	m3/day	9335	9336	9336	2887	2887	3215
Steam Rate	kg/hr	11172			3432		
Naphtha D86 95	C	110	110	110	177	177	177
Naphtha-Bottoms Gap	C	-16.9	-12.7	-36.7	-17.1	-3.8	-70.4
Naphtha Cutpoint	C	92	92	92	198	198	187
Naphtha in Bottoms	vol %	24.0	23.9	24.4	12.7	12.5	20.4
Naphtha Recovery	%	80.8	80.8	80.8	106.4	106.3	103.7
Heat Balance							
Fired, Feed	GJ/hr	93.3	58.4	131.2	176.1	143.7	189.2
Fired, Reboiler	GJ/hr		46.8			26.4	
Cold Train Condenser	GJ/hr	86.6	93.3	96.2	99.6	99.6	99.6
Cold Train Rundown	GJ/hr	9.4	6.3	3.4			
Cold Train Steam	GJ/hr	3.6					
Cold Train Naphtha Rundown	GJ/hr						
Cold Train Duty Total	GJ/hr	99.6	99.6	99.6	99.6	99.6	99.6
Hot Train Condenser	GJ/hr				69.0	71.4	69.4
Hot Train Rundown	GJ/hr		39.1	26.9		28.6	12.8
Hot Train Duty Total	GJ/hr		39.1	26.9	69.0	100.0	82.2
Condenser	GJ/hr	121.6	101.8	127.4	252.6	243.2	258.3
Condenser rejected	GJ/hr	35.0	8.5	31.2	84.0	72.2	89.3
Rundown rejected	GJ/hr	51.6	58.1	59.9	35.2	32.5	35.6
Light Naphtha Cooling	GJ/hr	9.7	9.4	10.8	35.8	35.8	34.6
Total Cooling Utility	GJ/hr	96.3	76.1	101.9	155.0	140.4	159.4
Above Pinch Lost	GJ/hr	2.0			3.1		
Total Fired	GJ/hr	311.2	314.4	360.8	441.0	438.9	469.5
Total Fired + Steam Fired	GJ/hr	279.6	386.1		402.0	579.2	
Feed Temperature	C	146	148	172	214	213	229
Bottoms Temperature	C	129	186		194	287	
Mechanical							
Stages Rectifying		14	14	14	14	14	14
Stages Stripping		3	9		3	9	
Trays Rectifying		18	18	18	18	18	18
Trays Stripping		12	12		12	12	
Diameter Rectifying	m	3.5052	3.3528	3.9624	4.7244	4.7244	4.8768
Diameter Stripping	m	2.7432	3.3528		1.524	2.286	

Table B-2 Cutpoint and Configuration Comparison – F3FB Condensate

Feed Operation Configuration		F3FB Low Cut Strip	F3FB Low Cut Reboil	F3FB Low Cut Preflash	F3FB High Cut Strip	F3FB High Cut Reboil	F3FB High Cut Preflash
Streams							
Feed gravity	SG	0.7260	0.7260	0.7260	0.7260	0.7260	0.7260
Bottoms gravity	SG	0.7703	0.7694	0.7665	0.8146	0.8151	0.8076
Gas Rate	m3/day	10	10	10	10	10	10
Nap Rate	m3/day	6958	6958	6958	12649	12649	12415
Btms Rate	m3/day	8933	8933	8933	3243	3243	3476
Steam Rate	kg/hr	10680			3858		
Naphtha D86 95	C	110	110	110	177	177	177
Naphtha-Bottoms Gap	C	-10.9	-6.0	-28.4	-7.4	5.1	-68.3
Naphtha Cutpoint	C	96	96	96	194	196	190
Naphtha in Bottoms	vol %	21.0	20.9	21.5	11.1	10.9	16.6
Naphtha Recovery	%	85.9	85.9	85.9	106.0	106.0	104.1
Heat Balance							
Fired, Feed	GJ/hr	100.2	61.9	134.6	168.9	132.4	197.0
Fired, Reboiler	GJ/hr		48.4			32.2	
Cold Train Condenser	GJ/hr	87.7	93.9	96.0	100.8	100.8	100.8
Cold Train Rundown	GJ/hr	6.8	6.9	4.7			
Cold Train Steam	GJ/hr	6.3					
Cold Train Naphtha Rundown	GJ/hr						
Cold Train Duty Total	GJ/hr	100.8	100.8	100.8	100.8	100.8	100.8
Hot Train Condenser	GJ/hr				64.1	66.7	64.9
Hot Train Rundown	GJ/hr		41.8	26.8	4.7	35.7	17.1
Hot Train Duty Total	GJ/hr		41.8	26.8	68.9	102.3	82.0
Condenser	GJ/hr	137.7	117.1	141.6	258.5	246.7	275.8
Condenser rejected	GJ/hr	50.0	23.2	45.6	93.6	79.2	110.1
Rundown rejected	GJ/hr	60.1	56.3	57.2	35.4	35.9	37.9
Light Naphtha Cooling	GJ/hr	1.5	1.4	2.2	19.6	19.6	18.9
Total Cooling Utility	GJ/hr	111.6	80.9	105.0	148.6	134.7	166.9
Above Pinch Lost	GJ/hr						
Total Fired	GJ/hr	314.4	315.4	362.9	437.8	434.7	486.4
Total Fired + Steam Fired	GJ/hr	281.7	396.7		402.0	591.9	
Feed Temperature	C	148	148	173	213	211	238
Bottoms Temperature	C	131	191		194	294	
Mechanical							
Stages Rectifying		14	14	14	14	14	14
Stages Stripping		3	9		3	9	
Trays Rectifying		18	18	18	18	18	18
Trays Stripping		12	12		12	12	
Diameter Rectifying	m	3.5052	3.5052	4.1148	4.7244	4.572	5.1816
Diameter Stripping	m	2.7432	3.3528		1.524	2.4384	

Table B-3 Inspection Comparison for Two-step Preflash Splitting

Feed		F3FB	Griffin	Two-Step	F3FB	F3FB	F3FB
Operation		Low Cut	High Cut		High Cut	High Cut	High Cut
Configuration		Preflash	Preflash		Strip	Reboil	Preflash
Streams							
Feed gravity	SG	0.7260	0.7595		0.7260	0.7260	0.7260
Bottoms gravity	SG	0.7665	0.8236		0.8146	0.8151	0.8076
Gas Rate	m3/day	10	5		10	10	10
Nap Rate	m3/day	6958	5284	12242	12649	12649	12415
Btms Rate	m3/day	8933	3647	12580	3243	3243	3476
Steam Rate	kg/hr				3858		
Naphtha D86 95	C	110	177		177	177	177
Naphtha-Bottoms Gap	C	-28.4	-42.9		-7.4	5.1	-68.3
Naphtha Cutpoint	C	96	176		194	196	190
Naphtha in Bottoms	vol %	21.5	16.0		11.1	10.9	16.6
Naphtha Recovery	%	85.9	99.6		106.0	106.0	104.1
Heat Balance							
Fired, Feed	GJ/hr	134.6	106.2	241	168.9	132.4	197.0
Fired, Reboiler	GJ/hr					32.2	
Cold Train Condenser	GJ/hr	96.0	58.0	154	100.8	100.8	100.8
Cold Train Rundown	GJ/hr	4.7		5			
Cold Train Steam	GJ/hr						
Cold Train Naphtha Ru	GJ/hr						
Cold Train Duty Total	GJ/hr	100.8	58.0	159	100.8	100.8	100.8
Hot Train Condenser	GJ/hr		44.3	44.3	64.1	66.7	64.9
Hot Train Rundown	GJ/hr	26.8	21.6	48	4.7	35.7	17.1
Hot Train Duty Total	GJ/hr	26.8	66.0	93	68.9	102.3	82.0
Condenser	GJ/hr	141.6	126.9	269	258.5	246.7	275.8
Condenser rejected	GJ/hr	45.6	24.6	70	93.6	79.2	110.1
Rundown rejected	GJ/hr	57.2	44.4	102	35.4	35.9	37.9
Light Naphtha Cooling	GJ/hr	2.2	19.8	22	19.6	19.6	18.9
Total Cooling Utility	GJ/hr	105.0	88.8	194	148.6	134.7	166.9
Above Pinch Lost	GJ/hr						
Total Fired	GJ/hr	134.6	106.2	241	168.9	164.6	197.0
Total Fired + Steam Fir	GJ/hr	134.6	106.2	241	177.9	164.6	197.0
Feed Temperature	C	173	263		213	211	238
Bottoms Temperature	C				194	294	
Mechanical							
Stages Rectifying		14	14		14	14	14
Stages Stripping					3	9	
Trays Rectifying		18	18		18	18	18
Trays Stripping					12	12	
Diameter Rectifying	m	4.1148	3.6576		4.7244	4.572	5.1816
Diameter Stripping	m				1.524	2.4384	

14. Appendix C – Glossary, Definitions, Abbreviations and Conventions

Abbreviations and terms in text and figures that may not have been defined or that may be non-standard usage.

These definitions are tailored to the most common industry use of the term in crude distillation and petroleum refining. Many plants use their own variations on these definitions and other terms in place of those listed. Other technical fields may have different meanings for these terms.

Term or abbreviation	Definition
AEBP	Atmospheric equivalent boiling point. The temperature at which a compound would have boiled at atmospheric pressure if it could be heated to that temperature without thermal decomposition.
AET	Atmospheric equivalent temperature. See AEBP.
ANS	Alaskan North Slope – crude oil type
API degrees, API°	See API Gravity
API Gravity	A measure of liquid density defined by the American Petroleum Institute. The scale was derived from the reading marks on a specific type of measurement instrument. $API^{O} = \frac{141.5}{SG} - 131.5$ $SG = \frac{141.5}{API^{O} + 131.5}$ Where SG is specific gravity at 60°F (15.6°C). Water has an API gravity of 10. An API gravity lower than 10 is denser than water, an API gravity higher than 10 is less dense than water.
ASTM Distillation	Old usage. ASTM Distillation quoted without a descriptive qualifier refers to an ASTM D86 distillation. This is a batch distillation without reflux.
Boiling range	See distillation range.
bpd	Barrels per day, standard volume barrels at 60°F (15.6°C)
Cold train	Heat exchangers upstream of the desalter in a crude unit.
Condensate	Low density, naturally occurring petroleum liquid mixture. By industry usage this has a API gravity of 49° or higher (SG 0.7839 or lower). Condensates may come from direct production from wells or as captured liquids from pipeline systems (often called retrograde condensates).
CP	Cutpoint
Cutpoint	The ASTM D2892 distillation temperature that would achieve the same split between overhead products and bottoms products that the unit achieves. The D2892 distillation basis is atmospheric equivalent

	temperature. Typical usage is volume percent.
Distillation range	Range of AEBP boiling temperatures for a stream. Typically expressed as an ASTM D-2892 equivalent (true boiling range) or an ASTM D-86 equivalent (ASTM boiling range) on a 5-95 volume percent basis. Other basis may be used.
FBP	Final boiling point
Flash zone	Feed entry point into a main fractionator.
Gap-overlap	A measure of fractionation performance between the distillation of a light product and the distillation of the next heavier product from a main fractionators
IBP	Initial boiling point.
hnap	Heavy naphtha
Hot train	Exchangers in a crude preheat sequence downstream of the desalter.
IBP	Initial boiling point.
IP	Institute of Petroleum, UK
ISO	International Standards Organization.
LSR, lsr	Light straight run naphtha, same as light naphtha
MB	Million btu/hr (abbreviated to fit in figures), mmbtu/hr
MBtu/hr	Million btu/hr, mmbtu/hr
mmbtu/hr	Million btu/hr, industry standard usage remains the old form using a lower-case m for thousand. 'mm' equals thousand X thousand to get million.
NBP	Normal boiling point, see AEBP
NWS	Northwest Shelf, crude (condensate) type.
Overlap	See Gap
Specific gravity	Density of a stream at 60°F (15.56°C) compared to the density of water at 60°F (15.56°C).
Stabilizer	Distillation tower with a main purpose to control the vapor pressure of the bottoms product.
Stove oil	Alternate name for kerosene
T_{XX}	Temperature at which XX volume (or weight) percent of liquid is distilled.
TBP	True boiling point. Distillation curve based on ASTM D2892 or an equivalent.
Transfer line	Pipe connecting the outlet of a fired heater to the downstream main

	frationator.
TVP	True vapor pressure.
Wild naphtha	Unstabilized naphtha. Naphtha that still has a high vapor pressure due to a high butane and lighter content (typically more than one volume percent).

Figure conventions;

- Lower case text and abbreviations in figures refer to a stream.
- Upper case text and abbreviations in figures refer to an equipment item.

Slippery Slopes:

How should we approximate the slope of the equilibrium line in multicomponent systems?

Emily Capek, Ross Taylor
Clarkson University
Potsdam, New York, USA

Markus Duss
Sulzer Chemtech
Winterthur, Switzerland

Emily Capek

- BS: Chemical Engineering (2015)
  Clarkson University
  - Minors: Mathematics
    Biomedical engineering
    Project Management
  - Most Outstanding Senior Award (2015)
- Undergraduate Work Experience
  - ExxonMobil
    - Business Optimization Co-op; Spring 2013
    - Process Engineering Intern; Summer 2014
  - Corning Incorporated
    - Operations Engineering Intern; Summer 2012
    - Process Engineering Intern; Summer 2013
- Future
  - Corning Incorporated, Lexington, KY
    - Process Engineering Support

Background: Tray Efficiency

For a binary mixture

$$E_{OV} = 1 - \exp(-N_{OV})$$

Slope of Equilibrium Line

$$\frac{1}{N_{OV}} = \frac{1}{N_V} + \frac{\Lambda}{N_L'} \qquad \Lambda = M\frac{V}{L}$$

$$\frac{E_{MV}}{E_{OV}} = \frac{\exp(E_{OV}\Lambda) - 1}{E_{OV}\Lambda}$$

Lewis Case I

Some of the well known equations for estimating first point and then tray efficiency are shown here. Lewis Case 1 for the tray efficiency calculation is included only as one way to estimate the tray efficiency. For other models see, For example, Distillation Tray Fundamentals by M.J. Lockett (Cambridge).

Note that the term representing the slope of the equilibrium line appears in two different places, first in the calculation of the point efficiency and then a second time in the calculation of the tray efficiency from the point efficiency.

Binary McCabe-Thiele

- Slope depends on height
- Definition of slope is unambiguous
- This does NOT mean it is always calculated accurately

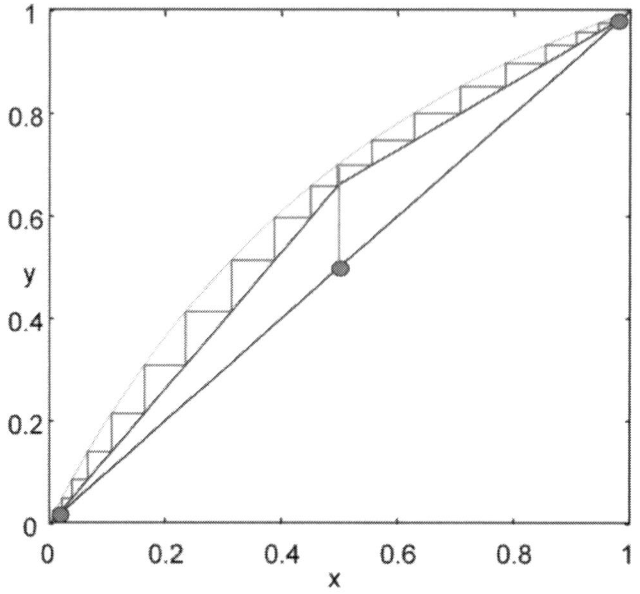

For binary systems there is no ambiguity about the definition of the slope of the equilibrium line. In principle the calculation of the slope can always be done accurately but that does not mean the slope is always calculated properly.

Multicomponent Efficiency Model

$$[E_{OV}]=[\mathrm{I}]-\exp\{-[N_{OV}]\}$$

[ ] = square matrix

$$[N_{OV}]^{-1}=[N_V]^{-1}+\frac{V}{L}[\mathrm{M}][N_L']^{-1}$$

Is this framework used?

Ognisty, T. P. and Sakata, M., "Multicomponent Diffusion: Theory vs. Industrial Data," *Chem. Eng. Prog.*, 83(3), 60-65 (1987).
Aittamaa, J., "Estimating Multicomponent Plate Efficiencies in Distillation," *Kem.-Kemi*, 8, 295-300, 317 (1981).

For multicomponent systems we must use a matrix formulation for the proper calculation of the point and tray efficiency. This formulation is decades old, but is rarely used (a couple of cases where the method is used are cited above).

Note that the slope is in fact a matrix of slope factors of dimenstion c-1 times c-1 where c is the number of compounds.

Pseudo-Binary Approach

$$E_{OV} = 1 - \exp(-N_{LK-HK})$$

Slope of Pseudo-Equilibrium Line

$$\frac{1}{N_{OV:LK-HK}} = \frac{1}{N_{V:LK-HK}} + \frac{\Lambda}{N'_{L:LK-HK}} \qquad \Lambda = M_{LK-HK}\frac{V}{L}$$

$$Y_{LK} = \frac{y_{LK}}{y_{LK} + y_{HK}} \qquad X_{LK} = \frac{x_{LK}}{x_{LK} + x_{HK}}$$

One way to calculate the slope in multicomponent mixtures is to use the equations for binary systems but use the properties for the two key components. Note the calculation of the apparent mole fractions as though the mixture contains only the keys.

Pseudo McCabe-Thiele Diagrams

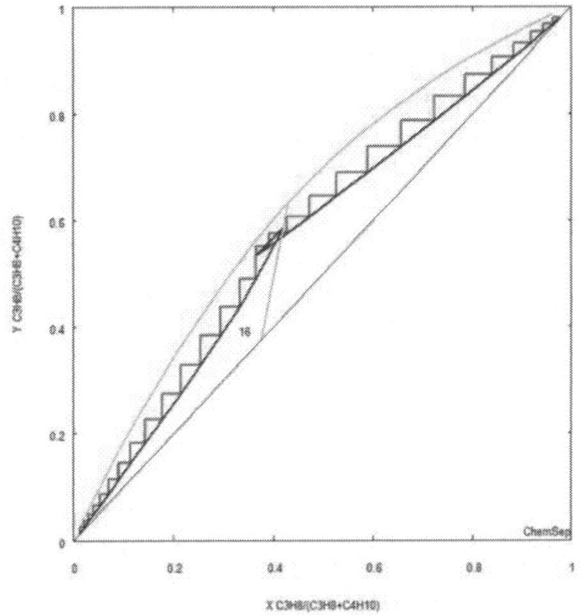

It is possible to construct McCabe-Theile type diagrams using these apparent mole fractions. The slope of the equilibrium line follows straightforwardly.

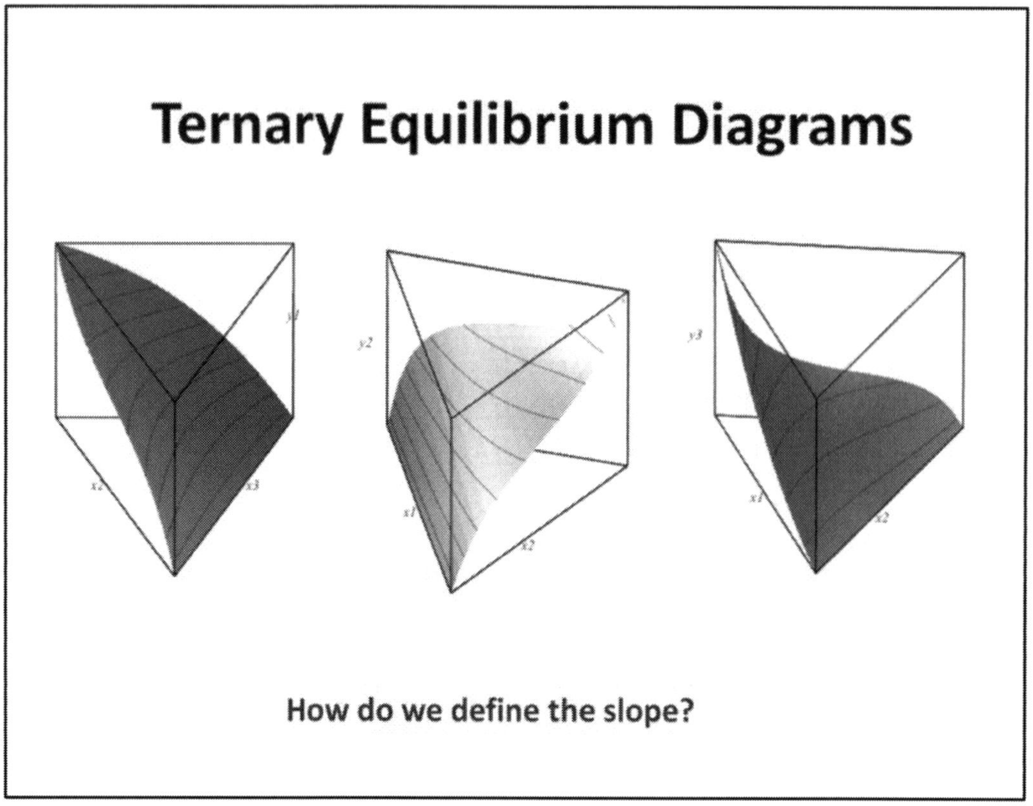

Here we show three different images of the equilibrium surface for a hypothetical ternary system. What is unusual about these images is that the vertical axis is one of the mole fractions in the vapor phase, the triangular axes are the mole fractions in the liquid phase. Rarely are these diagrams seen.

The key point to be made here is that there is no unique simple way to represent the slope when there are more than two compounds in the mixture.

Key Question

- How can we define the pseudo-binary slope in a multicomponent system?

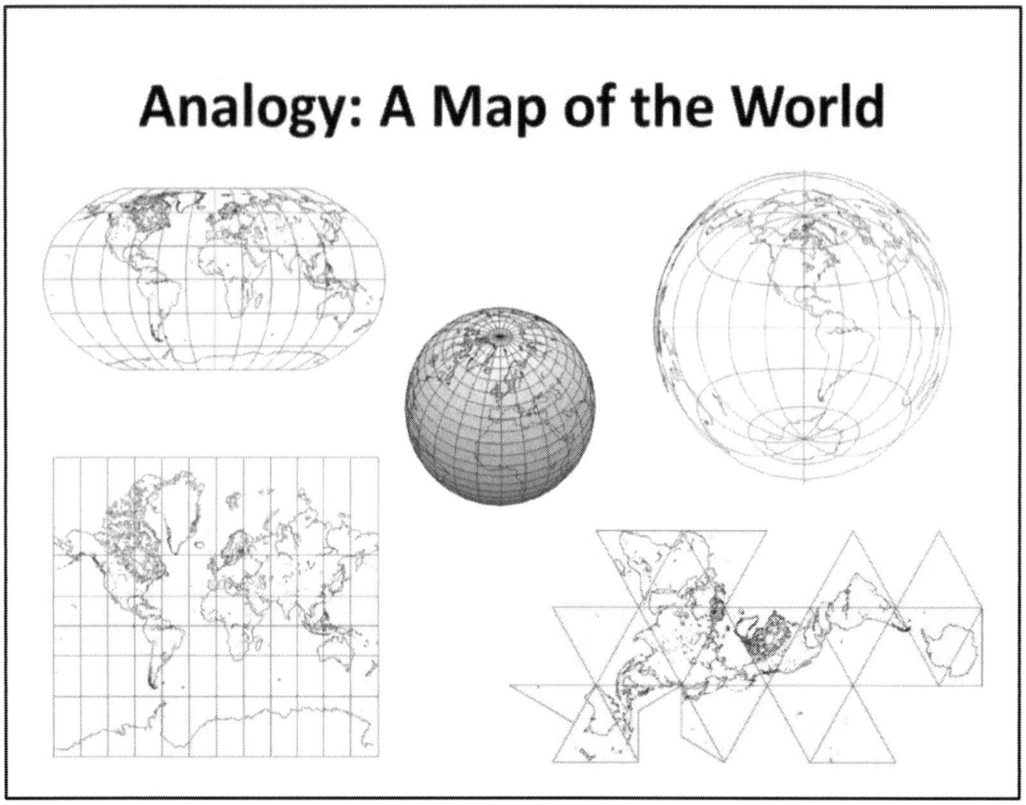

The problem of crafting a simple way to estimate the slope in a multicomponent system is related to the problem of making a two dimensional map of the world. There are more than 100 different map projections that have been proposed (four are shown here). All of them represent some sort of compromise because we are trying to reduce the dimensionality from 3 to 2.

The problem is arguably more complicated for distillation systems since we have to reduce the dimensionality from c-1 to 2.

Key Question

- How can we define the pseudo-binary slope in a multicomponent system?

- No unique answer
- Only approximations are possible

Method 1: Constant Relative Volatility

- Pseudo-binary VLE

$$Y_1 = \frac{\alpha_{12}\, X_1}{1 + X_1(\alpha_{12} - 1)}$$

- Slope

$$\mathrm{M} = \frac{\alpha_{12}\, X_1}{(1+X_1(\alpha_{12}-1))^2}$$

- where 1 = LK
  2 = HK

Method 2: LK-HK Binary Slope

- Set all mole fractions to zero except those of keys
- Make key fractions sum to 1
- Calculate $y^*(x+\Delta x)$ and $y^*(x)$
- Slope $$m = \frac{y^*(x+\Delta x) - y^*(x)}{\Delta x}$$

Method 3: LK-HK slope in mixture

- Calculate $y^*(x)$ using all species
- Increase slightly mole fraction of light key
- Make mole fractions sum to 1
- Calculate $y^*(x+\Delta x_{LK})$
- Slope

$$m=\frac{y^*_{LK}(x+\Delta x_{LK})-y^*_{LK}(x)}{\Delta x_{LK}}$$

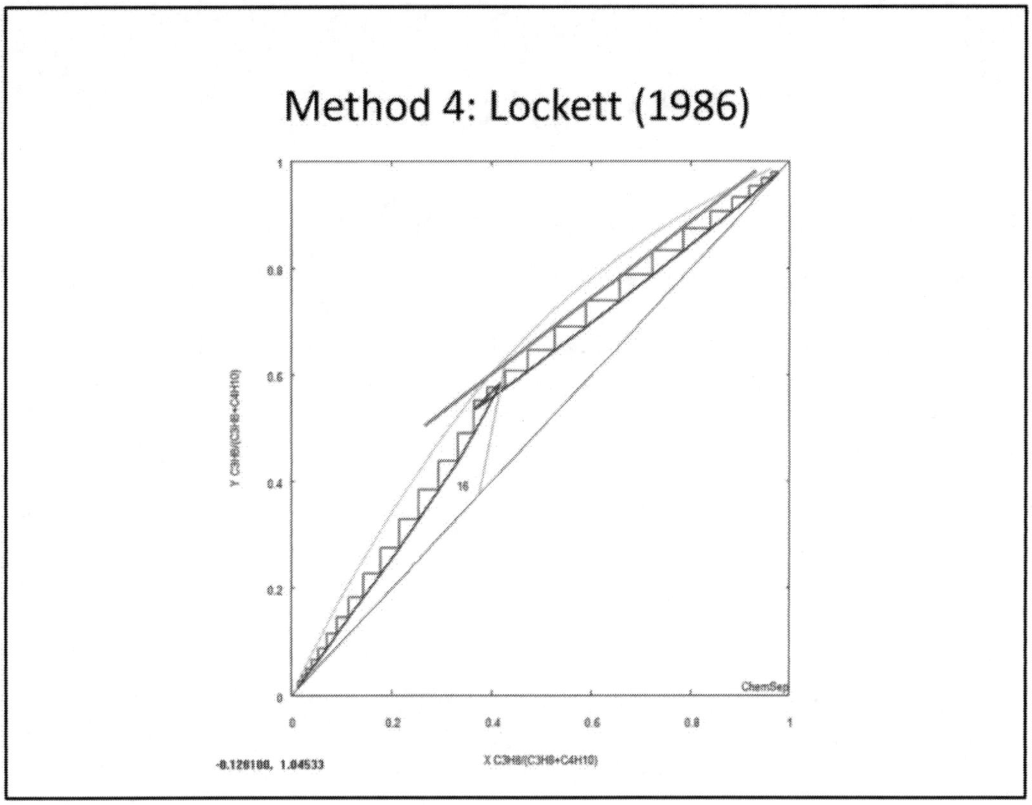

This method, described in the book by Lockett (1986) is to approximate the slope by using the apparent mole fractions on adjacent stages.

More Questions

- What slope approximation is "best"?
- What does "best" mean?
- How accurate is the slope approximation using only the relative volatility?

Some Examples

Slope approximations incorporated in ChemSep
Uses to model many different columns, including:

- Demethanizer from LNG flowsheet
- Depropanizer from LNG flowsheet
- Extractive distillation
- Azeotropic distillation
- Acetone absorber
- Gasoline stabilizer

The case studies we have used for our investigation come from the open literature.

The demethanizer and deethanizer come from a LNG flowsheet described by W.L. Luyben.

The extractive distillation example is from a book by C.D. Holland.

The azeotropic distillation column is a very famous example used by many in the past. Our base case is from e paper by Prokopakis and Seider.

The acetone absorber is adapted from several example calculations in Perry's Chemical Engineers Handbook.

Finally, the gasoline stabilizer is from a paper by Ragatz et al.

LNG Train (Luyben)

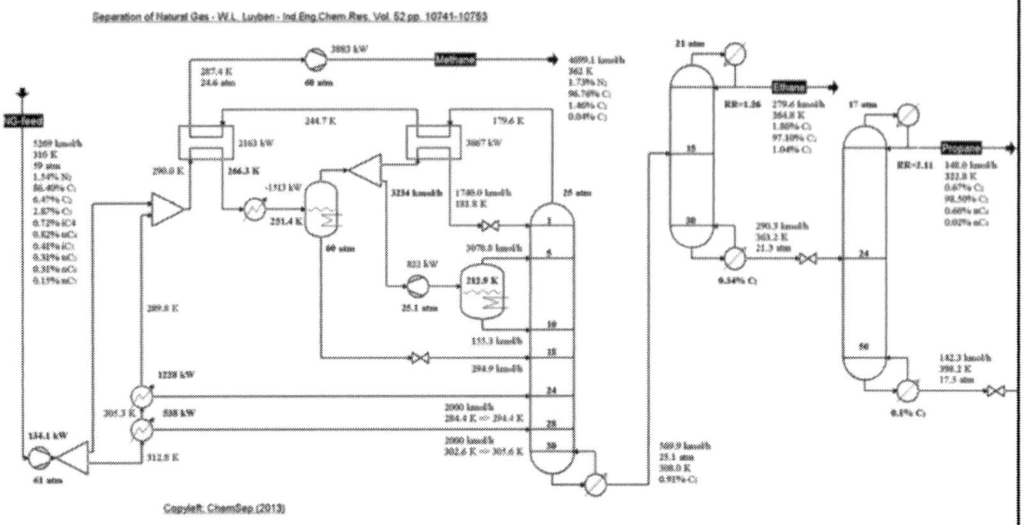

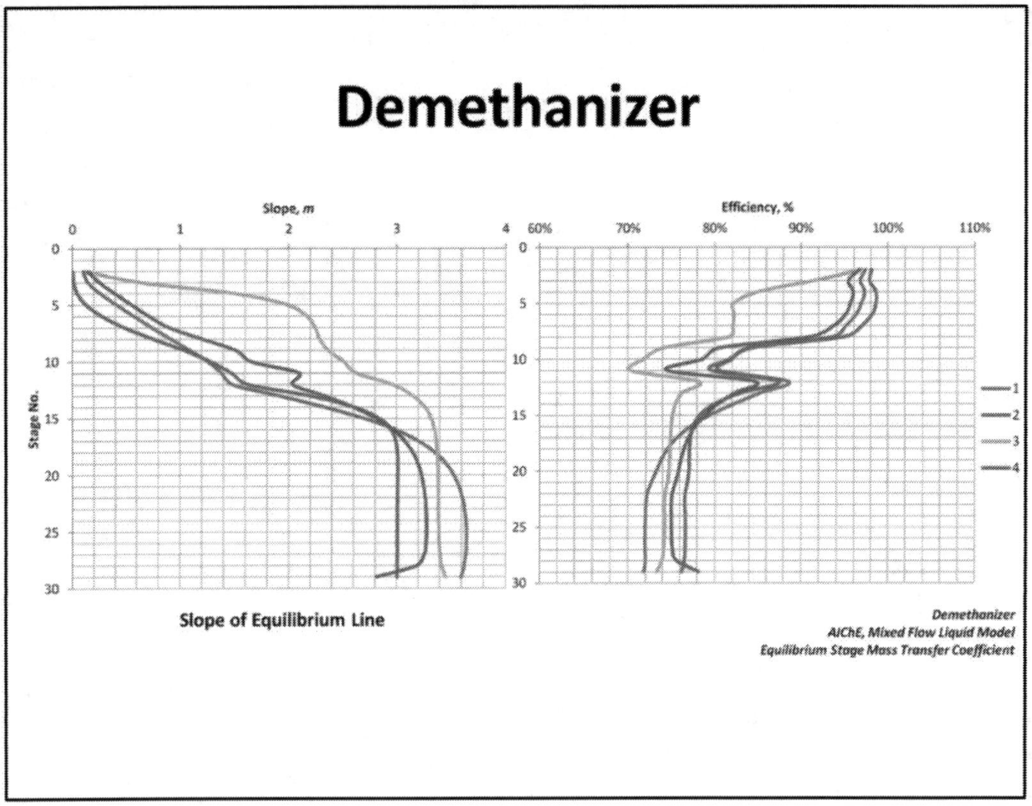

For the demethanizer three of the methods are reasonably close; only method 3 is somewhat different, especially in the upper section of the column.

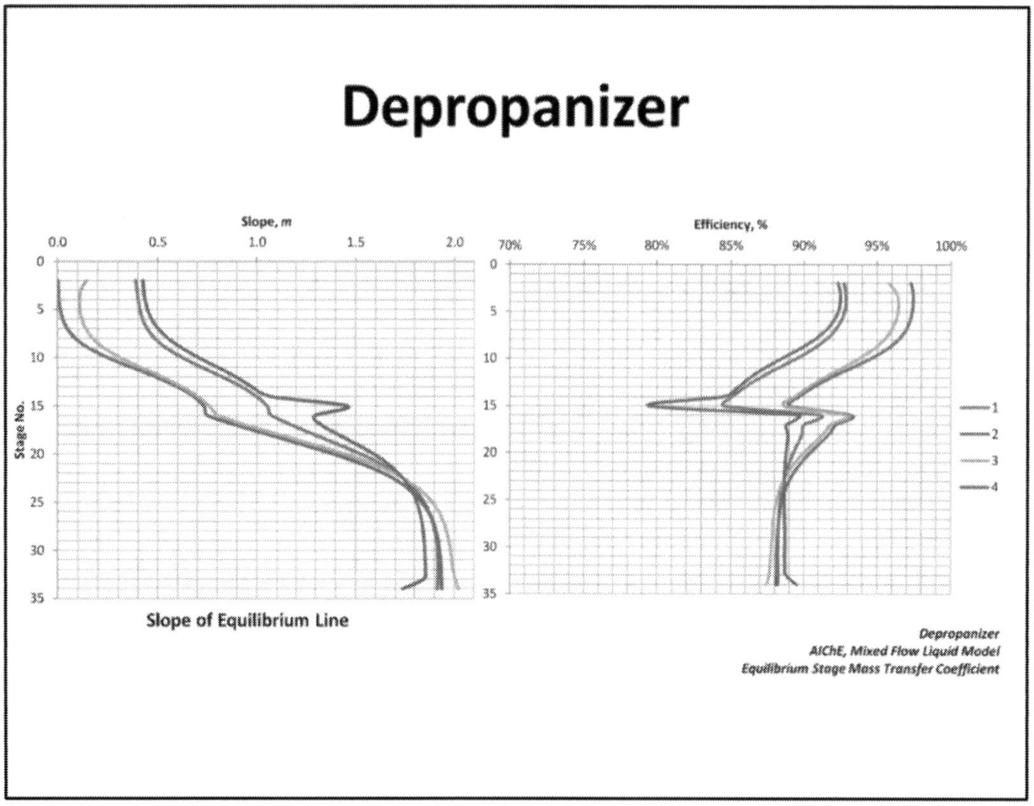

For the de-ethanizer there are two pairs of methods that are close to one anohter

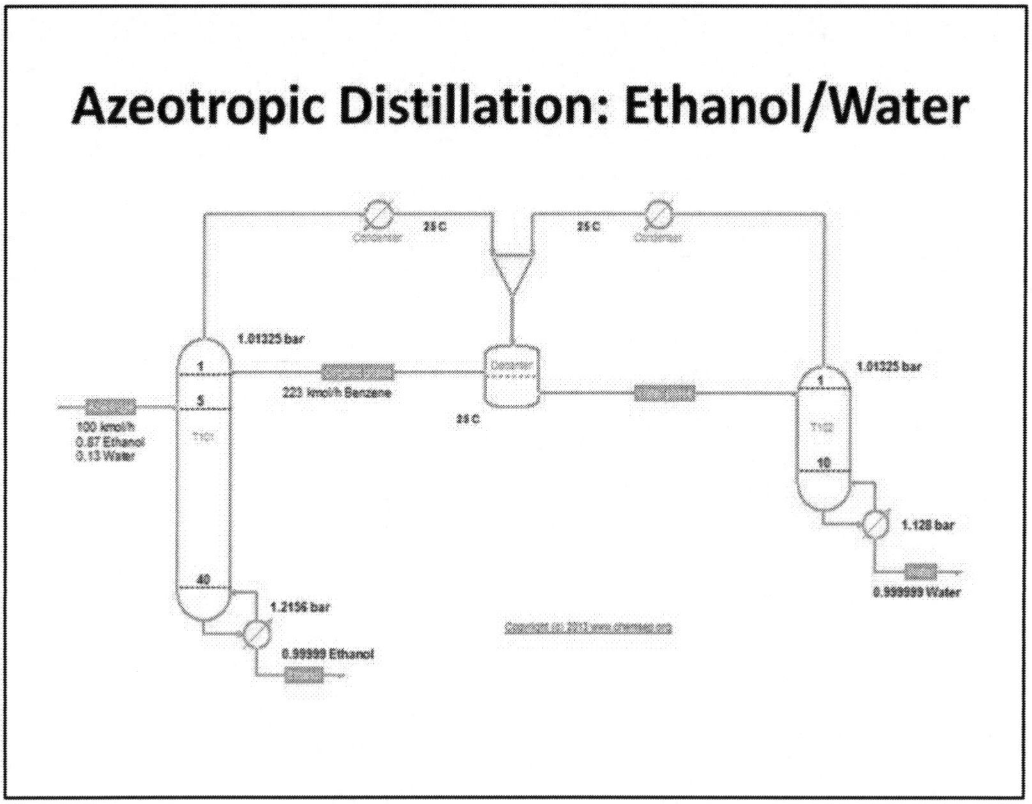
Azeotropic Distillation: Ethanol/Water
Condenser
25 C
25 C
Condenser
1.01325 bar
1
5
Decanter
223 kmol/h Benzene
100 kmol/h
0.87 Ethanol
0.13 Water
25 C
1.01325 bar
1
10
1.128 bar
0.999999 Water
40
1.2156 bar
0.99999 Ethanol

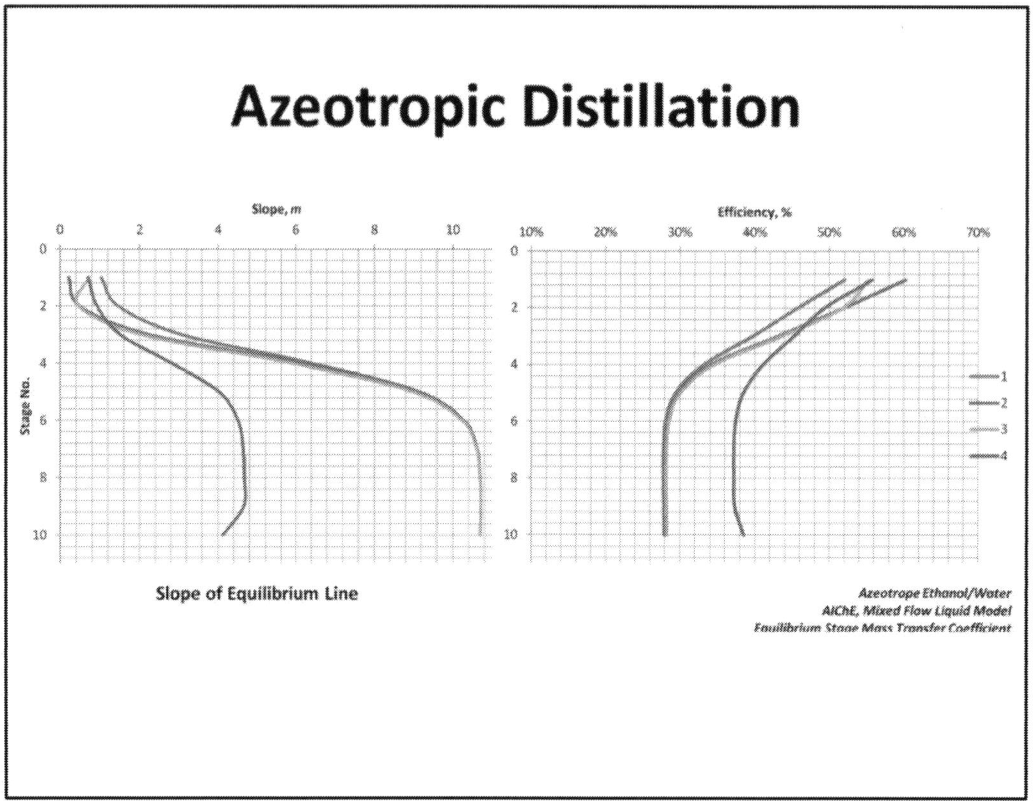

Here we show profiles for the azeotropic column (the first in the flowsheet on the prior page). Significant differences here for this highly non-ideal system.

Extractive Distillation

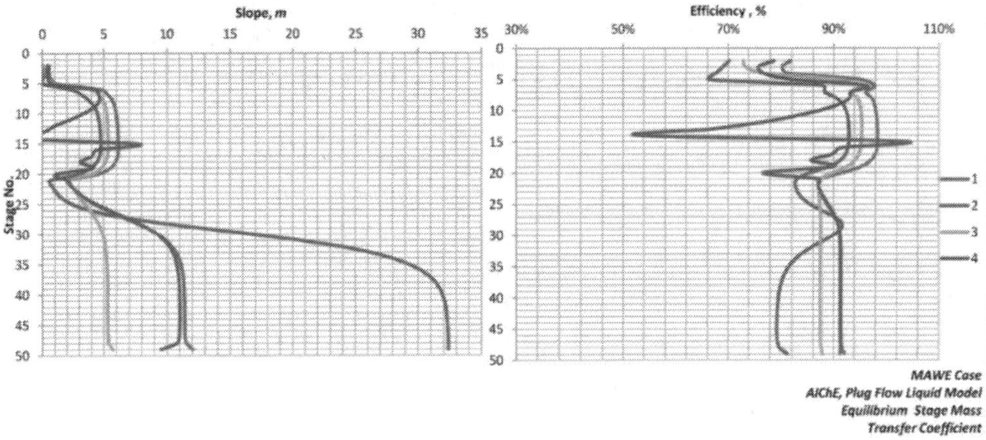

Similarly for this very nonideal extractive distillation; the methods show considerable differences.

More Open Questions

- Do dramatically different slopes matter?
- How can we answer the above?
  - Compare column profiles to data

It is not a simple matter to determine if these different approximations to the slope actually have an impact on the design and simulation of real columns.

To properly compare the methods we need composition profiles taken from operating columns. Such data is, unfortunately, relatively rare.

Gasoline Stabilizer (Ragatz)

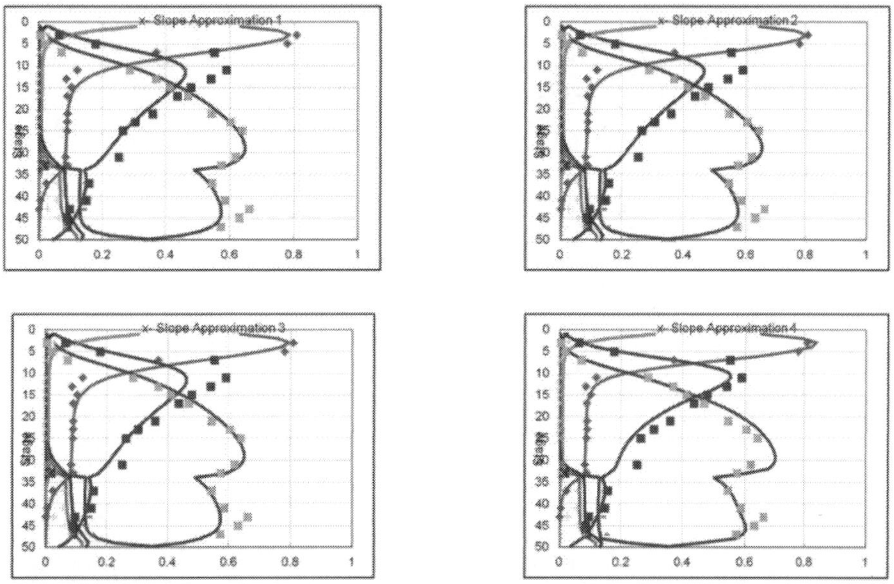

Here we show composition profiles for the four different methods to data for a gasoline stabilizer.

This is a system for which all methods should give reasonably close so it is not possible to draw definitive conclusions.

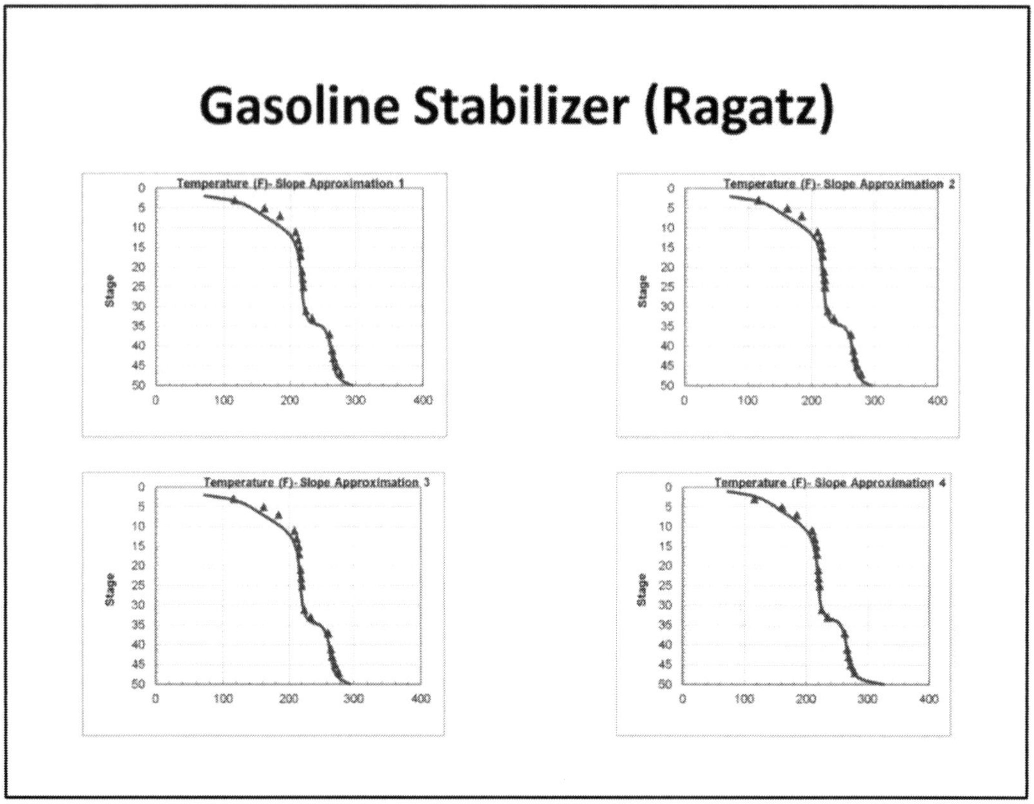

Temperature profiles for the gasoline stabilizer.

Acetone Absorber

- HK Water/ LK Acetone
  - High slope
- Complications with Method 4
  - Negative slope causing negative tray efficiency

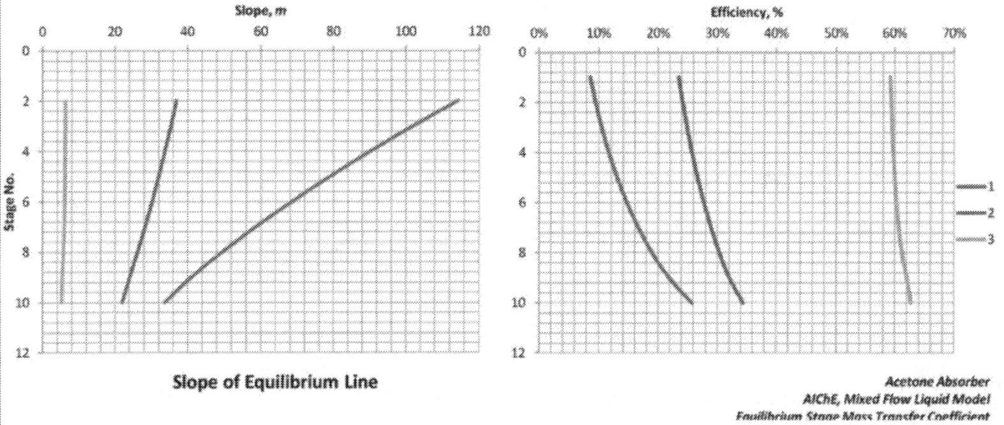

Significant differences for the acetone absorber.

Here we see that the choice of key components is of great importance when the relative volatility is so high.

Acetone Absorber

- HK Water/ LK Nitrogen
  - VERY HIGH slope: ~ 10^3 - 10^4 order
- Extremely low efficiency

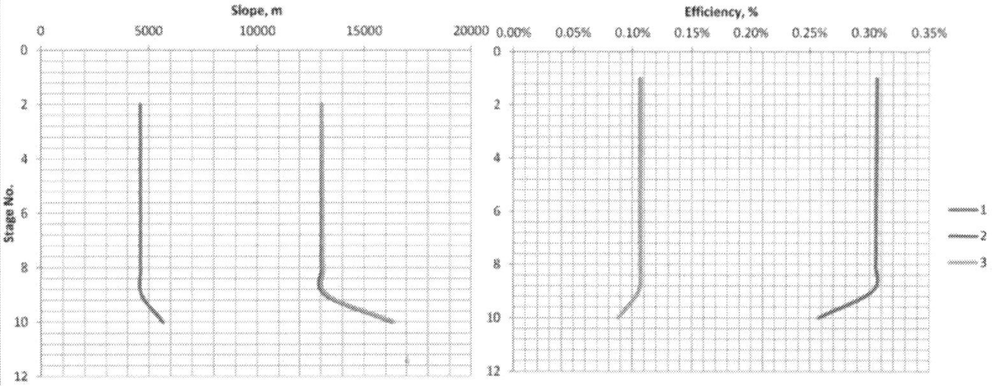

Slope of Equilibrium Line

Acetone Absorber, LKC=N_2
AIChE, Plug Flow Liquid Model
Equilibrium Stage Mass Transfer Coefficient

Slippery Slopes!

- Negative slopes
- Can lead to "absurd" efficiencies
- Methods 3 and 4 especially sensitive

Conclusions

- Many ways to approximate the slope of the pseudo-binary equilibrium "line" in multicomponent systems
- Constant alpha method OK for constant alpha systems
- Limited data makes it difficult to recommend a method
- But composition profiles not that sensitive to slope
- Key component selection matters
- The thermodynamic model matters
- Watch out for negative slopes!

More Open Questions

- What other methods exist for approximating the slope?
- Do you ever use multiple key pairs in a single column simulation?

Simulation of Batch Distillation with a Varying Alpha (α): The Ethanol-Water and Methanol-Water Systems at 1 Atm

Robert G. Kunz

RGK Environmental Consulting, L.L.C.
Hillsborough, North Carolina

Paper 99b for Presentation at

2015 AIChE Spring Meeting

April 26-30, 2015

Austin, Texas

Abstract

In a previous presentation [**1**], a newly developed technique was described that enables analytical integration of the Rayleigh Equation for batch distillation with a varying relative volatility (α). Heretofore, integration in closed analytical form has required that α be assumed constant. This new technique produces a continuous analytical function for the Rayleigh Equation integral for a varying α and reduces algebraically to the traditional function when α is constant. Detailed derivations are contained in the cited presentation.

Both the traditional expression at constant α and the new equation were evaluated against numerical integration via Simpson's Rule for a range of typical values of constant α, for a number of ideal and non-ideal binary vapor-liquid equilibrium (VLE) systems with varying α, and for an illustrative example contained in a chemical engineering textbook. The method was then utilized successfully in a computer-simulated batch distillation of an ethanol-water solution at 1 atm pressure. Results compared favorably with expected values.

The present paper reviews the ethanol-water case and extends the simulation procedure to methanol-water. Batch distillation of these systems at atmospheric pressure is a popular student laboratory exercise in colleges and universities throughout the world.

A summary of VLE relationships for ideal and non-ideal systems plus temperature-dependent methods for analysis of constituent composition specific to both the ethanol-water and methanol-water systems are discussed in separate appendices. These latter include, among others, estimates of liquid density for alcohol-water mixtures extended beyond the range of published data and refractive index measurements assembled from various sources.

The paper consists of new material supported by background information tutorial in nature.

Keywords. Alcohol "Proof"; Boiling Point Curves; Computer Simulation, Density, Refractive Indices, Gas Chromatography, Flash Points, and Auto-ignition Temperatures for Ethanol-Water and Methanol-Water Solutions; Numerical Integration; Rayleigh Equation; Simpson's Rule; Systems: Ethanol-Water, Methanol-Water, and Methanol-Ethanol; Vapor-Liquid Equilibrium.

Disclaimer
("Some restrictions apply; batteries not included; your mileage may vary")

The following caveat applies: The information contained herein is offered in good faith but without guarantee, warranty, or representation of any kind (expressed or implied) as to its usefulness, correctness, completeness, or fitness for any particular purpose. The user assumes all risk for its implementation and should seek independent professional verification of its accuracy. The author assumes no responsibility and shall not be liable for any loss of profit nor any special, incidental, consequential, or other damages which may result from the use of any of the information contained in this presentation, be it oral or written. Any statements concerning design, construction, operation, what constitutes regulatory compliance, and/or how to achieve such compliance should not be construed as recommendations on the part of the author and/or his organization.

Introduction

Distillation is a separation technique utilizing the difference in composition of a liquid and the vapor in equilibrium with that liquid [**2,3**]. Simple, or differential, distillation employs a single equilibrium stage to effect that separation. A batch distillation operates on an unreplenished charge of material being boiled off from a still pot and then condensed as product in another vessel. It is an unsteady-state process, where the amount and composition of feed and product vary with time. In a two-component (binary) system, the charge of feed material boils away and gradually becomes richer in the less volatile/higher-boiling/ heavy component. The condensed vapor product (distillate) increases in volume and is enhanced in the more volatile/lower-boiling/light component. This paper discusses the simple, differential batch distillation of a two-component (binary) system, in which none of the condensed vapor is returned, or refluxed, to the still pot. A typical laboratory setup is shown in Figure 1.

Simple batch distillation is not practiced extensively in industry. It is used commercially for separating small quantities of high-value chemicals, when high purities are not required, as a preliminary step to be followed by further processing, for separations where the equipment requires frequent cleanout, or where the feed mixture is very easy to separate [**2,3**]. A rule of thumb for ease of separation is that the feed components differ in atmospheric boiling point by at least 25°C [**4**, pp.167-168], 30-40°C [**5**, p.61], or 75°C [**6**, p.89; **7**, p.89], depending on the source. Whatever the precise numerical value, the intent is that the lighter component be much more easily vaporized in preference to the heavier component in order to distill a decent amount of high-purity product before too much of the feed material boils away.

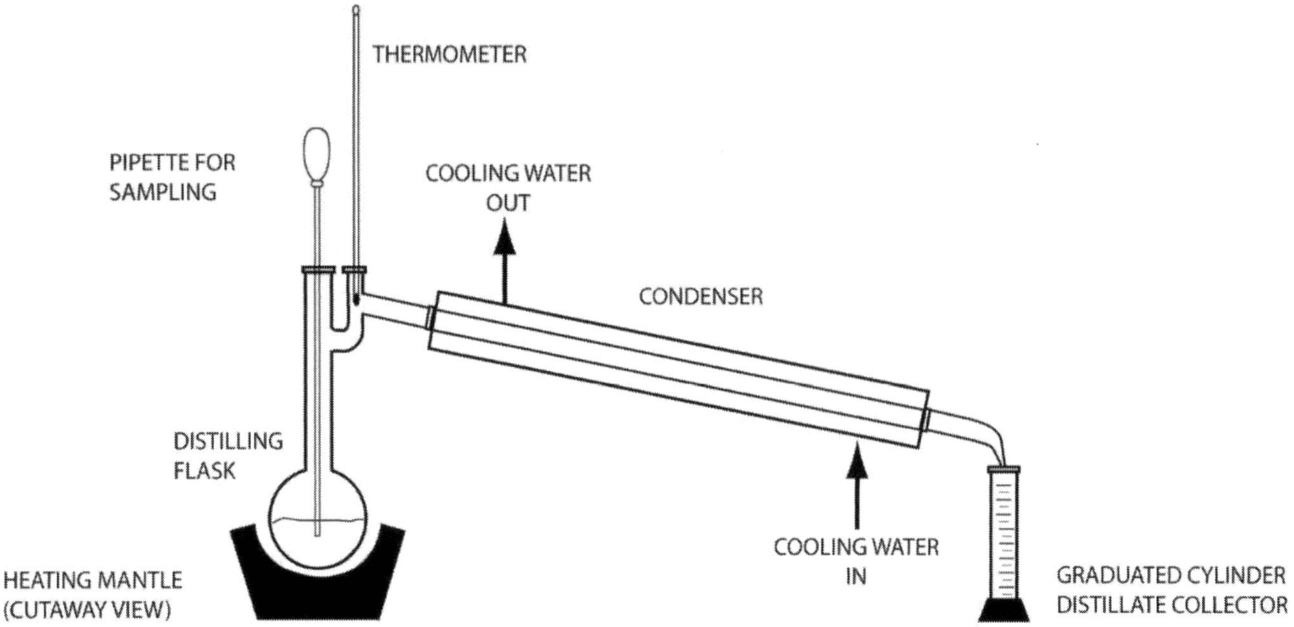

Figure 1. Typical Laboratory Setup for Simple, Differential Batch Distillation

The Rayleigh Equation

Simple batch distillation without reflux is described by the Rayleigh Equation, first advanced by Lord Rayleigh in 1902 [**8**]. A step-by-step derivation [**2**; **9**, p.580-581; **10**; **11**] is contained in a previous presentation.

$$\ln (W/F) = \int_{x_F}^{x_W} dx/(y - x) \tag{1}$$

where

F = total moles of feed in the still pot/distilling flask at starting time (t) = 0
W = total moles left in still pot (bottoms) at any time
D = total moles boiled off and condensed into the distillate at the same point in time
x = mol fraction of the more volatile component (MVC) in the liquid phase
y = mol fraction of the MVC in the vapor phase in equilibrium with x

Other quantities are determined by material balance or a component mass balance:

$$F = W + D \tag{2}$$

$$Fx_F = Wx_W + Dx_D \tag{3}$$

$$x_D = (Fx_F - Wx_W) / (F - W) \tag{4}$$

For the first drop of distillate, when W = F and x_W = x_F, the expression in Equation (4) becomes mathematically indeterminate. and x_D is provided by the x-y equilibrium curve, as is the instantaneous composition of the distillate at every moment during the distillation.

The Rayleigh Equation appears simple; yet its evaluation is somewhat involved. The integrand is not a simple function of x: at constant pressure, every value of x corresponds to a different temperature; except for the most ideal systems, y is not related to x in a simple manner. Important variables are temperature (t), total pressure (P_T), partial pressure (p), vapor pressure (p^{vap}), liquid-phase (x) and vapor-phase (y) mol fractions, plus so-called activity coefficients (γ's) when a given system cannot be considered to be ideal. Equations for low-pressure binary vapor-liquid equilibrium (VLE) are summarized in Appendix A.

The standard procedure is to evaluate the integral numerically or graphically,* or to come up with a suitable substitution in the hopes of producing an analytical solution in closed mathematical form. Numerical integration involves repeatedly fitting the numerical values of the integrand with a mathematical function over short ranges, integrating those functions mathematically, and then summing. This numerical fitting and summing process results in a series of discrete values, rather than a continuous function.

One analytical approach makes use of a parameter known as the *relative volatility*, denoted by the symbol *alpha* (α). It is defined as the ratio of the vapor-phase mol fraction of the more volatile component, denoted as *component 1,* to its liquid-phase mol fraction, this ratio being divided by a similar ratio for the less volatile *component 2.* For a successful separation by distillation, α must be greater than 1.0. In the case of a two component system

$$\alpha = (y_1/x_1) \,/\, (y_2/x_2) = (y_1/x_1) \,/\, [(1 - y_1) \,/\, (1 - x_1)] \qquad (5)$$

since $y_2 = (1 - y_1)$ and $x_2 = (1 - x_1)$.

Rearranging Equation (8), one obtains for $1 / (y_1 - x_1)$ to be substituted in Equation (1).

$$\frac{1}{(y_1 - x_1)} = \frac{1 + (\alpha - 1)\, x_1}{(\alpha - 1)\, x_1\, (1 - x_1)} = \frac{1}{(\alpha - 1)\, x_1\, (1 - x_1)} + \frac{1}{(1 - x_1)} \qquad (6)$$

The relative volatility (α) is evaluated at constant total pressure for any temperature or the corresponding value of composition at every point along the vapor-liquid equilibrium (VLE) x-y curve from a ratio of vapor pressures and, if necessary, activity coefficients:

$$\alpha = (y_1/x_1) \,/\, (y_2/x_2) = (p_1^{vap}\,\gamma_1) \,/\, (p_2^{vap}\,\gamma_2) \qquad (7)$$

The traditional approach has been to assume α to be constant and integrate Equation (6) to obtain Equation (8) below [**14**, #40, #29]:

* In previous times, in addition to counting boxes under the curve of the integrand or using a planimeter device [**12**], the curve was drawn carefully on high quality graph paper of uniform density and then was weighed and calibrated against a rectangular section of that same graph paper. This method is noted in Ref. [**13**, p.75].

$$\ln(W/F) = \frac{1}{(\alpha - 1)} \left\{ \alpha \ln[(1 - x_F)/(1 - x_W)] + \ln(x_W/x_F) \right\} \tag{8}$$

The percent of original charge remaining in the still pot is then given by:

$$100\,(W/F) = 100 \exp[\ln(W/F)] \tag{9}$$

However, even for the most ideal of systems, relative volatility is not exactly constant over the entire range of x.

A New Twist

Nevertheless, alpha need not necessarily be constant to obtain a closed-form analytical solution to Equation (1) as long as the factor $1/(\alpha - 1)$ can be fitted as a function of x over the range of interest, for example:

$$1/(\alpha - 1) = a x^2 + b x + c \tag{10}$$

This is much the same technique as the expression of gaseous heat capacities by an empirical power series in temperature.

With $1/(\alpha - 1)$ so represented, substitution in Equation (6) produces Equations (11) and (12), which can then be integrated analytically term by term [**14**, #32, #29, #40]:

$$\ln(W/F) = \int_{x_F}^{x_W} \frac{(a x_1^2 + b x_1 + c)\, dx_1}{x_1 (1 - x_1)} + \int_{x_F}^{x_W} \frac{dx_1}{(1 - x_1)} \tag{11}$$

$$\ln(W/F) = \int_{x_F}^{x_W} \frac{a x_1\, dx_1}{(1 - x_1)} + \int_{x_F}^{x_W} \frac{(1 + b)\, dx_1}{(1 - x_1)} + \int_{x_F}^{x_W} \frac{c\, dx_1}{x_1(1 - x_1)} \tag{12}$$

When the smoke clears and the dust settles, one is left with the following:

$$\ln(W/F) = a(x_F - x_W) + (1 + a + b + c) \ln[(1 - x_F)/(1 - x_W)] + c \ln[x_W / x_F] \tag{13}$$

The only source of error is the goodness of fit of the quadratic function in Equation (10). When α is constant for all values of composition, a = b = 0, c = 1/ (α − 1) identically, and Equation (23) reduces to Equation (13), the solution of the Rayleigh Equation with α constant. If necessary, an x^3 term can be added to Equation (10) but adds more terms to Equation (13).

Testing of the New Function against Numerical Integration

Simpson's Rule. Both the traditional expression at constant α and the new equation were evaluated against numerical integration via Simpson's Rule for a broad range of typical values of constant α, for a number of ideal and non-ideal binary vapor-liquid equilibrium (VLE) systems with varying α, and for a illustrative example contained in a chemical engineering textbook. A summary is given below; details have been presented previously [**1**].

It has been found that repeated application of Simpson's Rule in steps of 0.02 and an interval size of 0.01 mol fraction reproduces the exact integral at constant α for ln (W/F), even when exponentiated, to a degree better than can be measured experimentally. This situation applies down to a mol fraction bordering on 0.02. Narrowing the interval size at the lower end of the mol fraction range, where the integrand begins to increase rapidly, maintains the accuracy of the numerical integration even further. Eventually, however, the integrand grows without bounds as mol fraction remaining in the still pot approaches zero.

Numerous real vapor-liquid equilibrium systems were then chosen to test how well Equation (23) works in practice over the range of α from x_1 = 0.02 to x_1 = 0.5 when alpha is not exactly constant. Here, the previously validated Simpson's Rule integration is the standard, using the same interval size found suitable for the numerical integration with constant α. The overall conclusion is that agreement of the newly derived Equation (23) with numerical integration is as good as when α is constant. In addition, one has a continuous function in Equation (23) as opposed to numerical integration, which is valid only at discrete points.

Illustrative Example. The technique produced virtually identical results when tested against an illustrative example presented in McCabe and Smith [**11**] for batch/differential distillation of a 50/50 mol percent benzene-toluene solution at 1 atm pressure with 1/α taken as constant at 0.41. Calculated values for temperature and benzene compositions at various locations were plotted in the illustrative example against the mol fraction of distillate / charge distilled (D/F).

Testing of the New Function in Simulated Laboratory Experiment. Finally, to demonstrate the ease of use of the newly developed technique, batch distillation of a pure ethanol-water solution was simulated at a standard atmospheric pressure of 760 mmHg (1.013 bar). The methanol-water system has been added for discussion here. Results compare favorably with expected values. Both are favorite student laboratory exercises and constitute the main subject for the present paper.

Simulation of Laboratory Experiments

Experimental. An initial 200-mL charge of a pure ethanol-water or methanol-water solution, 20% alcohol by volume at 60 °F (15.56 °C) in an apparatus similar to Figure 1 at standard atmospheric pressure (760 mmHg, 1.013 bar) is typical [**6,7,15**]. This corresponds to 16 wt % approximately, and ~ 0.07-0.1 mol fraction alcohol (Tables 1 and 2). Volume of the distilling flask is 500 millimeters (mL). Boiling stones are added to prevent bumping [**4**, pp.146,170].

Samples of distillate and perhaps still bottoms are taken as every 5 mL of distillate is collected, and temperature is read at such times. The experiment is over when 50 mL of distillate has been collected. Sometimes the total collected distillate is subjected to further purification in a second step using a separate refluxed fractionating column. This experiment has also been conducted using denatured alcohol and water, fermented sugar solutions, and wine being distilled into brandy.

The student is cautioned to wait to read the temperature until droplets condense on the thermometer bulb [**4**, p.172]. The experiment then proceeds drop by drop. The recommended distilling rate varies, but numbers in the range of 2 [**6,7**] to 10 [**4**, p.171] drops per minute are stated. At an average 6 drops per minute (1 drop every 10 seconds), and 20 drops to the millimeter, collection of 50 mL would take just under 3 hours once equilibrium conditions are achieved, exclusive of analytical time. This would allow completion of the first step of the experiment within a typical standard laboratory period.

Quantitative analysis is done by refractive index, density, or possibly gas chromatography (Appendix B). Whatever the method of analysis, it is good practice to prepare one's own calibration curve(s) using the materials and analytical equipment employed rather than relying exclusively on literature data. A qualitative analysis for ethanol in the distillate is often conducted by attempting to ignite several drops of solution on a watch glass, since ethanol and its aqueous solutions greater than about 50 % by volume are combustible at room temperature [**16**]. This technique would also work for methanol, whose aqueous solutions containing ~ 40 % or more by volume also ignite [**17**]. (Refer again to Appendix B.)

Simulation. A computer simulation of the laboratory experiments described above was conducted using a personal computer (PC) spreadsheet. The simulated experiments are valid only for a binary mixture of pure alcohol and water, with an x-y VLE curve as in Figure A2 or Figure A3. It does not apply quantitatively to denatured ethanol containing methanol and other ingredients, fermented sugar solutions, or wine being distilled into brandy.

The simulation begins when equilibrium becomes established as the temperature in the laboratory experiment stabilizes and the first drop of liquid falls into the receiving vessel. The simulation also assumes that the distillation takes place slowly enough so as to maintain true vapor-liquid equilibrium with a drop of liquid condensed on the thermometer at all times.

Results from the simulations are summarized in Tables 1 and 2 for the first 50 mL of distillate collected at 20 °C. Temperature, amount of bottoms remaining and distillate formed, plus composition of bottoms and distillate are recorded at every 5 mL. There are two ways to conduct such an experiment – save each 5-mL increment separately for individual analysis, or let the distillate accumulate as a composite sample. Values are listed both ways in the tables.

Gram moles of bottoms (W) and mol fraction of alcohol in the bottoms (x_W) are related by Equation (13), using constants for $1/(\alpha - 1)$ from Equation (10) fitted in the narrow ranges of interest for x_W (a = 1.73317, b = 0.511067, c = 0.103302 for ethanol and a = 0.449813, b = 0.315463, c = 0.146141 for methanol). These constants provide a better fit for $1/(\alpha - 1)$ over the limited range of x_W in this "experiment" rather than the coefficients listed in Ref. [**1**] covering a larger span of x_W.

Moles of distillate (D) and mol fraction in the distillate (x_D) are computed from material balances [Equations (2) and (3), rearranged]. Moles are converted to grams (gm) using a molecular weight of 46.07 for ethanol, 32.04 for methanol, and 18.02 for water from Table A1:

$$gm\ C_2H_5OH = 46.07 \times mol\ C_2H_5OH,\quad gm\ CH_3OH = 32.04 \times mol\ CH_3OH,\quad gm\ H_2O = 18.02 \times H_2O \quad (14)$$

Mol fraction (m.f.) of ethanol or methanol is converted to wt % by:

$$\text{Ethanol:}\quad wt\ \% = 100 \times (m.f.)\,(46.07)\ /\ [(m.f.)\,(46.07) + (1 - m.f.)\,(18.02)] \quad (15)$$

$$\text{Methanol:}\quad wt\ \% = 100 \times (m.f.)\,(32.04)\ /\ [(m.f.)\,(32.04) + (1 - m.f.)\,(18.02)] \quad (16)$$

Mass is converted to volume by means of density at known temperature and composition, either at an assumed laboratory temperature of 20 °C or as estimated at the temperature of a distillate sample or that of the solution remaining in the distilling flask (Appendix B). Tables in Ref. [**18**, Tables 6.22 and 6.36], interpolated if necessary, relate wt % and % by volume for ethanol at the standard temperature of 60 °F (15.56 °C). For methanol, vol % at 60 °F was similarly determined from the density table in Perry's Handbook [**19**] and Equation (B2).

Graphical Results

Results of the simulation, extended past the entries in Tables 1 and 2, are plotted in Figures 2 and 3 following the same functional form suggested in the illustrative example for benzene-toluene in Ref. [**11**]. The alcohol composition in the distilling flask starts off at its initial value and approaches zero as the batch distillation proceeds (bottom curves in the figures). The last drop in the bottoms is pure water, although for safety reasons one would not actually want to distill the mixture to dryness. Composition of the distillate is listed in the last three columns of the tables. Three versions of distillate collection are shown. In Column 5, data are recorded at every 5 mL as the distillate is allowed to accumulate in the receiving vessel (top curves). In Column 6, the receiving vessel is changed as soon as each 5 mL of distillate is collected, and the individual distillate samples are kept separate for "analysis" (not plotted) Column 7 represents the instantaneous composition at each temperature and distillate volume noted (middle curves). For the first drop of distillate (0 mL accumulated), all three columns are exactly the same. Composition is also the same, but at a different value, in Columns 5 and 6 for the first 5 mL of accumulated sample. For other entries, compositions show a regular progression from top to bottom and from left to right in Columns 5, 6, and 7.

The atmospheric boiling temperature for a binary mixture depends on the vapor pressures, mol fractions, and activity coefficients of the pure components. Plotting distillate temperature against cumulative distillate from Tables 1 and 2 (Figures 2 and 3) results in the gentle s-shaped curve shown in Fig. 5.8 in Refs. [**6,7**] and Fig. 5.1 of Ref. [**20**]. The simulated experiment has been extended beyond the range of Tables 1 and 2 to collect extra distillate and provide additional temperature data points in the figures. The curves for ethanol-water and methanol-water cross at 100 °C for the last droplet remaining and extrapolate on the other end to the respective boiling points of pure ethanol and pure methanol.

The densities from Figures B1 and B2 in Appendix B can be used to estimate the hot still-pot liquid remaining. Bottoms volume in Column 2 at each boiling temperature is 3-5 % greater than the volume of the same liquid at 20 °C. Change in volume remaining is therefore not accounted for exactly by subtracting the volume of distillate removed, measured at 20 °C.

Table 1. Summary of Simulated Batch Distillation Experiment – Ethanol-Water

Temp (°C)	Bottoms (W) g moles mass (g) mL @ 20 °C *mL @ temp*	Distillate (D) g moles mass (g) mL @ 20 °C	Ethanol in Bottoms (x_W) mol fract wt % *% vol*	Ethanol in Distillate (x_D)		
				Cumulative mol fract wt % *% vol*	Individual mol fract wt % *% vol*	Instantaneous mol fract wt % *% vol*
89.1	9.7367 194.6994 **200.0** ***209.5***	0.0000 0.0000 **0**	0.0706 16.269 ***20.0***	0.3704 60.07 ***67.8***	0.3704 60.07 ***67.8***	0.3704 60.07 ***67.8***
89.6	9.5787 190.2358 **195.2** ***204.2***	0.1581 4.4636 **5**	0.0658 15.26 ***18.8***	0.3645 59.46 ***67.2***	0.3645 59.46 ***67.2***	0.3577 58.74 ***66.5***
90.0	9.4180 185.7561 **190.3** ***199.0***	0.3188 8.9433 **10**	0.0609 14.22 ***17.5***	0.3578 58.76 ***66.5***	0.3513 58.07 ***65.8***	0.3439 57.27 ***65.0***
90.5	9.2542 181.2583 **185.4** ***193.7***	0.4825 13.4410 **15**	0.0560 13.17 ***16.3***	0.3508 58.04 ***65.8***	0.3370 56.51 ***64.3***	0.3289 55.63 ***63.5***
91.0	9.0871 176.7406 **180.5** ***188.5***	0.6496 17.9588 **20**	0.0511 12.11 ***15.0***	0.3432 57.20 ***65.0***	0.3214 54.78 ***62.6***	0.3128 53.79 ***61.6***
91.6	8.9161 172.2004 **175.6** ***183.3***	0.8206 22.4989 **25**	0.0463 11.04 ***13.7***	0.3352 56.32 ***64.1***	0.3045 52.83 ***60.7***	0.2952 51.72 ***59.6***
92.2	8.7409 167.6352 **170.7** ***178.0***	0.9959 27.0641 **30**	0.0415 9.96 ***12.3***	0.3265 55.36 ***63.2***	0.2863 50.63 ***58.5***	0.2763 49.40 ***57.2***
92.8	8.5607 163.0420 **165.8** ***172.8***	1.1760 31.6573 **35**	0.0367 8.88 ***11.0***	0.3174 54.31 ***62.2***	0.2665 48.17 ***56.0***	0.2559 46.79 ***54.5***
93.5	8.3750 158.4176 **160.8** ***167.5***	1.3618 36.2818 **40**	0.0321 7.82 ***9.7***	0.3075 53.10 ***61.0***	0.2454 45.40 ***53.1***	0.2341 43.87 ***51.5***
94.2	8.1830 153.7584 **155.8** ***162.3***	1.5538 40.9410 **45**	0.0276 6.77 ***8.3***	0.2971 51.94 ***59.8***	0.2228 42.29 ***49.8***	0.2109 40.61 ***48.0***
94.9	7.9839 149.0609 **150.8** ***157.0***	1.7529 45.6385 **50**	0.0233 5.76 ***7.2***	0.2859 50.59 ***58.4***	0.1909 38.84 ***46.1***	0.1867 36.99 ***44.0***

Table 2. Summary of Simulated Batch Distillation Experiment – Methanol-Water

Temp (°C)	**Bottoms (W)** **g moles** **mass (g)** **mL @ 20 °C** ***mL @ temp***	**Distillate (D)** **g moles** **mass (g)** **mL @ 20 °C**	**Methanol in Bottoms (x_W)** **mol fract** wt % *% vol*	**Methanol in Distillate (x_D)**		
				Cumulative **mol fract** wt % *% vol*	**Individual** **mol fract** wt % *% vol*	**Instantaneous** **mol fract** wt % *% vol*
87.9	10.0386 194.8438 **200.0** ***206.7***	0.0000 0.0000 **0**	0.0991 16.359 ***20.0***	0.4169 55.97 ***63.8***	0.4169 55.97 ***63.8***	0.4169 55.97 ***63.8***
88.4	9.8485 190.3237 **195.1** ***201.7***	0.1902 4.5201 **5**	0.0931 15.26 ***18.7***	0.4100 55.27 ***63.1***	0.4100 55.27 ***63.1***	0.4029 54.54 ***62.3***
88.9	9.6560 185.7881 **190.1** ***196.6***	0.3826 9.0558 **10**	0.0871 14.50 ***17.8***	0.4027 54.52 ***62.3***	0.3957 53.79 ***61.6***	0.3880 52.99 ***60.8***
89.5	9.4610 181.2358 **185.2** ***191.6***	0.5776 13.6081 **15**	0.0810 13.55 ***16.6***	0.3951 53.74 ***61.5***	0.3807 52.22 ***60.0***	0.3722 51.32 ***59.1***
90.0	9.2634 176.6655 **180.3** ***186.5***	0.7753 18.1783 **20**	0.0750 12.60 ***15.5***	0.3872 52.90 ***60.7***	0.3647 50.51 ***58.2***	0.3554 49.51 ***57.2***
90.6	9.0628 172.0760 **175.3** ***181.5***	0.9758 22.7678 **25**	0.0690 11.64 ***14.3***	0.3788 52.02 ***59.8***	0.3478 48.67 ***56.3***	0.3376 47.54 ***55.1***
91.2	8.8590 167.4658 **170.4** ***176.4***	1.1796 27.3780 **30**	0.0630 10.68 ***13.2***	0.3701 51.09 ***58.8***	0.3299 46.68 ***54.2***	0.3188 45.42 ***52.9***
91.8	8.6517 162.8333 **165.4** ***171.4***	1.3869 32.0105 **35**	0.0571 9.72 ***12.0***	0.3610 50.11 ***57.8***	0.3112 44.55 ***52.0***	0.2989 43.12 ***50.5***
92.4	8.4407 158.1771 **160.5** ***166.3***	1.5979 36.6668 **40**	0.0513 8.78 ***10.9***	0.3514 49.06 ***56.7***	0.2920 42.31 ***49.6***	0.2780 40.64 ***47.8***
93.1	8.2256 153.4954 **155.5** ***161.2***	1.8130 41.3484 **45**	0.0457 7.85 ***9.7***	0.3414 47,96 ***55.6***	0.2721 39.93 ***47.0***	0.2562 37.98 ***44.9***
93.8	8.0061 148.7868 **150.5** ***156.1***	2.0325 46.0570 **50**	0.0402 6.94 ***8.6***	0.3310 46.80 ***54.4***	0.2517 37.43 ***44.3***	0.2335 35.14 ***41.7***

Density of the residual liquid (ρ_{BOTTOMS}) at each temperature and composition is plotted against temperature in Figure 5. The curves intersect at the liquid density of pure water boiling at 100 °C and 1 atm. "Experimental" points extrapolate smoothly to this datum along the curves drawn in the figure. At the other end, the curves extrapolate to the densities of pure methanol and pure ethanol at their respective atmospheric boiling points.

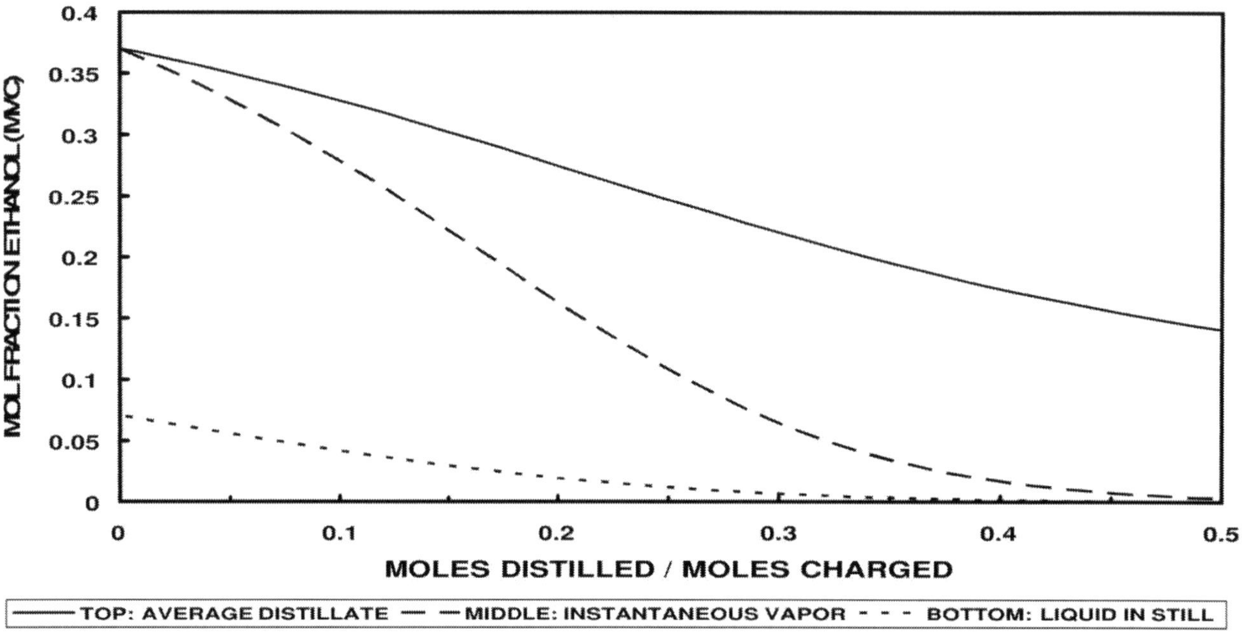

Figure 2. Batch Distillation of a 20 Volume % Ethanol-Water Solution at 1 Atm Pressure

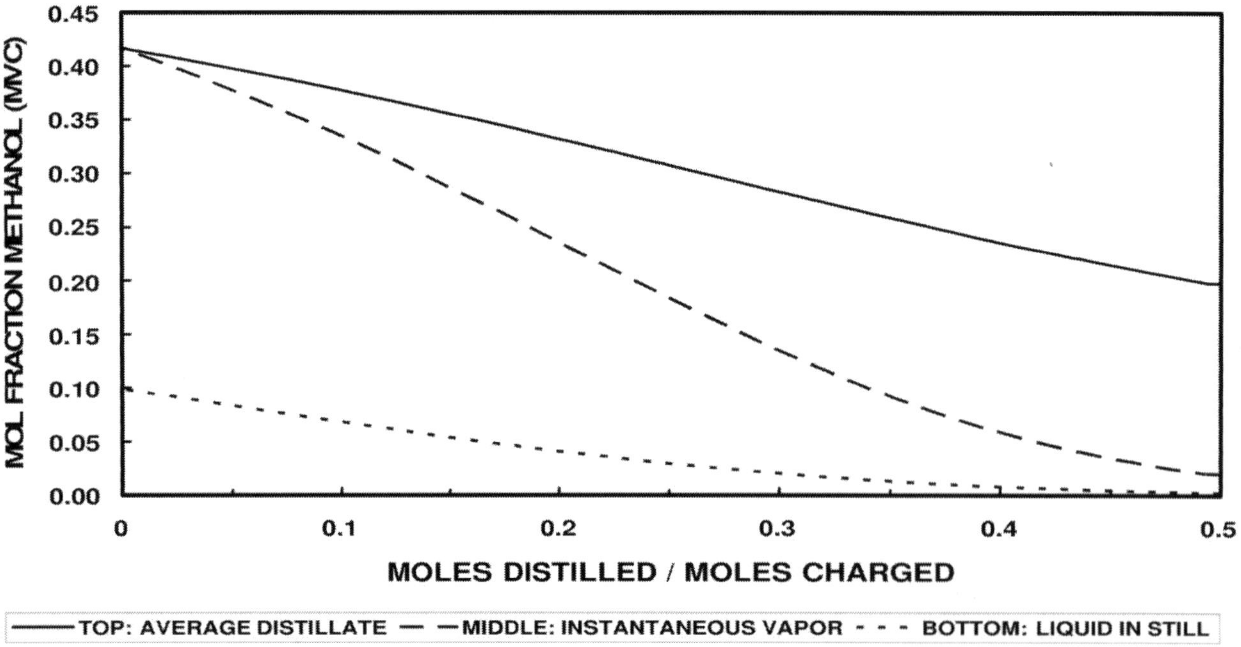

Figure 3. Batch Distillation of a 20 Volume % Methanol-Water Solution at 1 Atm Pressure

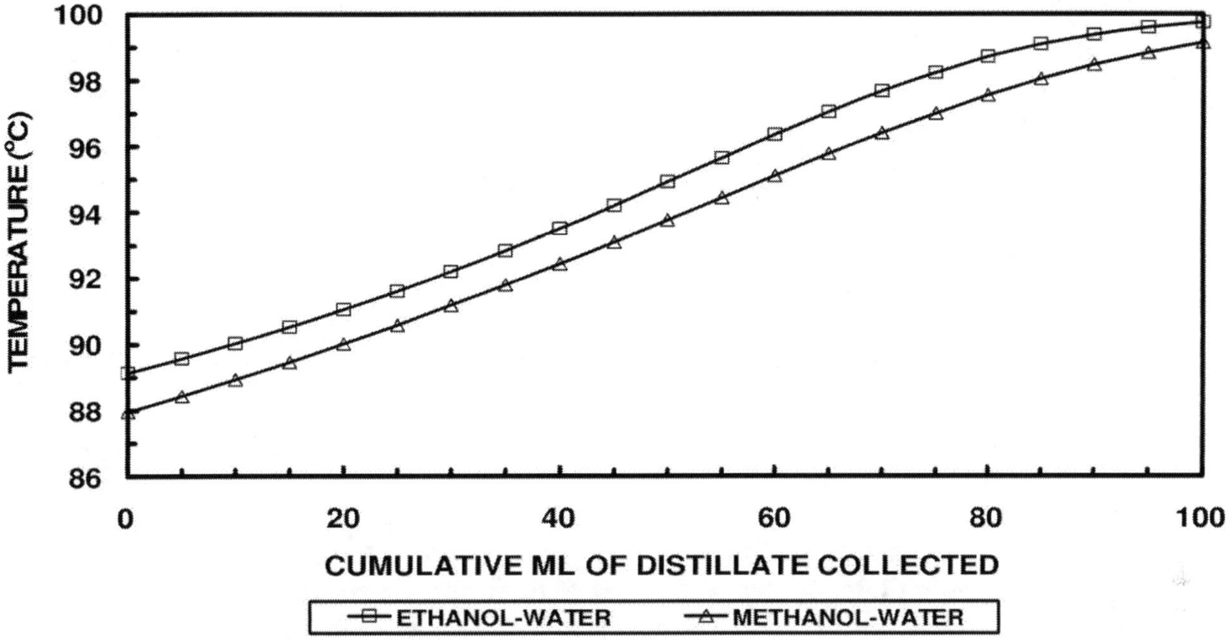

Figure 4. Still-Pot Temperature vs. Cumulative Volume of Distillate for Simulated Ethanol-Water and Methanol-Water Batch Distillations at 1 Atm Pressure

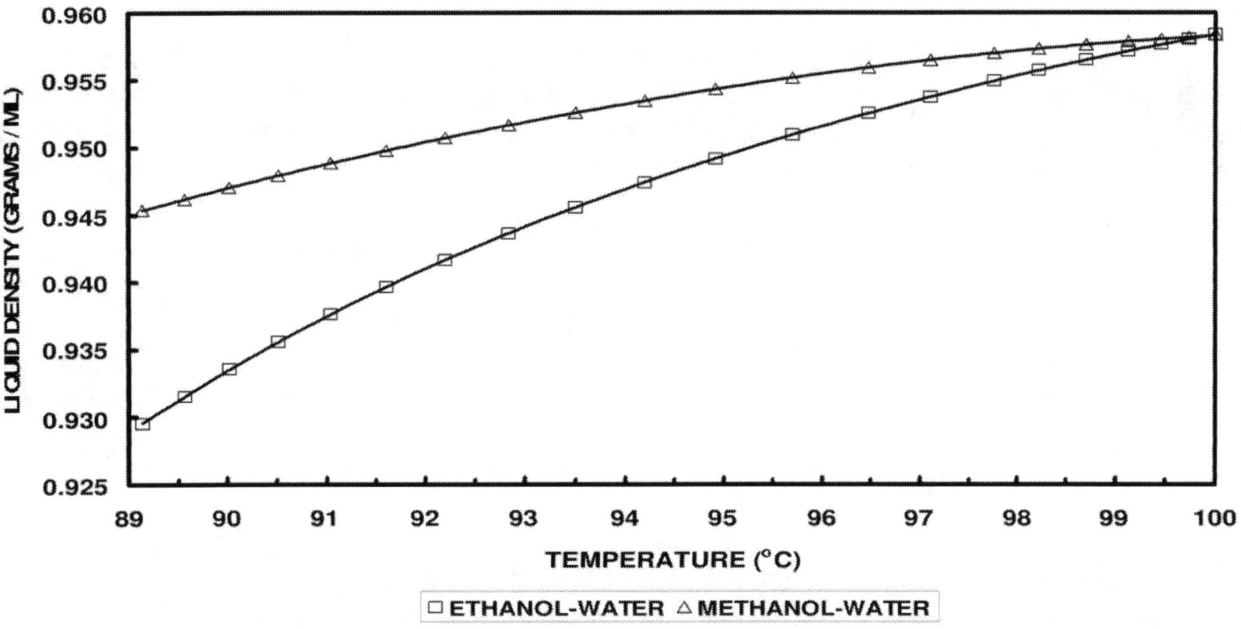

Figure 5. Estimated Density of Solution Remaining in the Still Pot at Boiling Temperature for Simulated Methanol-Water and Ethanol-Water Batch Distillations at 1 Atm Pressure

The empirical curve-fit function for ethanol-water, forced to pass through the 100 °C point:

$$\rho_{\text{BOTTOMS}} = 0.958380 + 1.35444 \times 10^{-3} (t - 100\ °C) - 8.1550 \times 10^{-5} [(t - 100\ °C)]^2 - 3.23156 \times 10^{-7} [(t - 100\ °C)]^4 \quad (17)$$

reproduces the points of the simulation with a correlation coefficient (r^2) of 100 % and a standard deviation (s^2) of 0.00002. For methanol-water, the function

$$\rho_{\text{BOTTOMS}} = 0.0201463 + 0.0147232 \times (t\ °C) - 0.00007155 \times (t\ °C)^2 \quad (18)$$

reproduces the points of the simulation with a correlation coefficient (r^2) of 100 % and a standard deviation (s^2) of 0.00008370. Calculated ρ_{WATER} at 100 °C is 0.95828 vs. 0.95838. These functions allow one to estimate the bottoms density (and therefore volume remaining) from temperature alone when batch-distilling without reflux a 20 vol % (60 °F, 15.56 °C) of ethanol-water or methanol-water solution. Division of mass by density gives volume.

In the laboratory, distillate solutions cooled to a standard temperature of perhaps 20 °C would be analyzed quantitatively by any of the methods discussed in Appendix B. For example, densities at laboratory temperatures are well known. Any samples extracted from the distilling flask should also be cooled to laboratory temperature before analysis by density.

On a qualitative basis, unheated ethanol-water distillate samples of about 50 % by volume or more would light off at laboratory temperature with a small flame and leave no residue on the watch glass (Appendix B). This qualitative ignition test should produce positive results at every 5 mL increment for the cumulative samples and for most of the individual samples (Table 1).

In some versions of the experiment, the composite ethanol-rich distillate is saved and further purified using a separate refluxed fractionating column. In a properly designed column, composition will approach pure water in the bottoms and the azeotrope in the overhead. A sample of that distillate would ignite on the watch glass as well, burning more vigorously with a brighter flame and again leaving no visible residue.

Based on flash-point data, anecdotal evidence from the Internet, and results of outdoor flammability tests (all discussed in Appendix B), all of the calculated methanol-water distillate compositions in Table 2 would likely be flammable when the ambient temperature is about 30 °C (86 °F) or above. In addition, the cumulative samples and most of the individual samples would be expected to ignite at a lower ambient temperature in the laboratory. Methanol fires are particularly dangerous because the flame is invisible in daylight to the naked eye [**21**]; combustible paper has been used to reveal the invisible flame [**17**].

Wrap-up

The newly developed equation has successfully facilitated the simulation of unrefluxed batch distillation experiments; however, modeling of any further purification of distillate samples in a fractionating column is beyond the scope of the present investigation.

Summary and Conclusions

- Batch, or differential, distillation without reflux is described by the Rayleigh Equation:
  - Relates composition and amount of material remaining in distilling flask
  - Other quantities determined by material balance.
- Numerical integration is required for the Rayleigh Equation in its basic form.
- Substitution of relative volatility (α) allows analytical integration for constant α.
- New equation derived here allows analytical integration whether α is constant or not.
- The new function was used in simulation of batch distillation of two alcohols and water.
- Ethanol-water and methanol-water results compare favorably with expectations, e.g.,
  - Boiling point vs. composition curve
  - Still pot temperature vs. cumulative volume of distillate collected
  - Density of solution remaining in still pot.
- Methods to analyze liquid composition are discussed for both systems:
  - Gas chromatography (GC).
  - Refractive index data compiled from various sources.
  - Densities of alcohol-water solutions extended beyond range of published data.
  - Qualitative ethanol-water and methanol-water ignition test / flash points.

About The Author

Robert G. Kunz was an environmental engineering manager at Air Products and Chemicals, Inc., Allentown, PA before retiring after 26+ years of service. He then joined Cormetech, Inc., Durham, NC as Technical Project Manager in support of sales and marketing efforts in the petroleum refining and petrochemical industries. He held engineering positions previously at Esso Research and Engineering Company, Florham Park, NJ and The M.W. Kellogg Company, New York, NY and is currently an independent environmental consultant. "Dr. Bob" has earned a BChE degree in Chemical Engineering from Manhattan College, a PhD in Chemical Engineering from Rensselaer Polytechnic Institute, an MS in Environmental Engineering from Newark College of Engineering, and an MBA from Temple University. He has contributed numerous publications to the technical literature, including topics in phase equilibria, and is a recipient of the Water Pollution Control Federation's Harrison Prescott Eddy Medal for noteworthy research in wastewater treatment. He is author or co-author of two books and holder of one U.S. patent. He is a member of the American Institute of Chemical Engineers (AIChE), the American Chemical Society (ACS), and the Air & Waste Management Association (A&WMA) and is a licensed professional engineer in several states.

Acknowledgement

The author is grateful to Roberta Kunz Fox, AIA, of fox^2 design for help with Figure 1.

References

1. Kunz, R.G., "The Rayleigh Equation Revisited: What to Do When Alpha (α) Isn't Constant," Paper 118b presented at the 2012 A.I.Ch.E. Spring Meeting, Houston, TX (April 1 - 5, 2012).

2. Kroschwitz, J.I. and M. Howe-Grant, editors, "Kirk-Othmer Encyclopedia of Chemical Technology," 4 ed., Vol. 8, pp.311-322, Wiley, New York (1993).

3. Coulson, J.M. and J.F. Richardson with J.R. Backhurst and J.H. Harker, "Chemical Engineering," 4 ed., Vol. 2, pp.436-440, Butterworth-Heinemann, Oxford (1999).

4. Zubrick,J.W.,"TheOrganicChemLabSurvivaManual,"7 ed., 350 pp., Wiley, Hoboken, NJ (2008).

5. Mayo, D.W., R.M. Pike, and D.C. Forbes, "Microscale Organic Laboratory with Multistep and Multiscale Syntheses," 5 ed., p.61, Wiley, Hoboken, NJ (2011).

6. Williamson, K.L., R. Minard, and K.M. Peters, "Macroscale and Microscale Organic Experiments," 5 ed., pp.88-104, Houghton Mifflin Company, Boston, MA (2007).

7. Williamson, K.L., R. Minard, and K.M. Peters, "Macroscale and Microscale Organic Experiments," Selected Material Special Edition for University of Massachusetts, Amherst, 5 ed., pp.88-104, Houghton Mifflin Company, Boston, MA (2007).

8. Lord Rayleigh, *Phil. Mag.*, "On the Distillation of Binary Mixtures," Series 6, **4(23)**, 521-537 (Nov. 1902).

9. Perry, J.H., editor, "Chemical Engineers' Handbook," 3 ed., 1942 pp., McGraw-Hill, New York (1950).

10. Badger, W.L. and J.T. Banchero, "Introduction to Chemical Engineering," pp. 245-258, McGraw-Hill, New York (1955).

11. McCabe, W.L. and J.C. Smith, "Unit Operations of Chemical Engineering," 3 ed., pp. 484-487, 535-540, 606, McGraw-Hill, New York (1976).

12. Thomas, G.B., Jr., "Calculus and Analytic Geometry," 3 ed., pp.385-387, Addison-Wesley Publishing Company, Inc., Reading, MA (1960).

13. Shoemaker, D.P., C.W. Garland, and J.W. Nibler, "Experiments in Physical Chemistry," 5 ed., 890 pp., McGraw-Hill, New York (1989).

14. Hodgman, C.D., editor-in chief, "Mathematical Tables from Handbook of Chemistry and Physics," p.249, Chemical Rubber Publishing Co., Cleveland, OH (1958).

15. Fleser, L.F. and K.L. Williamson, "Organic Experiments," 7 ed, pp.64-68, D.C. Heath and Company, Lexington, MA (1992).

16. Anon., "Ethanol," www.worldlingo.com/ma/enwiki/en/Ethanol (accessed on 3/8/2011).

17. "Water Meth is it Safe question asked.. Flammability etc." – SEAT Cupra.net – SEAT Forum www.seatcupra.net/forums/showthread.php?t=360226 (accessed on 5/8/2014).

18. Flick, E.W., editor, "Industrial Solvents Handbook," 5 ed., 994 pp., Noyes Data Corporation, Westwood, NJ (1998).

19. Green, D.W. and J.O. Maloney, "Perry's Chemical Engineer's Handbook,"6 ed,.p.3-88, McGraw-Hill, New York (1984).

20. "5. Distillation and Boiling Points," http://courses. chem. psu/chem36/ Experiments/PDF's_ for_ techniques/Distillation_&_BP.pdf (accessed on 4/5/2014).

21. "Water Meth is it Safe question asked.. Flammability etc." – SEAT Cupra.net – SEAT Forum www.seatcupra.net/forums/showthread.php?t=4015096 (accessed on 5/8/2014).

22. Dean, J.A., editor, "Lange's Handbook of Chemistry," 11 ed., McGraw-Hill, New York (1973).

23. Weast, R. C. and M. J. Astle, "CRC Handbook of Chemistry and Physics," 62 ed., CRC Press, Inc., Boca Raton, FL (1981).

24. Dean, J.A., editor, "Lange's Handbook of Chemistry," 12 ed., McGraw-Hill, New York (1978).

25. Dean, J.A., editor, "Lange's Handbook of Chemistry," 13 ed., p.4-59 (water) and Table 7-15 (organics), McGraw-Hill, New York (1985).

26. Kunz, R.G., "Environmental Calculations A Multimedia Approach," pp.41-52, Wiley, Hoboken, NJ (2009).

27. Carlson, H.C. and A.P. Colburn, "Vapor-Liquid Equilibrium of Nonideal Solutions – Utilization of Theoretical Methods to Extend Data," *Ind. Eng. Chem.*, **34**(**5**), 581-589 (May 1942).

28. Beebe, A.H., Jr., K.E. Coulter, R.A. Lindsay, and E.M. Baker, "Equilibria in Ethanol-Water System at Pressures Less Than Atmospheric," *Ind. Eng. Chem.*, **34**(**12**), 1501-1504 (Dec. 1942).

29. Otsuki, H. and F.C. Williams, "Effect of Pressure on Vapor-Liquid Equilibria for the System Ethyl Alcohol-Water," *CEP Symposium Series No. 6*, **49**, 55-67 (1953).

30. Othmer, D.F., W.P. Moeller, S.W. Englund, and R.G. Christopher, "Composition of Vapors from Boiling Binary Solutions," *Ind. Eng. Chem.*, **43**(**3**), 707-711 (Mar. 1951).

31. Speight, J.G., editor, "Lange's Handbook of Chemistry," 16 ed., p.1.95 and p.2.294, McGraw-Hill, New York (2005).

32. Department of Commerce, Bureau of Standards, "Standard Density and Volumetric Tables – Circular No. 19," 6 ed., 72 pp., U.S. Government Printing Office, Washington, DC(Oct.31, 1924).

33. Washburn, E.W., editor-in-chief, "International Critical Tables of Numerical Data, Physics, Chemistry and Technology," 1 ed., Vol. III, 444 pp., McGraw-Hill, New York (1928).

34. Carr, C. and J.A. Riddick, "Physical Properties of Methanol-Water System," *Ind. Eng. Chem.*, **43**(**3**), 692-696 (Mar. 1951).

35. Mikhail, S.Z. and W.R. Kimel, "Densities and Viscosities of Methanol-Water Mixtures," *J. Chem. Eng. Data*, **6**(**4**), 533-537 (Oct. 1961).

36. Sakuri, M. and T. Nakagawa, "Densities of Dilute Solutions of Water in Benzene and in Methanol at 278.15, 288.15, 298.15, 308.15, and 318.15 K. Partial Molar Volumes and Vw and Values of ∂Vw/∂T for Water in Benzene and in Methanol," *J. Chem. Thermodynamics*, 14, 269-274 (1982).

37. Shoemaker, D.P., C.W. Garland, and J.W. Nibler, "Experiments in Physical Chemistry," 5 ed., 890 pp., McGraw-Hill, New York (1989).

38. "The New Encyclopædia Britannica," Vol. 13, p.525, Encyclopædia, Inc., Chicago, IL (2002).

39. Kenkel, J., "Analytical Chemistry for Technicians," 3 ed., 554 pp., Lewis Publishers, Boca Raton, FL (2003).

40. Jones, C.A. E.M. Schoenborn, and A.P. Colburn, "Equilibrium Still for Miscible Liquids," *Ind. Eng. Chem.*, **35**(**6**), 666-672 (June 1943).

41. Schweppe, J.L. and J.R. Lorah, "Ternary System Ethyl Alcohol – *n*-Heptane – Water at 30 °C," *Ind. Eng. Chem.*, **46**(**11**), 2391-2392 (Nov. 1954).

42. Gray, D.E., editor, "American Institute of Physics Handbook, pp.2-141,2-142,2-152,2-153,2-157, McGraw-Hill, New York (1957).

43. "Ethyl Alcohol Handbook, 6 ed., Lyondell Chemical Company, Houston. TX, www.itecref.com/pdf/**Ethyl_Alcohol**_Handbook_Equistar.pdf (accessed on 3/8/2011).

44. Brochure "Technical Information & Safe Handling Guide for Methanol," Version 3.0, Methanex Corporation, Vancouver, B.C., Addison, TX (Sept. 2006).

45. Palleros, D.R., "Experimental Organic Chemistry," 833 pp., Wiley, New York (2000).

46. Dean, J.A., editor, "Analytical Chemistry Handbook," p.9.1, McGraw-Hill, New York (1995).

47. J. Ortega, J., "Densities and Refractive Indices of Pure Alcohols as a Function of Temperature," *J, Chem. Eng. Data*, **27**(**3**), 312-317(July 1982).

48. Andrews, L.W., "The Refractive Indices of Alcohol-Water Mixtures," *J. Am. Chem. Soc.*, **30**(**3**), 353-360 (1908).

49. Nowakowsha, J., "The Refractive Indices of Ethyl Alcohol and Water Mixtures," Masters Theses Loyola University Chicago, IL (1939). Paper 668. http://ecommons.luc.edu/luc_thesis/668.

50. "Distillation Course CHEE 4500 (Spring 2004)" https://www.myedu.com/UT-The-University-of-Toledo/CHEE-4500-Chemical-Engineering-Laboratory-I/course/ 290063/ (accessed 4/13/2014).

51. Balke, C., N. Butler, and K. Courtright, Distillation Column and Refractometer Calibration," Oklahoma State University CHE 4002 Project I, 26 pp. (Feb. 10, 2010). (accessed 4/16/2014).

52. Ho, V. and M. Butzow, "Methanol Recovery via Distillation -- Rotation 1," 36 pp. (Oct. 6, 2005). http://chem.engr.utc.edu/ench435/2005/CHBE-430-Distillation-Final-Report.htm. (4/17/2014).

53. Doroschewsky and Dvorsehantschik, *J. Russ. Phys. Chem. Soc.*, **40**, 908-931 (1908); *Chem. Zentr.*, **2**, 1569-1571 (1908), as cited in Simmonds, C., "Alcohol, Its Production, Properties, Chemistry, and Industrial Applications," Macmillan, London (1919).

54. Speight, J.G., editor, "Lange's Handbook of Chemistry," 16 ed., p.1.95 and p.2.294, McGraw-Hill, New York (2005).

55. Washburn, E.W., editor-in-chief, "International Critical Tables of Numerical Data, Physics, Chemistry and Technology," 1 ed., Vol. VII, 506 pp., McGraw-Hill, New York (1930).

56. Openhaim, G. and E. Grushka, "Temperature-dependent Refractive Index Issues Using a UV – visible Detector in High Performance Liquid Chromatography," *J. Chromatology A*, **942**, 63-71 (2002).

57. "Ethyl Alcohol Handbook," 6 ed., Lyondell Chemical Company, Houston. TX, www.itecref.com/pdf/**Ethyl_Alcohol**_Handbook_Equistar.pdf (accessed on 3/8/2011).

58. Anon., "Alcoholic Proof," http://www.bookrags.com/wiki/proof-alcohol (accessed on 3/3/2011).

59. U.S. Census Bureau, Foreign Trade Statistics, "2011 Schedule B – Chapter 22," http://www.census.gov/foreign-trade/schedules/b/2011/c22.html (accessed 3/8/2011).

60. Anon., "At What 'Proof' Will Spirits Burn? – Alcohol Fire Chemistry," http://ask.metafilter.com/95484/At-what-proof-will-spirits-burn (accessed on 3/8 2011).

61. Spencer, A.B. and G.R. Colonna, editors, "Fire Protection Guide to Hazardous Materials," 13 ed., p. 325-57, p. 325-81, p. 497-8, p. 497-10, NFPA, Quincy, MA (2002).

62. "Materials Safety Data Sheet Methanol," Terra Nitrogen Corporation, Sioux City, IA (April 1, 2001).

63. "Materials Safety Data Sheet 10% Methanol / 90% Water Volume / Volume," Honeywell, Burdick & Jackson, Muskegon, MI (Oct. 2005).

64. "Flammability of Methanol-Water Mixtures" – waterinjection.info forum www.aquamist.co.uk/vbulletin/showthread.php?t=856 (accessed on 5/7/2014).

65. "Water Meth is it Safe question asked.. Flammability etc." – SEAT Cupra.net – SEAT Forum www.seatcupra.net/forums/showthread.php?t=4015096 (accessed on 5/8/2014).

66. "Flash Points of Methanol based Water Solutions," http://www.engineeringtoolbox.com/methanol- water-d_987.html (accessed on 4/23/2014).

APPENDIX A

Summary of Vapor-Liquid Equilibrium (VLE) Relationships

Low-pressure binary vapor-liquid equilibrium is described by the following relationships involving temperature (t), total pressure (P_T), partial pressure (p), vapor pressure (p^{vap}), and liquid- and vapor-phase phase mol fractions (x and y) of each component, plus so-called activity coefficients (γ's) when a given system cannot be considered to be ideal. The subscript 1 denotes the more volatile component (MVC); 2 the less volatile component (LVC),

Ideal Systems

A system is said to be ideal when it follows Raoult's Law [Equations (A1) and (A2)]

$$p_1 = p_1^{vap} x_1 \tag{A1}$$

$$p_2 = p_2^{vap} x_2 \tag{A2}$$

and

$$P_T = p_1 + p_2 \tag{A3}$$

with

$$x_1 + x_2 = 1 \tag{A4}$$

If the Ideal Gas Law and Dalton's Law are assumed to apply to the vapor phase, vapor-phase mol fractions (y_1 and y_2) can be calculated from:

$$y_1 = p_1 / P_T = p_1^{vap} x_1 \ / \ P_T \tag{A5}$$

$$y_2 = p_2 / P_T = p_2^{vap} x_2 \ / \ P_T \tag{A6}$$

with

$$y_1 + y_2 = 1 \tag{A7}$$

Vapor Pressure. Pure-component vapor pressures in both ideal and non-ideal systems are conveniently calculated using the Antoine Equation:

$$\log_{10} p^{vap}\ (\text{mmHg}) = A - B \ / \ [t\ (°C) + C] \tag{A8}$$

Antoine constants A, B, and C are listed in Table A1 for the chemicals under consideration here.

Calculation of x-y Curve. By combining and rearranging the equations above, one obtains:

$$x_1 = (P_T - p_2^{vap}) \ / \ (p_1^{vap} - p_2^{vap}) \tag{A9}$$

Chiefly by means of this equation, the entire x-y equilibrium curve and temperature-composition curves at constant pressure for ideal systems can be mapped out in a straightforward manner. One first chooses a temperature, computes vapor pressures from Equation (16), then calculates x_1 from Equation (A9) and y_1 from Equation (A5). Only if a specific value of x_1 is desired, does the calculation become trial and error, in which a new value of t is repeatedly chosen until the desired value of x_1 is obtained to within whatever tolerance is satisfactory.

Table A1. Antoine Equation Constants of Interest [a]

Chemical Name and Synonyms	Formula	MW [b]	BP (°C) [c]	Applicable Range(°C)	A	B	C
Ethanol Ethyl Alcohol Grain Alcohol Methyl Carbinol	C_2H_5OH	46.07	78.3	—	8.04494	1554.3	222.65
Methanol Methyl Alcohol Wood Alcohol Carbinol	CH_3OH	32.04	64.7	–20 to +140	7.87863	1473.11	230.0
Water Hydrogen Oxide	H_2O	18.02	100.00	0 to 60 60 to 150	8.10765 7.96681	1750.286 1668.21	235.0 228.0

Notes:

(a) Antoine Constants from "Lange's Handbook of Chemistry", 11th Edition [**22**, Table 10-10].

(b) Molecular Weight (MW) from "Handbook of Chemistry and Physics," 62nd Edition [**23**, pp. C-65 to C-576], rounded for water [**23**, p.B-105] to 2 decimal places.

(c) Boiling point (BP) at 760 mmHg from "Lange's Handbook," 11th-13th Editions [**22**, p.4-59 and Table 7-4; **24**, p.4-59 and Table 7-4; **25**] or "Handbook of Chemistry and Physics," 62nd Edition [**22**, pp. C-65 to C-576, p.B105], chosen to be the most consistent with boiling point calculated using the Antoine Constants listed.

Methanol-Ethanol. Ideal systems occur with members of homologous series, isomers, and in certain other cases [**9**]. Because of the similarity of the two molecules, interactions between like and unlike species are nearly the same. Hence, each component behaves as if independent of the other. The classic system used to illustrate Raoult's Law is benzene-toluene. It is highlighted in many textbooks including Ref. [**11**]. Likewise, methanol and ethanol, of interest here, form an ideal binary system with each other. Its x-y curve, published previously [**26**], is drawn in Figure A1 along with experimental data from several investigators.

Non-Ideal Systems

For these, a so-called activity coefficient (γ) is introduced into Raoult's Law to yield:

$$p_1 = p_1^{vap}\,\gamma_1\,x_1 \qquad \text{(A10)}$$

$$p_2 = p_2^{vap}\,\gamma_2\,x_2 \qquad \text{(A11)}$$

and

$$P_T = p_1 + p_2 \qquad \text{(A3)}$$

with

$$x_1 + x_2 = 1 \qquad \text{(A4)}$$

Under the Ideal-Gas and Dalton's Law assumptions, vapor phase mol fractions are given by:

$$y_1 = p_1 / P_T = p_1^{vap} \gamma_1 x_1 / P_T \quad (A12)$$

$$y_2 = p_2 / P_T = p_2^{vap} \gamma_2 x_2 / P_T \quad (A13)$$

again with

$$y_1 + y_2 = 1 \quad (A7)$$

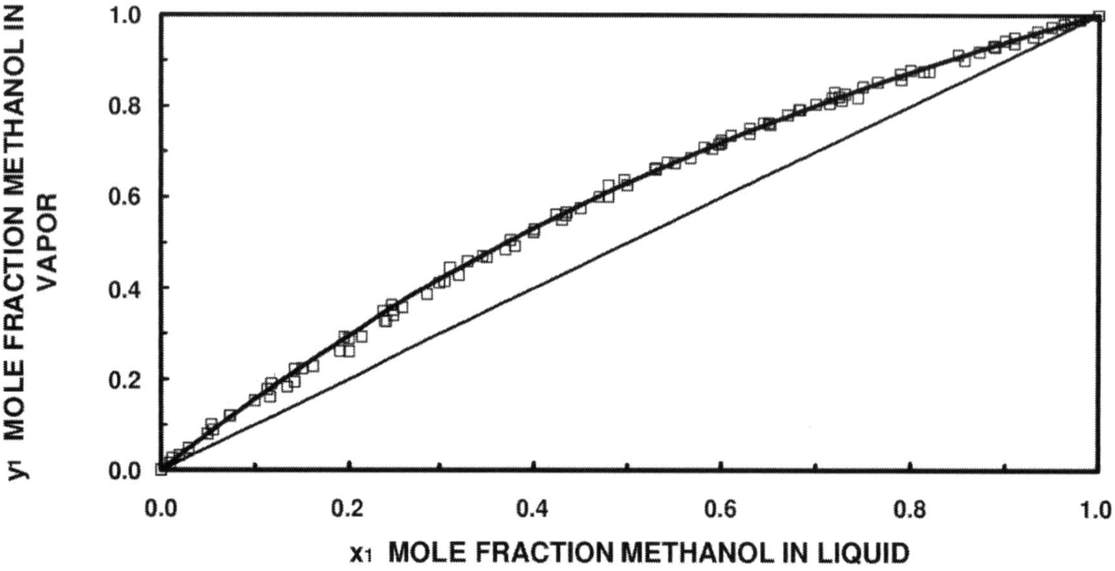

Figure A1. Vapor-Liquid Equilibrium Diagram Methanol-Ethanol at 1 Atm Pressure

Activity Coefficients. Activity coefficients (γ's) account for deviations from ideality. They are back calculated as point values from experimental data and then fitted to a special type of continuous function for use in VLE calculations. There are several empirical activity-coefficient formulations in widespread use; for any given application, some fit the deviations better than others across the range of data. The van Laar equations [**2**; **9**, p.527, **27**], for example:

$$\log_{10} \gamma_1 = \frac{A_{1\text{-}2}}{[1 + (A_{1\text{-}2}\, x_1) / (A_{2\text{-}1}\, x_2)]^2} \quad (A14)$$

and

$$\log_{10} \gamma_2 = \frac{A_{2\text{-}1}}{[1 + (A_{2\text{-}1}\, x_2) / (A_{1\text{-}2}\, x_1)]^2} \quad (A15)$$

fit the data well for the non-ideal systems considered here. These particular equations are also written in terms of natural logarithms (base e), using constants 2.302585... × $A_{1\text{-}2}$ and $A_{2\text{-}1}$.

Methanol-Water. Activity coefficients for methanol-water at 1 atm pressure derived from experimental data are given on p.528 in Perry's 3rd Edition [**9**]. Coefficients $A_{1\text{-}2}$ and $A_{2\text{-}1}$ of are 0.36 and 0.22 ($\log_{10}$ based) or 0.83 and 0.51 (ln based). The previously published [**26**] methanol-water x-y curve with accompanying experimental data from a number of investigators is reproduced in Figure A2. It resembles the curves for ideal systems. It uses the van Laar constants enumerated above and Antoine constants from Table A1, plus a trial-and-error† procedure to compute each point on the continuous curve at 1 atm total pressure.

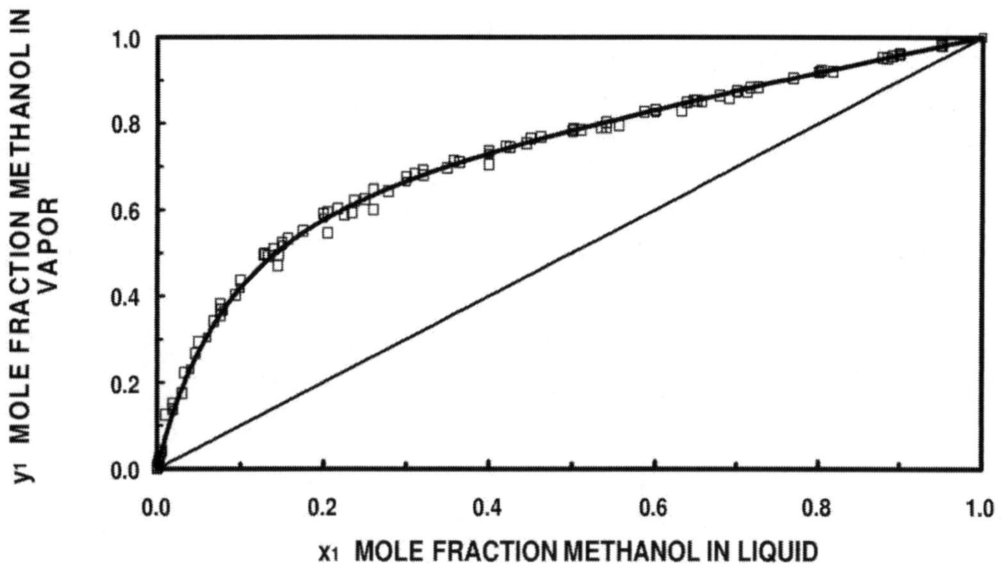

Figure A2. Vapor-Liquid Equilibrium Diagram for Methanol-Water at 1 Atm Pressure

Ethanol-Water. The typical example frequently cited to illustrate an azeotrope, or constant boiling mixture, is the ethanol-water system. This system is well investigated in the literature. Its predicted x-y diagram at atmospheric pressure, shown in Figure A3, was drawn by using the same trial-and-error procedure described above for methanol-water.

It exhibits an azeotrope composed of 95.6 wt % ethanol and 4.4 wt % water at 1 atm pressure (760 mmHg, 1.013 bar) and 78.15 °C; this corresponds to 89.43 mol % ethanol and 10.57 mol % water [**9**, pp.631,633]. This cannot be purified further to 100 % ethanol by distillation at atmospheric pressure. The azeotropic point is different at other total pressure levels [**9**, p.631]; water content decreases with decreasing pressure, and at least in theory becomes zero when distillation pressure is reduced low enough [**9**, p.631;**40**].

In the special case of an azeotrope, one can obtain the van Laar constants for use in Equations (A14) and (A15) from the composition of the azeotropic point [**26**]. For the ethanol-water system, coefficients of 0.68381, 0.41724 ($\log_{10}$ based) or 1.5745, 0.9607 (ln based)

† Constant-pressure calculations for non-ideal systems are always trial and error since the activity coefficients are functions of liquid composition. An efficient method to map out the x-y equilibrium curve is to select a value of x_1 and total pressure and then vary the temperature until the calculated total pressure from Equation (A3) agrees as closely as desired with the total pressure selected.

were determined from the azeotrope at 1 atm in conjunction with Antoine Equation vapor pressures calculated using the ethanol and water constants of Table A1. Since the fitting procedure to obtain them involves only the data at the azeotropic point, agreement with other experimental values along the curve is not guaranteed. However, the myriad of accompanying data points from multiple investigators in Figure A3 are so numerous as to obscure the calculated x-y equilibrium curve almost completely. Data were not selectively chosen to show agreement with the curve, as evidenced by the few stray points on the graph; rather, complete data sets were added one after another until the underlying computer file filled up.

Furthermore, the ethanol-water system at constant pressure has been investigated from below atmospheric pressure [**28**] up to at least 300 lb_f/in^2 absolute (psia) [**29**,**30**]. It is well described in Ref. [**29**] by van Laar constants only slightly different from those obtained here and differing among themselves but little for huge changes in pressure. Therefore, small day-to-day variations in barometric pressure encountered in the laboratory would alter only P_T in simulation calculations, and not require modification of the 760-mmHg van Laar constants.

Relative Volatility. The relative volatility (α) in Equation (5) can be calculated at constant pressure for any temperature or the corresponding value of composition (x_1) along the VLE curves for ideal systems from a ratio of vapor pressures [Equation (A16)], and for non-ideal systems through the addition of activity coefficients [Equation (7)]:

$$\alpha = (y_1/x_1) / (y_2/x_2) = p_1^{vap} / p_2^{vap} \tag{A16}$$

$$\alpha = (y_1/x_1) / (y_2/x_2) = (p_1^{vap} \gamma_1) / (p_2^{vap} \gamma_2) \tag{7}$$

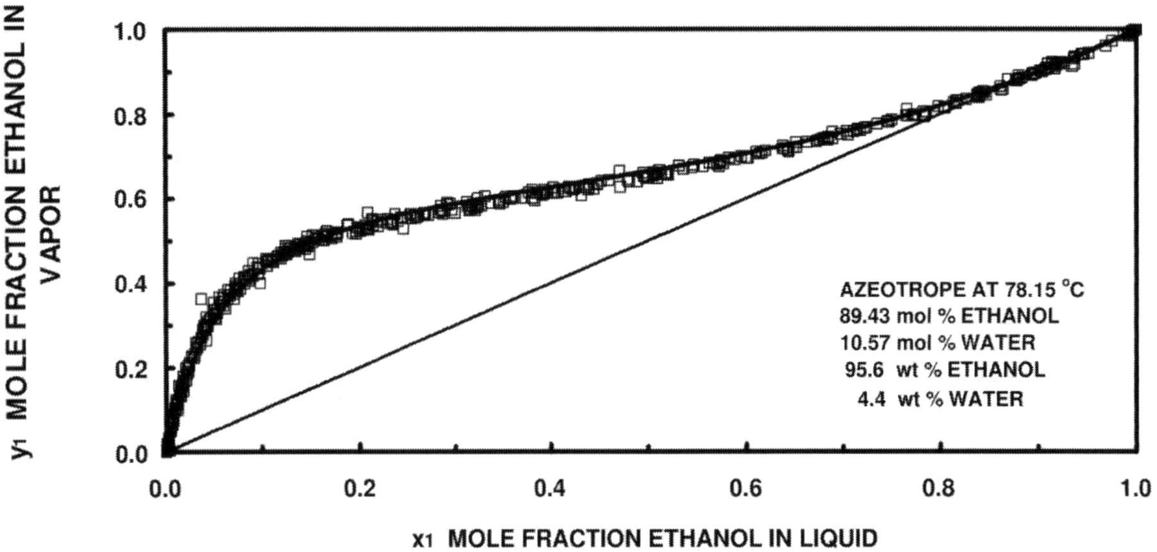

Figure A3. Vapor-Liquid Equilibrium Diagram for Ethanol-Water at 1 Atm

APPENDIX B

ANALYTICAL METHODS FOR ALCOHOL-WATER SOLUTIONS

This appendix reviews several ways to analyze for chemical composition in the ethanol-water and methanol-water systems. These include at least three quantitative methods and one qualitative technique. Quantitative analysis is done by refractive index (RI), density (ρ), or possibly gas chromatography (GC). A qualitative analysis for ethanol is often conducted by attempting to ignite several drops of distillate solution on a watch glass since aqueous solutions of ethanol greater than about 50 vol % are combustible at room temperature [**16**]. Methanol-water combustion characteristics are similar. Topics are discussed in turn below.

Analysis by Density

Published densities of solutions of ethanol and water [**9**,**22-25**,**31**-**33**] and methanol and water [**9**,**22-25**,**31**-**36**] facilitate chemical analysis by careful measurement of density at constant temperature. Unlike the refractive index of alcohol-water solutions discussed below, solution density at a given temperature corresponds to a unique value of composition. The simplest technique employs a pycnometer, or specific gravity bottle, a piece of calibrated glassware of precisely known volume [**37**, pp.191-193]. It is weighed first empty and then full of solution. Density is calculated by dividing solution weight by solution volume. Pycnometer volume is typically about 10 mL [**38**], but a 5-mL pycnometer is also available [**39**, p.434]. Related studies used a 10-mL [**28**,**40**] or 5-mL [**41**] pycnometer to analyze by density.

Advantage: Analysis by density has the advantage of using simple laboratory equipment in widespread use. It takes a minimum of time and provides a unique value for ethanol or methanol composition at a constant known temperature. Disadvantage: In some cases, the sample size needed may be greater than the amount of distillate collected over a short period.

Density-Composition Plots. Density data for ethanol-water and methanol-water mixtures across the entire gamut of compositions and sample temperatures are difficult to find. This is not so much of a problem for analysis by density, but rather for converting from mass to volume at any given temperature and vice versa. Density data (g/mL) for liquid water (0-102 °C) and ethanol-water mixtures are tabulated in NBS Circular 19 [**32**] and handbooks reproducing said tables. For pure ethanol, density or its reciprocal, specific volume, is listed in various tables up to 80 °C [**33**, p.41], and tabular entries and correlating equations are provided for density from 0-78 °C [**42**]. Density data in lb_m/gal is plotted either against °C [**18**, p.286] or °F [**43**, p.62] from 5 to 60 °C, parametric in vol % ethanol. Density data for liquid solutions outside this temperature range could not be found in the published literature.

However, the parametric plots [**18**, p.286; **43**, p.62] appear to be well behaved, consistently spaced, and fairly linear at the upper temperatures, suggesting that missing values might be interpolated or extrapolated with some degree of success. To that end, the density plot has been reconstructed here in units of g/mL between 0 and 100 °C, with wt % ethanol as parameter (Figure B1).

The newly developed curves of Figure B1 are based on the data from NBS Circular 19 [**32**], the equations of Ref. [**42**], and the aforementioned graph [**18**, p.286]. The fit is exact for all values for pure water, for pure ethanol up to 78 °C, and for tabulated mixture data from 10-40 °C [**32**], and in reasonable agreement with values read with difficulty from the curves of Ref. [**18**, p.286] between 5 and 60 °C. Two different methods produce the same extrapolated pure ethanol curve from 78 to 100 °C.

The top curve (0 % ethanol, 100 % water) contains the actual maximum point in the density for pure water at about 4 °C. Curve fitting predicts a maximum in density for 20 wt % ethanol at about 2 °C; this may or may not be the case. As drawn in Figure B1, the other curves exhibit a continuous decrease in density with increasing temperature.

Estimated densities for pure ethanol beyond 80 °C and for mixtures below 5 °C and above 60 °C are not based on data and should be used with caution, since extrapolated densities are most likely not exact. In addition, although values are shown in Figure B1 for pure ethanol and for some mixture compositions above their atmospheric boiling points, these solutions do not exist as stable liquids at 1 atm pressure. In that case, the predicted densities correspond to whatever slightly elevated pressure is necessary to maintain the solution in the liquid state. This slight difference in pressure should have a negligible effect on density since enormous pressure increases are necessary to effect a significant change in liquid density [**33**, pp.40-42]. Finally, while estimated values of density may be useful to calculate volume from mass or vice versa, density values outside the range of data should <u>not</u> be used for analytical purposes to determine solution composition.

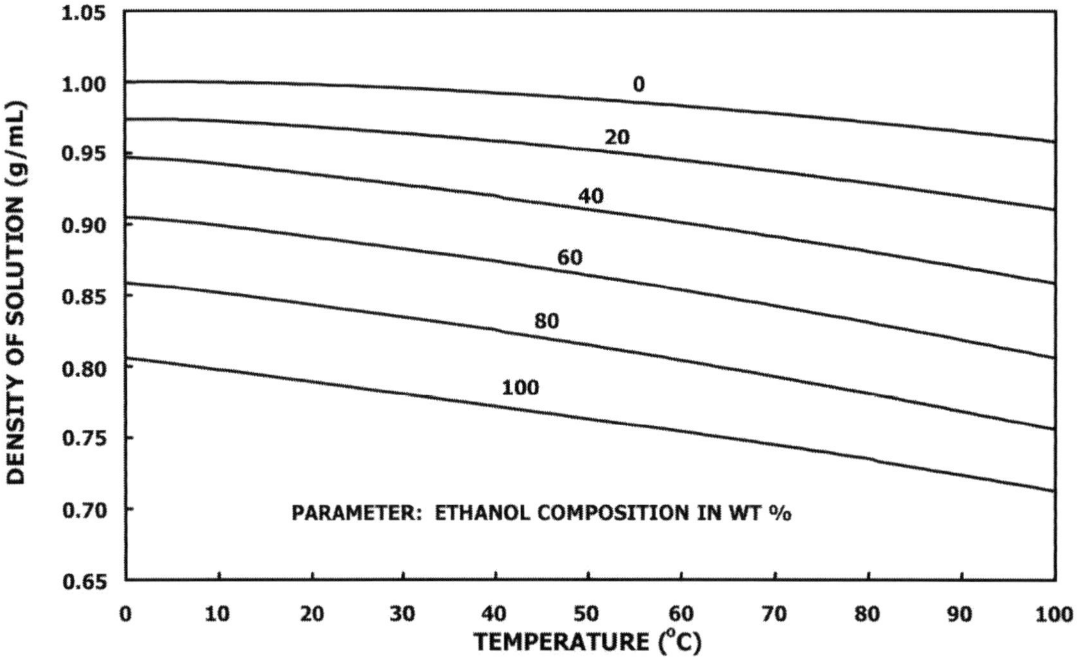

Figure B1. Estimated Liquid Density of Ethanol-Water Solutions

A similar plot of density vs. temperature from +30 °C to −40 °C for methanol-water solutions is available from a methanol manufacturer's website Ref. [**44**]. It too appears to be well behaved and can possibly be extended to the temperature range of interest with the help of other data. Such data exist from 0 °C to 100 °C for pure water as noted above, for 100% methanol [**42,44**] and for methanol-water solutions from 0 °C to 50 °C previously cited. The variation of solution density is said to be linear with mole fraction [**45**, pp.139-140].

Density-composition curves at constant temperature can then be fitted using only pure component densities without needing any data at intermediate compositions; such data where available simply provide a check that the curves are reasonably linear. The slope of a linear density-composition curve varies with temperature and, as it turns out, is well correlated by a quadratic function. The intercept of each density-composition line is the density of pure water at the given temperature, tabulated to 5 decimal places in NBS Circular 19 [**32**].

The resulting estimated density graph for methanol-water solutions from 0 °C to 100 °C is shown in Figure B2. The same cautions noted above for the ethanol-water curves of Figure B1 apply here also. Solution densities are good only to 2 or perhaps 3 decimal places. Pure component densities, all based on data, are better. Densities for water depicted at the top of the diagram follow the same curve as in Figure B1 (maximum at ~4 °C). Even after all the roundabout calculations, the straight line at the bottom corresponding to 100 % methanol reproduces the starting data, showing a slope of −0.0009+ g/ml per °C, as noted earlier[**34**].

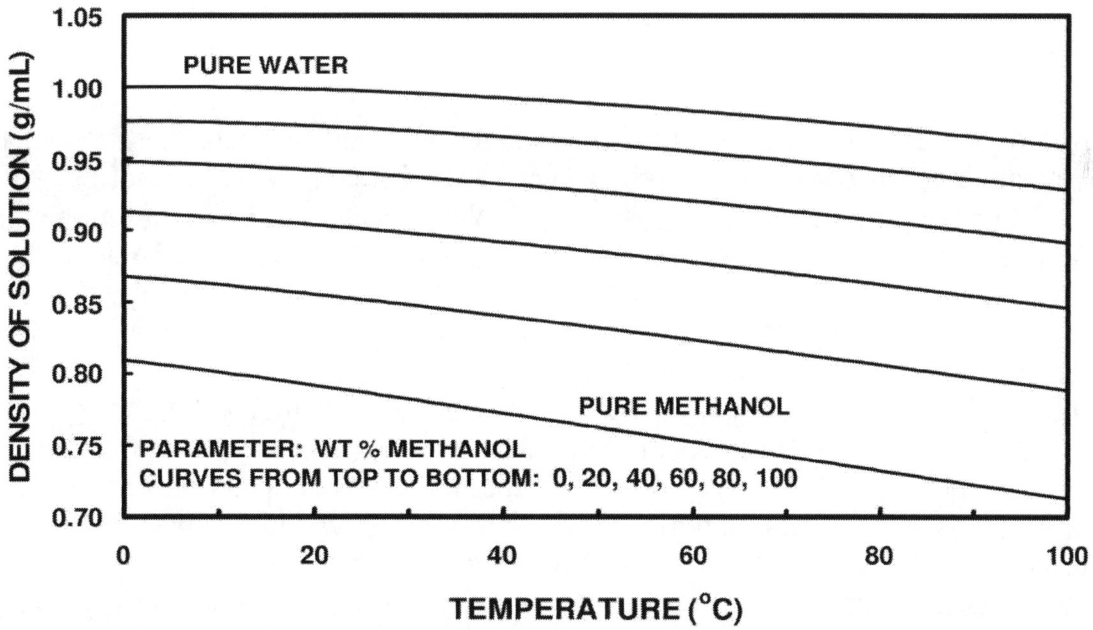

Figure B2. Estimated Liquid Density of Methanol-Water Solutions

Analysis Using Refractive Index

Refractive index (RI) data for mixtures of ethanol and water at various temperatures (n_D^t) are summarized in Table B1, assembled from a number of sources. Refractive index can theoretically be determined to 1 part in 10,000 [**4**, p.319; **6**, p.299] but is more likely to agree with literature values to 1 part in 1,000 because of the presence of impurities [**6**, p.279].

Refractive index is commonly measured at 20 °C (n_D^{20}) [**17**, pp.D-200,D207]. Minor excursions from 20 °C are corrected for by adding a mean value of 0.00045 refractive index units for every °C above 20 °C [**6**,p.299; **46; 47**].

$$n_D^{20} = n_D^t + 0.00045\,(t - 20\ °C) \qquad \text{(B1)}$$

As seen in Table B1, differences in refractive index for more appreciable differences in temperature are not exactly accounted for by the general factor.

Refractive index for ethanol-water solutions passes through a maximum point when plotted against composition. This occurs at 79.3 wt % ethanol and 25 °C, as reported by Andrews [**48**], and in the vicinity of 79-80 wt % ethanol at 60 °F (15.56 °C), as determined by curve-fitting the data in Ref. [**24**, p.250] combined with differential calculus. Another study of the maxima is also available [**49**]. A less rigorous inspection of the other entries in Table B1 detects maxima occurring somewhere in a wider range between 70-80 wt % ethanol.

Refractive index measurements for methanol-water solutions are shown in Table B2. Note that some entries in this table relate to volume % rather than to weight %.

$$\text{vol \% } CH_3OH = \text{wt \% } CH_3OH \times \rho_{\text{solution}} / \rho_{\text{pure methanol}}\ [\mathbf{34}] \qquad \text{(B2)}$$

$$\text{wt \% } CH_3OH = \text{vol \% } CH_3OH \times \rho_{\text{pure methanol}} / \rho_{\text{solution}} \qquad \text{(B3)}$$

Density (ρ) of solution and of pure methanol is at the given temperature, and in addition solution density is at the given concentration. Vol % CH_3OH is 100 × volume of pure methanol divided by volume of resulting solution, not by total volume of methanol plus water being mixed.

Maximum points in RI also occur for methanol-water solutions (Table B2). These maxima are also a subject of discussion by others [**50-52**].Refractive indices for methanol-water solutions have been fitted as a function of temperature using a 4th or 5th degree polynomial [**51-52**.]

In the affected range, therefore, more than one alcohol composition corresponds to the same refractive index when the axes of either curve are switched to make refractive index the independent variable for calibration. However, location on the curve can be resolved by adding alcohol or water to the sample and noting whether the refractive index goes up or down.

Advantage: Determination of refractive index requires only two or three drops of sample [**6**, p.300]. Once the procedure is set up and the proper technique is developed [**4**, pp.247-251], it is possible to analyze a new sample every minute [**13**, p.237]. RI should be measured as soon as possible after sampling lest the samples decompose on standing [**6**, p.300]. Disadvantage: Analysis by RI does, however, require specialized equipment [**6**, p.300].

Table B1. Refractive Index (n_D) Values of Ethanol-Water Solutions at Various Temperatures (°C)

Ethanol (wt %)	n_D^{15}	$n_D^{15.56}$	n_D^{20}	n_D^{25}	n_D^{30}	n_D^{40}	n_D^{50}	n_D^{55}
0	1.33345	1.33336	1.33300	1.33250	1.3318	1.3306	1.3290	1.3281
10	1.34020	1.3402	1.3395	1.3389	1.3384	1.3368	1.3349	1.3339
20	1.34778	1.3479	1.3469	1.3462	1.3450	1.3429	1.3406	1.3393
30	1.35470	1.3542	1.3535	1.3520	1.3510	1.3481	1.3452	1.3435
40	1.35948	1.3590	1.3583	1.3565	1.3550	1.3518	1.3484	1.3468
50	1.36290	1.3626	1.3616	1.3598	1.3578	1.3543	1.3506	1.3488
60	1.36505	1.3650	1.3638	1.3620	1.3597	1.3560	1.3522	1.3501
70	1.36645	1.3662	1.3652	1.363038	1.3608	1.3570	1.3528	1.3505
80	1.36690	1.3665	1.3658	1.36331	1.3611	1.3569	1.3525	1.3502
90	1.36626	1.3659	1.3650	1.36239	1.3603	1.3561	1.3515	1.3491
100	1.36332	1.36316	1.3614	1.35941	1.3573	1.3531	1.3487	1.3465

Notes:

(a) Values at 15°C, 30°C, 40°C, 50°C, and 55°C from Ref. [**53**].

(b) Values at 15.56°C (60°F) for pure water and ethanol from Ref. [**18**, p.250]. Other entries at 15.56°C from regression of Ref. [**18**, p.250] values at other compositions.

(c) Values at 20°C from Ref. [**23**, p.D-200 (water) and p.D207].

(d) Values at 25°C either reported by Andrews [**48**], assembled from entries on p.1.95, p.2.294, and Tables 10-71 and 10-72 of Lange's Handbook [**54**], or curve-fitted to interpolate between those values.

Table B2. Refractive Index (n_D) Values of Methanol-Water Solutions at Various Temperatures (°C)

Methanol (wt %)	n_D^{15}	n_D^{20}	Methanol (vol %)	n_D^{25}	n_D^{30}	n_D^{35}	n_D^{40}	n_D^{45}	n_D^{50}
0	1.33339	1.3330	0	1.3325	1.3321	1.3316	1.3309	1.3300	1.3292
10	1.33584	1.3354	10	1.3341	1.3336	1.3330	1.3321	1.3310	1.3303
20	1.33879	1.3381	20	1.3360	1.3357	1.3352	1.3339	1.3330	1.3318
30	1.34138	1.3407	30	1.3382	1.3378	1.3366	1.3352	1.3343	1.3337
40	1.34308	1.3425	40	1.3402	1.3393	1.3376	1.3369	1.3358	1.3342
50	1.34378	1.3431	50	1.3412	1.3400	1.3386	1.3374	1.3360	1.3348
60	1.34327	1.3426	60	1.3412	1.3400	1.3388	1.3372	1.3355	1.3342
70	1.34179	1.3411	70	1.3401	1.3389	1.3371	1.3359	1.3340	1.3318
80	1.33925	1.3385	80	1.3375	1.3356	1.3344	1.3330	1.3313	1.3298
90	1.33545	1.3348	90	1.3333	1.3316	1.3305	1.3290	1.3259	1.3244
100	1.33057	1.3290	100	1.3269	1.3248	1.3235	1.3212	1.3192	1.3168

Notes:

(a) Values at 15°C from Ref. [**55**, p.66]. ICT, Vol. 7, p.66 (1930).
The original reference includes additional entries at intermediate values of ***wt %*** methanol.

(b) Values at 20°C from Ref. [**23**, p.D-200 (water) and p.D217]. HCP, 62 ed., pp.D-200, D-217
The original reference includes additional entries at intermediate values of ***wt %*** methanol.

(c) Data entries between 25°C and 50°C corresponding to values of ***vol %*** methanol from Ref. [**56**].

(d) .All of the above data pass through a maximum point as a function of % methanol.

Analysis by Gas Chromatography (GC)

Gas chromatography is a well-documented separation and analysis technique used in a chemical laboratory. When a liquid sample is injected into the gas chromatograph instrument, it is picked up by a carrier gas, usually helium but possibly nitrogen, and passed through a heated column packed with adsorbent material, where different chemicals are adsorbed and desorbed at different rates. The chemicals being analyzed for arrive at a detector downstream as a series of peaks traced on a strip chart recorder. The time for the arrival of a peak is indicative of each particular chemical in the mixture. The width of each peak, or more properly the area under the peak, represents the amount of that chemical present. More information to supplement this highly simplified explanation is contained in the cited references [**4**, pp.255-274; **6**, pp.215-223; **7**, pp.165-173; **13**, pp.149-151; **45**, pp.144-169].

Advantages: A gas chromatograph can detect extremely low concentrations in an unknown sample. It requires a sample size on the order of 1-25 micro liters (μL), smaller or much smaller than the several drops (20 drops = 1 mL) needed for a refractive index determination.

Disadvantages: The GC instrument is exceptionally expensive to purchase and may have to be shared with other investigators, if it is available at all. It needs an hour or so warm-up time, and requires experience it setting its temperature controls for the particular system being analyzed and some technique/practice in handling the micro-liter syringe to pierce the septum at the instrument's injection port. Residence time for each sample to pass through the column and show up as peaks at the detector may be several minutes or longer.

Summary of Analytical-Quantitative

Properties of alcohol-water solutions from the literature are provided here for reference in conducting laboratory analyses of unknown solution compositions encountered during a distillation experiment. Nevertheless, it is advisable to prepare one's own calibration curves employing the same reagents and laboratory equipment being used to analyze the unknowns. A procedure for preparing standard methanol-water solutions is outlined in Ref. [**39**, p. 458].

Analytical-Qualitative

A qualitative analysis for ethanol in the distillate is often conducted by attempting to ignite several drops of solution on a watch glass. Like hydrogen, ethanol burns with a bluish flame not easily visible in ordinary daylight [**57**, p.101]. Ethanol and its aqueous solutions greater than about 50 % by volume are combustible at room temperature [**16**]. These will ignite and burn to CO_2 and water without leaving any residue.

Alcohol Proof. In former times, this phenomenon formed the basis of a test of the alcoholic spirits given to members of the British Royal Navy as part of their rations [**58**]. In that test, a sample lighting off with gunpowder constituted "proof" that the alcohol had not been watered down. In the U.S. today, *proof* is defined as exactly twice the alcoholic content by volume, measured at 60 °F (15.56 °C) [**18**, Table 6.36; **57**, p. 108; **59**].

Anecdotal evidence suggests that a certain brand of vodka at 100 proof (50 % ethanol by volume) lights off, whereas this same vodka at 80 proof (40 % ethanol by volume) does not [**60**]. That this should be so is explained by the flash point of ethanol-water solutions (Table B3, prepared from a single source [**61**]); flash point is plotted vs. temperature in Ref. [**18**, p.254]. The flash point is the lowest temperature at which just enough liquid is evaporated at the surface to create a combustible concentration in air of gas that will catch fire if an ignition source is present.‡ The table shows that an ethanol-water solution is combustible at "room temperature" at about the stated 50 % by volume. Heating the solution first, which generates additional combustible ethanol vapor (and thereby lowers the flash point), allows solutions of lesser ethanol content to ignite at the same room temperature [**60**].

Table B3. Flash Points of Ethanol-Water Solutions [**61**]

Ethanol Volume %	Temperature		Ethanol Volume %	Temperature	
	°F	°C		°F	°C
5	144	62	60	72	22
10	120	49	70	70	21
20	97	36	80	68	20
30	85	29	95	63	17
40	79	26	96	62	17
50	75	24	100	55	13

Although not specifically mentioned as an analytical test of distillate composition, some methanol-water mixtures are also combustible, burning with a flame that may not be visible to the naked eye in daylight [**62**]. Methanol solutions containing up to 74 % water are classified by the National Fire Protection Association (NPFA) as flammable [**44**]. This means that such mixtures with as little as 21 vol % methanol (25 wt %) are flammable liquids; waste methanol in concentrations greater than or equal to 24 % falls under the definition of an ignitable hazardous waste [**62**]. An aqueous solution of 10 % methanol by volume is stated to be stable at atmospheric pressure and a temperature greater than 21 °C (70 °F) [**63**].

Anecdotal information from Internet chat forums agrees with this basis. Methanol-water mixtures are used as a performance-boosting fuel additive in some cars, and the discussions found center on questions about the maximum concentration for the safe storage, handling, and use of methanol. Methanol fires are particularly dangerous because the flame is invisible in daylight to the naked eye [**21**]. One source relates "kitchen chemistry" experience that a solution with a ratio of 1/3 methanol to 2/3 water cannot be made to catch fire and that a 50 % / 50 % methanol-water solution ignites but burns poorly [**64**]. Elsewhere, a 50 / 50 methanol/water mix by volume is said to be on the edge of flammability [**65**]. An additional source states that a 30 % methanol in water solution just sits there and cannot be lighted in free air [**17**], and another quotes an industrial brochure as stating that a blend of 33 % methanol / 67 % water is the absolute maximum methanol concentration that will keep the

‡ The autoignition temperature, for which no external ignition source is necessary, is much higher [for example, 686 °F (363 °C) for 100 % ethanol (Ref. [**61**], p.325-57) and 867 °F (464 °C) for 100 % methanol (Ref. [**61**], p.325-81)].

mixture non-flammable [**64**]. Still another claims personally to use a 15-20 % methanol solution, which is said to work well, presumably without flammability problems [**64**].

Finally, flammability tests on methanol-water mixtures [**17**] were conducted outdoors in a metal trough, with results linked to videos posted on other Internet sites. Four methanol concentrations were investigated – 100 %, 50 %, 40 %, and 30 % (all most likely by volume). Combustible paper was used to show up the invisible flame. The burning paper revealed a rather robust combustion for the 100 % case, and the 50 % and 40 % solutions burning with increasing difficulty. The paper from the 30 % trial did not light off at all.

The foregoing observations and findings are consistent with the flash-point entries in Table B4. Flash points of methanol-water solutions are plotted vs. concentration in Ref. [**44**]. Based on the flash-point data, the anecdotal evidence, and the results of the flammability tests, all the calculated methanol-water distillate compositions in Table 2 would likely be flammable when the ambient temperature is about 30 °C (86 °F) or above. In addition, the cumulative samples and most of the individual samples would be expected to ignite at a lower ambient temperature in the laboratory (leaving no residue from combustion of the solution).

Table B4. Flash Points of Methanol-Water Solutions [**44, 61, 66**]

Methanol Weight %	Methanol Volume %	Temperature	
		°F	°C
0	0	—	—
10	13	130	54
20	24	110	43
30	35	95	35
40	46	85	29
50	56	75	24
60	66	70	21
70	75	60	16
80	83	54	12
90	92	54	12
100	100	54	12

Multicomponent Batch Distillation Optimal Control and Thermodynamic Uncertainties

Urmila Diwekar
Center for Uncertain Systems:
Tools for Optimization & Management
Vishwamitra Research Institute
Crystal Lake, IL 60012
Email: urmila@vri-custom.org

Claudio Rivier
Corporate R&D Division
Firmenich SA
Geneva,
Switzerland
E-mail: claudio.river@firmenich.com

Abstract:

Batch distillation is commonly used unit operation in specialty chemicals. In order to maximize the profitability and improve separation using batch distillation column, optimal reflux operation can be used. The mathematical and numerical complexities of the optimal control problem get worse when uncertainty is present in the formulation. However, in specialty chemical industries thermodynamic properties of not all chemical components of a mixture to be separated are known. These uncertainties in the problem formulation can result in sub-optimal solutions. These static thermodynamic uncertainties result in dynamic uncertainties due to unsteady state nature of batch distillation operation. In this work, we derive new optimality conditions based on Ito's lemma derived from financial literature where time dependent uncertainties are common. We use stochastic maximum principle for formulation of the resulting stochastic optimal control problem. This algorithm is implemented in the MultiBatchDS batch distillation process simulator. Finally, a numerical case-study derived from a real world batch distillation separation problem encountered at Firmenich is presented to show the scope and application of the proposed approach.

1. Introduction

Batch distillation is commonly used unit operation in specialty chemicals and pharmaceutical industries. One of the major problem in optimizing batch distillation operation for a particular product specification is that the number of components whose thermodynamic properties known are relatively small as compared to the total chemicals in the mixture. In order to circumvent this problem the components are merged and a mixture of these merged component is used for simulation and optimization. For example, Figure 1 shows the gas chromatograph of several chemicals in a mixture used for separation from Firmenich, a world renowned specialty chemical industry. There are more than 10 components present in this mixture but they are merged into 5 major chemical components whose properties are known.

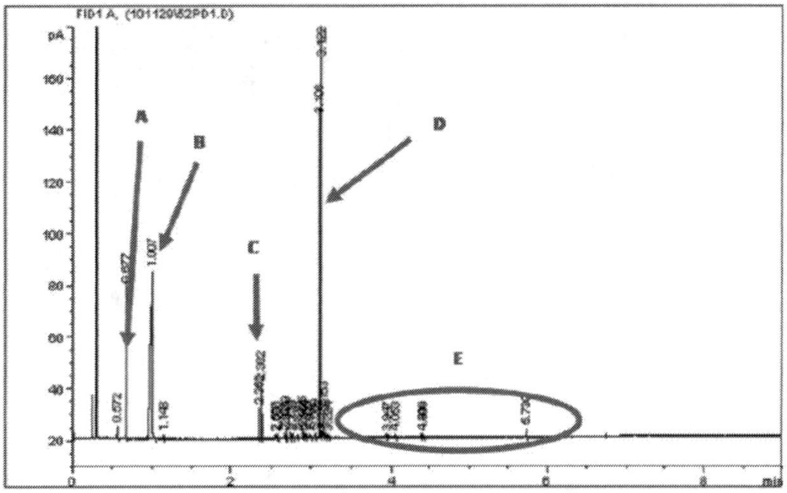

Figure 1: Gas Chromatograph of One of the Firmenich Mixtures

In order to study this problem, we derived an example mixture similar one of the Firmenich mixtures. Table 1 shows the proposed mixture relative volatility and composition. From the table, it can be seen that relative volatility and moles of component 1, 3, 5, and 7 are known exactly but for other components only ranges are known. The common approach to simulate and optimize the operation of batch column is to merge components. Table 2 shows the relative volatility and feed composition of the merged mixture. Section 2 presents the optimization and optimal control study carried out with this merged mixture. Since we do not know where exactly the example mixture lies in the relative volatility and composition range, we created 15 random sample mixtures from the ranges. We assumed that the uncertainty can be represented by normal distribution spanning these ranges. In order to select these random samples, we have taken samples from the normal distribution as well as added some samples which represent the end points of the distribution (low probability events). This analysis shows that clearly taking merged mixtures and optimizing the column performance is suboptimal. We then used a new approach for optimization and optimal control considering uncertainty. This is presented in sections 3 and 4, along with the results obtained using this approach. Finally section 5 presents the summary of the paper.

Components	Relative Volatility	Feed Mixture (moles)
1	2.5488	9.03
2	1.8-2.3	5-10
3	1.5744	22.68
4	1.2-1.4	5-10
5	1	29.4
6	0.4-0.7	5-10
7	0.128	8.89
8	0.1-0.12	5-10

Table 1: Example mixture ranges for relative volatility and feed composition

Components	Relative Volatility	Feed Mixture
1	2.5488	9.03
2	1.5744	30.18
3	1	36.9
4	0.128	23.89

Table 2: Merged components for the example mixture

2. Deterministic Optimization & Optimal Control

The optimization problem in batch distillation mostly involve optimal control problem where time dependent reflux profile is determined by optimizing some performance index. These problems are described below.

- Maximum Distillate Problem - where the amount of distillate of a specified concentration for a specified time is maximized (Converse and Gross, 1963; Keith and Brunet, 1971; Murty et al., 1980; Diwekar et al., 1987; Logsdon et al., 1990; Farhat et al., 1990)

- Minimum Time Problem - where the batch time needed to produce a prescribed amount of distillate of a specified concentration is minimized (Coward, 1967a, b; Hansen and Jorgensen, 1986; Mujtaba and Macchietto, 1988)
- Maximum Profit Problem - where a profit function for a specified concentration of distillate is maximized (Kerkhof and Vissers, 1978; Logsdon et al., 1990)

Here in this work, we are considering the maximum profit problem described by Kerkohf and Vissers (1978). Kerkohf and Vissers were the first one to use profit function for maximization. The profit function they have used is given below.

$$\max_{T,\ R_t} \quad Profit = \frac{P_r D - C_0 F}{T + t_s}$$

where D = amount of product distilled [mol]

P_r = sales value of the product [$/mole]

F = amount of feed [mol]

C_0 = cost of feed [$/mol]

T = batch time [hr]

t_s = set up time for each batch [hr]

R_t = reflux ratio profile with respect to time

Diwekar (1992) provided a unified approach to solving various optimization and optimal control problems in batch distillation. In this approach the problem can be converted as:

$$\max_{T} \quad Profit = \frac{P_r \left(\max\limits_{R_t} \ D\right) - C_0 F}{T + t_s}$$

The inner problem is the maximum distillate optimal control problem which can be solved using a modified formulation using maximum principle while the outer problem can be solved as a nonlinear programming problem.

The inner maximum distillate problem can be formulated as:

$$\max_{R_t} \quad Profit = \int_0^T \frac{dD}{dt}\, dt = \int_0^T \frac{V}{R_t + 1}\, dt$$

Subject to the following purity constraint

$$x_{Dav} = \frac{\int_0^T \frac{V}{R_t + 1} x_D(1)\, dt}{\int_0^T \frac{V}{R_t + 1}\, dt}$$

where V = reboiler vapor rate [mol/hr]

x_{Dav} = average distillate composition of component 1

$x_D\,(1)$ = distillate composition of component 1 at time t

In the coupled approach only two differential equations are considered for time dynamics with respect to component 1. Rest of the components are assumed to be in quasi-steady state and short cut model equations are used to obtain relation between bottom composition and top composition (using the value of reflux ratio at time t and number of plates). These two differential equations are given below.

$$\frac{dx_1}{dt} = \frac{dB}{dt} = \frac{-V}{R_t + 1} \quad x_1\,(0) = F$$

$$\frac{dx_2}{dt} = \frac{dx_B(1)}{dt} = \frac{V}{R_t + 1} \frac{(x_B(1) - x_D(1))}{B} \quad x_2\,(0) = x_F(1)$$

where B = amount remaining in the bottom
$x_B(1)$ = bottom composition of component 1

$x_D(1)$ = top composition of component 1

$x_F(1)$ = feed composition of component 1

In the maximum principle approach to solving optimal control number of adjoint variables and adjoint differential equations added to the formulation equals that of state variables, x (2 in this case). Additionally the purity constraint is added in the objective function with Lagrange multipliers. This complicates the problem and adds several loops to the problem. In the modified approach proposed by Diwekar(1992) the two adjoint variables are merged into one and purity constraint is used to calculate initial value for this adjoint variable. This results in the following adjoint equation for adjoint variable, z, and analytical expression for optimal reflux, R_t.

$$\frac{dz}{dt} = -z\frac{V}{R_t + 1}\frac{(1 - \frac{dx_D(1)}{dx_1})}{x_1} - z^2\frac{V}{R_t + 1}\frac{(x_2 - x_D(1))}{x_1^2}$$

$$R_t = \frac{B - z(x_B(1) - x_D(1))}{z\frac{dx_D(1)}{dR_t}} - 1$$

The coupled NLP and optimal control problem is solved to obtain the value of T and R_t. Figure 2 shows the optimal reflux profile obtained for the merged mixture for the following batch distillation column.

- Feed is 100 moles to start with.
- Batch column Number of Plates=14
- Product Specification is 0.6 for component 1 at the top
- Constant reflux to obtain the purity with batch time =1.1 is 7.0
- Cost of raw materials, C_0 = $100 / mole
- Cost of product P_r = $ 3000/mole

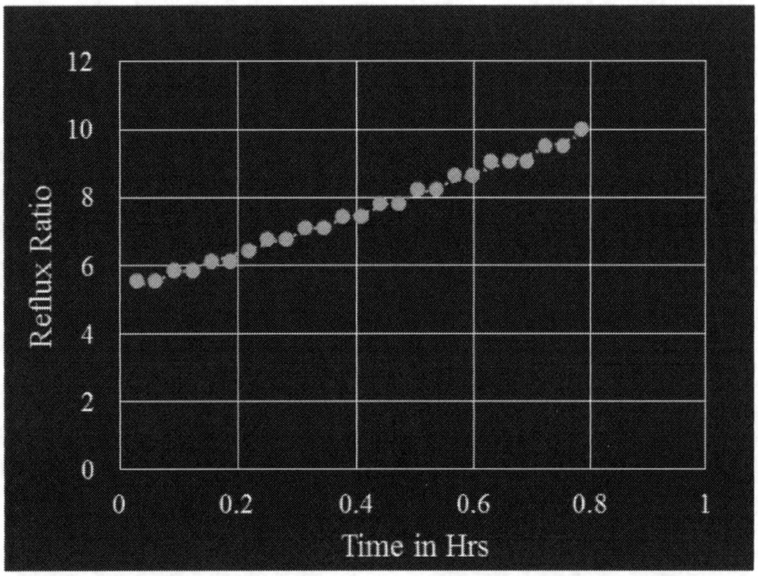

Figure 2: Deterministic Optimal Reflux Profile for the Merged Mixture

The deterministic optimal profit found to be $2211 per batch as compared to the constant reflux profit of $1300 per batch.

3. Uncertainties due to Missing Components

Since we do not know where exactly the example mixture lies in the relative volatility and composition range, we created 15 random sample mixtures from the ranges. We assumed that the uncertainty can be represented by normal distribution spanning these ranges. In order to select these random samples, we have taken samples from the normal distribution (please see Table 3) as well as added some samples which represent the end points of the distribution (low probability events). These mixtures are shown in Table 4. The last four mixtures represent low probability events.

Components	Relative Volatility	Feed Mixture
1	2.5488	9.03
2	Normal 1.8-2.3	Normal 5-10
3	1.5744	22.68
4	Normal 1.2-1.4	Normal 5-10
5	1	29.4
6	Normal 0.4-0.7	Normal 5-10
7	0.128	8.89
8	Normal 0.1-0.12	Normal 5-10

Table 3: Uncertainties due to missing data

Rigorous model in MultiBatchDS are used as a proxy to experiment and these mixtures along with the deterministic optimal control profile is used to simulate various possible scenarios. The profits calculated are presented in Table 5. It can be seen that the deterministic optimal control profile actually shows loss in many scenarios. This calls for stochastic optimal control instead of deterministic optimal control.

Sample Mixture		Comp 1	Comp 2	Comp 3	Comp 4	Comp 5	Comp 6	Comp 7	Comp 8
1	**α(i)**	2.5488	2.1387	1.5744	1.2847	1	0.5137	0.128	0.1155
	xF(i)	0.08925465	0.0741318	0.2241745	0.08765044	0.2905965	0.07597024	0.08787086	0.070351
2	**α(i)**	2.5488	2.1868	1.5744	1.3242	1	0.5388	0.128	0.1045
	xF(i)	0.091931415	0.0782486	0.2308975	0.07446139	0.2993116	0.06732169	0.09050612	0.0673217
3	**α(i)**	2.5488	2.1105	1.5744	1.3355	1	0.6321	0.128	0.1107
	xF(i)	0.090501366	0.0662745	0.2273058	0.06910376	0.2946556	0.08406003	0.08909824	0.0790008
4	**α(i)**	2.5488	2.0882	1.5744	1.2758	1	0.5612	0.128	0.1076
	xF(i)	0.086637238	0.0850801	0.2176005	0.08047093	0.2820747	0.08508007	0.08529402	0.0777624
5	**α(i)**	2.5488	2.0118	1.5744	1.2926	1	0.6032	0.128	0.1135
	xF(i)	0.090994011	0.0816729	0.2285431	0.07745072	0.2962596	0.06179431	0.08958325	0.0737021
6	**α(i)**	2.5488	1.9613	1.5744	1.3074	1	0.55	0.128	0.1065
	xF(i)	0.090122908	0.0729966	0.2263552	0.07867041	0.2934234	0.07299656	0.08872565	0.0767093
7	**α(i)**	2.5488	1.9895	1.5744	1.3153	1	0.4968	0.128	0.1115
	xF(i)	0.090735895	0.0842779	0.2278948	0.06161902	0.2954192	0.08144124	0.08932914	0.0692828
8	**α(i)**	2.5488	2.0314	1.5744	1.2645	1	0.4679	0.128	0.1085
	xF(i)	0.090735895	0.061619	0.2278948	0.07536204	0.2954192	0.07536204	0.08932914	0.0842779
9	**α(i)**	2.5488	2.0686	1.5744	1.2453	1	0.5863	0.128	0.1093
	xF(i)	0.091817094	0.0723708	0.2306104	0.06723797	0.2989394	0.07237078	0.09039357	0.07626
10	**α(i)**	2.5488	2.05	1.5744	1.3547	1	0.5271	0.128	0.1124
	xF(i)	0.089616495	0.0782284	0.2250833	0.07063626	0.2917746	0.0684281	0.08822709	0.0880058
11	**α(i)**	2.5488	1.9132	1.5744	1.3	1	0.5729	0.128	0.11
	xF(i)	0.091198487	0.0696361	0.2290567	0.08185645	0.2969253	0.07960931	0.08978456	0.0619332
12	**α(i)**	2.5488	1.8	1.5744	1.2	1	0.4	0.128	0.12
	xF(i)	0.0903	0.075	0.2268	0.075	0.294	0.075	0.0889	0.075
13	**α(i)**	2.5488	1.8	1.5744	1.2	1	0.4	0.128	0.12
	xF(i)	0.100333	0.055556	0.252	0.055556	0.326667	0.055556	0.098778	0.055556
14	**α(i)**	2.5488	2.3	1.5744	1.4	1	0.7	0.128	0.1
	xF(i)	0.0903	0.075	0.2268	0.075	0.294	0.075	0.0889	0.075
15	**α(i)**	2.5488	2.3	1.5744	1.4	1	0.7	0.128	0.1
	xF(i)	0.100333	0.055556	0.252	0.055556	0.326667	0.055556	0.098778	0.055556

Table 4: Sample mixtures from the probability distributions

Mixture	Profit (det)X 10^2
1	-200
2	-200
3	-200
4	-200
5	-200
6	-37.8605
7	-200
8	-13.06
9	-37.864
10	-200
11	-23.1286
12	4.229
13	23.8766
14	-200
15	4.228

Table 5: Deterministic profits for sample mixtures

It has been shown by Dr. Diwekar's group that static uncertainties (Table 3) in batch distillation result in dynamic uncertainties in composition as shown in Figure 3. These uncertainties can be captured using stochastic processes called Ito processes (Rico-Ramirez et al, 2003; Rico-Ramirez and Diwekar, 2004; Ulas and Diwekar, 2004). Figure 3 shows uncertainties in multiple components. The Ito process which fits this uncertainties is found to be simple Brownian motion. However, when we analyzed the spread of the uncertain region, it has been found that the Ito process is significant for the first component. This allows for quasi-steady state assumption to be valid for other components and discrete Ito processes can be used for those. This is shown in following equations.

$$\frac{dx_2}{dt} = \frac{dx_B(1)}{dt} = \frac{V}{R_t + 1} \frac{(x_B(1) - x_D(1))}{B} + \sigma_1 \, dz \;\; x_2\,(0) = x_F(1)$$

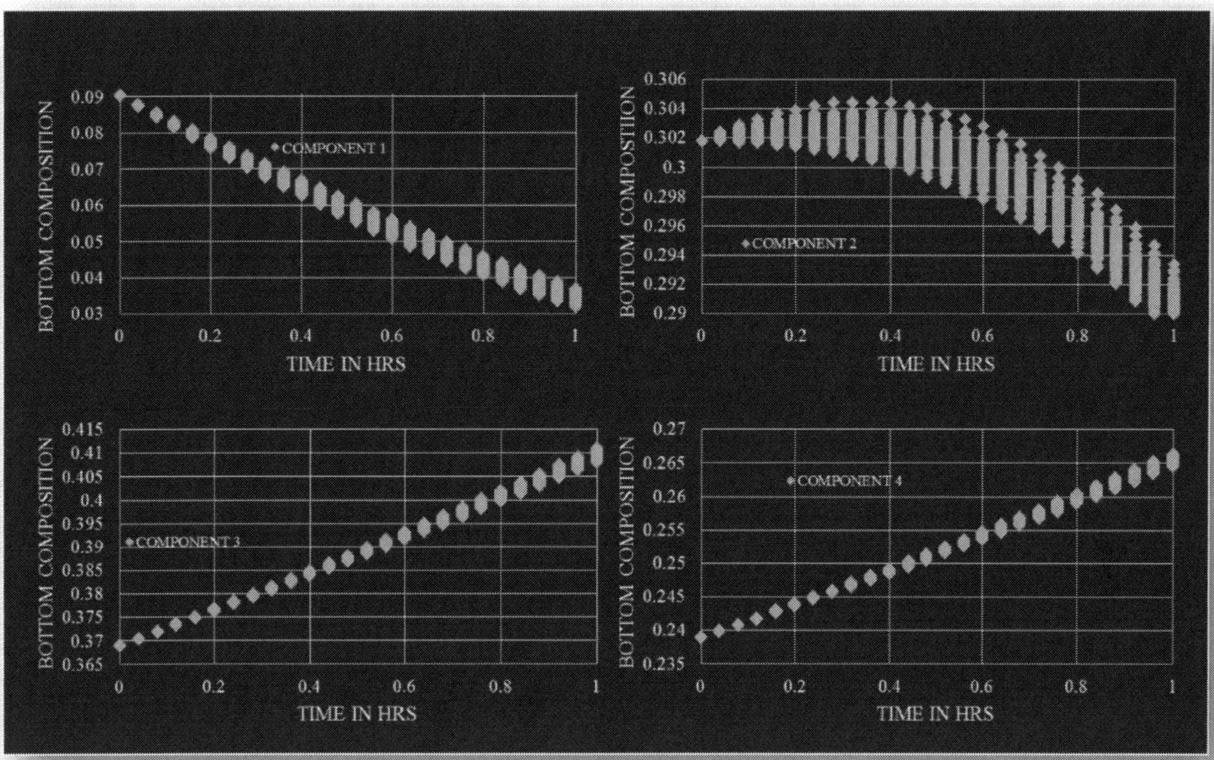

Figure 3: Dynamic uncertainties resulted from static uncertainties (Table 3)

Component	**Max. % Change**	σ_i
1	33.333	0.001572721
2	1.369863	0.010815823
3	0.6024	0.004440619
4	0.625	0.003057177

Table 6: Maximum % change in Ito processes and $\sigma_i s$

where σ_i = standard deviation of the Ito process for component i

$dz = \varepsilon\sqrt{dt}$, where ε is random number derived from unit normal distribution

The quasi-steady state for other components is given by

$$x_{B,t}(i) = x_{B,\ t-1}(i) + \frac{\Delta x_B(i)\,(x_B(i) - x_D(i))}{(x_B(1) - x_D(1))}|_{t-1} + \sigma_i \Delta z,\ \ i = 2,..n$$

We normalize it with the following equations.

$$x_{B,t}(i) = \frac{x_{B,\ t}(i)}{\sum_{j=2}^{n} x_{B,t}(j)} \left(1 - x_B(1)\right)\ \ i = 2, \ldots n$$

4. Stochastic Optimization & Optimal Control

Since the differential equations now include terms with Ito processes, the coupled optimization & optimal control problem becomes stochastic optimization and stochastic optimal control problem with the following objective function.

$$\max_{T} \quad E(Profit) = \frac{P_r\ (\max\limits_{R_t} E(D)) - C_0 F}{T + t_s}$$

where E = represents the expected value of the function

Subject to following differential equations for the two state variables

$$\frac{dx_1}{dt} = \frac{dB}{dt} = \frac{-V}{R_t + 1} \quad x_1\,(0) = F$$

$$\frac{dx_2}{dt} = \frac{dx_B(1)}{dt} = \frac{V}{R_t + 1}\,\frac{(x_B(1) - x_D(1))}{B} + \sigma_1\, dz\ \ x_2\,(0) = x_F(1)$$

The coupled approach presented earlier can be used to solve this problem. However we have to use stochastic maximum principle (Rico-Ramirez and Diwekar, 2004) to solve the underlying optimal control problem. The modified stochastic maximum principle adds the merged adjoint variable as well as a variable representing second order terms. The analytical expression of reflux includes additional stochastic terms as given below.

$$\frac{dz}{dt} = -Z \frac{V}{\frac{d\sigma_1}{dR_t}} \frac{(1-\frac{dx_D(1)}{dx_1})}{x_1} - z^2 \frac{V}{R_t+1} \frac{(x_2 - x_D(1))}{x_1^2} \cdot \sigma_1^2 \in$$

$$R_t = \frac{B-z(x_B(1)-x_D(1))}{z\frac{dx_D(1)}{dR_t}} + \frac{B(\sigma_1 \in \frac{d\sigma_1}{dR_t})}{z\frac{dx_D(1)}{dR_t}V} (R_t+1)^2 - 1$$

The coupled approach is used to solve the maximum expected profit problem. The optimal reflux profile obtained is shown in the Figure 4. The expected profit is found to be $1840 per batch. It can be seen from figure 4 that the stochastic profile is significantly different than the deterministic profile. Again the same mixtures were used with the new profile to obtain profits for the mixtures. This is presented in Table 7. It can be seen that for the stochastic cases most of the mixtures show significant profit for which the deterministic case shows losses. The last few cases where deterministic profile provides profit are the cases with low probability scenarios.

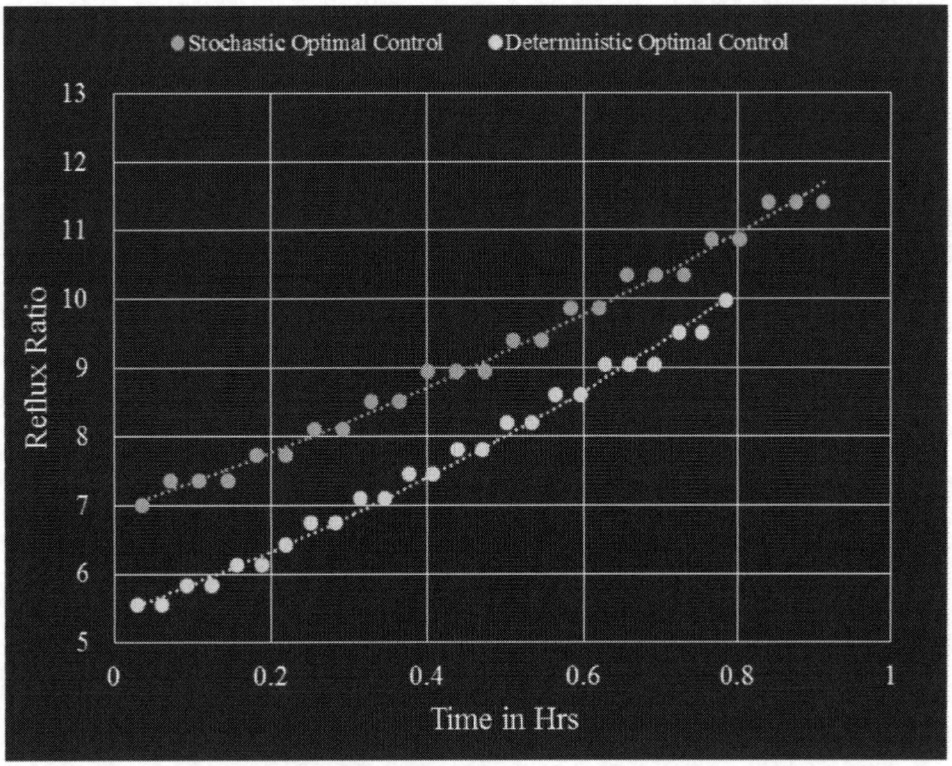

Figure 4: The deterministic and stochastic optimal control profiles.

Mixture	Profit (det)X10^3	Profit (stocha)X10^2
1	-200	16.7329
2	-200	16.1453
3	-200	19.1652
4	-200	11.1569
5	-200	18.4327
6	-37.8605	19.1652
7	-200	18.7117
8	-13.06	18.0778
9	-37.864	19.3476
10	-200	18.0778
11	-23.1286	17.2752
12	4.229	17.9064
13	23.8766	20.8965
14	-200	-21.2815
15	4.228	20.5372

Table 7: Sample mixture profits deterministic as well as stochastic

5. Summary

Mixtures with some components with no or little thermodynamic data is a commonly observed in specialty chemical industries. Optimizing merged components result in suboptimal solutions and many times losses instead of profits. A robust approach is to characterize the uncertainties as Ito processes and use stochastic optimal control approaches. This approach makes sure that high probability of having profits with a particular mixture. In the future, this approach will be applied to a real case study and results will be validated with real experiments.

Bibliography

Boltyanskii V. G., R. V. Gamkrelidze, and L. S. Pontryagin (1956), On the theory of optimum processes (in Russian), *Doklady Akad. Nauk SSSR*, **110**, no. 1.

Bonny L., P. Floquet, S. Domench and L. Pibouleau (1996), Optimal strategies for batch distillation campaign of different mixtures, *Chemical Engineering and Processing*, **35**, 349.

C. Flemming M. and S. B. Jorgensen (1987), Optimal control of binary batch distillation with recycled waste cut, *The Chemical Engineering Journal*, **34**, 57.

Converse A. O. and G. D. Gross (1963), Optimal distillate policy in batch distillation, *Industrial Engineering Chemistry Fundamentals*, **2**, 217.

Coward I. (1967a), The time optimal problems in binary batch distillation, *Chemical Engineering Science*, **22**, 503.

Coward I. (1967b), The time optimal problems in binary batch distillation, a further note, *Chemical Engineering Science*, **22**, 1881.

Diwekar U. M. (1992), Unified approach to solving optimal design-control problems in batch distillation, *AIChE Journal*, **38**, 1551.

Diwekar U. (2008), *Introduction to Applied Optimization, 2nd edition*, Springer.

Diwekar U. (2012), *Batch Distillation: Simulation, Optimal Design and Control, 2nd edition*, CRC Press.

Diwekar U. M. and K. P. Madhavan (1991), Multicomponent batch distillation column design, *Industrial and Engineering Chemistry Research*, **30**, 713.

Diwekar U. M., R. K. Malik, and K. P. Madhavan (1987), Optimal reflux rate policy determination for multicomponent batch distillation columns, *Computers and Chemical Engineering*, **11**, 629.

Egly H., V. Rubby, and V. Seid (1979), Optimum design and operation of batch rectification accompanied by chemical reaction, *Computers and Chemical Engineering*, **3**, 169.

Fan L. T. (1966), *The Continuous Maximum Principle*, John Wiley & Sons, NY.

Farhat S., M. Czernicki, L. Pibouleau, and S. Domench (1990), Optimization of multiple-fraction batch distillation by nonlinear programming, *AIChE Journal*, **36**, 1349.

Furlonge H. I., C. C. Pantelides, E. Sorensen (1999), Optimal operation of multivessel batch distillation columns, *AIChE J.*, **45**, 781.

Hansen T. T and S. B. Jorgensen (1986), Optimal control of binary batch distillation in tray or packed columns, *Chemical Engineering Journal*, **33**, 151.

Keith F, and M. Brunet (1971), Optimal operation of a batch packed distillation column, *Canadian Journal of Chemical Engineering*, **49**, 291.

Kerkhof L. H. and H. J. M. Vissers (1978), On the profit of optimum control in batch distillation, *Chemical Engineering Science*, **33**, 961.

Li P., G. Wozny, and E. Reuter (1997), Optimization of multiple-fraction batch distillation with detailed dynamic process model,*Institution of Chemical Engineers Symposium Series*,**142**, 289.

Logsdon J. S. and L. T. Biegler (1993), Accurate determination of optimal reflux policies for the maximum distillate problem in batch distillation, *Industrial and Engineering Chemistry Research*, **32**, 692.

Logsdon J. S., U. M. Diwekar, and L. T. Biegler (1990), On the simultaneous optimal design and operation of batch distillation columns, *Transactions of IChemE*, **68**, 434.

Mayur D. N. and R. Jackson (1971), Time optimal problems in batch distillation for multicomponent mixtures columns with holdup, *Chemical Engineering Journal*, **2**,150.

Mujtaba I. M. and S. Macchietto (1988), Optimal control of batch distillation, *IMAC World Conference*, Paris.

Mujtaba I. M. and S. Macchietto (1992), An optimal recycle policy for multicomponent batch distillation, *Computers & Chemical Engineering*, **16**, S273.

Mujtaba I. M. and S. Macchietto (1996), Simultaneous optimization of design and operation of multicomponent batch distillation column–Single and multiple separation duties, *Journal of Process Control*, **6(1)**, 27.

Murty B. S. N., K. Gangiagh, and A. Husain (1980), Performance of various methods in computing optimal control policies, *Chemical Engineering Journal*, **19**, 201.

Pontryagin L. S. (1957), Basic problems of automatic regulation and control (in Russian), *Izd-vo Akad Nauk SSSR*.

Rico-Ramirez V. and U. Diwekar (2004), Stochastic maximum principle for optimal control under uncertainty, *Computers & Chemical Engineering*, **28**, 2845.

Rico-Ramirez V., B. Morel, and U. Diwekar (2003), Real option theory from finance to batch distillation, *Computers and Chemical Engineering*, **27**, 1867.

Robinson E. R. (1970), The optimal control of an industrial batch distillation column, *Chemical Engineering Science*, **25**, 921.

Ulas S. and U. Diwekar (2004), Thermodynamic uncertainties in batch processing and optimal control, *Computers and Chemical Engineering*, **28**, 2245.

Ulas S., U. Diwekar, and M. Stadtherr (2005), Uncertainties in parameter estimation and optimal control in batch distillation, *Computers and Chemical Engineering*, **29**, 1805.

Design, Fabrication, and Operation of a Distillation System to Recycle Nuclear Waste

Andrew J. Eckles (Jack)
NATIONAL THERMODYNAMIC LABORATORY

15 AIChE
Spring Meeting, Austin, TX

Pre-Introduction

15 AIChE
Spring Meeting, Austin, TX

Pre-Introduction 2

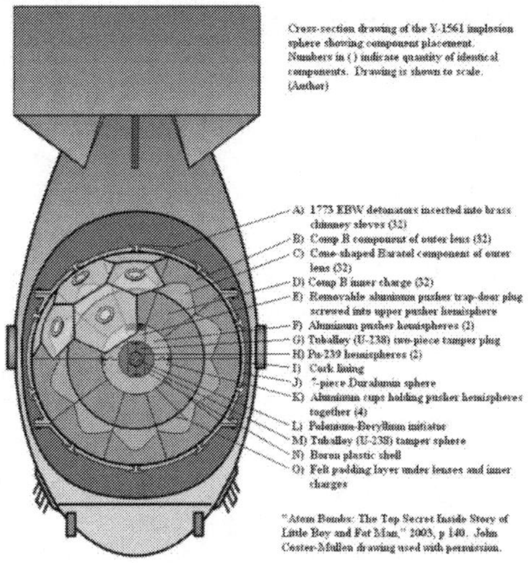

15 AIChE
Spring Meeting, Austin, TX

Firing Mechanism is Complex

- Because of its complicated firing mechanism, and the need for previously untested synchronization of explosives and precision design, it was thought that a full test of the concept was needed before the scientists and military representatives could be confident it would perform correctly under combat conditions. On 16 July 1945, a Y-1561 model Fat Man known asthe gadget for security reasons was detonated in a test explosion at a remote site in New Mexico, known as the "Trinity" test. It gave a yield of about 20 kilotonnes (84 TJ).[24] Some minor changes were made to the design as a result of the Trinity test.[25] Philip Morrison recalled that "There were some changes of importance... The fundamental thing was, of course, very much the same."

15 AIChE
Spring Meeting, Austin, TX

Introduction

- Typical: Model, Pilot Test, Design, Assemble, Startup, ~~Fix minor bugs/problems found during startup~~

- **<u>Atypical</u>**: Increased Importance of Pilot Testing, Very Limited Access After Startup, Radiation Hazards, Nuclear QA/QC Program, Potential Criticality Concerns, Material Degradation in High Radiation Environment, Potential Limitations on Operating Hours

15 AIChE
Spring Meeting, Austin, TX

Nuclear QA/QC Program

- Design
- Change Management
- Inspection/Qualification of all Vendors and Subcontractors
- Documentation
- Continuous Monitoring
- Segregation of all materials, equipment, & supplies

15 AIChE
Spring Meeting, Austin, TX

Pilot Testing

15 AIChE
Spring Meeting, Austin, TX

Radiation Embrittlement

In the typical nuclear environment, the average energy of a neutron is about 2MeV while the thresheold energy to displace an atom from its lattice position in metals is just 20-40 eV; this means that about 50,000 atoms are displaced in a typical collision. In most cases, 90% of the displaced atoms recombine with vacated lattice positions and move back to its original position. However, the remaining non-correct but stable defects and rearrangements from radiation damage will affect material's mechanical properties.

15 AIChE
Spring Meeting, Austin, TX

Electronics and Sensors

- Dramatically reduced life expectancies
- Must design for change out with remote manipulation
- Dual sensors used wherever practical

15 AIChE
Spring Meeting, Austin, TX

Criticality

- Design for worst case
- Shape is EXTREMELY important (most vessels are long, thin cylinders carefully spaced
- Do NOT attempt to perform calculations yourself (unless you are expert yourself – and then I still recommend against it)

15 AIChE
Spring Meeting, Austin, TX

Nitric Acid with Plutonium

* **Nitric Acid Recovery System** -- A Los Alamos team devised a unique way to eliminate acid waste at the Lab's plutonium facility. The team was recognized for its Nitric Acid Recovery System with a White House Closing the Circle Award. The Nitric Acid Recovery System also won a Department of Energy Pollution Prevention Award.

15 AIChE
Spring Meeting, Austin, TX

Piping

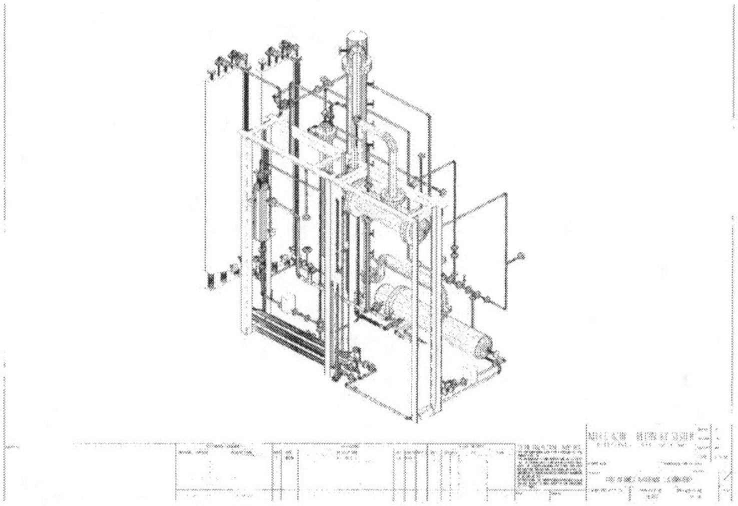

15 AIChE
Spring Meeting, Austin, TX

Operating Time

- Continuous Systems most efficient and cost effective
- Qualified Operators May Not Be Available 24/7
- Automated Operation Without Operator Supervision Not Allowed

15 AIChE
Spring Meeting, Austin, TX

Materials Of Construction

- Many Plastics Subject to Severe Degradation In High Radiation Environments
- Hydrogen Embrittlement
- Radiation Hardening & Eventual Embrittlement
- Potential Enhanced Corrosion
- Hastelloys

15 AIChE
Spring Meeting, Austin, TX

Material Certificates

- Easy and Free If Done In Advance

15 AIChE
Spring Meeting, Austin, TX

Counterfeit Bolts

Substandard materials are known within DOE as suspect/counterfeit items (S/CI). They pose immediate and potential threats to the safety of DOE and contractor workers, the public and the environment. Failure of a safety system due to an S/CI could also have security implications at the DOE facilities. Fortunately, there have been no major accidents or any loss of life within the complex resulting from S/CIs. However, over the years S/CIs have entered the procurement system(s) at DOE sites. Left unidentified and uncorrected these items do present potential risks. A wide variety of counterfeit items have been identified and documented.

15 AIChE
Spring Meeting, Austin, TX

Half Life of Plutonium 240 is
6,280 years (Pu 239 is 24,100 years)

15 AIChE
Spring Meeting, Austin, TX

Novel methods to separate azeotropic butanol-water mixtures without the need for extractive distillation or pervaporative membrane separation

Authors

Kareemuddin Shaik, Zhonglin Zhang, Hiren Shethna, Kadhim Mohammed

Saudi Aramco, Dhahran 31311, Saudi Arabia

Abstract

Conventional methods to separate azeotropic mixtures containing isomers of butanol i.e., 2-butanol and tert-butanol, require either multi-column extractive distillation or an integrated pervaporation membrane-distillation configuration to effectively break the azeotropes. While the azeotropic mixture of 2-butanol – water is easy to separate as the aqueous and organic phase compositions lie on the opposite side of the azeotrope, the presence of tert-butanol and iso-octene in the mixture makes the separation complicated as its aqueous and organic phase compositions lie on the same side of the azeotrope, making it difficult to cross the distillation boundary to achieve the required separation.

In this work, various process configurations to effectively bring the mixed butanol product containing 2-butanol, tert-butanol and iso-octene on-specification in-terms of water content were developed without the need for above mentioned cost-intense conventional separation schemes. An optimized separation scheme along with the hydration process was then scaled- up to a 60 kg/day integrated pilot plant and successfully demonstrated the production of on-spec product.

1. Introduction

While it is easier to develop feasible separation scenarios to ideal and near ideal product compositions, it can get really challenging when dealing with non-ideal mixtures, specifically when a combination of homo and hetero azeotropes are present in a product mixture due to the non-linear vapor-liquid equilibrium behavior, which complicates the prediction of feasible separation sequences further. In order to effectively develop feasible separation scenarios for efficient separation of azeotropic mixtures, a tool that is known as thermodynamic topological analysis of distillation (in Russian Literature) or residue curve (or distillation line) map analysis is employed (1) for three different scenarios listed in this publication.

2. Conventional azeotropic separation techniques utilizing residue curves

A residue curve represents the liquid residue composition with time as a result of simple, one stage batch distillation. The results, when plotted on a triangular graph, are called a residue curve because the plot follows the liquid residue composition in the still. Different residue lines result from different starting compositions. A collection of these curves for a given ternary system is called a residue curve map (2).

a. Solvent Extraction

The extractive distillation is a very efficient and proven technique for the separation of azeotropic mixtures. Residue curve maps using process modelling software such as Aspen Plus expedite the process of selecting effective solvents for the required separation of azeotropic mixtures. Fig. 1 represents a typical scheme employed to separate the azeotropic mixtures.

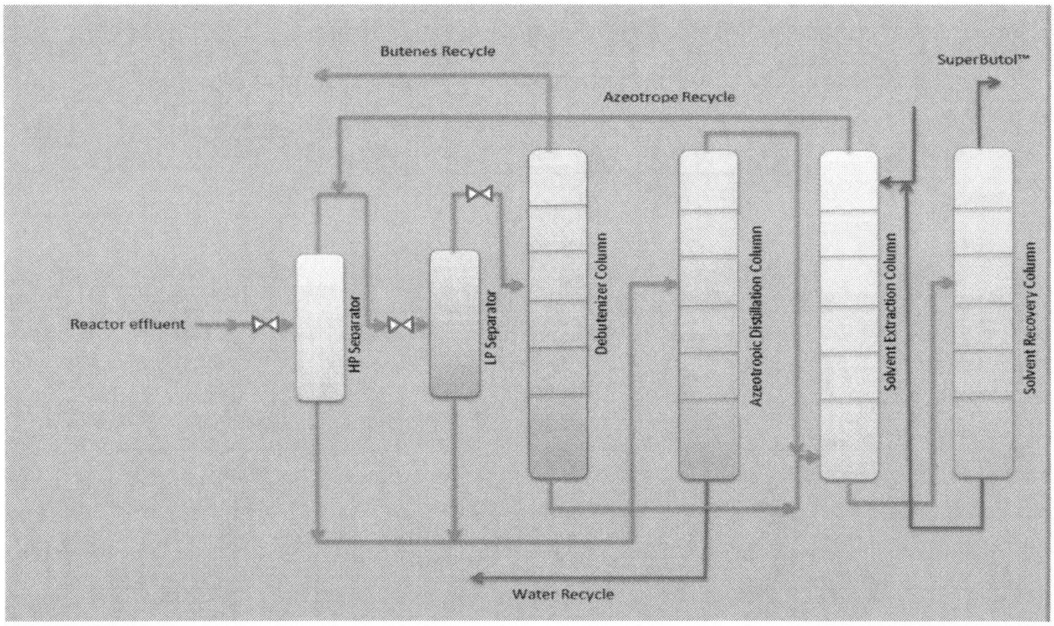

Fig. 1 Solvent extraction scheme at the backend of a separation process

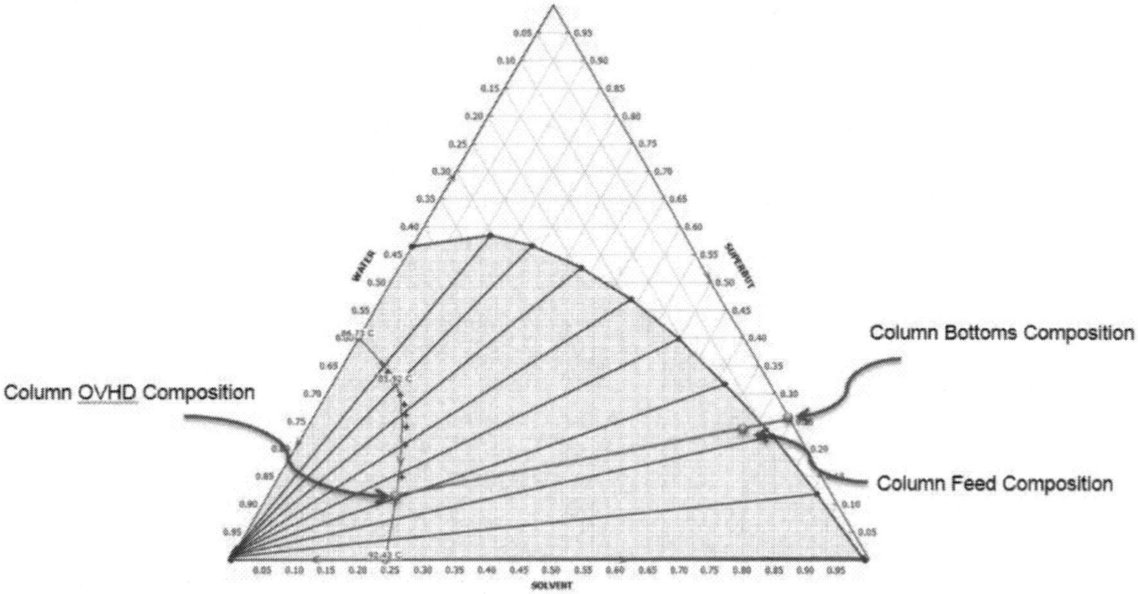

Fig. 2 Tertiary Plot depicting the compositions of the feed, overhead and bottoms of the solvent extraction column showed in Figure 1

The solvent (p-Xylene in this scenario) extracts the complex mixture of butanol isomers and iso-octene in the solvent extraction column from the azeotropic mixture and the solvent is then recovered in the solvent recovery column and recycle to the solvent extraction column in a closed loop. Several solvents were studied and the closest to the acceptable separation is achieved was by using p-Xylene as a solvent. The composition profile of solvent extraction column based on the feed composition follows the trend showed in Fig. 2. It can be inferred from the Fig. 2 that the overhead composition, which is supposed to be pure water, is stopped at the distillation boundary that it can't cross due to the complexing of the azeotropic mixture resulting in inefficient separation of the mixture and loss of product into the recycle stream. Table 1 lists the compositions of the feed, overhead and bottoms stream of the solvent extraction column with less than optimum separation obtained through this conventional solvent extraction scheme.

	Feed, mol%	BTMS, mol%	OVHD, mol%
SuperButol™	23.75%	25.6%	10.32%
Water	8.14%	-	76.94%
Solvent (pX)	68.11%	74.4%	12.74%

Table 1: Solvent Extraction Column composition profile

b. Pervaporation Membrane Technology

Pervaporation is based on the use of a membrane which is permeable to certain substances but not to others. The membranes are able to absorb certain substances which have chemical affinity for the membrane and reject substances which have less affinity. For instance, hydrophilic membranes are used to separate water from solvents and organophilic membranes are used to remove organics from water. Since the main driving force for separating the mixture is the partial pressure difference, this technique can be applied regardless of the VLE of the mixture and the azeotropes can be broken without adding additional substances and component with same volatility can also be separated (3). Fig. 3 depicts one such scheme that utilizes pervaporation membrane technology.

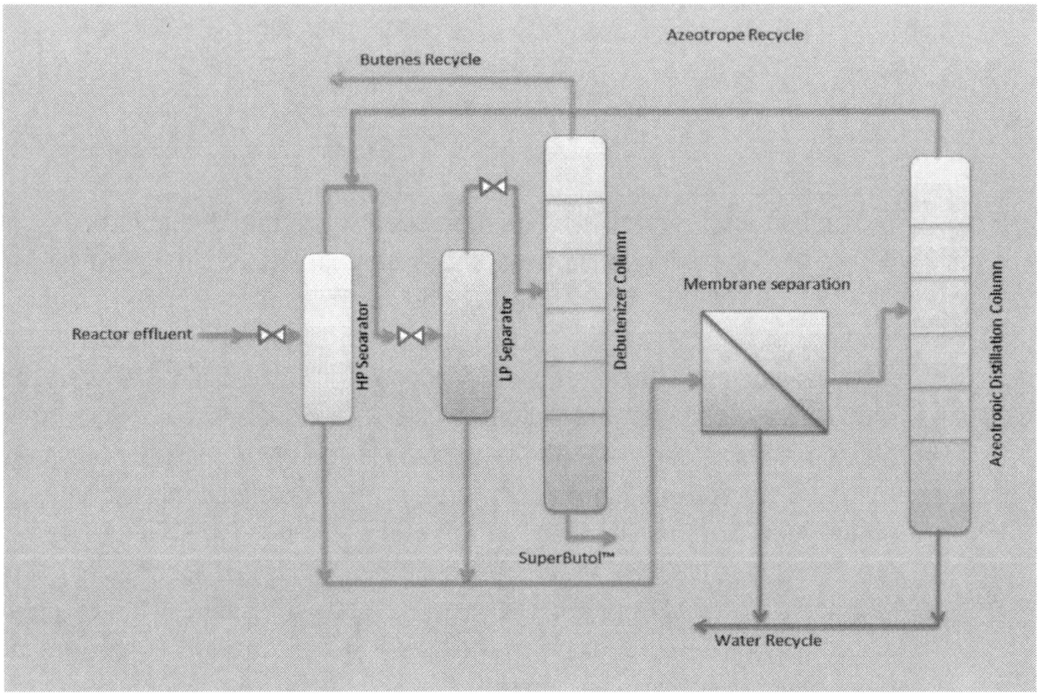

Fig. 3: Pervaporation scheme at the backend of a separation process

As it can be inferred from Fig. 3, the impact of incorporating pervaporation membrane set-up may not have any significant impact on how the debutanizer column operates, resulting in inefficient separation as shown in Fig. 3 and Table 2. The separation is again limited by the fact the feed composition and the required product composition lie on the opposite side of the distillation boundary despite employing the pervaporative membrane technology into the separation scheme.

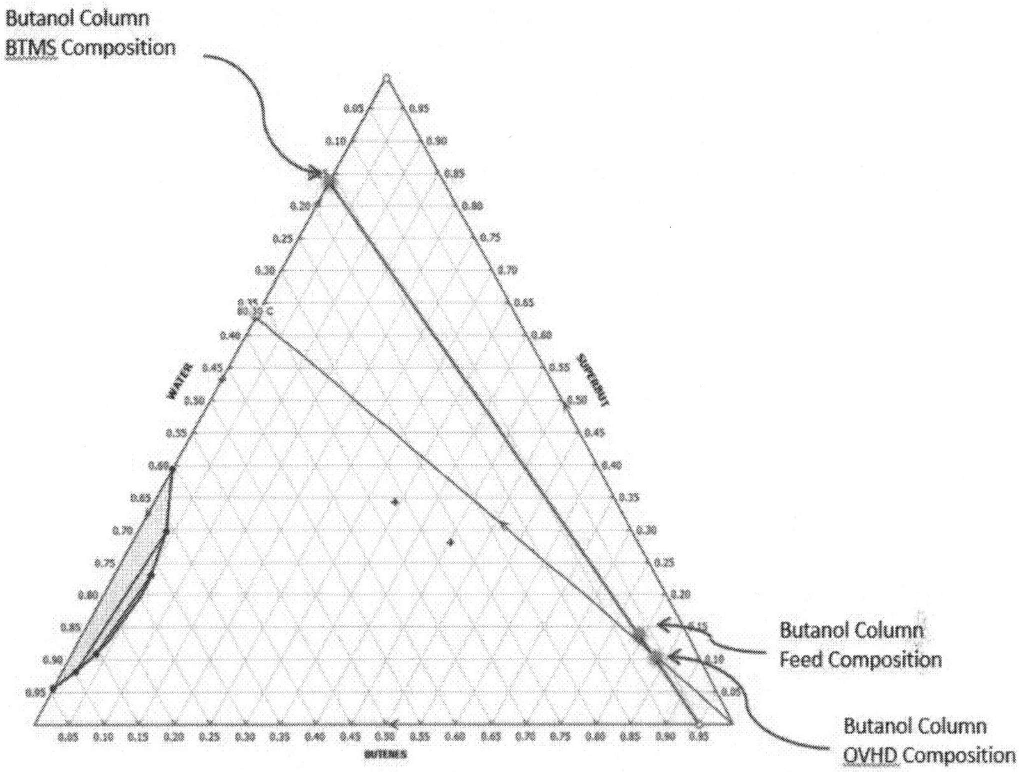

Fig. 4: Tertiary Plot depicting the compositions of the feed, overhead and bottoms of the debutenizer column showed in Figure 3

	Feed, mol%	BTMS, mol%	OVHD, mol%
SuperButol™	14%	85.2%	11.1%
Water	7%	14.8%	6.8%
Butenes	79%	-	82.2%

Table 2: Debutenizer Column composition profile

3. **SuperButol™ Separation Technology**

A novel process to effectively and efficiently separate the complex azeotropic mixture to produce an on-spec mixture of butanol isomers and iso-octene with less than 500 ppm of water has been patented by Saudi Aramco. This process utilizes unreacted mixed butenes feed as a solvent to extract all the organic components from the product mixture. Fig. 5 depicts a simplified process flow diagram with internal and external recycle of unreacted mixed butenes so as to increase the butenes to water ratio such that the extraction of organic phase into mixed butenes is maximized.

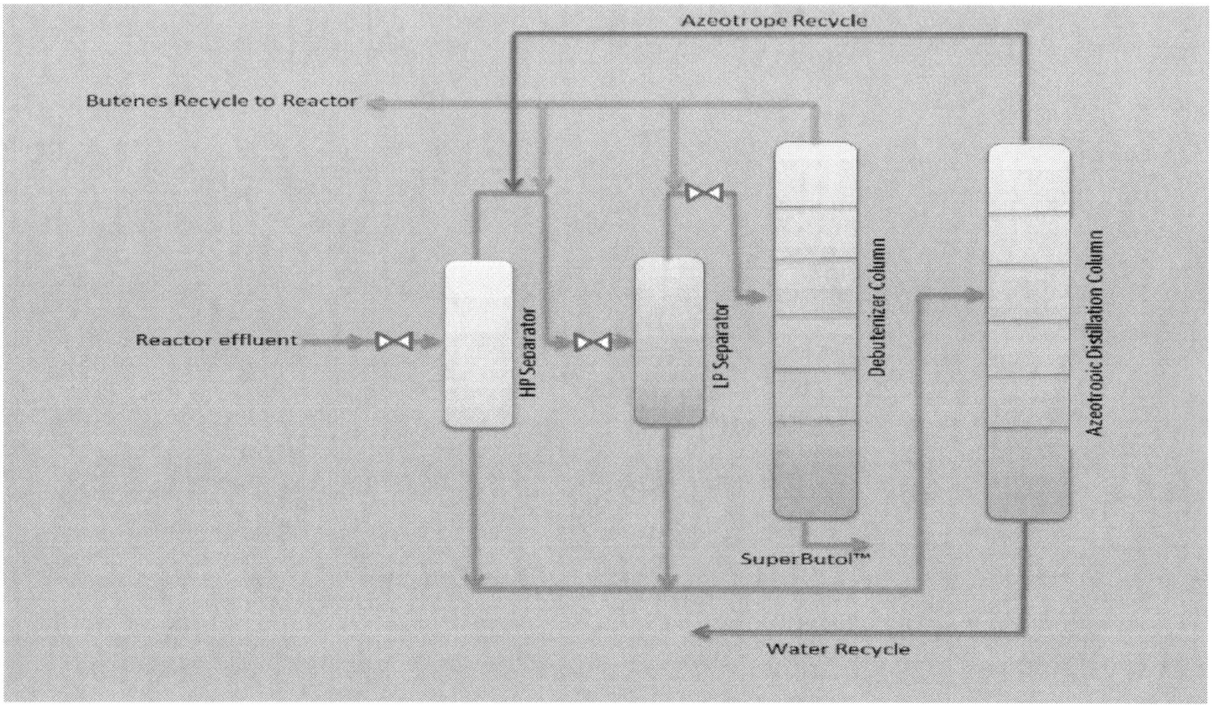

Fig. 5: SuperButol™ Separation Technology

The above set-up has been tested at an integrated pilot plant at 60 kg/day capacity and produced2.5 tons of on-spec SuperButol™ product consisting of butanol isomers and iso-octene with a targeted water content of less than 500 ppm.

The high pressure separator (HP separator) shown in Fig. 5 receives the unreacted mixed butenes feed such that the feed composition of the HP separator falls on the aqueous side of the azeotrope. The HP separator is designed such that the residence time will enable the composition of the organic layer to lay close to the distillation boundary as shown in Fig. 6. The resulting composition of the organic and aqueous layers of the HP separators are shown in Table 3. The organic layer from the HP separator is subsequently sent to the low pressure separator (LP separator) along with the partial recycled unreacted mixed butene feed so as to get the feed composition as shown in Fig. 7. The intent here is to push the LP separator organic composition away from the aqueous phase so as to further reduce its water content. The organic and aqueous layer compositions are tabulated in table 4.

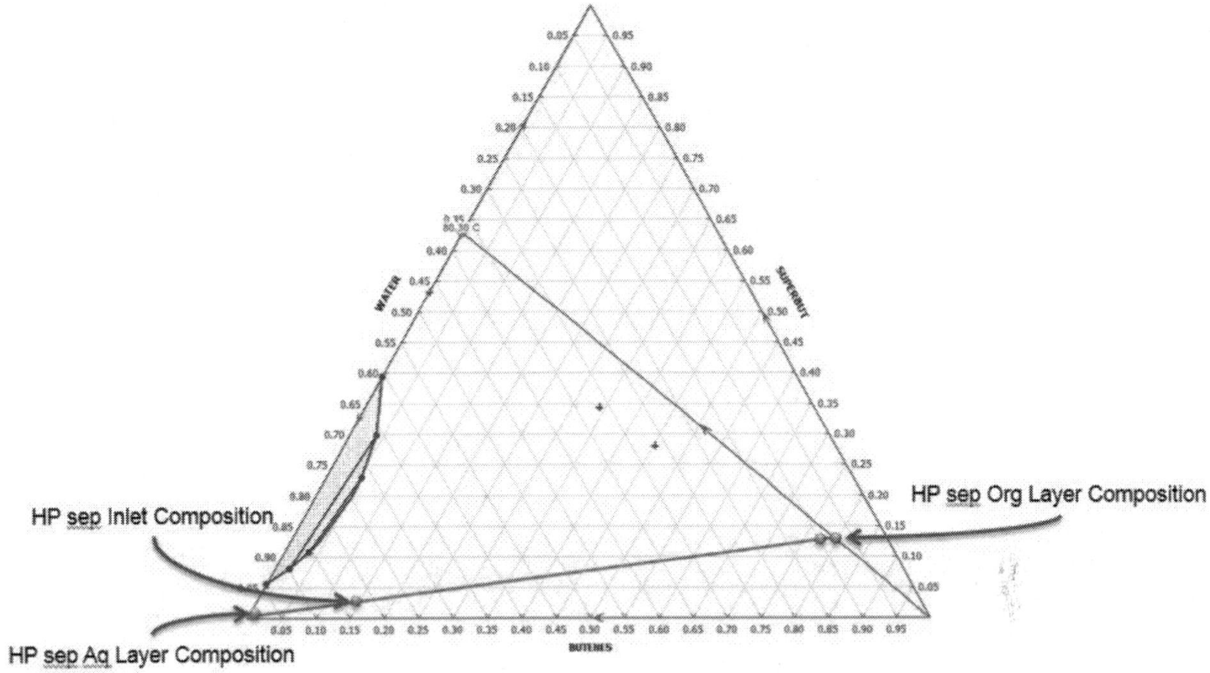

Fig. 6: Composition profile across the HP separator

HP Separator	Feed, mol%	Aqueous Layer mol%	Org Layer, mol%
SuperButol™	2.9%	0.6%	13%
Water	83.2%	99%	9.2%
Butenes	14.5%	0.4%	77.8%

Table 3: HP separator composition profile

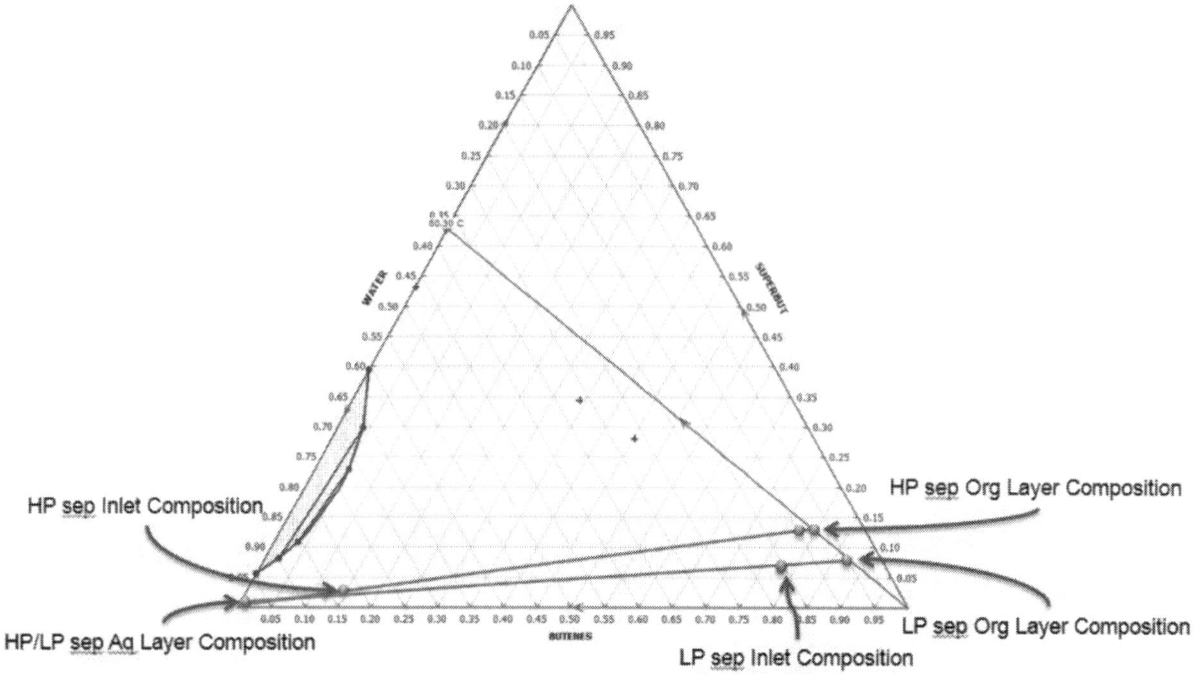

Fig. 7: Composition profile across the LP separator

LP Separator	Feed, mol%	Aqueous layer, mol%	Org Layer, mol%
SuperButol™	6.8%	0.7%	7.7%
Water	15.2%	99%	4.3%
Butenes	78%	0.3%	88%

Table 4: LP separator composition profile

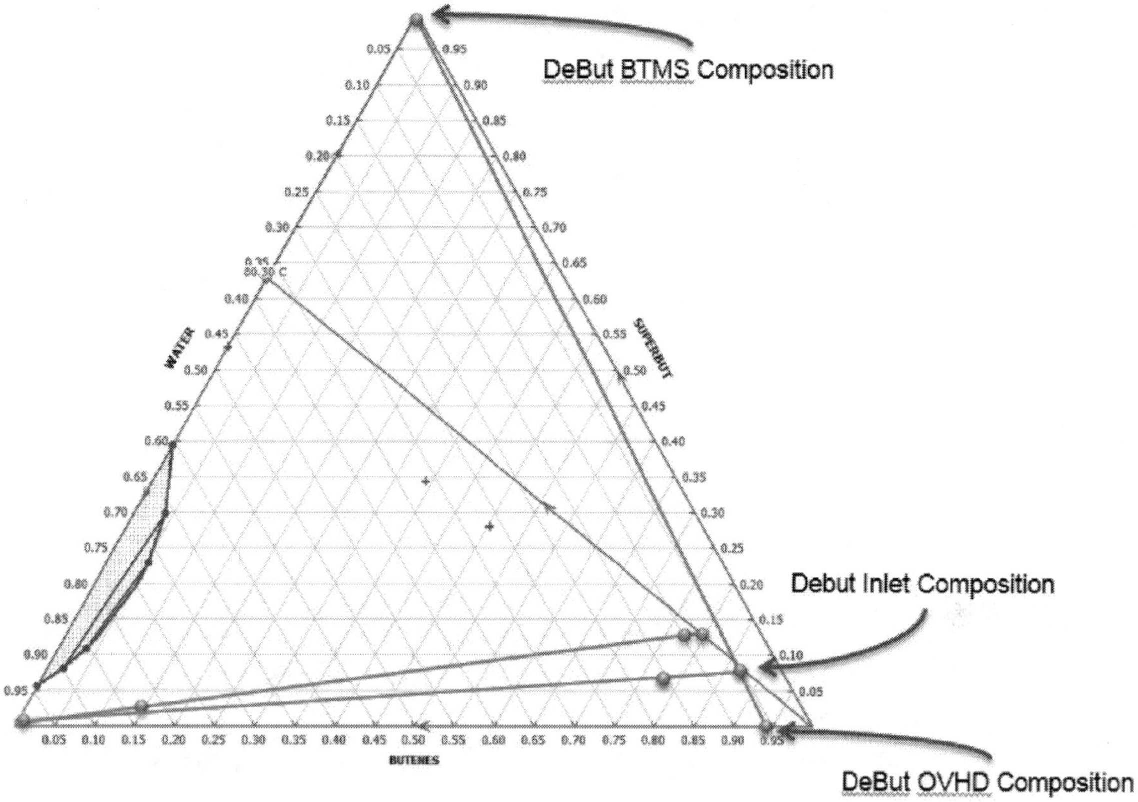

Fig. 8: Composition profile across the Debutenizer Column

The organic layer from the LP separator thus becomes the feed to the Debutenizer with the inlet composition shown in Fig. 8. From the Fig. 8 it may appear that in order to achieve the required SuperButol™ composition at the Debutenizer bottoms, it must cross the distillation boundary as the mass balance would dictate the composition of Debutanizer overheads to be as shown in Fig. 8 and Table 5. Upon careful evaluation of Fig. 9, which depicts the residue curves of the azeotropic mixture, it can be inferred that all the compositions i.e., Debutenizer inlet composition (On the distillation boundary), bottoms composition and overhead composition fall on the aqueous side of the azeotropic, thereby facilitating the required product separation to achieve the targeted product composition.

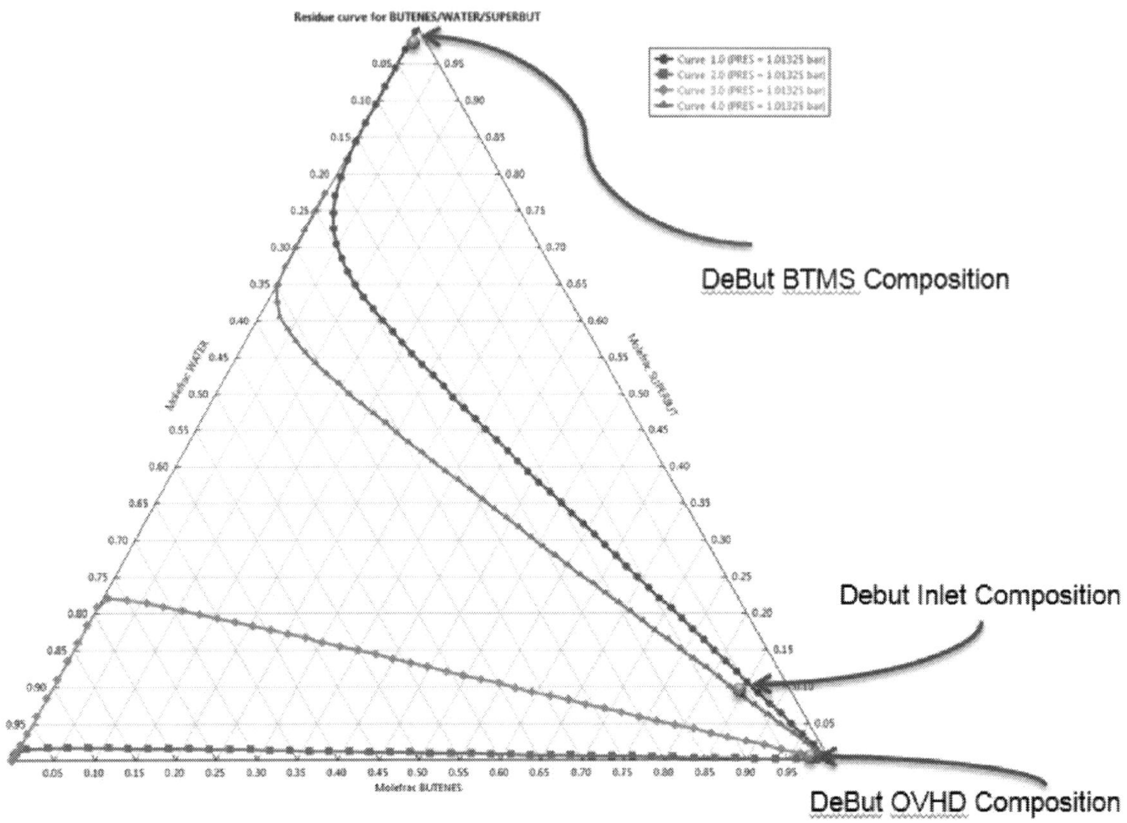

Fig. 9: Residue curves conferring that the Debutenizer composition profile didn't actually cross the distillation boundary

DeButenizer	Feed, mol%	BTMS, mol%	OVHD, mol%
SuperButol™	7.7%	99.5%	0.5%
Water	4.3%	0.5%	4.6%
Butenes	88%	-	94.9%

Table 5: Debutenizer composition profile

4. Results and Discussion

Table 6 summarizes the product compositions obtained using the three schemes employed in the study i.e, solvent extraction, integrated pervaporation/distillation set-up and SuperButol™ separation technology. It can be inferred from the results that for this particular azeotropic mixture, SuperButol™ separation technology has been effective in obtaining desired product composition. The residue curve maps greatly enhanced our ability to understand the system so as to efficiently design a separation scheme, there-by significantly reducing the capital and operating costs. A preliminary economic analysis was conducted to gauge the capital operating costs of these various schemes and the results are tabulated in Table 7 with SuperButol™ technology being an optimizal scheme.

Component	Product Composition based on Separation Scheme, wt%		
	Solvent Extraction	Integrated pervap/Distillation	SuperButol™ Technology
2-Butanol	55.11%	61.91%	63.89%
Tert-Butanol	25.08%	27.40%	29.67%
DIB (Iso-Octene)	5.78%	5.48%	5.93%
Water	3.09%	5.21%	0.50%
Solvent	10.94%	-	-

Table 6: Comparison of effectiveness of various separation technologies

Separation Scheme	Equipment Cost, $MM	Energy Costs, kW
Solvent Extraction	1.525	16.92
Integrated Pervapration/Distillation	2.31	7.42
SuperButol™ separation scheme	0.89	16.93

Table 7: Comparison of capital/operating costs of various separation technologies

5. Conclusion

Separation of Azeotrpic mixture consisting of 2-Butanol/tert-Butanol/iso-Octene/Water has been successfully demonstrated in an Integrated Pilot Plant using Saudi Aramco patented technology. While conventional separation schemes were inadequate and capital intensive, SuperButol™ process separation scheme has been proven to be simple, robust and cost effective.

6. References

1. V. N. Kiva et al. Chemical Engineering Science 58 (2003) 1903-1953
2. Content from Chemistations website
3. Norbert Martin, Sulzer Chemtech Review Paper 3/98

A Total Correlation Evaluating User Fitting UNIFAC Determined from VLE Data Covering 5000 Binaries (ID#400309)

Satoru Kato[1,2], *Hiroshi Tachibana*[2], *Hidetoshi Kuramochi*[3], *Ichiro Minami*[4]
[1] *TC Lines Co., Ltd., Yokohama, Japan,* [2] *Tokyo Metropolitan University, Hachiohji, Tokyo, Japan,* [3]*National Institute for Environment Studies, Tsukuba, Japan,* [4] *Epsilon Co., Ltd., Tokyo, Japan*

1. INTRODUCTION

Vapor-liquid equilibria (VLE) are of practical importance for the rational design and control of distillation towers. In the present investigation, a high-precision correlation, called the TC line (Thermodynamic Consistency line)[1-3], is applied to the propylene + propane distillation separations, benzene + cyclohexane + acetone azeotropic distillation and user fitting UNIFAC. The TC line enables us to remove systematic errors from the VLE experimental data. It is a long-standing problem for process simulator users to eliminate inaccurate VLE data predicted using parameters integrated into the process simulators. The TC line is advantageous for eliminating these inaccurate data due to its highly precise convergence. In the following analysis, a total correlation for VLE data covering 5000 binaries[4] is used to evaluate the data consistency of the binary VLE data.

2. MODELS

2.1 Correlation of VLE data using activity coefficient models

The combination of the equation of state and a mixing rule may not satisfy the Gibbs-Duhem equation; therefore, the present investigation uses the activity coefficient models. The vapor-liquid equilibrium equation for component i is formulated as follows [5]:

$$Py_i = \gamma_i x_i p_{si} \quad (i = 1, 2) \tag{1}$$

where P, x_i and p_{si} denote the system pressure, the liquid-phase mole fraction of component i and the vapor pressure of component i, respectively. The activity coefficient, γ_i, is formulated using the following models:

(1) the Margules equation

$$\ln \gamma_1 = x_2^2[A + 2(B - A)x_1], \quad A = \ln \gamma_1^{\infty} \tag{2}$$

$$\ln \gamma_2 = x_1^2[B + 2(A - B)x_2], \quad B = \ln \gamma_2^{\infty} \tag{3}$$

(2) the van Laar equation

$$\ln \gamma_1 = A[1 + Ax_1/(Bx_2)]^{-2} \qquad A = \ln \gamma_1^{\infty} \tag{4}$$

$$\ln \gamma_2 = B[1 + Bx_2/(Ax_1)]^{-2} \qquad B = \ln \gamma_2^{\infty} \tag{5}$$

(3) the UNIQUAC equation

$$\ln \gamma_i = \ln \frac{\Phi_i}{x_i} + \frac{z}{2} q_i \ln \frac{\theta_i}{\Phi_i} + \Phi_i \left(l_i - \frac{r_i}{r_j} l_j \right) - q_i \ln(\theta_i + \theta_j \tau_{ji}) + \theta_j q_i \left(\frac{\tau_{ij}}{\theta_i + \theta_j \tau_{ji}} - \frac{\tau_{ij}}{\theta_j + \theta_i \tau_{ij}} \right) \tag{6}$$

$$\Phi_i = \frac{x_i r_i}{x_1 r_1 + x_2 r_2} \tag{7}$$

$$\theta_i = \frac{x_i q_i}{x_1 q_1 + x_2 q_2} \tag{8}$$

$$l_i = \frac{z}{2}(r_i - q_i) - (r_i - 1) \quad (9)$$

$$\ln \tau_{ij} = -\frac{u_{ij}}{RT} \quad (10)$$

where the molecular volume, r_i, and the molecular surface area, q_i, of component i are given in ref. 6. In Eq. (10), R and T denote the gas constant and system temperature, respectively.

2.2 An Example of VLE data correlation

In **Fig. 1**, P is plotted versus x_1 at 373.15 K for the methanol (1) + water (2) binary. Any of the activity coefficient equations can satisfactorily correlate the non-azeotropic mixtures, from which the γ_i^∞ values can be determined.

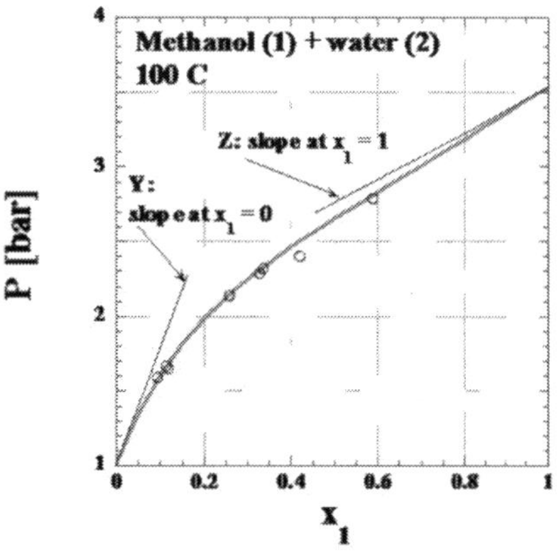

Fig. 1 P vs. x_1 at 373.15 K for the methano (1) + water (2) binary system, (o) experimental data[7], (—) correlated using the Margules equation with $A = 0.882$ and $B = 0.539$.

2.3 Faults in conventional correlations

In **Fig. 2**, the UNIQUAC binary parameter, u_{12}, is plotted versus $T - 273.15$ for the methanol (1) + water (2) binary. In **Fig 3**, the margules binary parameter, A, is plotted versus $T - 273.15$. These figures show that data scatterings are notable, and a line representing the data cannot be fixed. Similar faults occour even if other activity coefficient equations are used. Furthermore, the scattering is more notable, if azeotropic mixtures are analyzed.

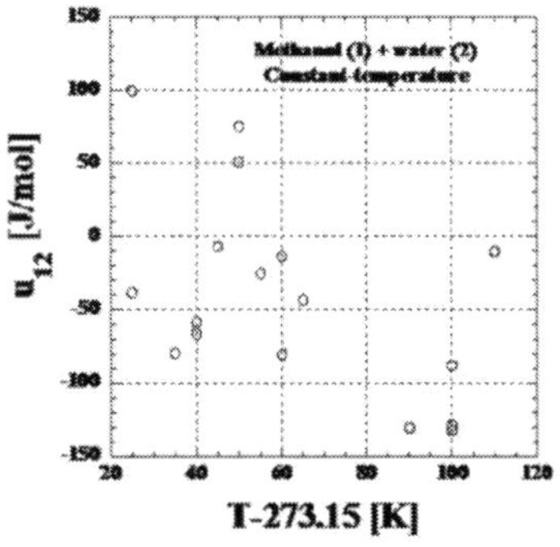

Fig. 2 u_{12} vs. $T - 273.15$ for the methanol (1) + water (2) binary, (o) UNIQUAC parameters calculated from constant temperature data[7-15].

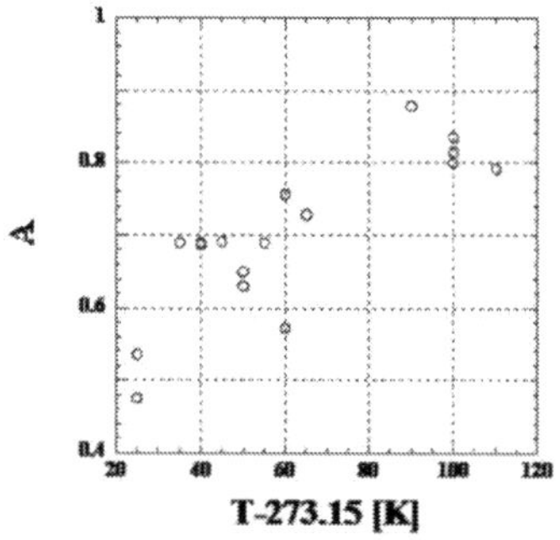

Fig. 3 A vs. $T - 273.15$ for the methanol (1) + water (2) binary, (o) Margules parameters calculated from constant temperature data[7-15].

3. TC LINE

3.1 Definition and advantages of TC line

The TC line is high-precision correlations that occur between X versus Y and X versus Z. The quantities, X, Y and Z, are defined as follows[2]:

$$X = (p_{s1} + p_{s2})/(P_{c1} + p_{s2}) \tag{11}$$

$$Y = (p_{s1}\gamma_1^{\infty} - p_{s2})/(P_{c1} - p_{s2}) \tag{12}$$

$$Z = (p_{s1} - p_{s2}\gamma_2^{\infty})/(P_{c1} - p_{s2}) \tag{13}$$

where X and P_{c1} denote the proximity ratio to the critical point and critical pressure, respectively. As shown in Fig. 1, Y and Z denote the non-dimensional infinite dilution pressure gradients at x_1=0 and 1, respectively.

Examples of the TC lines are shown in **Figs. 4 a and b**. In Fig. 4, the following X, Y and Z were used, because, at a low pressure VLE, the effect of the denominators in Eqs. (11)-(13) are negligible.

$$X = (p_{s1} + p_{s2}) \tag{14}$$

$$Y = (p_{s1}\gamma_1^{\infty} - p_{s2}) \tag{15}$$

$$Z = (p_{s1} - p_{s2}\gamma_2^{\infty}) \tag{16}$$

The TC line is formulated as Eq. (17) or Eq. (18).

$$Y, Z = aX^b \tag{17}$$

$$Y, Z = aX + bX^2 + c\ln(X+1) \tag{18}$$

The great advantages of the TC line are that it passes through the origin, and constant-temperature VLE data and constant-pressure VLE data converge to the same line, as shown in Figs. 4 (a) and (b).

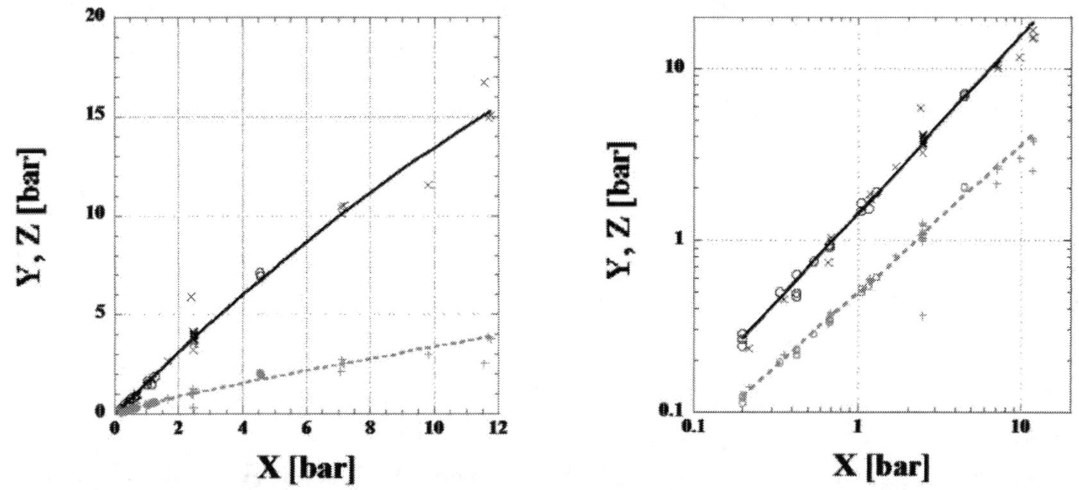

(a) (b)

Figs. 4 (a) and (b) Y and Z vs. X for the methanol water bianry, (o) X vs. Y calculated from constant temperature data, (x) X vs. Y calculated from constant-pressure data, (—) X vs. Y TC line with $Y = 1.73X^{1.04}$ as shown in ref. 3, (●) X vs. Z calculated from constant-temperature data, (+) X vs. Z calculated from constant pressure data, (- - -) X vs. Z TC line with $Z = 0.283X^{0.871}$ as shwon in ref. 3; exp. data were cited from ref. 6

3.2 x, y Relationships predicted from the TC line

In **Fig. 5**, y_1 is plotted versus x_1 for the benzene (1) + 1-propanol (2) binary at 1.013 bar, in which the van Laar equation was used for the prediction of y_1. The van Laar parameters were determined using the TC lines shown in **Fig. 6.** In **Fig. 7**, y_1 is plotted versus x_1 for the 2-butanone (1) + water (2) binary at 1.013 bar. The van Laar equation was also used, and the van Laar parameters were determined using the TC lines shown in **Fig. 8**. Figs. 5 and 7 show that accurate x, y relationships are predicted using the TC line, even if the azeotropic and the two-liquid phase forming binaries are analyzed.

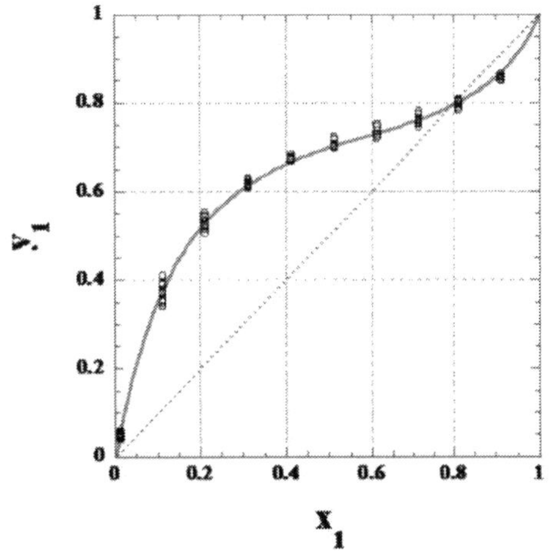

Fig. 5 y_1 vs. x_1 for the benzene (1) + 1-proapnol (2) binary at 1.013 bar, (o) exp. data, (—) predicted using TC lines.

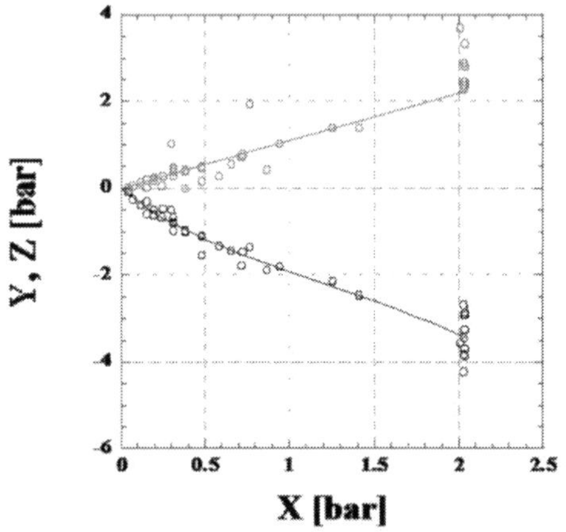

Fig. 6 Y and Z vs. X for the benzene + 1-propanol binary, (o) X vs. Y calculated from exp. data, (—) X vs. Y TC line, (o) X vx. Z calculated from exp. data, (—) X vs. Z TC line, exp. data were cited from ref. 6.

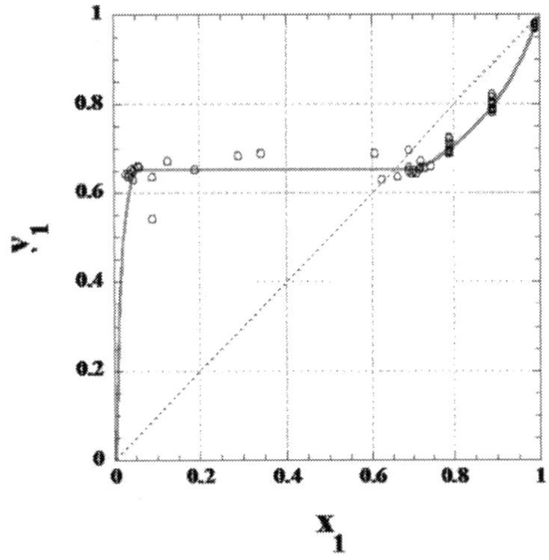

Fig. 7 y_1 vs. x_1 for the 2-butanone (1) + water (2) binary at 1.013 bar, (o) exp. data, (—) predicted using TC lines.

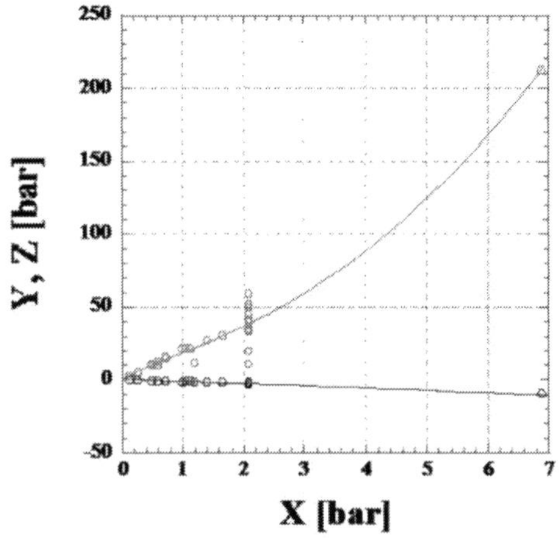

Fig. 8 Y and Z vs. X for the 2-butanone + water binary, (o) X vs. Y calculated from exp. data, (—) X vs. Y TC line, (o) X vx. Z calculated from exp. data, (—) X vs. Z TC line, exp. data were cited from ref. 6.

4. TOTAL CORRELATION

In **Fig. 9**, I_{exp} is plotted versus I_{cal} using the TC lines satisfying $n \geq 7$, $AADY < 0.2$ and $AADZ < 0.2$, where n denotes the number of data sets, and $AADY$ and $AADZ$ are the average deviations from the Y and Z TC lines, respectively. The total correlation function, I, includes universal constants, which are independent of the system. Fig. 9 shows that $I_{exp} = I_{cal}$ holds, if reliable TC line data are used. In **Fig. 10**, I_{exp} is plotted versus I_{cal} for 5000 binaries. Of the 5000 binaries, 92 % are reliable, because they satisfy the relationship, $I_{exp} = I_{cal}$. Most of the binaries reflecting low quality data have only one dataset.

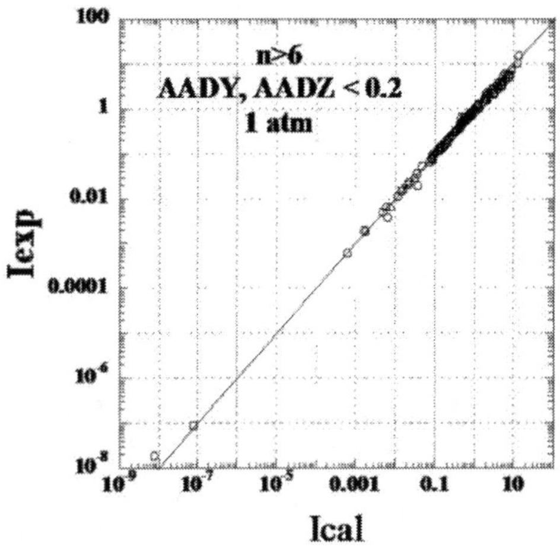

Fig. 9 I_{exp} vs. I_{cal}, plotted using reliable TC line data, as $n \geq 7$, $AADY < 0.2$ and $AADZ < 0.2$.

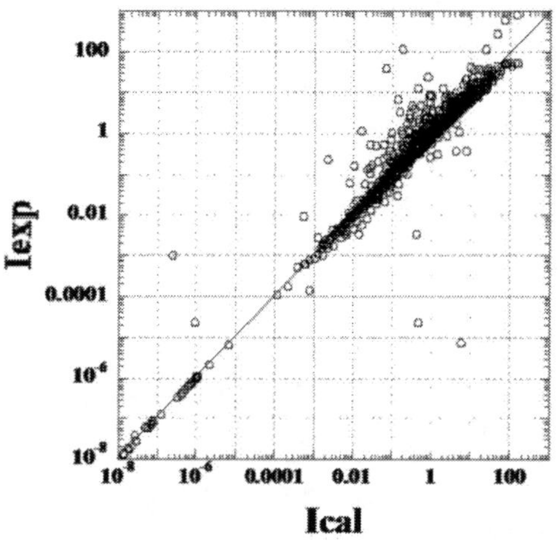

Fig. 10 I_{exp} vs. I_{cal}, plotted using the TC line data of 5000 binaries.

5. APPLICATIONS

5.1 Propylene + propane distillation separation

To show the usefulness of the TC line and the total correlation, the number of theoretical stages operated at total reflux satisfying three-nine purities was calculated using the Ponchon-Savarit method. The method of Rowlinson was used for the prediction of the pure liquid heat capacities. The partial molar excess enthalpies were calculated from the temperature dependence of the TC line. This procedure is a notable advantage that originated in the TC line. In **Fig. 11**, the TC lines for the propylene + propane binary are plotted. The infinite dilution activity coefficient values were calculated from the Margules equation determined from the P, x data, as shown in **Fig. 12**. As shown in **Table 1**, the number of theoretical stages, N, calculated at $T = 223.75$ K was 80. The TC lines at 223.75 K satisfy the total correlation; therefore, they are reliable. Meanwhile, the experimental data at 223.75 K[16], shown in Fig. 11 using the symbols ▼ and ▲, resulted in $N = 66$. Conventional consistency tests cannot eliminate inaccurate data. In contrast to this fault, the TC line can provide accurate VLE.

Table 1 Number of theoretical stages at total reflux satisfying three-nine purities.

System	Purpose	N from TC line	N from exp. data
Propylene + propane	C3 distillation	80	66
Benzene + cyclohexane + acetone	Azeotropic distillation	17	15
Benzene + cyclohexane + acetone	User fitting UNIFAC	17	15

N: number of theoretical stages calculated using the Ponchon-Savarit method.

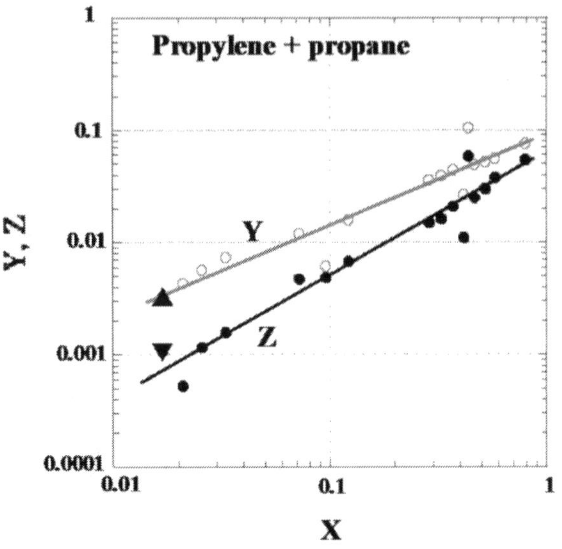

Fig. 11 TC lines for the propylene (1) + propane (2) binary, where $T < T_{c1}$ is satisfied.

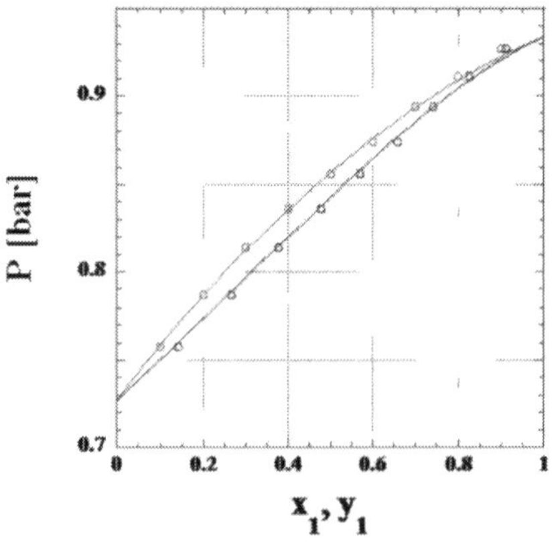

Fig .12 P vs. x_1 and y_1 for the propylene (1) + proapne (2) binary at 223.75 K, (o) P vs. x_1 data[16], (o) P vs. y_1 data[16], (—) correlation of P, x data using the Margules equation, and (—) P, y predicted using the Margules equation.

5.2 Azeotropic distillation and an application to model predictive control of distillation towers

Benzene is separated from a mixture with cyclohexane by azeotropic distillation with acetone. The TC lines of the three binaries consisting of the ternary system, benzene + cyclohexane + acetone, satisfy the total correlation. The UNIQUAC equation was used for the calculation of the activity coefficients in the ternary system. As shown in Table 1, $N = 17$ was obtained using the TC lines. Meanwhile, if the TC lines of one of the binaries are replaced with experimental data, the worst case was $N = 15$. As shown in **Fig. 13**, the partial molar excess enthalpy cannot be determined from a conventional correlation. Meanwhile, the TC lines, as shown in **Fig. 14**, can provide an accurate liquid phase enthalpy. In **Fig. 15**, (L/Q) divided by its value at the bottom is plotted versus N, where L and Q denote the liquid flow rate and heat load, respectively. Fig. 15 shows that, by using the TC line method, McCabe-Thiele and athermal approximations, respectively, provide 3% and 2% differences in the (L/Q) values from that calculated using the TC lines. It is estimated that the differences increase with the increasing gap in solute volatility. The TC line method will be integrated into a process simulator using a pseudo binary method (PBM)[17, 18] for calculating the real-time model predictive control of distillation towers.

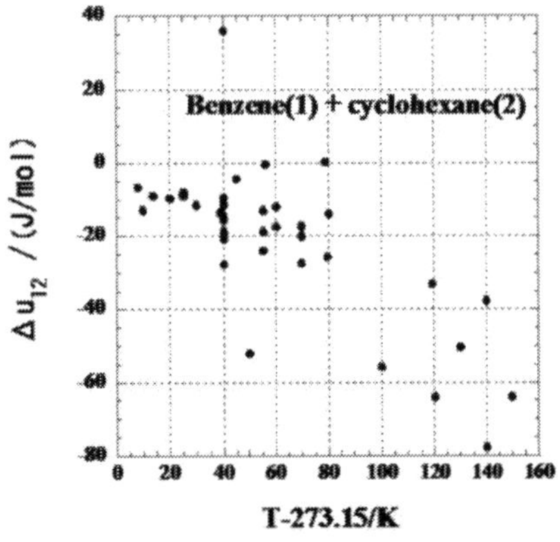

Fig. 13 u_{12} vs. $T - 273.15$ for the benzene (1) + cyclohexane (2) binary, (•) exp. data determined using the UNIQUAC equation.

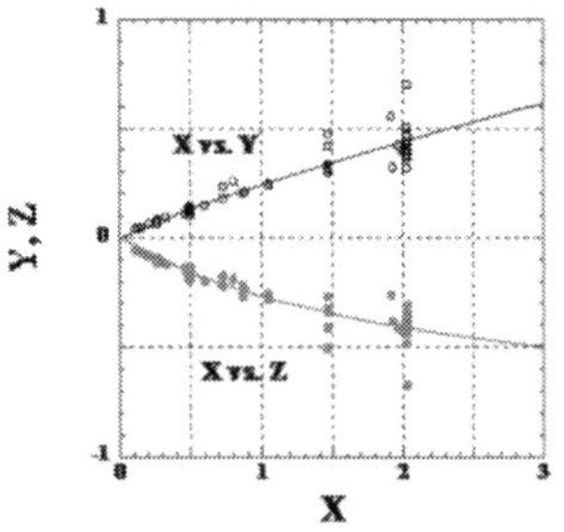

Fig 14. Y and Z vs. X for the benzene (1) + cyclohexane (2) binary, (o) X vs. Y calculated from exp. data using the UNIQUAC equation, (—) X vs. Y TC line with $Y = 0.239X^{0.831}$, (•) X vs. Z calculated from exp. data using the UNIQUAC equation, (—) X vs. Z TC line with $Z = -0.255\ X^{0.692}$.

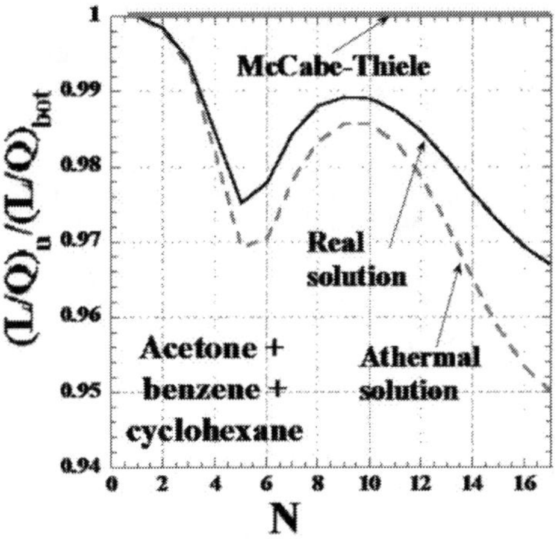

Fig. 15 $(L/Q)/(L/Q)_{bot}$ vs. N for the benzene + cyclohexane + acetone system, (—) McCabe-Thiele, (- - -) athermal solution, and (**—**) real solution calculated from TC lines.

5.3 User fitting UNIFAC

a. Establishment of group interaction parameters

UNIFAC group interaction parameters can be obtained from the Dortmund Data Bank. However, accurate UNIFAC parameters can be determined using the TC lines by user's own fitting, in which the UNIFAC group interaction parameters can be optimized using the following objective function:

$$O.F. = (1/n)\Sigma|(\ln\gamma_i^{\infty}{}_{TC} - \ln\gamma_i^{\infty}{}_{UNI})/\ln\gamma_i^{\infty}{}_{TC}| \qquad (19)$$

Interaction parameters between the CH_3OH and $ACNO_2$ groups are lacking in the original UNIFAC (ver. 5)[5]. In **Table 2**, optimized parameters are listed determined from the user fitting UNIFAC using the TC lines of the methanol + nitrobenzene binary.

Table 2 Interaction parameters, a_{nm} and a_{mn}, from the methanol + nitrobenzene binary

Group pair (m, n)	a_{mn}	a_{nm}
CH_3OH - $ACNO_2$	139.1 K	978.7 K

b. Prediction of VLE for binaries lacking experimental data

In **Fig. 16**, the x, y relationship for the water (1) + 4-methylpyridine (2) binary is predicted. The UNIFAC group interaction parameters were optimized using the TC lines of the following six binaries: pyridine +, 2-methylpyridine +, 3-methylpyridine +, 2,6-dimethylpyridine +, 2,4-dimethylpyridine+ and 2-methyl-5-ethylpyridine + water. In **Fig. 17**, y_1 is plotted versus x_1 for the acetone (1) + dodecane (2) binary. Five acetone + alkane binaries were used except for the acetone + dodecane binary. Figs. 16 and 17 show that the user fitting UNIFAC is useful for the prediction of VLE for a binary, in which experimental data are lacking.

c. Improvement of group interaction parameters

Accurate group interaction parameters can be determined from the user fitting UNIFAC. In **Fig. 18**, x, y relationships for the acrolein + water binary is plotted, as determined using the TC lines of four binaries: acrolein + water, 1-butanol + octen-4-al, 1-pentene + acetaldehyde and crotonaldehyde + vinyl acetate. In **Fig. 19**, y_1 is plotted versus x_1 for the vinyl acetate (1) + chrotonaldehyde (2) binary. **Table 3** demonstrates the improvement of UNIFAC interaction parameters realized by the user-fitting UNIFAC for the group pairs including CH_2=CH. Table 3 shows that the $\Delta\ln\gamma_i^{\infty}$ values of the original UNIFAC decrease by 70 % in average, if the user-fitting UNIFAC is used. Figs 18, 19 and Table 3 show that the user fitting UNIFAC is useful for improving group interaction parameters. Table 1 shows the improvement of the N values, as improved from $N = 15$ obtained using original UNIFAC to $N = 17$ obtained using the user-fitting UNIFAC.

d. Strength of molecular interactions acting on chemical groups

The strength of molecular interactions acting on a chemical group is formulated by the partial molar excess enthalpy of the group, H_i, as follows:

$$H_i = \frac{\partial \ln \Gamma_i}{\partial (1/T)} \qquad (20)$$

In Eq. (20), Γ_i denotes the activity coefficient of chemical group *i*. In **Table 4**, partial molar excess enthalpy, H_i/RT, and entropy, S_i/R, of CH_2=CH group in the acrolein + water binary and CH_3 group in the 1-butanol +water binary are shown as calculated using the TC lines at 333.15 K. Table 4 clearly shows that the CH_2=CH group is instable in water, because $\Delta H_i/RT >> 0$ and $\triangle S_i/R >> 0$ hold. One of the great advantages of the TC line is that it can provide the accurate temperature dependence of Γ_i., from which the strength of molecular interactions can be quantitatively identified. This fact can be applied to chemical and pharmaceutical material development procedures.

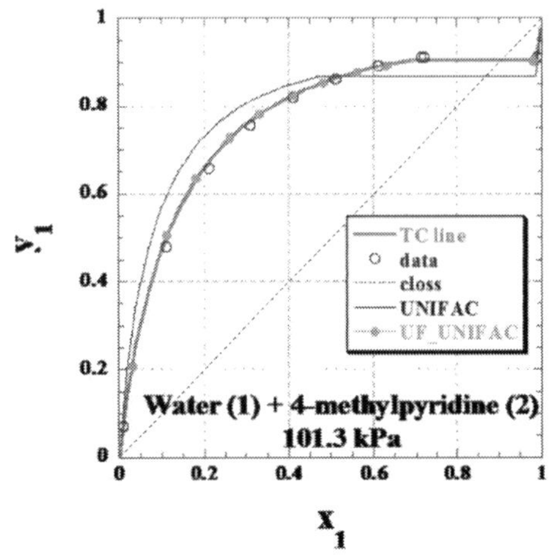

Fig. 16 y_1 vs. x_1 for the water (1) + 4-methylpyridine (2) binary at 1.013 bar, (o) exp. data, (—) predicted using TC lines, (—) predicted using UNIFAC (Ver. 5), and (—●—) predicted using the user fitting UNIFAC.

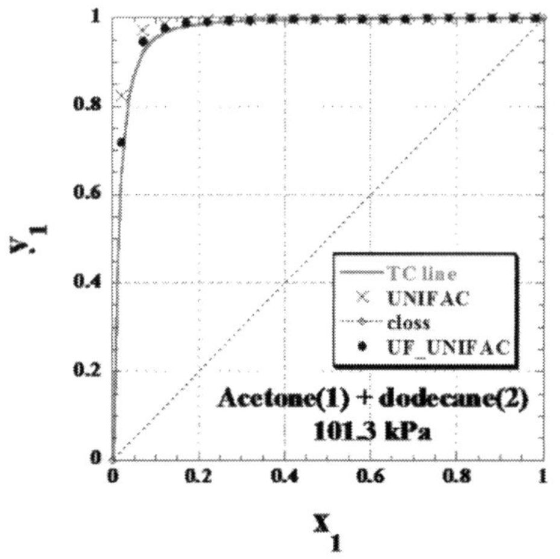

Fig. 17 y_1 vs. x_1 for the acetone (1) + dodecane (2) binary at 1.013 bar, (—) predicted uaing TC lines, (x) predicted using UNIFAC (Ver. 5), and (●) predicted using the user fitting UNIFAC.

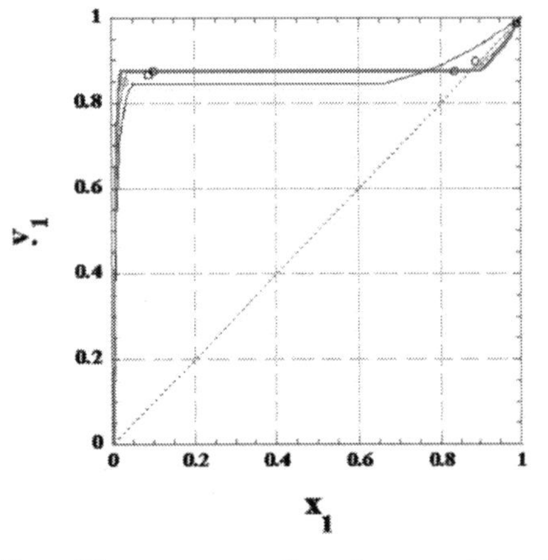

Fig. 18 y_1 vs. x_1 for the acrolein (1) + water (2) binary at 1.013 bar, (o) exp. data, (—) predicted using TC lines, (—) predicted using UNIFAC (Ver. 5), and (—●—) predicted using the user fitting UNIFAC with optimized parameters.

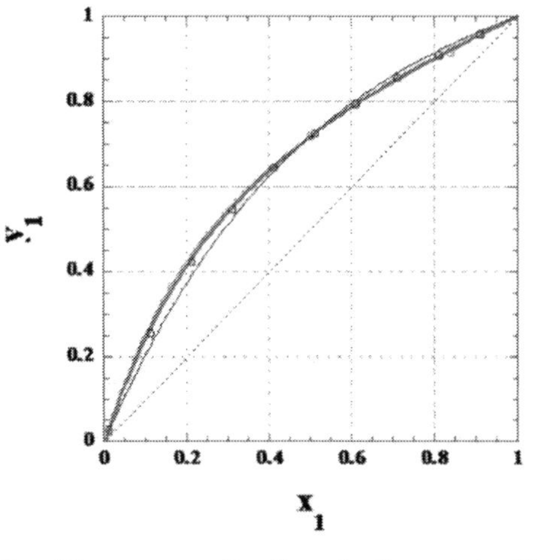

Fig. 19 y_1 vs. x_1 for the vinyl acetate (1) + croton aldehyde (2) binary at 1.013 bar, (o) exp. data, (—) predicted using TC lines, (—) predicted using UNIFAC (Ver. 5), and (—●—) predicted using the user fitting UNIFAC with optimized parameters.

Table 3 Improvement of UNIFAC interaction parameters using the user-fitting UNIFAC

Group num.	Interacting groups	$(\Delta \ln\gamma_i^{\infty})_{UNIFAC}$	$(\Delta \ln\gamma_i^{\infty})_{UFU}$	s	m
2,8	CH_2=CH, ACOH	0.13	0	0.00	1
2,9	CH_2=CH, CH_3CO	0.18	0.09	0.50	6
2,10	CH_2=CH, CHO	0.31	0.051	0.16	6
2,11	CH_2=CH, CH_3COO	0.34	0.16	0.47	7

$(\Delta \ln\gamma_i^{\infty})_{UNIFAC} = 1/(2m)\Sigma|\ln\gamma_i^{\infty}{}_{UNIFAC} - \ln\gamma_i^{\infty}{}_{TC}|$

$(\Delta \ln\gamma_i^{\infty})_{UFU} = 1/(2m)\Sigma|\ln\gamma_i^{\infty}{}_{UFU} - \ln\gamma_i^{\infty}{}_{TC}|$

$s = (\Delta \ln\gamma_i^{\infty})_{UNIFAC}/(\Delta \ln\gamma_i^{\infty})_{UFU}$

m: number of binaries used for the user-fitting UNIFAC (UFU).

Table 4 Infinite dilution partial molar excess enthalpies and entropies of chemical groups

	H_i/333.15R	S_i/R		H_i/333.15R	S_i/R
CH_2=CH in water	6.6	3.9	CH_3 in water	-0.83	-2.4
CH_2=CH in acrolein	-0.27	-0.6	CH_3 in 1-butanol	-0.13	-0.4
H_2O in acrolein	2.3	0.33	H_2O in 1-butanol	0.2	-1.1

6. CONCLUSION

The TC line accompanied with the total correlation is useful for predicting the accurate number of theoretical stages of distillation towers. It was shown that the user-fitting UNIFAC improves UNIFAC group interaction parameters as notable as a 70 % decrease in $\ln \gamma_i^{\infty}$ prediction errors.

Acknowledgement

The authors are grateful to the Tokyo Metropolitan University for the approved use of part of the TC lines applied for patents.

References

1 Patent: US20140288905A1, EP276218A1, KR20140079425A, CN103906553A, WO2013047494A1.
2 S. Kato, J. P. Schmelzer, P. Gostomski, K. N. Marsh, J. Chem. Eng. Data, 56 (2011) 4927-4934.
3 S. Kato, P. Gostomski, A. Oda, J. Freitag, S. Hirohama, D. Bluck, Fluid Phase Equilib., 357 (2013) 36-42.
4 S. Kato, Kagaku Kogaku, 77 (2013) 474-477, http://tc-lines.com
5 B.E. Poling, J.M. Prausnitz, J.P. O'Connell, The Properties of Gases and Liquids, 5th ed. McGraw-Hill, New York, 2001.
6 J. Gmehling, U. Onken, Vapor–Liquid Equilibrium Data Collection, Chemistry, Data Series, vol. 1, parts 1–8, 1981, Dechema.
7 M. Albert, I. Hahnenstein, H. Hasse, G. Maurer, AIChE J., 42 (1996) 1741-1752.
8 M.L. MC Glashan, A.G. Williamson, J. Chem. Eng. Data, 21 (1976) 196-199.
9 G.A. Ratcliff, K.C. Chao, Can. J. Chem. Eng., 47 (1969) 148-153.
10 M.S.H. Bader, K.A.M. Gasem, Chem. Eng. Commun., 140 (1995) 41-72.
11 K. Kurihara, T. Minoura, K. Takeda, K. Kojima, J. Chem. Eng. Data, 40 (1995) 679-684.
12 M.J. Lee, S.L. Chen, C.H. Kang, H.M. Lin, Ind. Eng. Chem. Res., 39 (2000) 4383-4391.
13 P. Sentenac, Y. Bur, E. Ruzy, C. Berro, J. Chem. Eng. Data, 43 (1998) 592-600.
14 J. Yao, H. Li, S. Han, Fluid Phase Equilib., 162 (1999) 253-260.
15 J. Zielkiewicz, P. Oracz, Fluid Phase Equilib., 59 (1990) 279-290.
16 M. Hirata, T. Hakuta, T. Onoda, Int. Chem. Eng., 8 (1968) 175.
17 I. Minami, Kagaku Souchi, 10 (2003) 92. 11 (2003) 202, 1 (2005) 87, 2 (2005) 87, 4 (2005) 88, 8 (2005) 99.
18 Y. Ishii, F.D. Otto, Compt. Chem. Eng. ,32 (2008) 1842.

MODIFICATION OF THE BILLET AND SCHULTES-MODEL FOR THE HYDRAULIC MODELLING OF RANDOM AND STRUCTURED PACKING

Paper 4a

Verena Wolf, Markus Lehner, Montanuniversitaet Leoben, Leoben, Austria
Karin Hoffmann, RVT Process Equipment GmbH, Steinwiesen, Germany

Abstract

If a new packing is developed and introduced to the market, a reliable calculation of the properties is hard to perform as the known correlations are not able to consider all the special influences of the geometrical properties of a packing. Therefore, the published hydraulic models only work acceptable for validated packing types.

To ensure a good rating of a newly developed packing for different application areas, a reliable forecast of the hydraulic properties has to be guaranteed. Therefore, the following models from literature have been examined on their ability to model a newly developed grid-structured packing: SRP-Model, Delft-Model, NNA-Model of Piché et al., Billet and Schultes-Model, Maćkowiak-Model and Stichlmair-Engel-Model. Unfortunately, none of them is able to calculate the hydraulic properties of the grid structured packing satisfactorily. The best modelling results could be reached with the model of Billet and Schultes. As a result of that and the fact that the Billet and Schultes-Model is applicable for structured as well as for random packing, it was examined more in more detail. Due to a relatively simple but physically justified modification of the Billet and Schultes-Model good modelling results can be reached.

Furthermore, it can be shown that this adjustment also leads to a better prediction of the wet pressure drop of further examined packing types. Moreover, the area of validity of the Billet and Schultes-Model of 0.61 – 60 $m^3/(m^2*h)$ can be extended to an irrigation density of 120 $m^3/(m^2*h)$ for the calculation of the wet pressure drop.

Introduction

The historical development of random and structured packings is characterised by a continuous improvement of the hydraulic capacity as well as the mass transfer properties. Already decades ago it was ascertained that a structured arrangement of the packing is highly advantageous in order to reduce the pressure loss. Due to the closed structure of conventional plate packings, the exchange of gas and liquid in horizontal direction is limited. Furthermore, particularly in terms of plastic structured packings, the production costs are high compared with random plastic packings. RVT Process Equipment developed a new grid structured plastic packing, called Hiflow Plus (Figure 1). This packing is optimized for the material consumption and the open structure enables a free horizontal exchange of gas and liquid. This new packing type aims to close the gap between random and structured packings.

The fields of industrial application of this grid structured packing are separation processes with high volume flow rates and moderate temperatures as for example sea water scrubbers for SO_2 removal from exhaust gases of power plants, amine scrubbers for carbon capture units or ground water strippers, just to mention a few. (1)

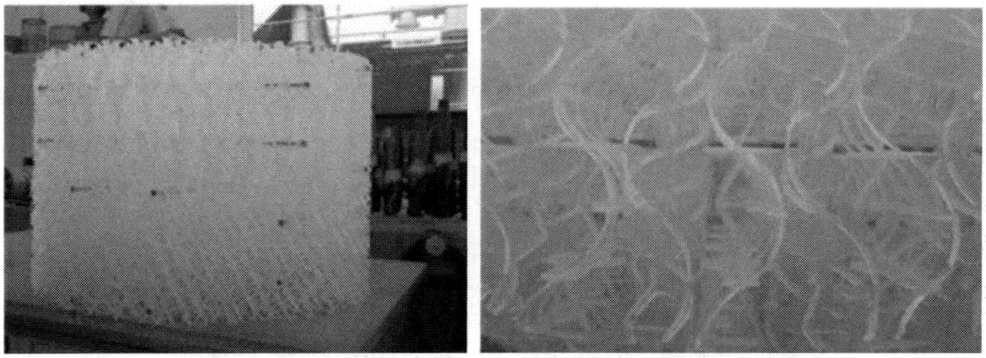

Figure 1: Plastic grid structured packing Hiflow Plus

For the introduction of a newly developed packing to the market, a reliable calculation of the hydraulic properties is needed. With the known correlations from literature this is hard to perform as these models are not able to consider all the special influences of the geometrical properties of a packing. Consequently, the published hydraulic models only work acceptable for validated packing types. Therefore, detailed and reliable experimental measurements are necessary in order to validate the results of existing models, or to develop a new correlation for the calculation of the packing properties.

Experimental Set-Up

The hydraulic properties (dry and wet pressure drop, hold-up, loading and flooding point) have been examined at an experimental plant of the Chair of Process Engineering of Industrial Environmental Protection at Montanuniversitaet Leoben. In Figure 2 the flow sheet of the used experimental set-up is depicted. The plant consists of a DN 600 saturation column and a DN 450 measurement column, both made of polypropylene. The air is sucked from the environment and flows through the saturation column where it is saturated with water. After passing a tube type gas distributor, the gas streams through the measurement column. Afterwards, it enters a radial blower (frequency converted gear, max. 5000 m^3/h, max. 5000 Pa pressure increase) and is then finally released to the environment. The gas flow is measured by a flow grid (airflow) downstream of the blower. Furthermore, the temperature of the gas is measured at every in- and outlet of both columns (Endress & Hauser, PT 100 type Omnigrad M TR 13; WIKA, type 55) which have their own closed liquid circulation. The packing is irrigated by a trough type liquid distributor (128 dripping points per m^2, free cross sectional area of 41%, three available operating ranges: 5-30 $m^3/(m^2h)$, 5-60 $m^3/(m^2h)$ and 30-120 $m^3/(m^2h)$). The used radial pumps are able to handle a volume flow of 50 m^3/h. The liquid flow is measured by a magnetic inductive flow meter (Krohne, type optiflux 4000) and the temperature of the inlet stream at the measurement column is gauged by a resistance thermometer (Voltcraft, type K).

The hydraulic properties of a packing are examined at different gas and liquid loads. At the beginning of a test run the adjustments of the selected loads are made and it is necessary to wait until steady state conditions have been reached. After five minutes of stable operation all necessary values can be read off. Then the liquid level in the bottom tank of the measuring column is raised to a defined height by opening the water supply line which is necessary for the hold-up measurement described below. Afterwards, the pumps and the blower are shut down simultaneously. During the following 30 minutes the liquid can drop off from the packing completely and leave the column bottom tank before the next operating point (same irrigation density but higher F-factor) can be adjusted. Once the flooding point for the chosen irrigation density is reached, the test series can be accomplished for the next irrigation density.

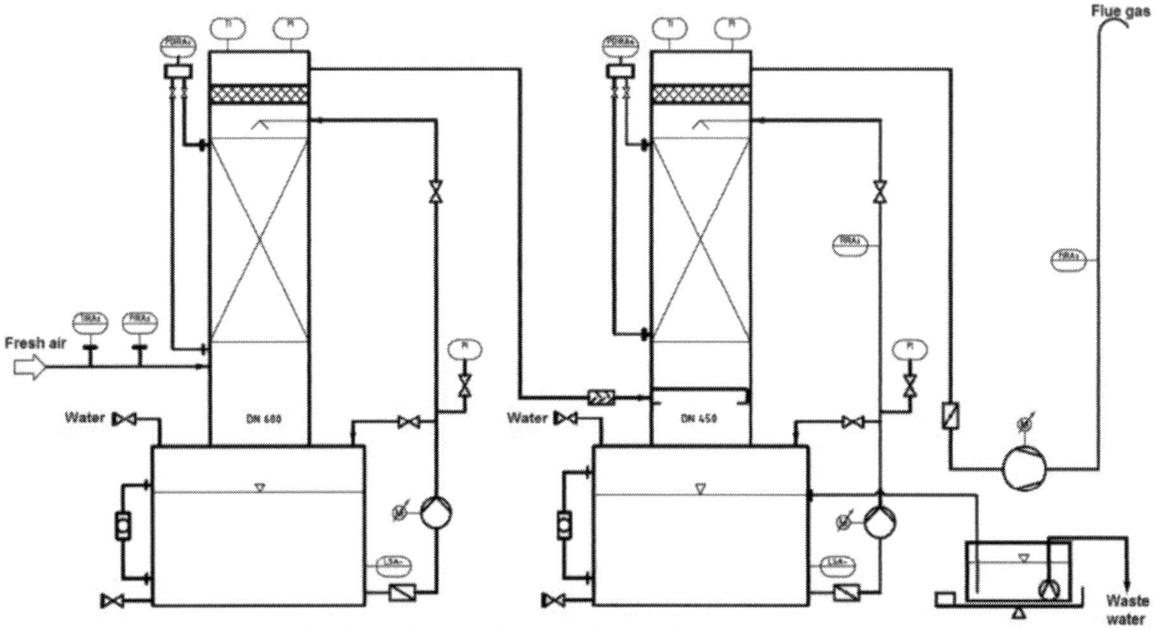

Figure 2: Flow sheet of the experimental plant "Packed column"

Pressure Loss Measurement

The pressure loss can be examined by calculating the difference of the pressure measurements which are taken by a differential pressure transmitter (Endress & Hauser, Deltabar S PMD75) situated at the upper and the lower region of the packing. Due to the experimental set up the static pressure is measured at these positions, and therefore it needs to be ensured that the dynamic pressure is equal on both pressure measurement positions. This can be guaranteed if the gas velocity is constant wherefore an equal available cross-section for the gas stream is required. Therefore, both pressure measurement positions have to be either located "inside" of the packing, or below and above of the packing. To be able to neglect the influence of the support grid and a hold-down grid, which is used in some cases, the pressure measurement positions are located at a column height where packing is situated in the column. For the design of the pressure measuring taps the findings of Cai and Resetarits (2) were considered.

Hold-up Measurement

The measurement of the hold-up is achieved by collecting and weighing the amount of water which is in the column during operation. Therefore, the water level in the bottom tank of the measuring column is adjusted by adding water from the tap to a defined height during operation. This height is given by the lower end of a nozzle in the tank, from which a pipe is leading to a tub standing on a scale

(Figure 3). Afterwards, the column is shut down and the operating liquid is flowing out of the bottom tank into the tub on the scale, where it can be weighed. To exactly calculate the hold-up of the packing itself, a calibration of the column without packing is needed to be able to subtract the amounts of liquid which are situated in the pipe, in the liquid distributor and as droplets below and above of the packing (Figure 3).

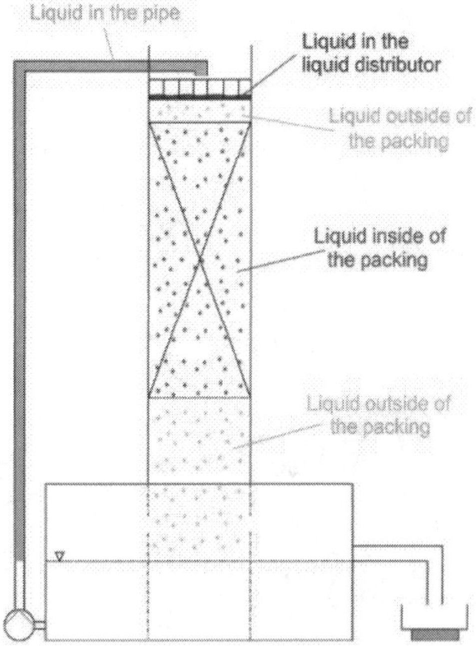

Figure 3: Setup for hold-up measurements with different parts of liquid situated in the column during operation

Loading and Flooding Point Measurement

The loading point can be identified by the change of the slope of the pressure loss line, when the pressure loss is plotted over the F-factor in a double logarithmic diagram. For this reason the measurement results for each operating point are immediately entered into a spread sheet including such a diagram. Furthermore, the increase of the gas load is done in smaller steps after reaching the loading point for a special irrigation density to ensure a better resolution of the hydraulic properties in the loading zone.

The flooding point can be detected if it is not possible to reach stationary conditions. At this point the pressure loss over the packing is not stable any more but is increasing although the gas and the liquid loads are constant. Furthermore, the flooding point can also be identified in the above mentioned diagram when the slope of the pressure loss line approaches the value of one.

Hydraulic Modelling

Following models from literature have been examined on their ability to rate the newly developed grid-structured packing: SRP-Model (3-10), Delft-Model (10-12), NNA-Model of Piché et al. (13-18), Billet and Schultes-Model (19-28), Maćkowiak-Model (29-31) and Stichlmair-Engel-

Model (32-34). Unfortunately, none of them is able to calculate the hydraulic properties of the grid structured packing satisfactorily. The best modelling results could be reached with the model of Billet and Schultes. As a result of that, and the fact that the Billet and Schultes-Model is applicable for structured as well as for random packing, it was examined in more detail.

Billet and Schultes-Model

The Billet and Schultes-Model assumes the packing as a bundle of parallel channels, and that the liquid trickles down as a laminar flowing film with constant film thickness in the inside of these channels while the gas phase flows upwards in counterflow. By neglecting the inertia forces and assuming that the liquid film is in equilibrium state, gravity and shear forces in the liquid film are held in equilibrium with the frictional forces by the shear stress in the vapour phase at the surface of the film (Figure 4). (23)

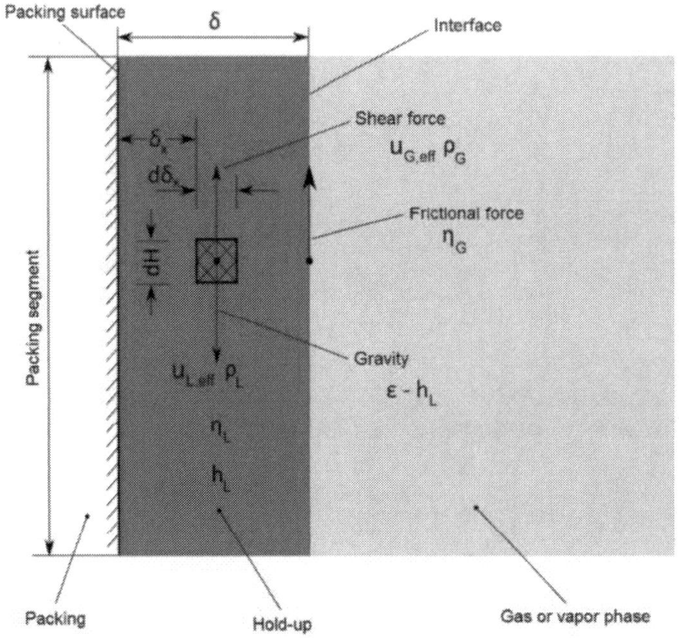

Figure 4: Theoretical assumptions of the Billet and Schultes-Model (26)

Based on this theoretical background the equations for the calculation of the loading and the flooding point as well as the liquid hold-up **at** respectively **below** of the loading point are established. The correlation for the hold-up at the flooding point is derived from empirical observations and the calculation equation for the hold-up in the loading zone is an interpolation of the 13th order between the hold-up at the loading point and the one at the flooding point. The calculation of the dry pressure drop (Equations (1)-(3)) is based on the Ergun-Equation (35) which is in turn a further development of the Darcy-Weisbach Equation.

$$\frac{\Delta p_{tr}}{H} = \psi_{tr} * \frac{a_{geo}}{\varepsilon^3} * \frac{F^2}{2} * \frac{1}{K} \tag{1}$$

$$\frac{1}{K} = 1 + \frac{2}{3} * \frac{1}{1-\varepsilon} * \frac{d_P}{d_S} \tag{2}$$

$$\psi_{tr} = C_{P,0} * \left(\frac{64}{\mathrm{Re}_G} + \frac{1{,}8}{\mathrm{Re}_G^{0{,}08}} \right) \tag{3}$$

The wet pressure drop (Equation (4)) is again calculated similar to the dry pressure drop but the herein used resistance coefficient is combined by four terms (Equation (5)).

$$\frac{\Delta p}{H} = \psi_L * \frac{a_{geo}}{(\varepsilon - h_L)^3} * \frac{F^2}{2} * \frac{1}{K} \tag{4}$$

$$\psi_L = \psi_{tr} * \left(\frac{\varepsilon - h_L}{\varepsilon} \right)^{1{,}5} * \left(\frac{h_L}{h_{L,S}} \right)^{0{,}3} * B_f \tag{5}$$

- Dry resistance coefficient ψ_{tr},
- one term, who takes into account that the free cross sectional area is reduced through the down trickling fluid $\left(\frac{\varepsilon - h_L}{\varepsilon} \right)^{1{,}5}$,
- the term $\left(\frac{h_L}{h_{L,S}} \right)^{0{,}3}$, which was found through empirical adaptions and describes the huge increase of the pressure loss in the loading zone and
- the wetting factor B_f.

The wetting factor describes the dependency of the pressure loss from the liquid load and was found through empirical adaption of Billet and Schultes in the year of 1991 (23) as given in Equations (6) and (7).

$$B_f = \exp\left(\frac{\mathrm{Re}_L}{200} \right) \qquad \text{(Billet and Schultes 1991 (23))} \tag{6}$$

$$\mathrm{Re}_L = \frac{u_L * \rho_L}{a_{geo} * \eta_L} \tag{7}$$

In the year 1999 Billet and Schultes (28) introduced a new version of the wetting factor, which can be seen in Equations (8) and (9), as this one is able to describe the experimental results at high pressures and at very low liquid viscosities better.

$$B_f = \exp\left(\frac{13300}{a_{geo}^{\frac{3}{2}}} * \sqrt{Fr_L} \right) \qquad \text{(Billet und Schultes 1999 (28))} \tag{8}$$

$$Fr_L = \frac{u_L^2 * a_{geo}}{g} \tag{9}$$

As reality deviates from the presented theoretical assumptions and the geometrical shape of all the different packing types cannot be described by surface area and void fraction, four empirical packing specific shape constants ($C_{P,0}$, C_h, C_S, C_{Fl}) are needed for the equations of the hydraulic part of the Billet and Schultes-Model (28). These constants are given for a lot of packing types in the literature (28) or can be calculated with the help of experimental measurement results.

Modelling of the Structured Packing Hiflow Plus

The Billet and Schultes-Model is able to predict the loading and the flooding point as well as the hold-up and the dry pressure drop satisfactorily. Furthermore, the wet pressure drop prediction of the packing Hiflow Plus #2 is acceptable as depicted in Figure 5.

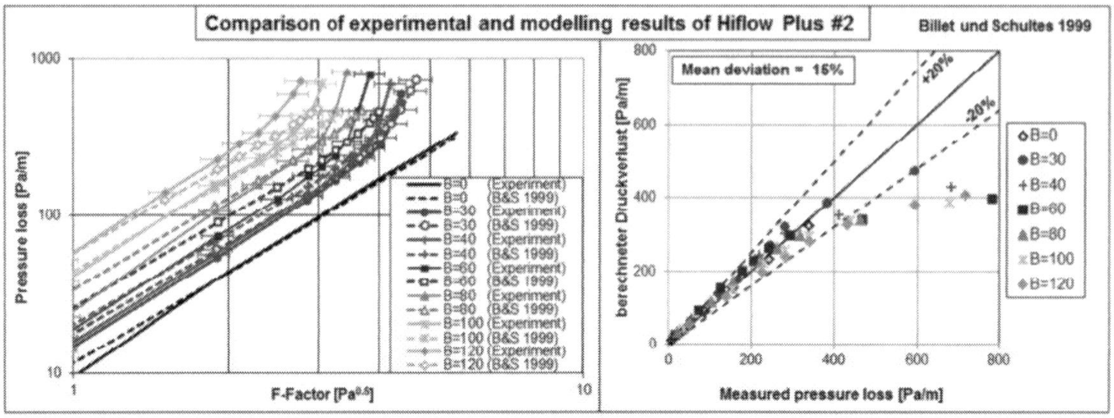

Figure 5: Modelling results by the Billet and Schultes Model 1999 for the packing Hiflow Plus #2

Nevertheless, the wet pressure drop forecast for the packing Hiflow Plus #1 does not work very well which also leads to a mean deviation over the whole considered range of 35% as can be seen in Figure 6.

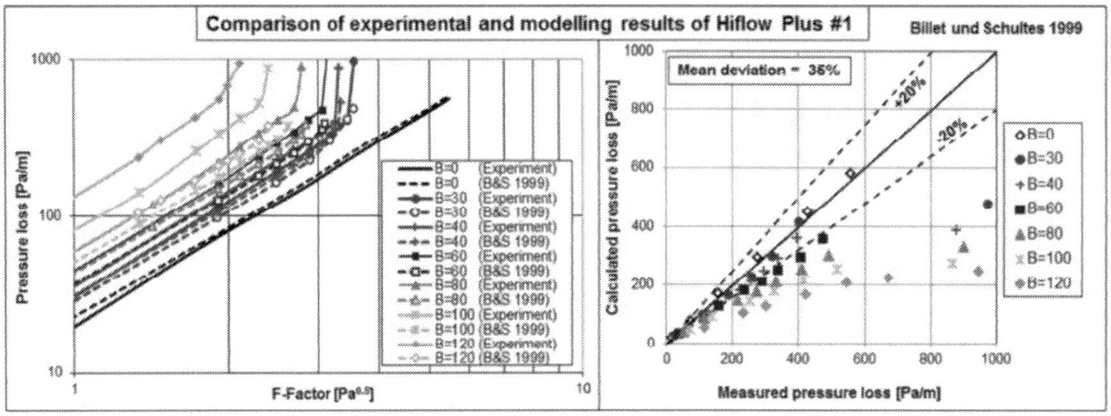

Figure 6: Modelling results by the Billet and Schultes Model 1999 for the packing Hiflow Plus #1

As the two different packings Hiflow Plus #1 and Hiflow Plus #2 have a similar geometrical structure and only differ in their packing properties (Table 1), the influence of these packing properties on the calculation of the wet pressure loss was examined as a function of gas and liquid load.

Table 1: Packing properties of Hiflow Plus

Packing	a_{geo} [m^2/m^3]	ε [m^3/m^3]
Hiflow Plus #1	180	0.93
Hiflow Plus #2	100	0.95
Hiflow Plus #3	80	0.96

In particular, the influence of the geometrical surface area on the wet pressure drop as a function of the liquid load is remarkable as shown in Figure 7, calculated by the equations given by Billet and Schultes 1991 as well as the one of 1999 for a constant gas load (F-factor = 1 $Pa^{0,5}$). It can be seen that the curve for an irrigation density of 60 $m^3/(m^2*h)$ has already a different slope than the one at an irrigation density of 30 $m^3/(m^2*h)$. At an irrigation density of 100 or 120 $m^3/(m^2*h)$ the Billet and Schultes-Model calculates a higher pressure loss for a packing with a geometric surface area of 90 m^2/m^3 as for instance for a packing with 120 m^2/m^3.

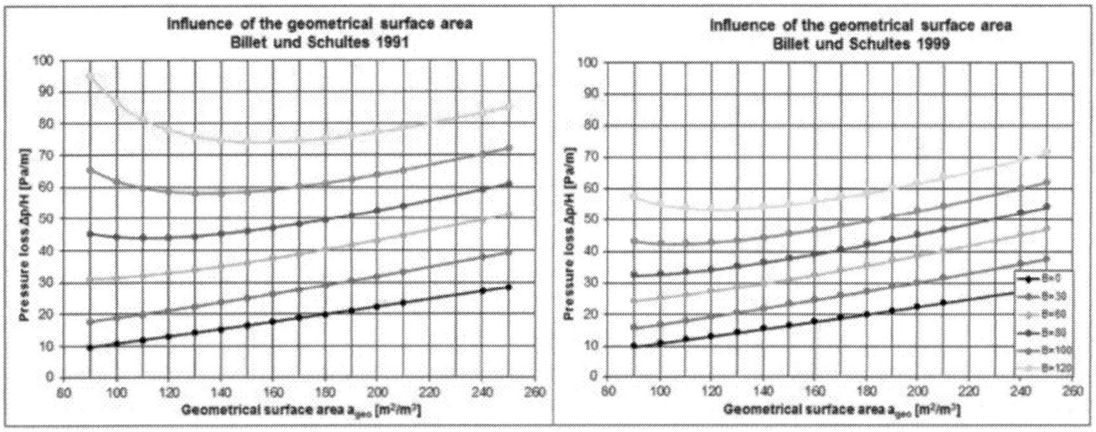

Figure 7: Influence of the geometrical surface area on the calculation of wet pressure loss

For this reason a closer consideration of the influence of the geometrical surface area a_{geo} as well as the liquid load u_L on the calculation of the wet pressure drop was accomplished in Equations (10) – (12). It shows that these two variables can both be found in the wetting factor. Furthermore, it is obvious that the influence of these two variables is similar in the two different definitions from 1991 and 1999.

$$\frac{\Delta p}{H} = \left\{ \psi_{ir} * \left(\frac{\varepsilon - h_L}{\varepsilon} \right)^{1,5} * \left(\frac{h_L}{h_{L,S}} \right)^{0,3} * B_f \right\} * \frac{a_{geo}}{(\varepsilon - h_L)^3} * \frac{F^2}{2} * \frac{1}{K} \qquad (10)$$

$$B_f = \exp\left(\frac{\mathrm{Re}_L}{200} \right) = \exp\left(\frac{1}{200} * \frac{\rho_L}{\eta_L} * \frac{u_L}{a_{geo}} \right) \qquad \text{(Billet and Schultes 1991 (23))} \qquad (11)$$

$$B_f = \exp\left(\frac{13300}{a_{geo}^{\frac{3}{2}}} * \sqrt{Fr_L} \right) = \exp\left(\frac{13300}{g^{0,5}} * \frac{u_L}{a_{geo}} \right) \qquad \text{(Billet and Schultes 1999 (28))} \qquad (12)$$

Furthermore, this relationship can be rearranged, see Equation (13): the wetting factor actually depends on the wettable length $L_{wettable}$ which is the product of the cross sectional area of the column and the specific geometrical surface area of the packing. This length gives kind of a wettable perimeter provided by the packing which the liquid can use for trickling downwards.

$$\frac{u_L}{a_{geo}} = \frac{\dot{V}_L}{A * a_{geo}} = \frac{\dot{V}_L}{L_{wettable}} \qquad (13)$$

Modification of the Billet and Schultes-Model

As the wettable length used in the Billet and Schultes-Model is a constant value over the whole operation range of a column, a term which is able to describe the condition of the liquid film as a function of the packing geometry as well as of the operating conditions has to be defined. With the equations given from Billet and Schultes for the liquid hold-up as well as for the hydraulic surface area, a film thickness can be defined which depends on the Reynolds- and on the Froude-Number as described in Equation (14).

$$\delta = \frac{h_L}{a_h} = \frac{h_L = f_1(\mathrm{Re}, Fr)}{a_h = f_2(\mathrm{Re}, Fr)} \Rightarrow \delta = f(\mathrm{Re}, Fr) \qquad (14)$$

By introducing this new term, and with adaption to the experimental results, a new wetting factor can be defined as shown in Equation (15). The new defined wetting factor depends on a Reynolds-Number, $\mathrm{Re}_{L,\text{film thickness}}$, which specific length is defined by the film thickness δ given in Equation (14).

$$B_f = \exp\left(\frac{1}{16} * \frac{\rho_L * u_L * \delta}{\eta_L}\right) = \exp\left(\frac{\mathrm{Re}_{L,\text{film thickness}}}{16}\right) \qquad (15)$$

In Figure 8, the influence of the geometrical surface area on the calculation of the wet pressure drop is examined in the same way as shown in Figure 7. The results of the Billet and Schultes-Model 1999 and of the introduced modification of this work are compared in Figure 8. As it can be seen, the correlation of this work expresses the wet pressure loss dependence of the geometrical surface area more realistic than the Billet and Schultes-Model 1999.

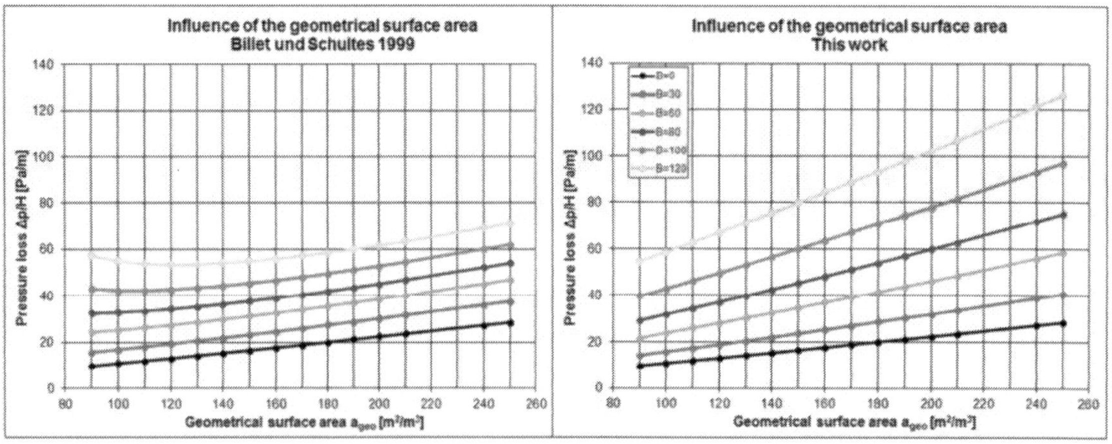

Figure 8: Influence of the geometrical surface area on the calculation of wet pressure loss

Results

In Figures 9 and 10 the experimental wet pressure loss results are compared with the calculated values of the modified Billet and Schultes-Model. It can be seen that the modelling results match better than with the original Billet and Schultes-Model, as the mean deviation for the packing Hiflow Plus #1 decreases from 35% to 15%. The deviations for the packing Hiflow Plus #2 are almost equal (15% for the Billet and Schultes-Modell 1999 and 14% for the modification).

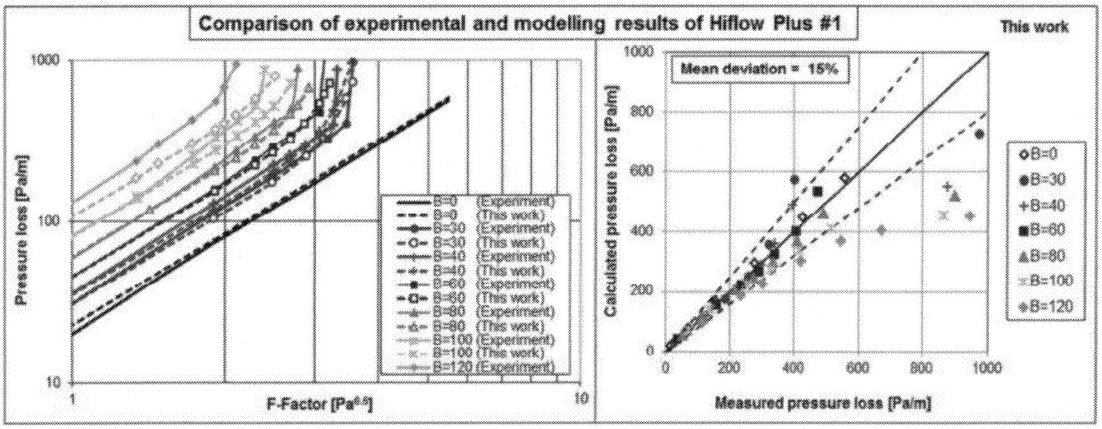

Figure 9: Modelling results by the modification of the Billet and Schultes Model for the packing Hiflow Plus #1

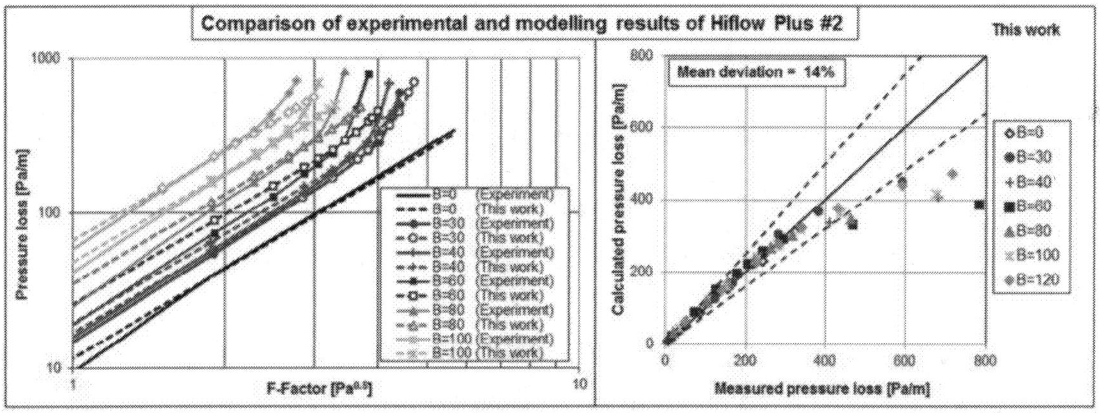

Figure 10: Modelling results by the modification of the Billet and Schultes Model for the packing Hiflow Plus #2

In order to assess the performance of this modification of the Billet and Schultes-Model, further packing types have been examined. In a first step of this validation process, two different structured plastic packings, four different random plastic packings (Table 2) as well as five different metal random packings (Table 3) have been investigated. All presented measurements were done with the system air/water under ambient conditions.

As Billet and Schultes (28) state that their wet pressure drop correlation is only valid up to an irrigation density of 60 $m^3/(m^2*h)$, the mean deviation between the measured and calculated values for this operating region was considered separately (Figure 12 and 14). Evaluating the results in the Figures 11-14, it can be summarised that the modification presented in this paper leads to significantly better forecasting results particularly for those packings which the Billet and Schultes-Model does not rate really well, as for instance for the packing RMSR 70-5. If the Billet and Schultes-Model already predicts acceptable wet pressure drops, the modification shows comparable results, e.g. packing RMSR 25-3. Furthermore, as the modification is able to predict the wet pressure drops up to an irrigation density of 120 $m^3/(m^2*h)$ with acceptable mean deviations (max. 25% for Raflux 25-5), the area of validity can be raised from an irrigation density of 0.61-60 $m^3/(m^2*h)$ up to 120 $m^3/(m^2*h)$.

Table 2: Properties of the examined plastic packing types

Packing	Material	Void Fraction [m³/m³]	Geom.Surface Area [m²/m³]
Hiflow Plus #1	PP	0.93	180
Hiflow Plus #2	PP	0.95	100
Hiflow 50-0	PP	0.94	110
Hiflow 50-6	PP	0.94	90
Hiflow 38-1	PP	0.95	150
Hiflow 90-7	PP	0.97	76

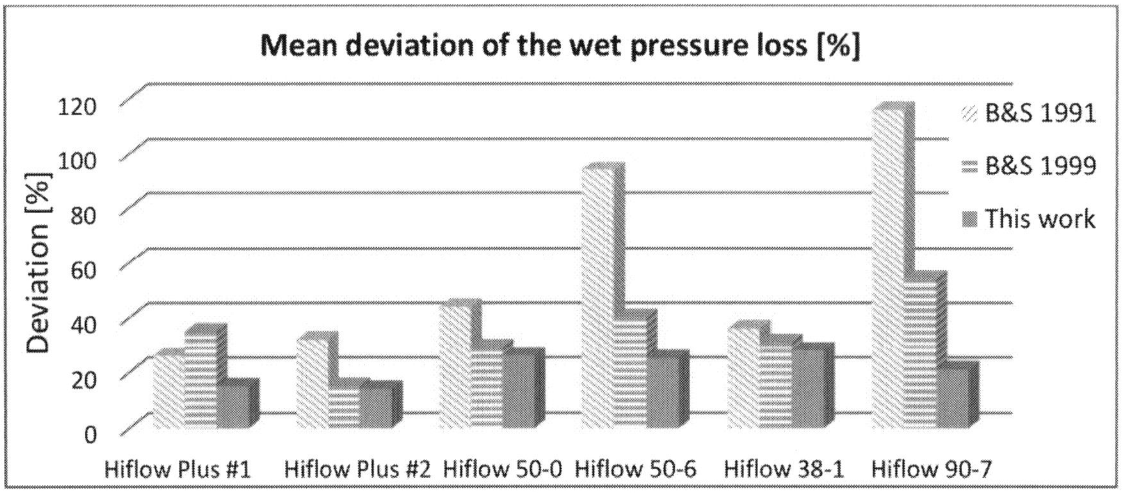

Figure 11: Mean deviation of the wet pressure drop for plastic packings for the whole examined load region

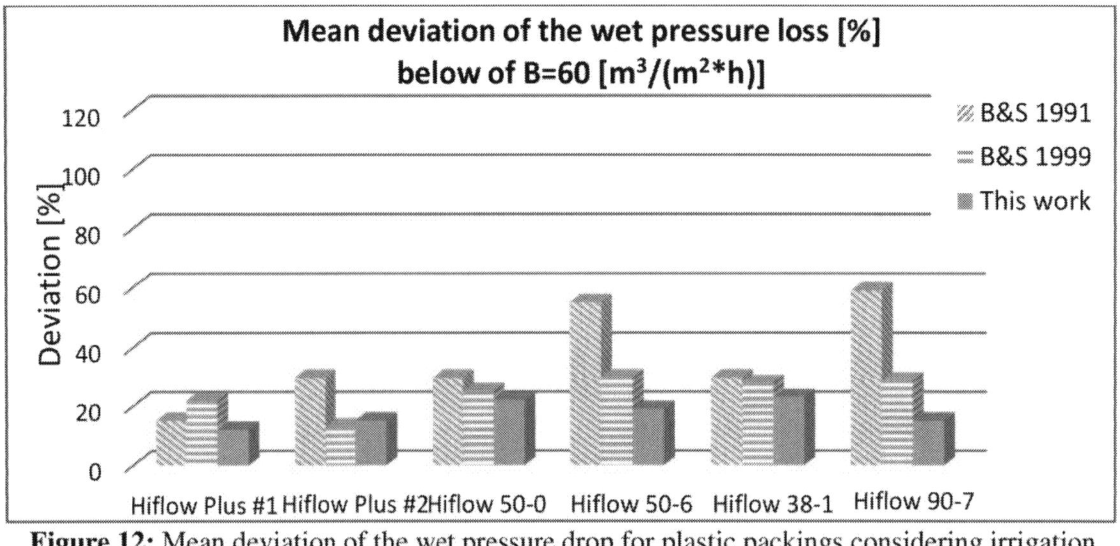

Figure 12: Mean deviation of the wet pressure drop for plastic packings considering irrigation densities only up to 60 $m^3/(m^2*h)$

Table 3: Properties of the examined metal packing types

Packing	Material	Void Fraction [m^3/m^3]	Geom.Surface Area [m^2/m^3]
Raflux 25-5	1.4404	0.95	393
Raflux 50-5	1.4306	0.97	112
RMSR 25-3	1.4000	0.97	235
RMSR 50-4	1.4301	0.98	115
RMSR 70-5	1.4000	0.98	67

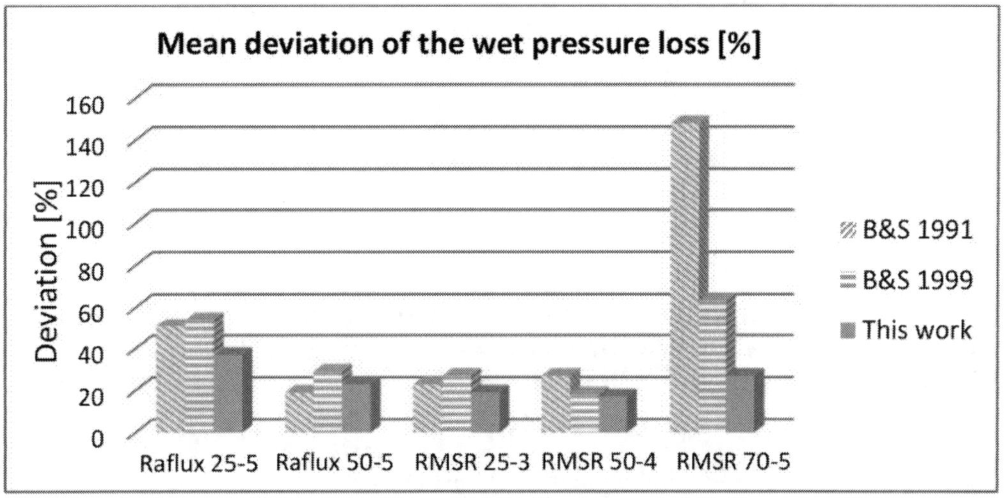

Figure 13: Mean deviation of the wet pressure drop for plastic packings for the whole examined load region

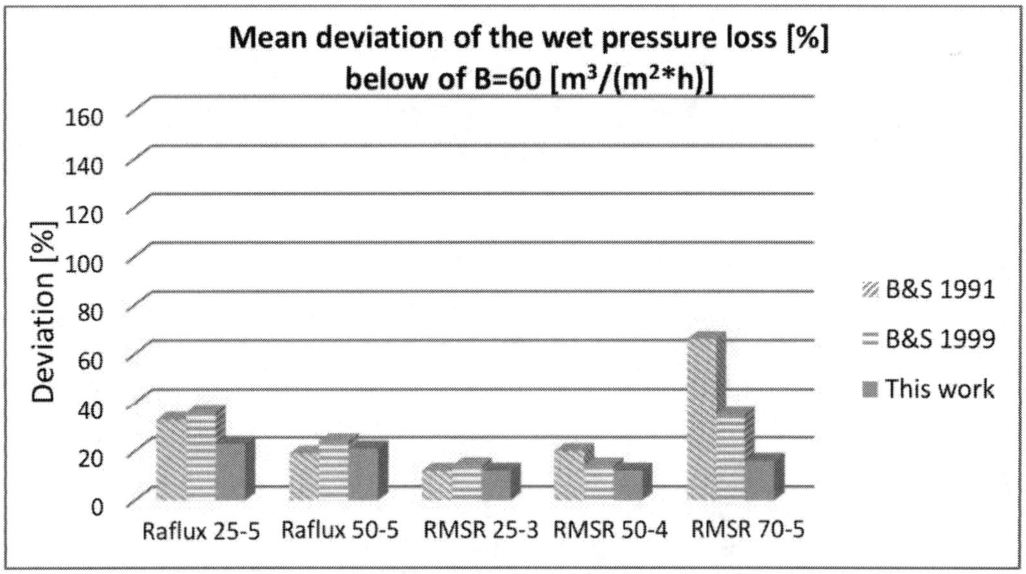

Figure 14: Mean deviation of the wet pressure drop for metal packings considering irrigation densities only up to 60 $m^3/(m^2*h)$

Conclusion and Outlook

Due to a relatively simple but physically justified modification of the Billet and Schultes-Model a better prediction of the wet pressure drop of the examined packing types (a_{geo} = 67 – 393 m^2/m^3; ε = 0.93 – 0.98 m^3/m^3) could be reached. The modification could be done by a new combination of equations given by the original model so that no new parameters have to be added. Moreover, the area of validity of the Billet and Schultes-Model of 0.61 – 60 $m^3/(m^2*h)$ can be extended to an irrigation density of 120 $m^3/(m^2*h)$ for the calculation of the wet pressure drop.

In the future further packing types, especially structured plate packings, will also be examined, and the validation will also be done with other test systems than water and air to investigate additionally the influence of the fluid properties.

Symbols

A	Column cross section [m^2]
a_{geo}	Geometrical packing surface [m^2/m^3]
a_h	Hydraulic surface area [m^2/m^3]
B_f	Wetting factor
B	Irrigation density [$m^3/(m^2*h)$]
DN	Diameter Nominal
d	Diameter [m]
d_P	Particle diameter [m] $d_P = 6 * \frac{(1-\varepsilon)}{a_{geo}}$
d_S	Column inner diameter [m]
F	Gas loading factor resp. F-factor [$Pa^{0.5}$]
f()	Function of
Fr	Froude-Number [-]
g	Gravity [N]
H	Height [m]
h_L	Hold-up [m^3/m^3]
K	Wall factor [-]
kg	Kilogram
m	Meter
Pa	Pascal
Re	Reynolds-Number [-]
s	Second
u_L	Liquid surface velocity [m/s]
$\dot{V}$	Volume flow rate [m^3/s]
Greek Symbols	
δ	Liquid film thickness [m]
Δ	Difference
Δp	Pressure loss [Pa]
ε	Void fraction [m^3/m^3]
η	Dynamic viscosity [kg/(m*s)], [Pa*s]
ρ	Density [kg/m^3]
Ψ	Resistance coefficient

Indices

ft	Wet
L	Liquid phase
S	At the loading point
tr	Dry

References

1. Wolf, V.; Lehner, M.; Hoffmann, K. (2012): Hydrodynamic and mass transfer processes of the grid-structured plastic packing Hiflow Plus. Filtech Exhibitions Germany (Hg.): 11. World Filtration Congress – Proceedings. Meerbusch: FILTECH, p. G14-01-059.
2. Cai, T. J.; Resetarits, M. R. (2011): Pressure drop measurements on distillation columns. Chinese Journal of Chemical Engineering 19 (5), p. 779–783.
3. Bravo, J. L.; Rocha, J. A.; Fair, J. R. (1985): Mass transfer in gauze packings. Hydrocarbon Processing 64 (1), p. 91–95.
4. Bravo, J. L.; Rocha, J. A.; Fair, J. R. (1986): Pressure drop in structured packings. Hydrocarbon Processing 56 (3), p. 45.
5. Fair, J. R.; Bravo, J. L. (1987): Prediction of mass transfer efficiencies and pressure drop for structured tower packings in vapor/liquid service. IChemE: Distillation and Absorption 87, Brighton, September 1987. London, (EFCE Publication Series 62), p. A183-A201.
6. Fair, J. R.; Bravo, J. L. (1990): Distillation columns containing structured packing. Chemical Engineering Progress 86 (1), p. 19–29.
7. Bravo, J. L.; Fair, J. R.; Rocha, J. A. (1992): A comprehensive model for the performance of columns containing structured packing. IChemE: Distillation and Absorption 1992, 2 (additional papers). Birmingham, September 1992. Rugby: (IChemE Symposium Series 128), p. A489-A507.
8. Rocha, J. Antonio; Bravo, Jose L.; Fair, James R. (1993): Distillation columns containing structured packings: a comprehensive model for their performance. 1. Hydraulic models. Ind. Eng. Chem. Res. 32 (4), p. 641–651.
9. Gualito, J. J.; Cerino, F. J.; Cardenas, J. C.; Rocha, J. A. (1997): Design method for distillation columns filled with metallic, ceramic, or plastic structured packings. Ind. Eng. Chem. Res. 36 (5), p. 1747–1757.
10. Verschoof, H.-J.; Olujić, Ž.; Fair, J. R. (1999): A general correlation for predicting the loading point of corrugated sheet structured packings. Ind. Eng. Chem. Res. 38 (10), p. 3663–3669.
11. Olujić, Ž. (1997): Development of a complete simulation model for predicting the hydraulic and seperation performance of distillation columns equipped with structured packings. Chemical and Biochemical Engineering Quaterly 11 (1), p. 31–46.
12. Olujić, Ž.; Behrens, M.; Spiegel, L. (2007): Experimental characterization and modeling of the performance of a large-specific-area high-capacity structured packing. Ind. Eng. Chem. Res. 46 (3), p. 883–893.
13. Piché, S.; Iliuta, I.; Grandjean, B. P.A.; Larachi, F. (2001a): A unified approach to the hydraulics and mass transfer in randomly packed towers. Chemical Engineering Science 56 (21-22), p. 6003–6013.
14. Piché, S.; Larachi, F.; Grandjean, B.P.A. (2001b): Improved liquid hold-up correlation for randomly packed towers. Chemical Engineering Research and Design 79 (1), p. 71–80.
15. Piché, S.; Larachi, F.; Grandjean, B.P.A. (2001c): Loading capacity in packed towers – database, correlations and analysis. Chemical Engineering & Technology 24 (4), p. 373–380.

16. Piché, S.; Larachi, F.; Grandjean, B. P. A. (2001d): Flooding capacity in packed towers: database, correlations and analysis. Ind. Eng. Chem. Res. 40 (1), p. 476–487.
17. Piché, S. R.; Larachi, F.; Grandjean, B. P. A. (2001e): Improving the prediction of irrigated pressure drop in packed absorption towers. Can. J. Chem. Eng. 79 (4), p. 584–594.
18. Ortiz-Arroyo, A.; Larachi, F.; Iliuta, I. (2003): Method for inferring contact angle and for correlating static liquid hold-up in packed beds. Chemical Engineering Science 58 (13), p. 2835–2855.
19. Billet, R. (1987): Modelling of fluiddynamics in packed columns. IChemE: Distillation and Absorption 87, Bd. 1. Brighton, 07.-09. September 1987. London, (EFCE Publication Series 62), p. A171-A182.
20. Billet, R. (1988): Relationship between residence time, fluid dynamics and efficiency in countercurrent flow equipment. Chem. Eng. Technol. 11 (1), p. 139–148.
21. Billet, R.; Schultes, M. (1987a): Capacity studies on two-phase countercurrent-flow columns. IChemE: Distillation and Absorption 87, Bd. 2. Brighton, 07.-09. September 1987. London, (EFCE Publication Series 62), p. B255-B266.
22. Billet, R.; Schultes, M. (1987b): Determination of liquid hold-up in gas-liquid two-phase countercurrent mass transfer columns. IChemE: Distillation and Absorption 87, Bd. 1. Brighton, 07.-09. September 1987. London, (EFCE Publication Series 62), p. A159-A170.
23. Billet, R.; Schultes, M. (1991): Modelling of pressure drop in packed columns. Chem. Eng. Technol. 14 (2), p. 89–95.
24. Billet, R.; Schultes, M. (1992): Advantage in correlating packed column performance. In: IChemE: Distillation and Absorption 1992, Bd. 1. Birmingham, 07.-09. September 1992. European Federation of Chemical Engineering. Rugby, (IChemE Symposium Series 128), p. B129-B136.
25. Billet, R.; Schultes, M. (1993): A physical model for the prediction of liquid hold-up in two-phase countercurrent columns. Chem. Eng. Technol. 16 (6), p. 370–375.
26. Billet, R. (1995): Packed towers in processing and environmental technology. 1. ed. Weinheim, VCH.
27. Billet, R.; Schultes, M. (1995): Fluid dynamics and mass transfer in the total capacity range of packed columns up to the flood point. Chem. Eng. Technol. 18 (6), p. 371–379.
28. Billet, R.; Schultes, M. (1999): Prediction of mass transfer columns with dumped and arranged packings. Chemical Engineering Research and Design 77 (6), p. 498–504.
29. Maćkowiak, J. (1990): Bestimmung des Druckverlustes berieselter Füllkörperschüttungen und Packungen. Staub - Reinhaltung der Luft 50, p. 455–463. (german)
30. Maćkowiak, J. (2003): Fluiddynamik von Füllkörpern und Packungen. Grundlagen der Kolonnenauslegung. 2nd edition Berlin, Springer. (german)
31. Maćkowiak, J. (2006): Neue Aspekte der Modellierung bei der Fluiddynamik von Packungskolonnen. Chemie Ingenieur Technik 78 (8), p. 1079–1086. (german)
32. Stichlmair, J.; Bravo, J. L.; Fair, J. R. (1989): General model for prediction of pressure drop and capacity of countercurrent gas/liquid packed columns. Gas Separation & Purification 3 (1), p. 19–28.
33. Engel, V. (1999): Fluiddynamik in Packungskolonnen für Gas-Flüssig-Systeme. PhD Thesis. TU München, München. Chair of Fluiddynamics. (german)
34. Engel, V.; Stichlmair, J.; Geipel, W. (2001): Fluid Dynamics of Packings for Gas-Liquid Contactors. Chemical Engineering Technology 24 (5), p. 459–462.
35. Ergun, S. (1952): Fluid flow through packed columns. Chemical Engineering Progress 48 (2), p. 89–94.

How Sensitive is Your Treating Plant to Operating Conditions?

Part2

By

Ralph Weiland[1], Michael Schultes[2], Mark Pilling[3], Juan Portillo[3],
Dan Summers[3], Glenn Shiveler[3], Markus Duss[4] , Erik Stewart[5], Attilio Praderio[6]

[1] Optimized Gas Treating Inc., Houston, Texas
[2] Raschig GmbH, Ludwigshafen, Germany
[3] Sulzer Chemtech, Tulsa, Oklahoma
[4] Sulzer Chemtech, Winterthur, Switzerland
[5] INEOS Americas LLC, Houston, Texas
[6] ConocoPhillips, LNG Process Group, Houston, Texas

Presented at the Distillation Topical Conference
AIChE Spring Meeting, Austin, Texas, April, 2015

How Sensitive is Your Treating Plant to Operating Conditions? — Part2

Ralph Weiland[1], Michael Schultes[2], Mark Pilling[3], Juan Portillo[3],
Dan Summers[3], Glenn Shiveler[3], Markus Duss[4], Erik Stewart[5], Attilio Praderio[6]

[1] Optimized Gas Treating Inc., Houston, Texas
[2] Raschig GmbH, Ludwigshafen, Germany
[3] Sulzer Chemtech, Tulsa, Oklahoma
[4] Sulzer Chemtech, Winterthur, Switzerland
[5] INEOS Americas LLC, Houston, Texas
[6] ConocoPhillips, LNG Process Group, Houston, Texas
6

Abstract

One of the most neglected areas in gas treating is the proper and methodical selection and comparison of tower internals. The performance of tower internals has been studied predominantly from the perspective of hydraulic capacity. Apart from efficiency measurements, notably by FRI and almost exclusively in the context of distillation, mass transfer performance has tended to receive relatively scant attention.

In distillation the dominant resistance to mass transfer is almost always in the gas or vapor phase. In gas absorption, on the other hand, the absorption rate of some components under many different conditions of operation lies in the liquid phase. Furthermore, in acid gas removal, chemical reactions play a decisive role in determining the separation. These factors alone set absorption apart from distillation. Another dissimilarity is the usually large density difference between phases. Hence, distillation data do not apply very well to absorption.

When parts-per-million specifications on product gases must be met, selecting the wrong packing size or bed depth can result in a failed design. For many years, packing has had a bad reputation in absorption and distillation at high pressure. Part of the reason is inattention to proper distributor design. Another is the persistent and still unresolved difficulty in translating ideal stages into actual packed bed depths and the selection of a particular commercial packing, especially in amine treating.

In this article a case study is used to reveal the important tower-internals parameters that determine an absorber's mass transfer performance. This is done in the context of an LNG project and it compares a structured packing with three random packings of different size and type, and with two types of commercially proven valve tray.

Back to Basics: Understanding the Difference

Tower internals have natural hydraulic capacity limits imposed by such parameters as jet flooding, downcomer backup, downcomer choke flood, packed column flooding, etc. that are generally very straightforwardly assessed through software freely available from each internals provider or through the Design and Rating Program available from Fractionation Research Inc. (FRI). All the various tools use different input and output data panels, reflecting the conformational and mechanical differences that pertain to the different types of internals that can be selected for a particular service.

All tower internals use hydraulics to promote contact across the interface between counter-currently flowing vapor and liquid phases, but from the point of view of mass transfer, the rates of transfer of components between phases depend on the interfacial area available, the turbulence level created in each phase, the physical and transport properties of the phases, their flow rates, and concentration differences between phases and across phase boundaries. Of great underlying importance is the fact that mechanically different internals create different levels of turbulence for every type and size of internal even at the same flow rates using the same fluids. Turbulence levels determine phase mass transfer coefficients just like turbulence levels on the shell and tube sides of heat exchangers determine heat transfer coefficients. Equally important is the interfacial area between the gas and liquid through which the transferring species must be transported to effect a separation. An example is the behavior of packing.

As the solvent flow rate to a packed bed is increased, the wetted interfacial area rises as well, and the mass transfer rate therefore increases with solvent flow. This is shown by the right side of Figure 1 where increasing liquid rates cover more and more of the packing surface with a film of liquid. Treating improves until the tower operation runs up against another limitation, in this case a lean end pinch in which the treated gas reaches an equilibrium state with the lean solvent. Beyond that point, more solvent does not result in any change in absorber mass transfer performance. On the left side of this plot, yet another limitation is encountered, namely a solvent capacity limit where insufficient solvent is being provided to remove enough CO_2, and breakthrough occurs. However, in the central region of the plot, mass transfer area is responding to changes in solvent flow. With trays, gas-liquid interfacial area for mass transfer is only a relatively weak function of liquid rate but a strong function of gas rate.

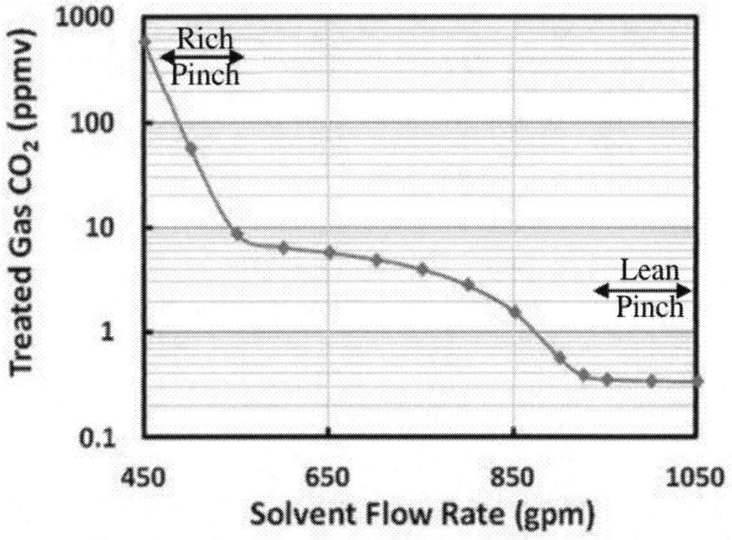

Figure 1. Performance of a Packed LNG Absorber with 0.015 Lean Loading

The case study involves the processing of 745 kNm^3/h of a gas containing 11.5% of CO_2 supplied to an onshore contactor at a pressure of 55 barg and a temperature of 31°C. The treated gas is to meet a specification of < 50 ppmv carbon dioxide. INEOS GAS/SPEC CS-1120® solvent at 50% weight concentration is used in this assessment. Tower diameter was calculated to achieve 60% of flood[1] or, in the case of Sulzer internals, 60% of maximum operating capacity. Simulations of the full treating plant were run using the ProTreat® mass transfer rate based gas treating simulator (Version 5.3.1.007). In each case, the absorber contains enough internals (packing volume or trays) to achieve <50 ppmv CO_2 in the treated gas when using the same lean solvent CO_2 loading (0.017 moles of CO_2 per mole of total amine in the CS-1120 solvent) throughout. The simulator uses established generalized correlations for mass transfer coefficients and specific effective interfacial area for structured and random packings and for trays. It has been thoroughly benchmarked against a wide range of commercial data on amine contactors using both random and structured packings as well as trays. Table 1 shows the key results of this simulation study.

Table 1 Treating with Various Packings and Trays

		Packings				Trays	
		Structured	Random			Valve	
		2nd Gen	4th Gen	3rd Gen	3rd Gen	2nd Gen	3rd Gen
		Mellapak 2X	Raschig SR #2	Nutter #2	Nutter #2.5	Sulzer BDH	Sulzer UFM
Dry Specific Area or Weir Height	(m^2/m^3) / (m)	210.0	97.6	96.0	85.0	0.076	0.076
Lean Amine Circulation	m^3/hr	1,817	1,817	1,817	1,817	1,817	1,817
	gpm	8,000	8,000	8,000	8,000	8,000	8,000
% Flood or % Capacity (at Bulge)		60.60%	60.20%	60.20%	60.20%	60.00%	60.00%
Packing Hydraulic Load	m^3/m^2-hr	88.60	81.40	70.71	74.97		
	gpm/ft^2	36.24	33.30	28.92	30.67		
Treated Gas - CO2 Content	ppmv	40.44	44.14	43.31	43.8	33.7	38.16
Rich Amine Load	mol CO2 / mol Amine	0.473	0.473	0.473	0.473	0.478	0.478
Interfacial Area - Total	m^2	24650	29930	31570	33160	29700	24800
- Relative to M2.X		1.00	1.21	1.28	1.35	1.20	1.01
Interfacial Area (m^2/m^3) Based on Tower Volume		238.0	137.5	132.5	118.0	97*	97*
Wetted/Dry Area Ratio		1.17	1.41	1.38	1.39		

Table 1 shows that despite the much higher specific dry area of Mellapak M2X™ packing, all the packings require roughly similar total wetted area to achieve the < 50 ppmv CO_2 treating specification. Total wetted or interfacial area is a key parameter. Nevertheless, the random packings all require slightly more total packing area than Mellapak M2X™, with Raschig SuperRings® No. 2 being closest in performance to the structured packing. The reason for the slight difference between the two types of packing is subtle differences in gas- and liquid-side mass transfer coefficients. And both gas and liquid coefficients are important. Liquid-side coefficients are more important *below* the temperature bulge because increasing acid-gas loading slows the CO_2 reaction rate. But gas-side coefficients are more important *above* the temperature bulge because the very fast kinetics of the lightly-loaded GAS/SPEC CS-1120 solvent moves the dominant mass transfer resistance from the liquid into the gas phase. Another cause

[1] Raschig uses the approach to flood concept while Sulzer uses capacity such that 100% capacity is defined to occur when the pressure drop reaches 12 mbar/m of packing depth.

of differences is temperature dependent transport properties that change in response to differing temperature profiles. There are many influences.

As shown in Table 1, Mellapak M2X^{TM} structured packing has more than twice the dry specific area of any of the random packings, and almost twice the specific wetted area. One might well ask the question then, if the physical volume of structured packing in an absorber can be made so much lower than random packings, why isn't structured packing used all the time? The answer isn't necessarily in mass transfer performance; rather, it lies in the hydraulics. Amine treating tends to be a fairly dirty service (especially in refineries) so structured packing, particularly small crimp packing, should be used only if the operator is prepared to give undivided attention to ensuring excellent amine hygiene. One might recommend, for example, full flow particulates filtration accompanied by charcoal filtration to remove surface active agents. The deposition of solids and gelatinous substances means early death to structured packing and the smaller the crimp, the earlier its demise. Figure 2 shows photographs provided by INEOS GAS/SPEC Analytical Lab of amine samples from an operating facility illustrating solids that can be present in plant solutions.

Figure 2. Amine Samples immediately after sampling and after settling

Some practitioners claim that structured packing is also more prone to initiate and stabilize foam—perhaps this occurs at the joints between packing layers where the gas flow undergoes sudden changes in direction. These joints are also where flooding often initiates, perhaps lending some credence to the foaming conjecture. It also possibly caused by localized disturbances to L/G ratios within the beds. Furthermore, structured packing may be somewhat more susceptible to liquid maldistribution. Figure 3 shows a foam test done on an operating plant solution by INEOS GAS/SPEC Laboratory, indicating a probable foaming issue in this plant and a potential operating issue with the contactor internals if not addressed. In any event, one may want to be a little cautious in giving an across-the-board structured packing recommendation without due regard for the particulars of the application. Close attention must be paid to installation of distributors, wall wipers and other design/installation issues recommended by structured packing vendors. In other words, it is necessary to have excellent design and installation of liquid and vapor distributors in order to obtain the best performance from structured packing.

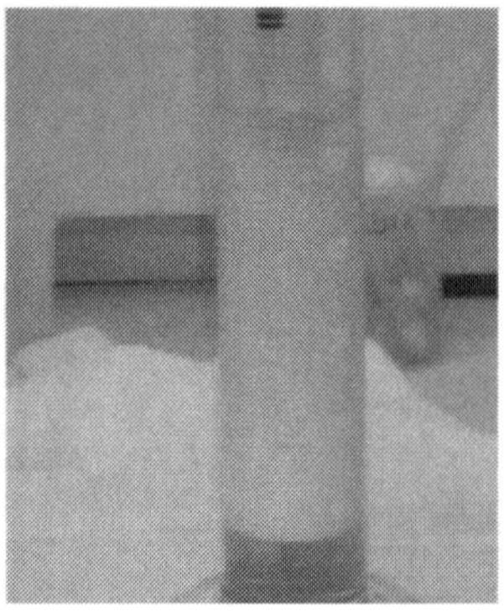

Figure 3. Foam Test Amine Samples immediately after sampling and after settling

This type of analysis can be carried out for any type and brand of packing contained within the ProTreat® database and, of course, findings will differ from packing to packing. It should be noted that the analysis for a contactor located offshore or on a floating facility will include additional considerations and, therefore, any conclusion from this work cannot directly be extrapolated to those situations. For example, on FPSOs rocking motion is a serious consideration—even here there are applications where structured packings are preferred but there are also applications where random packings are favored. Flow maldistribution must be considered in such applications. However, even greater attention to distributor design is mandatory. ProTreat can be used to assist in troubleshooting performance issues in contactors. INEOS uses an Infrared Camera to check for distributor issue in a packed column after plant simulation indicates poorer than expected treating performance. Figure 4 shows a thermal image (cf. Shiveler et al.[1] and Teletzke and Bickham[2]) of the liquid flow and reaction (indicated by the red region) occurring down the wall of contactor but with very little reaction in the packed bed. This indicates poor distribution of amine solution.

Figure 4. Infrared Camera Photo of Distribution Issue in Packed Tower

The purpose of this writing is to point out the controlling, underlying physical and chemical factors to which one should pay attention in selecting packed tower internals from a mass transfer performance standpoint. In that spirit, no endorsement of one packing over another is being made. There are always multiple choices, and indeed, multiple good choices. The best choice will depend on many factors, and in some cases mass transfer performance may not even be decisive. In any event, there is no one better qualified to help in making the best packing decisions than a reputable internals supplier.

Turning now to trays, Table 1 also shows a comparison with Sulzer BDH™ and Sulzer UFM™ (Umbrella Floating Mini-Valve) trays where it is immediately apparent that the total interfacial areas needed to achieve a specified separation using trays are very similar to those needed for packings. Note also that the interfacial area is based on the total tower volume over the active deck or tray bubbling area (i.e., exclusive of downcomers). In the case of the UFM tray deck, a total interfacial area similar to structured packing is sufficient to treat satisfactorily under these conditions because the gas-side mass transfer coefficient in the froth over the trays is a little higher than for random packing. Considering that the cost of a high-pressure vessel has more to do with its diameter than its length, it is worth noting that these trayed towers have diameters that are about 4 % larger than the structured packing option for the classical BDH rectangular floating valve and about 5% smaller than the structured packing option for the newest Sulzer UFM™ valve. For this specific application, the use of a four-pass layout is demanded by the high liquid weir loads (heavily loaded downcomers); however, this could limit the turndown ratio when compared with the packing options because multi-pass trays tend to have more limited turndown capabilities. The turndown limitation on packed beds, on the other hand, generally depends more on the distributors rather than on the packings themselves.

The Need and Importance of Performing a Sensitivity Analysis

A normal expectation in the course of operating an amine plant is that small changes in operating conditions should result in correspondingly small changes in plant performance. However this expectation is not always well founded[3]. Therefore, an analysis of performance sensitivity to all significant operating parameters should always be performed. This is especially recommended for amine systems operating near what is known as temperature pinch conditions where there can be stable operating regions demarcated by steep boundaries outside of which performance degrades precipitously. In other words, operationally *mass transfer* instabilities manifest as a sudden loss in treating, but they have little or no effect on hydraulic performance *per se*.

The main cause of sudden loss in treating is attempting to operate with the bare minimum of solvent absorptive capacity, whether from too low a solvent flow rate, too high a gas flow rate, too high temperature, or too high a CO_2 (or H_2S) concentration in the raw gas. With fast reactions, the reaction zone in the absorber extends only over a narrow band around the maximum temperature found in the absorber (bulge). Under normal operation, this reaction zone with its telltale bulge lies close to the bottom of the absorber. These profiles are illustrated in Figure 5. But, unlike slowly reacting solvents like DEA where the reaction zone is broader and more gradually emerges from the top of the tower (with decreasing solvent flow rate for example), a narrow reaction zone gives no warning of its impending emergence until it suddenly pops out the top of the column, and treating goes wildly off specification. Temperature bulges are also where towers find themselves hydraulically limited.

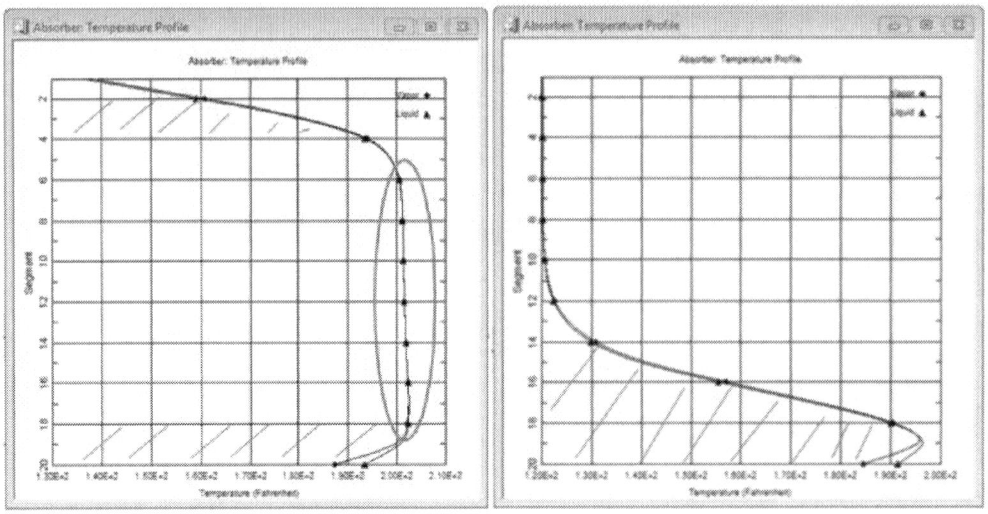

Figure 5 Bulge Pinched and Lean-end Pinched Temperature Profiles

For example, a treated gas exhibiting a carbon dioxide concentration of 10 ppmv can suddenly go to 1000s of ppmv with less than a 0.1% drop in solvent flow rate[4]. Figure 6 shows an example of such a stability boundary. The design point in this particular plant is 115°F but a temperature excursion of 10°F will throw treating into a tailspin. Note too, that although 50 ppmv CO_2 may be the targeted treating level, stable operation at that value is impossible, at least by manipulating lean amine temperature. In fact, in this case treating degenerates very rapidly from 10s of ppmv CO_2 to several 1000s of ppmv with a temperature increase less of than 1°F. The reasons for the shape of the curve in Figure 6 (and Figure 1) are complex and are discussed in detail by Hatcher et al.[5]

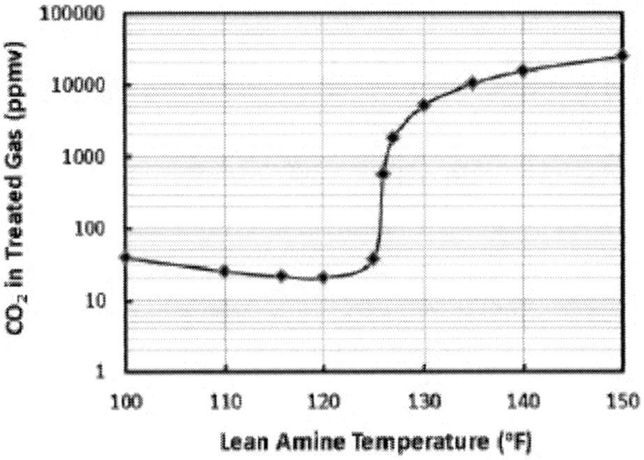

Figure 6: Extreme Sensitivity to Lean Amine Temperature in an LNG Absorber

The ability of the solvent to "load up" with carbon dioxide is compromised as it becomes too hot, and what cannot be absorbed must leave in the existing gas. Figure 7 shows contactor scans by INEOS for operating plants with a bulge pinch (left) versus a bulge illustrating the reaction zone in bottom of contactor (right).

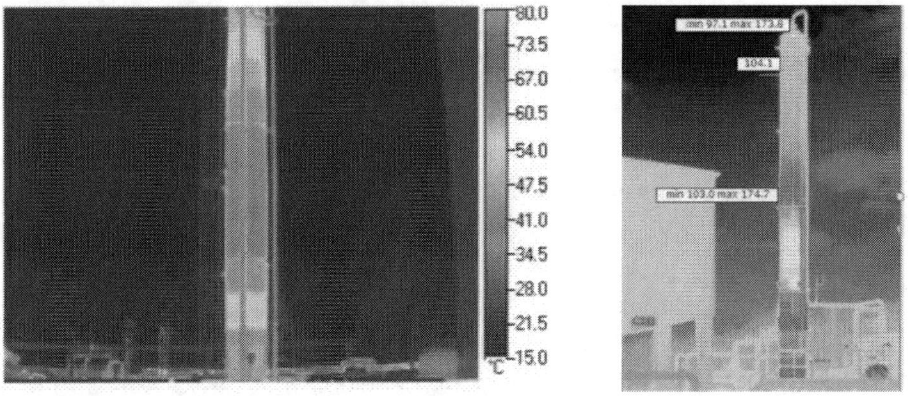

Figure 7. Infrared Scan of Contactor Profiles with Broad Bulge and Bottom Bulge

This is but one example pointing to the need for a careful sensitivity study in a new gas treating unit, especially one in which a fast reacting solvent is used for deep CO_2 removal as is often the case in LNG facilities.

Finally it should be noted that operation of the contactor with a hot narrow temperature profile may result in equipment corrosion due to breakout or flashing of CO_2. Figure 8 shows photos illustrating corrosion on contactor walls and in downcomer areas caused by CO_2 flashing and release in the tray liquid holdup area. In the hottest region of a contactor, turbulence can cause sudden short-term pressure fluctuations sufficient for CO_2 to flash from solution.

Figure 8. Contactor Wall Corrosion in Tray Liquid holdup Area and Downcomer

Concluding Remarks

There are many aspects that need to be considered when selecting between different internals for the same service, but ultimately the selection can only be made *logically* by using mass transfer rate-based simulation. Unlike ideal stages with efficiencies, the mass transfer rate approach recognizes that the tower is not a black box—it really does contain internals— and it accounts for the differences between various internals and the characteristics of each. Put another way, if your simulation tool doesn't understand how one packing behaves differently from another as a mass transfer device, how can it possibly be of any assistance in making the best internals selection? Without a mass transfer rate-based simulator, the comparative table presented here could not have been generated. Indeed, quite likely a non-optimal internals selection would have been made, with higher performance uncertainty leading to a design with excessive safety margin.

Naturally, mass transfer performance is not the only factor that must be considered. The final selection decision must also include the total installed cost, the operating costs, the availability of an experienced crew to install the selected internal at the specific location where the plant will be located, and the final technical-economical terms and conditions offered by every individual equipment supplier. Also, it must be noted that what was the best choice for one application will not necessarily be the best choice for another, regardless of apparent similarities.

The expedient selection of a particular column internal that fits within a current tower and extends hydraulic capacity is often the very decision that defeats the mass transfer treating performance. Rate-based simulation strongly emphasizes the fact that absorbers are truly mass transfer equipment—not ideal stages of hydraulic contact—and simulating this way provides important support for engineers working on industrial tower designs and retrofits.

This paper has focused on the sensitivity analysis that should be performed for the design of a new amine absorber. However, a sensitivity analysis should also be included in retrofitting projects: no amine treating plant should ever consider a revamp from trays to packing before fully understanding the limitations already existing in the plant, including current design and operating problems. For example, being unaware the effect of heat stable salts (HSSs) on performance can lead to a failed revamp because the HSSs may be the real cause of poor performance, not the tower internals. In another example, when moving from trays to packing to relieve vapor hydraulics, care must be taken to selecting the correct packing type to compensate for the tower volume now occupied by support and distribution internals, without serious loss of mass transfer performance. A typical problem relates to moving from trays to packing in order to relieve vapor hydraulics. Making the wrong choice of packing type, packing support, and liquid distribution for the conversion can result in serious loss of mass transfer performance. Thus, the retrofit of any absorber or mass transfer equipment must also focus on the mass transfer aspect, too, even though originally the exercise may have been driven to relieve a hydraulic capacity limitation.

Finally, by working in a collaborative environment, sharing common proven simulation tools and the non-substitutable value of the experience of "lessons learned", equipment and solvent providers, internals suppliers, process engineers, process designers, troubleshooters and plant operators, an initial design can be made much more solid with correspondingly less uncertainty, less design margin, better plant operability, and lower risk of unpleasant surprises.

There are many different opinions on designing tower internals for use in amine services, and because absorption and distillation are partly art and partly science, differing opinions can be shown to be serviceable in the field. Trays and packings can both be utilized for amine services. The key to a successful design or retrofit is the proper selection and application of the tower internals to the specific project conditions been assessed. The selection must be done from both a hydraulic and a mass transfer viewpoint. This paper is offered as an illustration of key facts that need to be considered in order to develop practical, serviceable column design for amine treating systems, where operating experience must be included as one of the key factors. Consequently, this paper is a compilation of the views of authors from a variety of backgrounds, and it does not necessarily represent completely the views of the five corporations which the authors represent professionally.

References

1. Shiveler, G., Solis, G. S., Gonzales, G. H. P., Bueno, M. L., *Retrofit of a H_2S Selective Amine Absorber Using MellapakPlus™ Structured Packing*, paper presented at Spring AIChE Meeting, Atlanta, Georgia, USA, April 10–13, 2005.

2. Teletzke, E., Bickham, C., *Troubleshoot Acid Gas Removal Systems*, Chemical Engineering Progress, pp. 47–52, July, 2014.

3. Hatcher, N. A., Alvis, R. S., Weiland, R. H., *Simulate Gas Treating Holistically: Part One*, Hydrocarbon Engineering, December, 2012. [Available from http://www.protreat.com/files/publications/Treating_Holistically_3_part_Series_HE.pdf]

4. Weiland, R. H., and Hatcher, N. A., *Foundations of Failure*, Hydrocarbon Engineering, December, 2011.

5. Hatcher, N. A., Alvis, R. S., Weiland, R. H., *Simulate Gas Treating Holistically: Part Three*, Hydrocarbon Engineering, May, 2013. [Available from http://www.protreat.com/files/publications/Treating_Holistically_3_part_Series_HE.pdf]

ExxonMobil

April 28, 2015

Experimental Investigation of the Effects of Tower Motion on Random Packing Efficiency

Energy lives here

Norman K. Yeh and Edward J. Grave
ExxonMobil Upstream Research Company

Presented at the Kister Distillation Symposium
2015 AIChE Spring Meeting, Austin, TX

Agenda

Motivation

Internals Selection and Design Considerations

Experimental Test Program

Results and Interpretation

Conclusions

Questions?

Motivation

- Floating processing facilities such as FLNG may be attractive options to develop offshore natural gas resources
- Technical challenges include proper design of process equipment for moving conditions
- Tower services typically required for gas processing facilities include
  - Absorption/stripping, e.g. H_2S and/or CO_2 removal, dehydration
  - Distillation of light hydrocarbon mixtures
- Challenges or uncertainties in these designs can lead to
  - Excessive overdesign of equipment
  - Inflated cost and therefore reduced economics for the project

ExxonMobil 3

Towers in Motion

- Motion imparted by wind, waves, and platform ballasting
  - Degrade tower performance due to poor vapor/liquid distribution

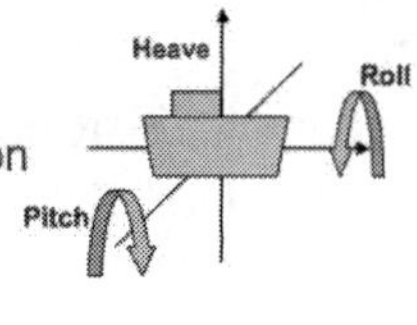

- Trayed Towers
  - Weeping, blowing, and sloshing
  - Dramatic decrease in efficiency
  - Not suitable for operation in motion

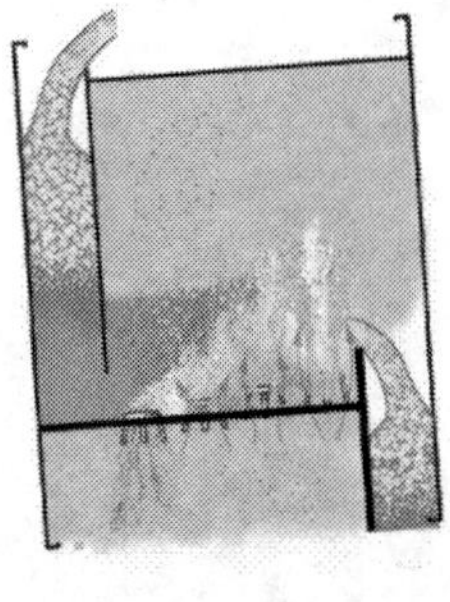

- Packed Towers
  - Maldistribution from the liquid distributor
  - Vapor and liquid channeling in the packing

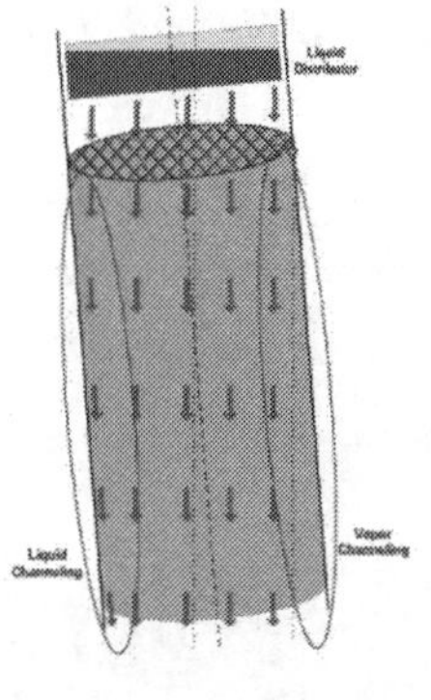

ExxonMobil 4

Internals Selection

- Packed beds are more suitable than trays, but packed towers are still subject to maldistribution from the liquid distributor and vapor/liquid channeling in the packing
- Random packing vs. structured packing
  - Some studies (e.g. Waldie, 2002) indicate that liquid distribution in structured packing is less susceptible to tilt effects than random packing
  - However, structured packing is not acceptable for high pressure distillation services within ExxonMobil
  - Random packing is preferred and commonly used in many upstream applications (high pressure distillation; higher liquid rate services such as gas treating towers)

ExxonMobil 5

Internals Design Considerations

- Tower efficiency requires a de-rating (marinization factor) based on operation at tilt or in motion
- Publicly available data does not provide an appropriate de-rating factor
  - Safety factors for packing height of 1.25 – 2.0 have been suggested
  - A single marinization factor is not appropriate for all towers and all services
- Conservative de-rating can significantly affect topsides design
  - Splitting of tall towers
  - >10% increase in length / width

ExxonMobil 6

ExxonMobil Tower Motion Test Program

- Objective of this work is to limit unnecessary overdesign and allow cost-effective tower designs customized for expected conditions
  - Understand the effects of tilt and motion on tower operation
  - Develop methods for collecting, interpreting, and applying data to project design
- EMURC conducted an experimental test program at Heriot-Watt University to quantify packing hydraulics and mass transfer with simulated wave motion

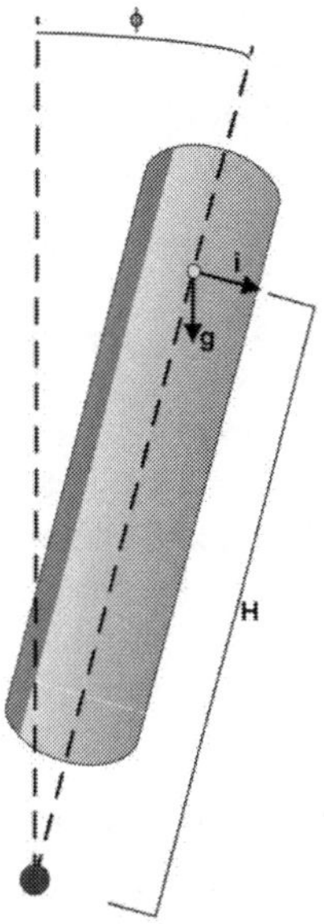

ExxonMobil

Experimental Test System

- Apparatus
  - Column: 1 m ID
  - Packing type: Random (Raschig Super Ring #2)
  - Pressurized liquid distributor
  - Bed heights (2 m and 4 m)
  - Motion: tilt angles (0-9°), periods (static, 10-30 sec)
  - Liquid flux rates (17-130 m^3/h per m^2 [7-53 gpm/ft^2])
- Mass transfer efficiency was measured with CO_2 absorption into caustic solution to obtain effective wetted area for mass transfer

ExxonMobil

Data Analysis & Interpretation

- Dry / under-wetted packing volume increases and effective mass transfer area decreases with
  - Increasing tilt angle
  - Increasing bed height/diameter ratio (L/D)
- Higher liquid fluxes also reduced the effective mass transfer area significantly
  - Liquid maldistribution causes vapor maldistribution so pressure drops in each section are important
  - Both liquid and vapor maldistribution can persist along the column unless measures are taken to mix and redistribute between beds
  - Effect of vapor channeling is a function of pressure balance between dry and wet areas
- Empirical parameters fitted for model coefficients and effects of motion period and amplitude

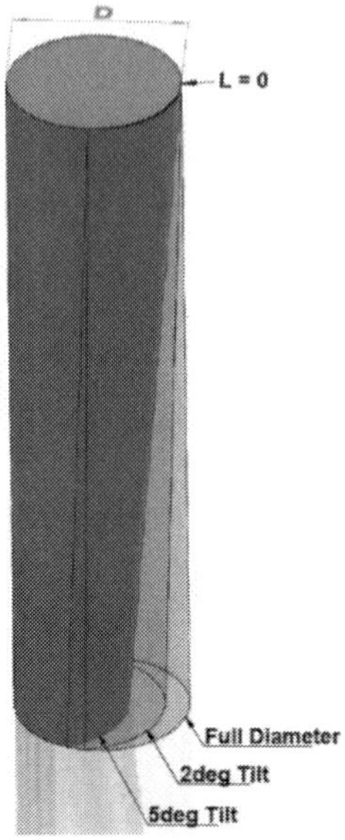

ExxonMobil

Selected Test Results

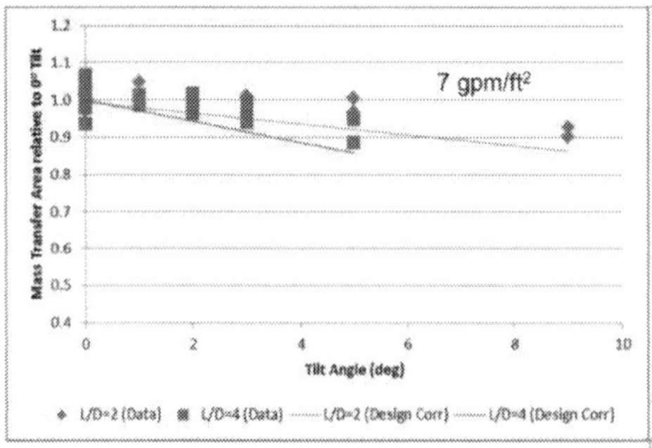

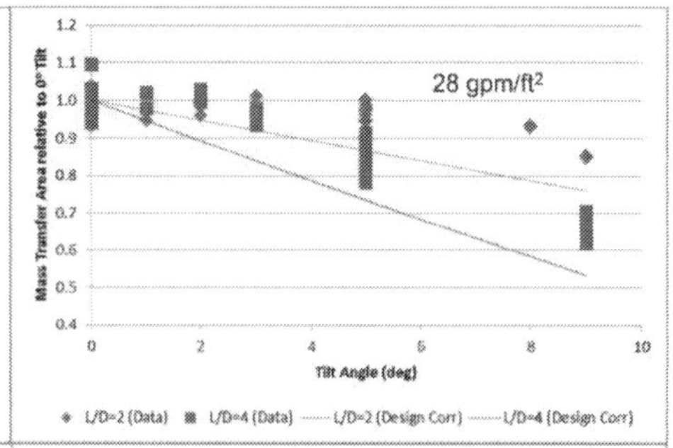

- Relative mass transfer area decreased with (L/D) ratios, tilt angles
- At low tilt angles (< 2°), impact on mass transfer area is relatively small
- At higher liquid rates, relative mass transfer area exhibited stronger dependence on both tilt angle and L/D
- Correlation parameters were tuned to conservatively predict majority of data for design purposes

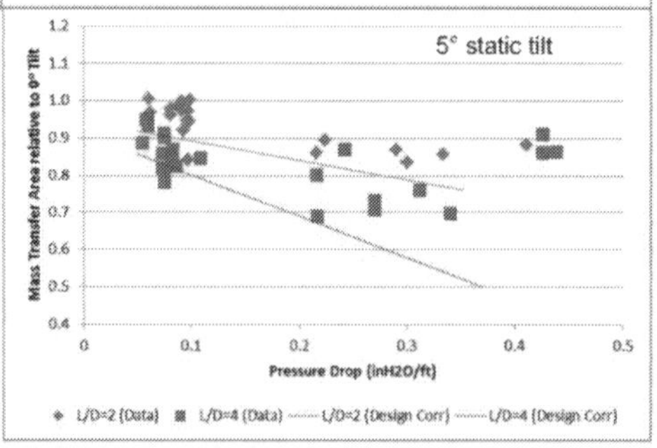

ExxonMobil

Conclusions

- Mass transfer performance has been measured for tilt and motion impacts on a modern random packing at industrial scale
  - Enable fit-for-purpose designs to deliver reliable performance
  - Contribute to minimizing overall cost of floating production facilities
- Significant impacts of L/D ratio, liquid flux, and vapor distribution – highlight importance of considering number/size of beds, internals for redistribution, etc. when designing columns for motion
- Concepts of dry/under-wetted packing volume and vapor channeling have successfully been used as the basis for model development
- Correlation for effective area may be incorporated into HETP calculation methods to improve accuracy of packing efficiency predictions under moving conditions and give fit-for-purpose designs

ExxonMobil 11

Acknowledgments

ExxonMobil colleagues who contributed over the course of the project:
- J. Tim Cullinane
- Andrew Steinhauser
- Berne Stober
- Graeme White, Heriot-Watt University

ExxonMobil 12

Packing Characterization for Post Combustion Capture: Model Development

By Chao Wang, Gary T. Rochelle
Department of Chemical Engineering, the University of Texas at Austin
Phone: (512) 228-9648
Email: wangchao@utexas.edu

Micah Perry, Frank A. Seibert
Separation Research Program, the University of Texas at Austin
Email: fseibert@mail.utexas.edu

Presented in the Kister Distillation Symposium of the 2015 Spring AIChE Meeting
Austin, TX, April 26- 30, 2015

Session: New Frontiers in Packed Tower Design and Application, Paper 144d

Part of the content is from paper submitted to I&EC Research

Abstract

Packing is widely used in post-combustion CO_2 capture because of its low pressure drop, good mass transfer efficiency, and ease of installation. In the CO_2 capture process, absorber and stripper performance are highly dependent on the effective mass transfer area of the packing (a_e). The stripper performance also depends on the liquid film mass transfer coefficient (k_L). Gas cooler and water wash performance depends on the gas film mass transfer coefficient (k_G).

In this work, eleven structured packings with surface area (a_p) from 125 to 500 m^2/m^3 and corrugation angle (θ) from 45 to 70 degrees were studied to explore the effects of packing geometry on mass transfer. A new concept, the mixing point density, was proposed to represent the effects of packing geometries. A generalized method to calculate mixing point density from a_P and θ was developed. Three dimensionless mass transfer models are developed to predict the effective area, liquid film and gas film mass transfer coefficients, with Mixing number and Reynolds number imbedded into the correlations. Compared with literature models, the models developed in this work capture the effects of packing geometry, and predict the mass transfer properties in aqueous systems accurately.

The content of this paper is partly from the paper submitted to Industrial & Engineering Chemistry Research. For more details, please refer to my paper in I&EC Research.

Experimental

The experiments are conducted at the Separations Research Program (SRP) at UT-Austin. A pilot-scale PVC column with an inner diameter of 0.428 m (16.8 in) and a total column height of 7.62 m (25 ft) capable of a maximum packed height of 3.05 m (10 ft) was utilized to measure the effective area, gas and liquid film mass transfer coefficients. Absorption of atmospheric CO_2 by 0.1 gmol/L NaOH (reaction film controlled) was used to measure the mass transfer area. Air stripping toluene from water (liquid film controlled) was used to measure the liquid film mass transfer coefficient. Absorption of SO_2 from air by 0.1 gmol/L NaOH (gas film controlled) was used to measure the gas film mass transfer coefficient. The inlet and outlet gas CO_2 was measured by a CO_2 analyzer (Horiba® VIA-510) with a typical range of 0-500 ppm. The inlet and outlet liquid toluene was measured by a Gas Chromatograph (Hewlett Packard® 5890A Gas Chromatograph) with a typical range of 0-1000 ppm. Two ambient level pulsed fluorescence analyzers (Thermo Scientific® Model 43i) were used to measure the inlet and outlet gas SO_2 at 0-100 ppm SO_2 and 0-2000 ppb SO_2. Detailed description of experimental methods and apparatus can be found in previous papers.[1-6]

Effective Area model

The effective mass transfer area model was developed based on the experimental data. The stainless steel structured packings were manufactured by Sulzer ChemTech, GTC Technology, and Raschig. The packing surface area varied from 125 to 500 m^2/m^3 while the corrugation angle varied from 45 to 70 degrees. Three random packings in the Raschig Super Ring family were also included in the database.

The effective mass transfer area model was developed based on the dimensionless model by Tsai (Equation 1).[7, 8] According to Tsai's experiments as well as effective area measurements conducted in this work, the effective area is a function of liquid flow rate, liquid density and surface tension, and is independent of gas flow rate and liquid phase viscosity. This model is supported by the majority of experimental data, although at some conditions the effective area changes slightly with gas flow rate.

$$\frac{a_e}{a_P} = 1.34[(We_L)(Fr_L)^{-1/3}]^{0.116} \quad (1)$$

Where,

We_L is the liquid phase Weber number, $\rho_L u_L^2 \delta_L/\sigma$;

Fr_L is the liquid phase Froude number, $u_L^2/g\delta_L$.

However, with a larger scope including random packing and hybrid packing such as the Raschig Super-Pak family, the original form of the Tsai model is not applicable. In those situations where channel dimensions are not known or hardly defined, using liquid superficial velocity over packing total area (u_L/a_P) instead of (Q/L_P) is a good alternative. The mass transfer area model in this work is developed based on the Tsai model, utilizes u_L/a_p as the liquid flow rate per wetted perimeter. The experimental coefficient is changed from 1.34 to 1.41 which provides a better fit of the larger database.

$$\frac{a_e}{a_p} = 1.41[(\frac{\rho_L}{\sigma})g^{1/3}(\frac{u_L}{a_p})^{4/3}]^{0.116} \tag{2}$$

Figure 1 shows the fractional mass transfer area as a function of the dimensionless number group $(We_L)(Fr_L)^{-1/3}$ at a moderate gas velocity (0.99 m/s). The mass transfer area model predicted by Equation (2) is shown by the black solid line in the middle. The database includes 14 packings measured in this work and contains a large variety of packing types (structured, random, and hybrid). The model shows a good fit with most data. GT-PAK™ 500Y shows a lower effective area than predicted due to the difficulty of wetting the large surface area effectively. The hybrid packing RSP200X shows a higher value than predicted. The model accurately captures the dependence of mass transfer area on liquid rate. The average deviation of this area model is 10.5%, which is quite acceptable considering the broad scope of the packing type.

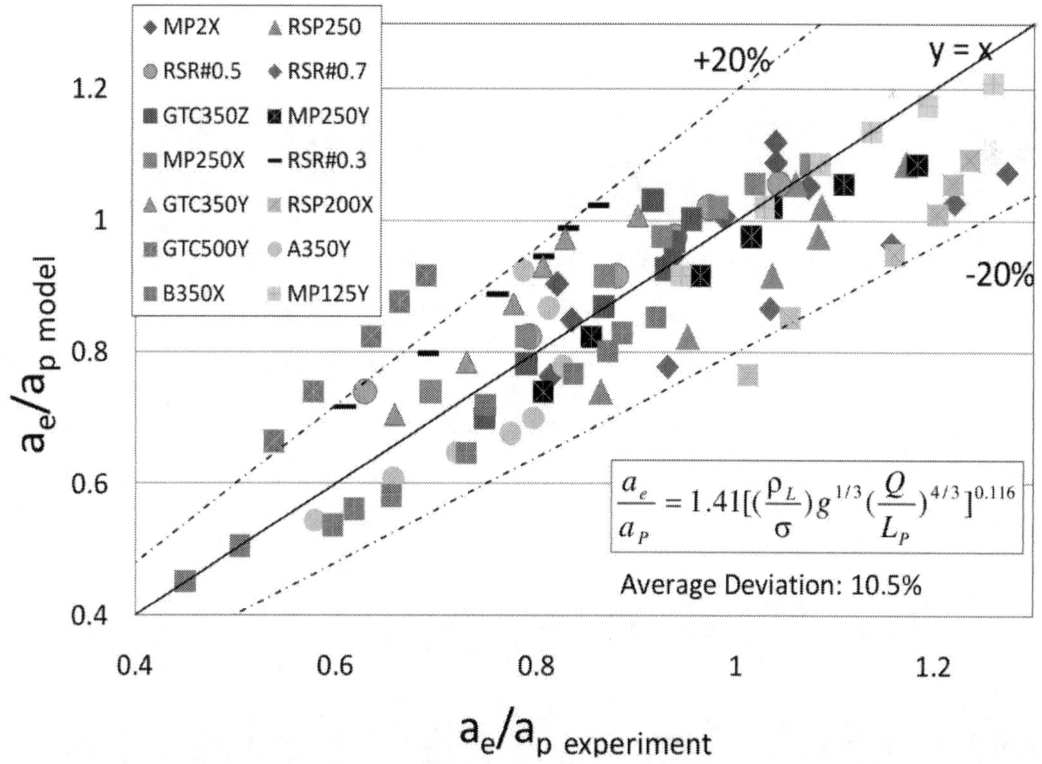

Figure 1. Mass transfer area of structured packing

Liquid film and Gas film mass transfer coefficient model

The dimensionless form of liquid the gas film mass transfer coefficient models are developed in this work. The models utilize the dimensionless form of velocity (Reynolds number, Re), the dimensionless form of liquid or gas phase physical properties (Schmidt number, Sc), and the dimensionless form of packing geometries (Mixing number, Mi) as the independent variables. The dimensionless form of k_L or k_G (Sherwood number, Sh) is used as the dependent variable. Thus, the model can be written as:

$$Sh_{L/G} = C * \mathrm{Re}_{L/G}^{m} Sc_{L/G}^{n} Mi^{p} \tag{3}$$

Conclusions of other researchers are used for the effect of Schmidt number on Sherwood number since the Schmidt number influence is not yet explored in this work. For the gas phase, Mehta's[9] conclusion is used in this model, which is that Sh_G depends on Sc_G to the power of 0.5. For the liquid phase, Mangers'[10] conclusion is used with a dependence of Sh_L on Sc_L to the power of 0.5.

The dimensionless k_L and k_G models for structured packings are:

$$Sh_L = 1.79\,\mathrm{Re}_L^{0.74} Mi^{0.42} Sc_L^{0.5},\ k_L = Sh_L a_P D_L \tag{4}$$

$$Sh_G = 0.83\,\mathrm{Re}_G^{0.58} Mi^{0.3} Sc_G^{0.5},\ k_G = Sh_G a_P D_G \tag{5}$$

In the model development, a new concept, mixing point density, was introduced to account for the effect of packing geometry on k_L and k_G. Mixing point density is the number of mixing points per m^3 in the packing. The dimensionless form of mixing point density, Mixing Number Mi, is used in the mass transfer correlations in this work. Mi is a transformation of the corrugation angle that should be better behaved in quantitative correlations. Mi can be calculated by Equation (6).

$$Mi = \frac{3\sin\theta\cos\theta}{16(\sin^2\theta + 1)^{3/2}} \tag{6}$$

Sh, Re, and Sc are defined as:

$$Sh = \frac{k * l_{eq}}{D} = \frac{k}{D * a_p} \quad (7)$$

$$\mathrm{Re} = \frac{\rho u l_{eq}}{\mu} = \frac{\rho u}{\mu a_p} \quad (8)$$

$$Sc = \frac{\upsilon}{D} = \frac{\mu}{\rho D} \quad (9)$$

Table 1 shows the standard error for the regressed power of Re and Mi.

Table 1. Standard error

	Re	Mi
k_L model	0.74 ± 0.045	0.42 ± 0.086
k_G model	0.58 ± 0.031	0.3 ± 0.046

The comparisons between experimental data and values predicted by dimensionless k_L and k_G models are shown in Figures 2 and 3. The k_L model shows good prediction for most packings, except for MP2X which shows a larger value than predicted. The mass transfer models accurately capture the effects of liquid or gas velocity, and packing geometry on mass transfer coefficients. The dependence of Sh_L on liquid rate is 0.74, an average of dependences for each individual packing. This dependence is larger than previous models developed from penetration theory, which will be discussed in the model comparison section. The dependence of Sh_G on gas rate is 0.58, which falls in the range of literature models (dependence varies from 0.5 to 1).

The deviation between experimental data and model value is 22% for Sh_L while the deviation between experimental data and model value is 12% for Sh_G, showing good consistence between model developed in this work and experimental data.

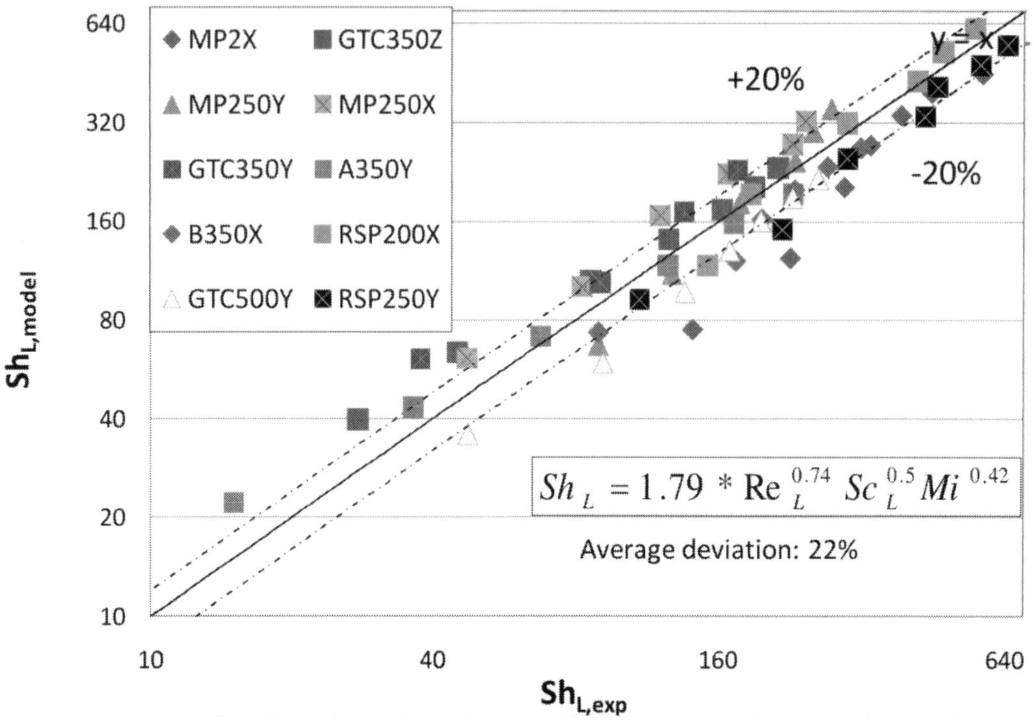

Figure 2. Comparison between experimental Sh_L and Sh_L predicted by dimensionless model

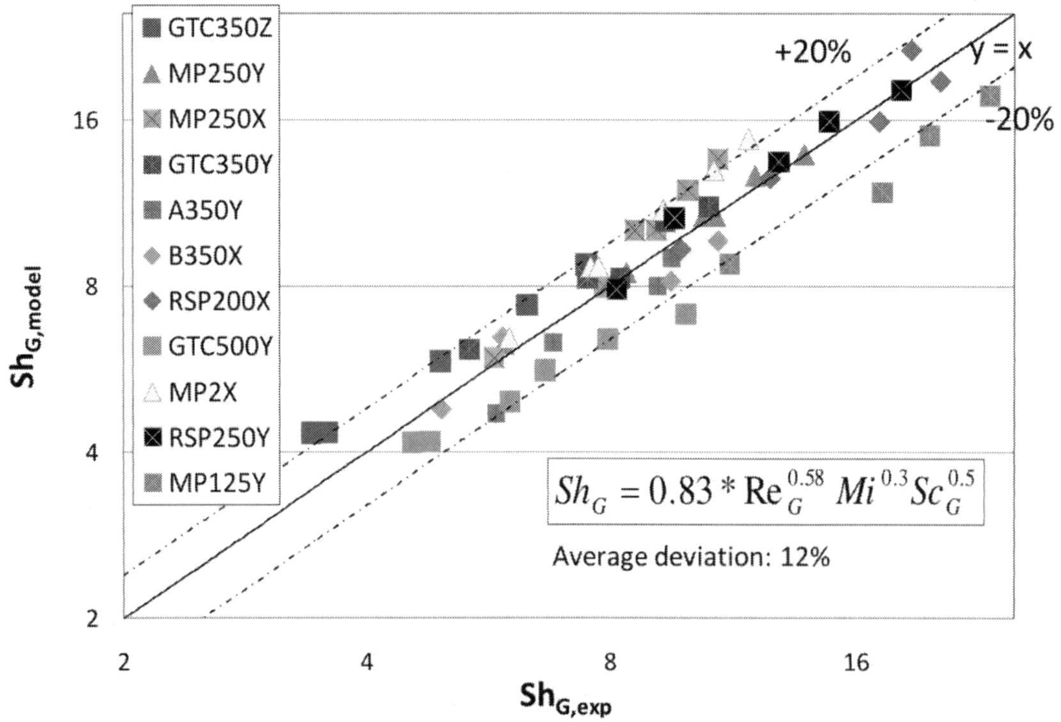

Figure 3. Comparison between experimental Sh_G and Sh_G predicted by dimensionless model

Conclusions

In this paper, three dimensionless mass transfer models are developed. The database includes eleven structured packings with surface area ranging from 125 m^2/m^3 to 500 m^2/m^3 and corrugation angle from 45 degree to 70 degree, and three random packings from the Raschig Super Ring family. The experimental systems use the absorption/desorption from aqueous solvents with liquid physical properties close to those of pure water.

The three dimensionless mass transfer correlations developed in this work are:

$$\frac{a_e}{a_P} = 1.41[(\frac{\rho_L}{\sigma})g^{1/3}(\frac{u_L}{a_P})^{4/3}]^{0.116}$$

$$Sh_L = 1.79 * \mathrm{Re}_L^{0.74} Mi^{0.42} Sc_L^{0.5}, k_L = Sh_L a_P D_L$$

$$Sh_G = 0.83 * \mathrm{Re}_G^{0.58} Mi^{0.3} Sc_G^{0.5}, k_G = Sh_G a_P D_G$$

$$Mi = \frac{3\sin\theta\cos\theta}{16(\sin^2\theta + 1)^{3/2}}$$

The effective area model uses liquid superficial velocity over packing total area (u_L/a_P) as the liquid flow rate per wetted perimeter. Thus the applied range of this area model is extended to include hybrid packings and random packings. The average deviation is 10.5% for the entire data bank, with a total of 14 packings including hybrid packings and random packings.

The liquid film and gas film mass transfer coefficients models are built in dimensionless form. The mass transfer coefficients are dependent on liquid or gas superficial velocity (u_L or u_G), packing corrugation angle (θ), and packing surface area (a_P). Through combination, two dimensionless numbers (Re and Mi) are used to represent these three variables. The average deviation is 22% for k_L model, and 12% for k_G model.

References

(1) Wilson, I. *Gas-Liquid Contact Area of Random and Structured Packing.* The University of Texas at Austin. M.S. Thesis. **2004.**

(2) Tsai, R.E.; Schultheiss, P.; Kettner, A.; Lewis, J.C.; Seibert, A.F.; Eldridge, R.B.; Rochelle, G.T. Influence of Surface Tension on Effective Packing Area. *Ind. Eng. Chem. Res.* **2008**, 47, 1253-1260.

(3) Wang, C.; Micah, P.; Rochelle, G.T.; Seibert, A.F. Packing characterization: Mass Transfer Properties. *Energy Procedia.* **2012**, 23, 23-32.

(4) Wang, C.; Micah, P.; Seibert, F.; Rochelle, G.T. Characterization of novel structured packing for CO_2 capture. *Energy Procedia.* **2013**, 37, 2145-2153.

(5) Wang, C.; Micah, P.; Seibert, F.; Rochelle, G.T. Packing characterization for post combustion CO_2 capture: mass transfer model development. *Energy Procedia.* **2014**, 63, 1727-1744

(6) Wang, C. *Mass Transfer Coefficients and Effective Area of Packing.* The University of Texas at Austin. Ph.D Thesis. **2015**.

(7) Tsai, R.E.; Seibert A.F.; Eldridge, R.B.; Rochelle, G.T. A Dimensionless Model for Predicting the Mass-Transfer Area of Structured Packing. *AIChE Journal.* **2011**, 57, 1173-1184.

(8) Tsai, R.E. *Mass Transfer Area of Structured Packing.* The University of Texas at Austin. Ph.D. Thesis. **2010**.

(9) Mehta, V.D.; Sharma, M.M. Effect of diffusivity on gas-side mass transfer coefficient. *Chemical Engineering Science.* **1966**, 21, 361-365.

(10) Mangers, R.J.; Ponter, A.B. Effect of viscosity on liquid film resistance to mass transfer in a packed column. *Ind. Eng. Chem. Process Des. Dev.* **1980**, 19, 530-537.

HIGH PRESSURE DISTILLATION WITH STRUCTURED PACKING – FRIEND OR FOE

By

Simon Chambers - Formerly of Fractionation Research Inc.

Ken McCarley - Fractionation Research Inc.

Tony Cai - Fractionation Research Inc.

Anand Vennavelli - Fractionation Research Inc.

Presented at the AIChE Spring National Meeting

Topical 8:

Kister Distillation Symposium

Paper 144e

April 28, 2015

Austin, Texas

High Pressure Distillation with Structured Packing – Friend or Foe

Simon Chambers**, Ken McCarley*, Tony Cai*, Anand Vennavelli*,
* *Fractionation Research Inc,, 424 Squires Blvd, Stillwater, OK 74074*
***Formerly of Fractionation Research Inc.*

Abstract

Throughout the 1980's and 90's, the industry reported catastrophic failures when structured packings were employed in light ends high pressure distillation. In 2001 Andrew Sloley, of The Distillation Group, set out criteria where structured packing should be avoided in high pressure distillation. The purpose of this paper is to quantify different physical properties where structured packing is most and least favored. Data were taken from FRI published Mellapak 250.Y data at ten different operating pressures. It will be shown that low surface tensions and high vapor densities are major indicators where structured packings should be avoided.

Background and Introduction

In the 1980's corrugated sheet metal structured packings enjoyed a broad range of success in commercial columns especially in vacuum distillations such as ethylbenzene-styrene monomer separation. Structured packing's greatest advantage is the ability to handle very high vapor rates relative to low liquid rates typical of vacuum conditions. Compared to trays and random packings, the thin turbulent liquid film flows or rivules on the packing surface afforded significantly lower pressure drops, higher capacities and favorable bed separation efficiencies. Growing in confidence and to penetrate the trays market share, the industry initiated deployment of structured packings in light ends high pressure distillation services (C2 to C4 separations). However this proved to be unsuccessful. The encouraging pilot plant results of Spiegel and Meier, 1987[1], and Billet, 1995[2], with structured packings, were not replicated in high pressure industrial large diameter columns. Efficiencies and useful capacities fell well short of expectations. It was speculated that maldistribution, created by back-mixing of the vapor or liquid phases at high pressure, led to the deterioration in performance.

With these discoveries being made, the membership of Fractionation Research Inc. (FRI) approved distillation testing of structured packing on a commercial scale from deep vacuum (0.20 bara) up to high pressure (27.6 bara). The purpose of this paper is to utilize the FRI published Sulzer Mellapak distillation results[3] to identify those physical properties that require serious consideration when selecting structured packing for a specific service. The test results will serve as a reference for the industry in evaluating specific applications and whether they corroborate two discussed mal-performance theories or not.

Fractionation Research Inc. Test Program on Structured Packing

Since 1987, FRI has been actively engaged in the hydraulic-efficiency characterization of structured packings. Sulzer Mellapak 250.Y was employed for distillation testing under total reflux conditions. The test systems employed were para-/ortho-xylene, cyclohexane/n-heptane (C_6/C_7) and iso-/normal-butane (iC_4/nC_4) at pressures ranging from 0.20 bara up to 27.6 bara. With the wide operating pressure range and flow rates, it was necessary to employ two 1.22 m

FRI test columns: vacuum to 11.4 bara in the low-pressure column and 6.9 to 27.6 bara in the high pressure column. In each column, the Mellapak 250.Y was packed to a depth of 3.78 m. A tubed drip pan liquid distributor (Shariat and Kunesh, 1995)[4] with appropriately sized drip tubes was employed for all but the low liquid rate deep vacuum tests. With the latter, a specially designed narrow arm-channel drip tube distributor was employed. A complete description of the tests is provided by Fitz et. al. (1999)[3]. *Figure 1* shows the setup of the low and high pressure columns complete with photographs of the Mellapak 250.Y and liquid distributors employed.

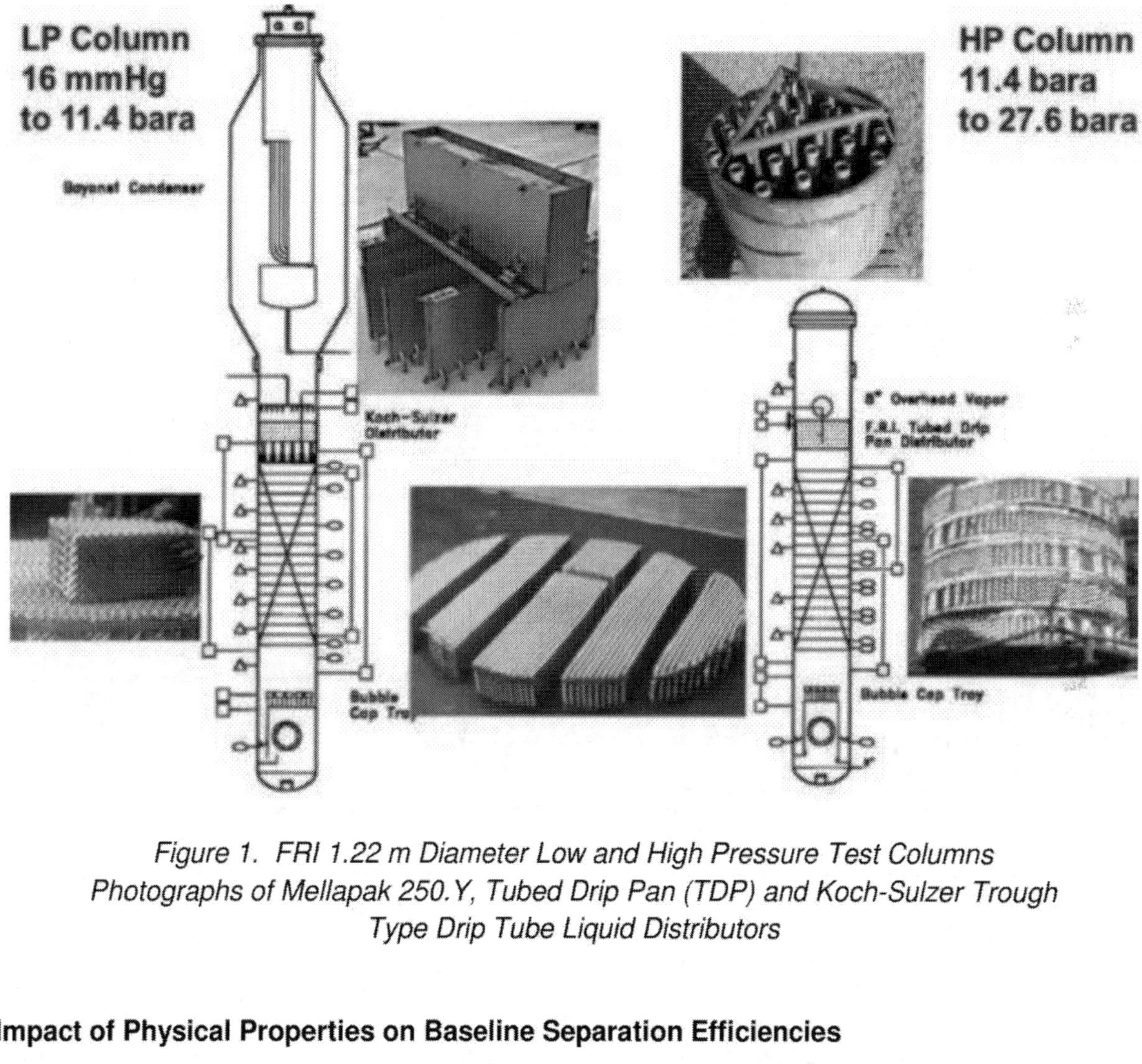

Figure 1. FRI 1.22 m Diameter Low and High Pressure Test Columns
Photographs of Mellapak 250.Y, Tubed Drip Pan (TDP) and Koch-Sulzer Trough Type Drip Tube Liquid Distributors

Impact of Physical Properties on Baseline Separation Efficiencies

Separation efficiencies (HETP's) under total reflux conditions were obtained using the Fenske equation. The number of theoretical stages is calculated first from the more volatile liquid compositions of the binary test system, obtained from in-bed samplers, and average relative volatilities (α_{avg}).

$$\text{No. of Theoretical Stages} = \ln[(x/(1-x))]_{Top} \times ((1-x)/x)_{Btm}]/\ln \alpha_{avg}$$

Thus: $$\text{HETP} = \text{Bed Height / No. of Theoretical Stages}$$

From the measured HETP's, one can determine the baseline HETP by ignoring obvious outliers such as packing under-wetting at the lowest rates and hydraulic flooding at the highest rates. *Figures 2* to *5* show examples of baseline HETP's determined for four out of ten system pressures investigated.

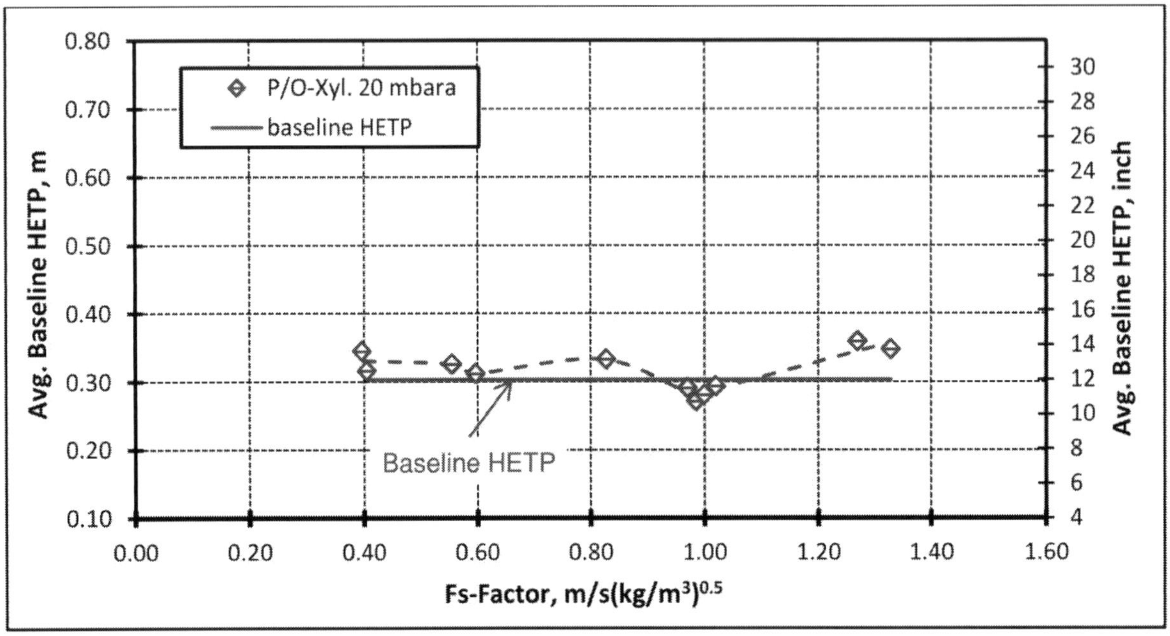

Figure 2. Sulzer Mellapak 250.Y HETP's with p/o-Xylene System at 0.02 bara

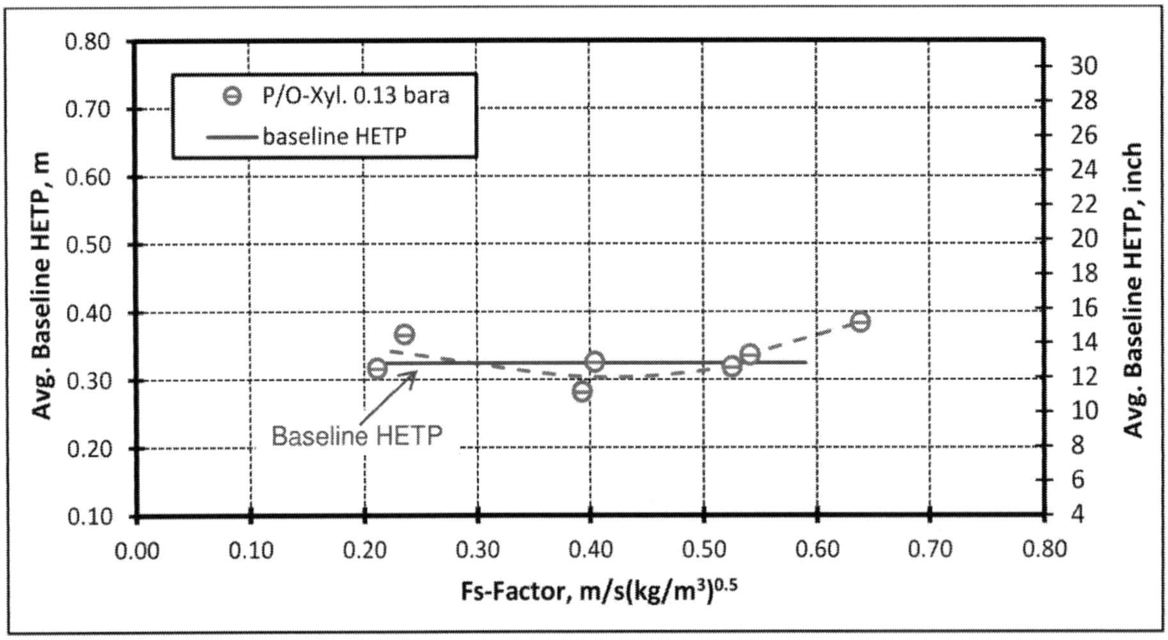

Figure 3. Sulzer Mellapak 250.Y HETP's with p/o-Xylene System at 0.13 bara

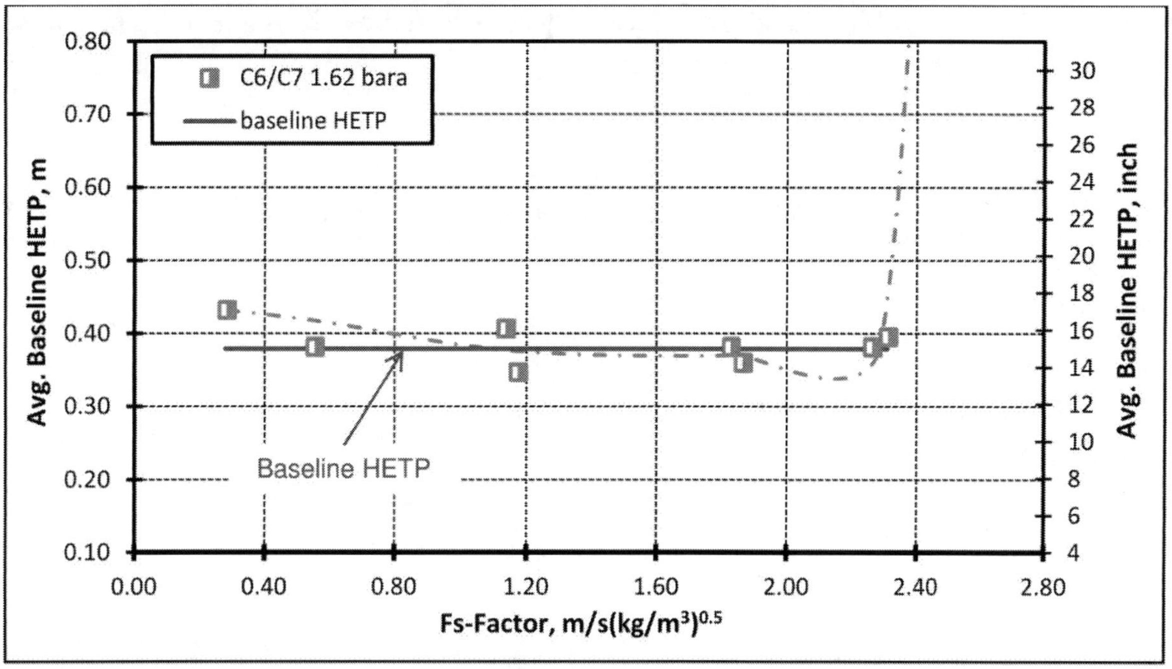

Figure 4. Sulzer Mellapak 250.Y HETP's with Cyclohexane/n-Heptane (C_6/C_7) System at 1.62 bara

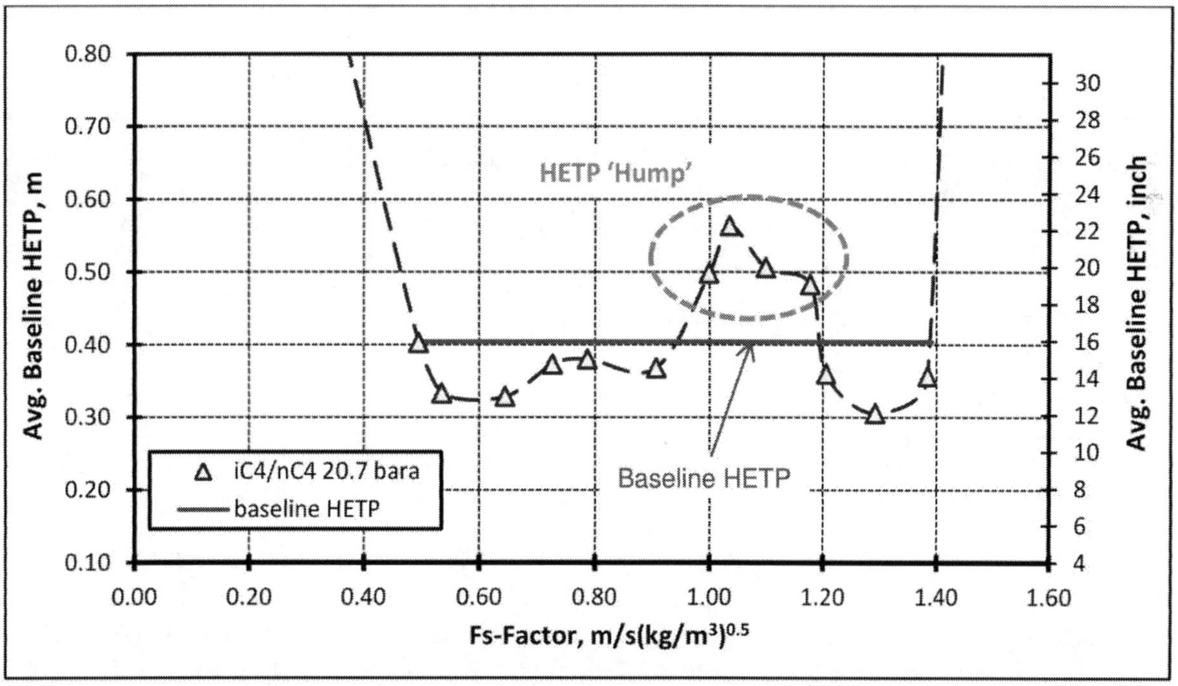

Figure 5. Sulzer Mellapak 250.Y HETP's with iso-/normal butane (iC_4/nC_4) System at 20.7 bara

The average baseline HETP's for all ten pressure system pressures are summarized as follows.

Table 1. Average Baseline HETP's at Ten Operating Pressures

System	Pressure bara	Baseline HETP meters	Comments
P/O-Xylene	0.02	0.303	
P/O-Xylene	0.033	0.301	
P/O-Xylene	0.10	0.324	
C6/C7	0.31	0.387	
P/O-Xylene	1.01	0.360	
C6/C7	1.62	0.379	
iC4/nC4	6.89	0.329	
iC4/nC4	11.4	*0.315	HETP 'Hump' 0.368 m
iC4/nC4	20.7	*0.412	HETP 'Hump' 0.512 m
iC4/nC4	27.6	*0.584	HETP 'Hump' 0.695 m

From those data, one can quantify which physical properties have a significant impact on structured packing performance from deep vacuum up to high pressure. Physical properties were obtained from data tables in two FRI AIChE papers presented in 1995 and 1996[5,6]. For the cyclohexane/n-heptane system, sources of physical property data were obtained from the "Standardization of Structured Packing Efficiency Measurements - A Work in Progress Version 1' article by Zarko Olujic[7] and from FRI published progress reports located in the library of Oklahoma State University.

The average baseline HETP's were plotted against the following:

a) Flow Parameter (FP), $L/V(\rho L/\rho V)^{0.5}$; [For Total reflux, L/V = 1 , thus FP = $(\rho L/\rho V)^{0.5}$]
b) Liquid to Vapor Density ratio
c) Vapor Density
d) Surface Tension
e) Liquid Flux

The resultant graphs are presented in *Figures 6* to *9*. At deep vacuum up to approximately 6.9 bara, the HETP's were approximately 0.30 to 0.38 m. However for pressures greater than 6.9 bara there was a noticeable decline of bed separation performance (higher HETP's) in the loading regime at approximately 60 to 90% of the flooding vapor rates. This became known as the efficiency 'hump' and is circled red in each of the figures. The actual measured 'hump' HETP's at the three highest pressures with the butanes system are summarized in *Figure 10.*

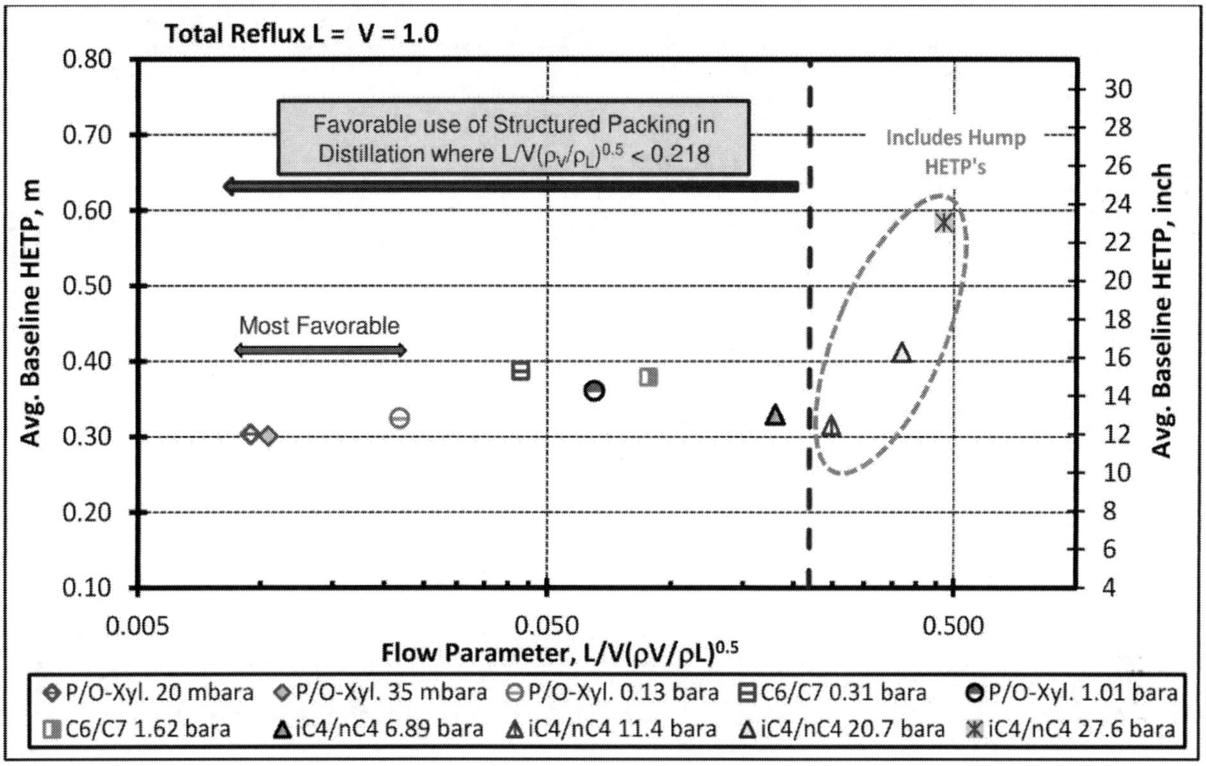

Figure 6. Influence of Flow Parameter Sulzer Mellapak 250.Y Baseline HETP's

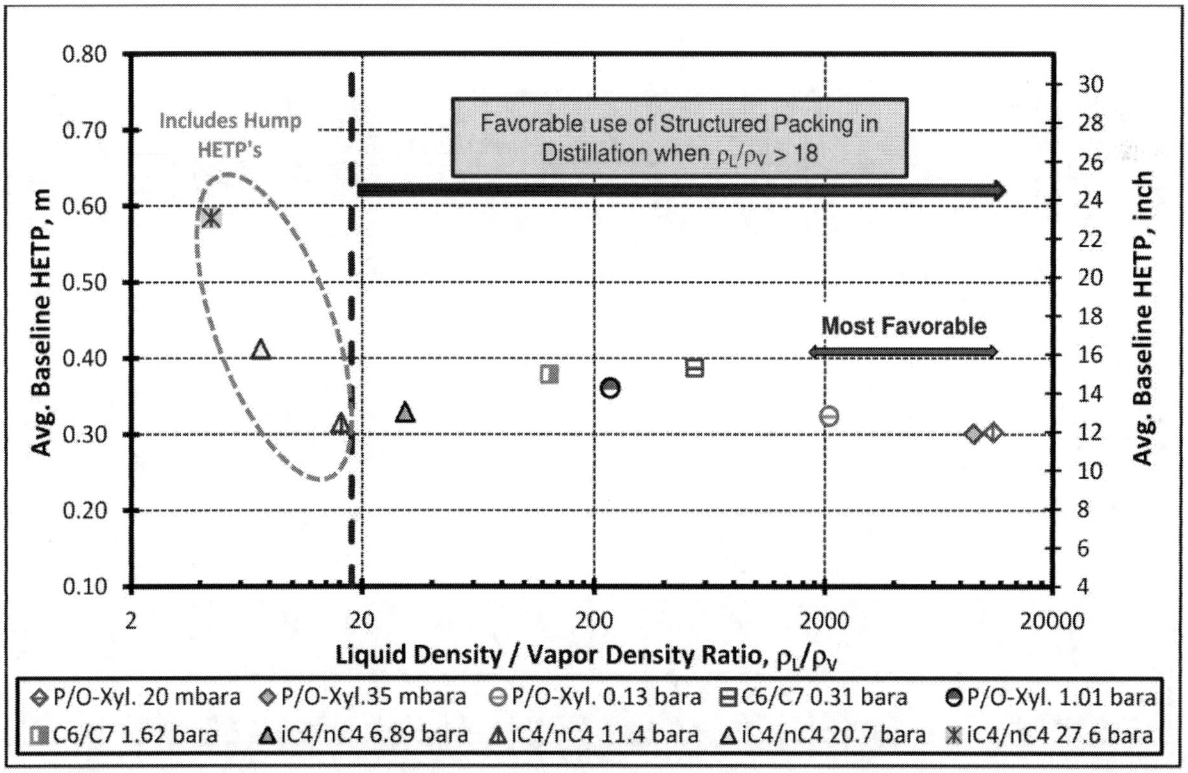

Figure 7. Influence of Phase Density Ratio on Sulzer Mellapak 250.Y Baseline HETP's

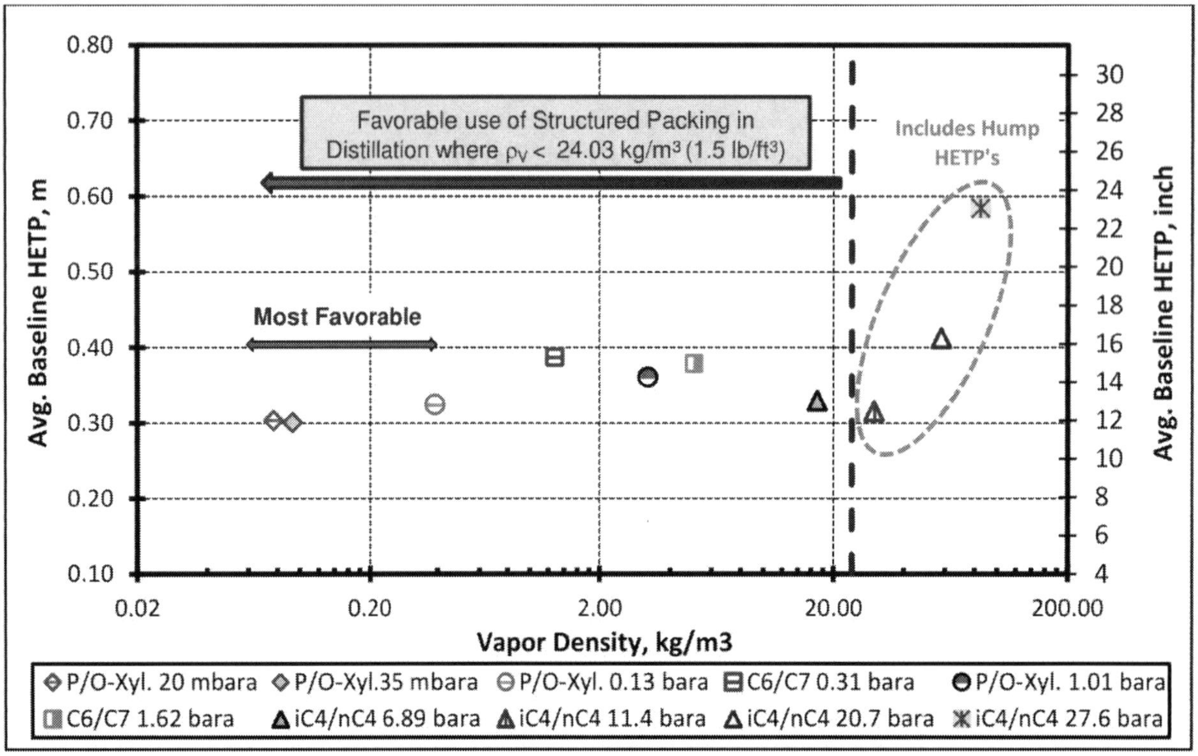

Figure 8. Influence of Vapor Density on Sulzer Mellapak 250.Y Baseline HETP's

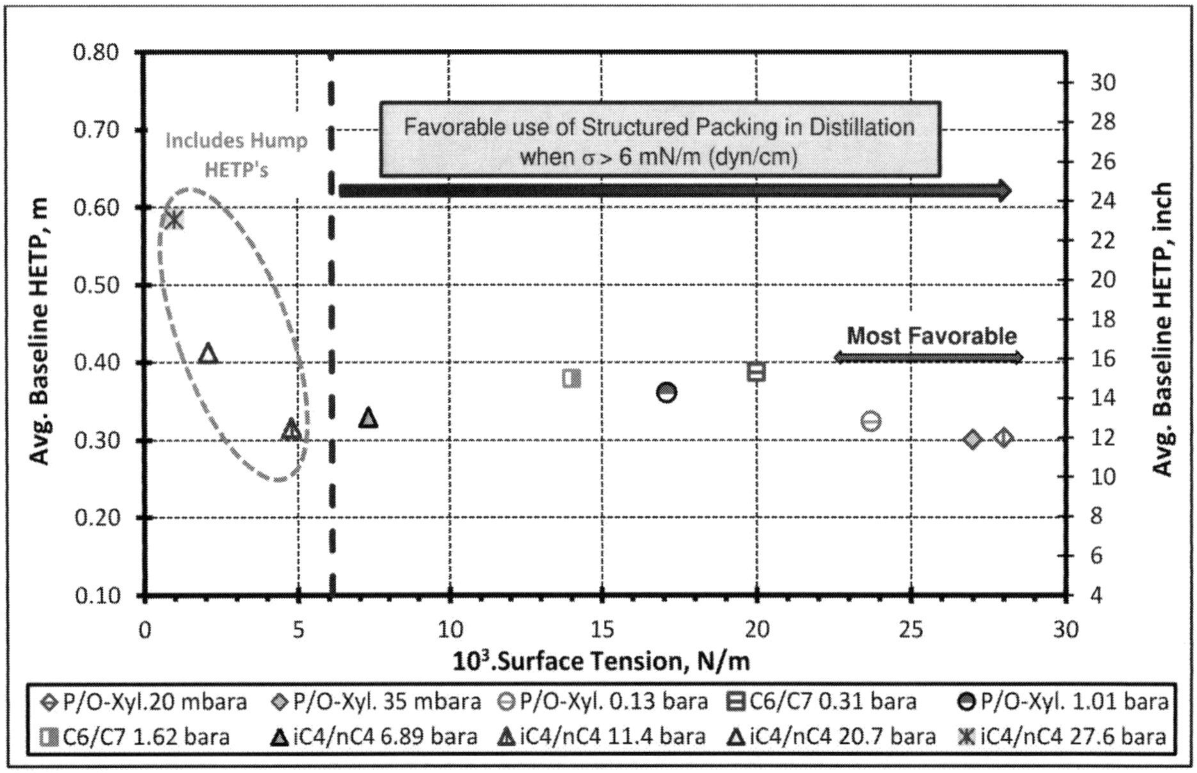

Figure 9. Influence of Surface Tension on Sulzer Mellapak 250.Y Baseline HETP's

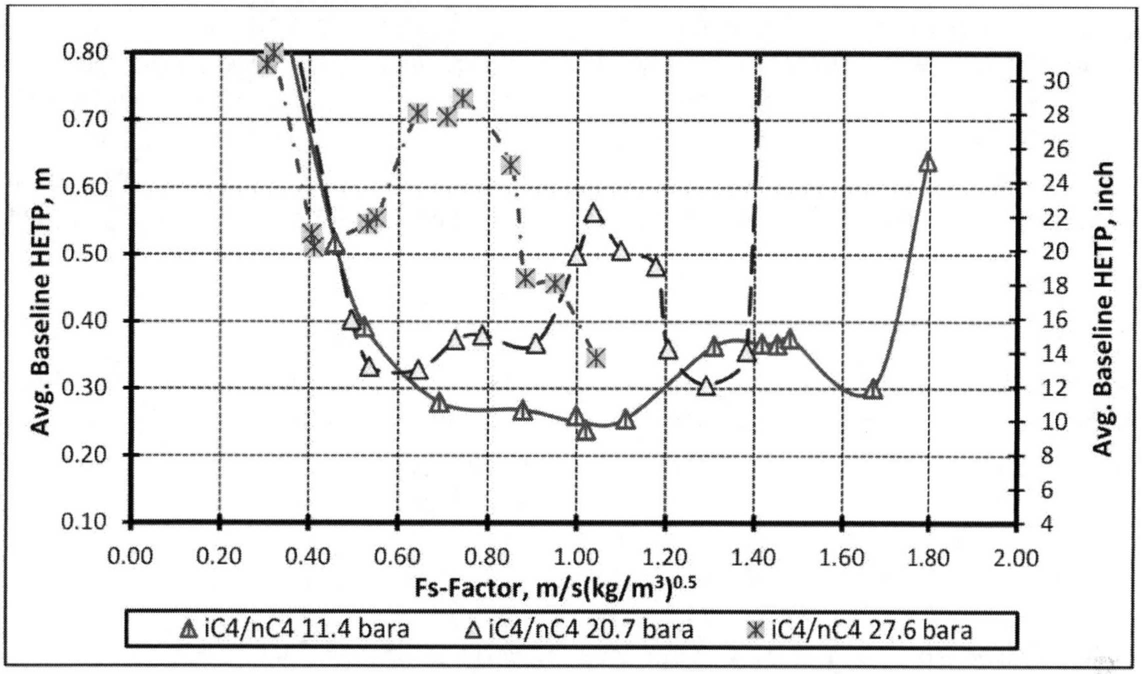

Figure 10. Sulzer Mellapak 250.Y HETP's with the iC4/nC4 System at High Pressures

Taking *Figures 6* to *9* together, a 250 m^2/m^3 specific area structured packing is more desirable in applications where the flow parameter is less than 0.218, the liquid-to-vapor density ratio is less than 18, the vapor density is less than 24.03 kg/m^3, and the surface tension is greater than 0.006 N/m.

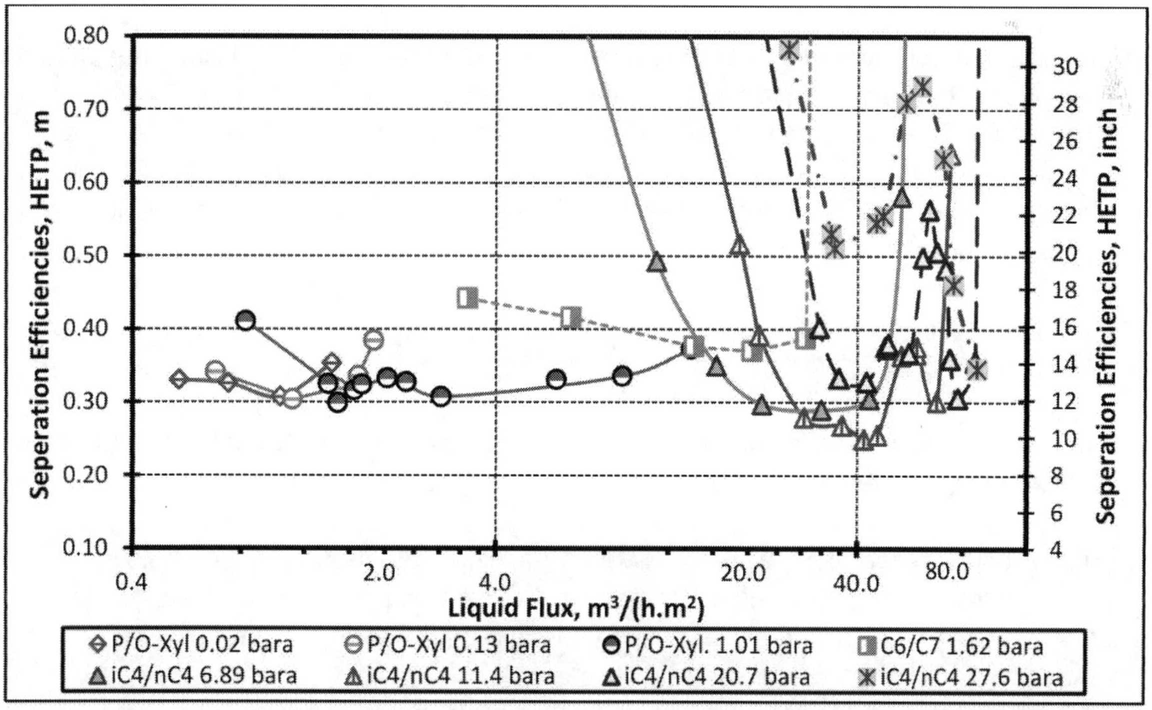

Figure 11. Influence of Liquid Flux on Mellapak 250.Y HETP's at Low to High Pressures

A further characteristic associated with high pressure distillation is high liquid loads. A high liquid flux in association with a low density difference and very low surface tension, leads to a fluid flow regime that deviates markedly from the preferred film flow. The influence of liquid flux on the Mellapak 250.Y HETP's from vacuum up to high operating pressures is presented in *Figure 11*. From those data a general heuristic is that common size structured packings with 250 m^2/m^3 specific area, tend to perform best when the liquid flux is less than 30 to 50 $m^3/m^2/h$.

Discussion of Results

Since the FRI binary test mixtures are close boiling, low relative volatility systems, the physical properties resemble the commercial application light hydrocarbon de-ethanizers, de-propanizers and de-butanizers. As a result, it was FRI's data that was the foundation of industrial guidelines as to which applications are most suitable for structured packings. To support those guidelines, hypotheses to explain potential reasons for bed performance deterioration (HETP 'hump') were put forward in the 1990's. For instance Zuiderweg(8), in 1997, postulated liquid-back-mixing as the reason for elevated HETP's at high pressure. In addition, large scale vapor back-mixing was cited by Nooijen et. al. (1997)(9) as the cause of bed performance deterioration.

Vapor Back-mixing Theory

Following the publication of the FRI Mellapak 250.Y structured packing data in (1999), Andrew Sloley of The Distillation Group consultancy set out criteria where structured packing should be avoided in light hydrocarbon distillations(10). This was in response to an Asian Refiners query and is illustrated in *Figure 12.* Vapor/liquid back-mixing theory (Zuiderweg(8), Kister(11)), can be summarized as follows:

a) Ideal structured packing behavior is the handling of thin flowing liquid films that promotes optimal interfacial area and the necessary mass transfer driving forces.
b) High pressures are associated with high liquid loads, low density difference and very low surface tensions that tend to cause a shift in fluid flow dynamics away from film flow.
c) Vapor pushes back against the direction of liquid flow, thus transforming thin films into thick liquid streams since the influence of packing surface texture for liquid spreading becomes negligible.
d) Interfacial area for mass transfer is diminished and corrugated channels 'choke' with liquid thus reducing efficient vapor capacity. High vapor velocities through a reduced packing cross-sectional area increases the likelihood of vapor back-mixing.
e) Due to minimized axial/radial mixing, vapor back-mixing extends over the height of a packing element leading to a pronounced bed performance deterioration

There are of course exceptions where structured packing was success fully employed at high pressures in which the liquid density is independent of system pressure and remains high at all pressures (10). Examples include glycol dehydration and amine sour gas scrubbing.

Figure 13 is an illustration of vapor back-mixing within the structured packing corrugations.

Question

Can Structured Packing be used in Light Ends Distillation up to 30 barg?

Answer

For light hydrocarbons avoid Structured Packing when:

- P ≥ 11.4 bara
- $\rho_V \geq 24$ kg/m³
- $\rho_L/\rho_{V\ ratio} \leq 18$

Exceptions - SP proven where:

- ρ_L independent System Pressure
- ρ_L high regardless of pressure e.g. Glycol dehydration, Sour Gas Scrubber

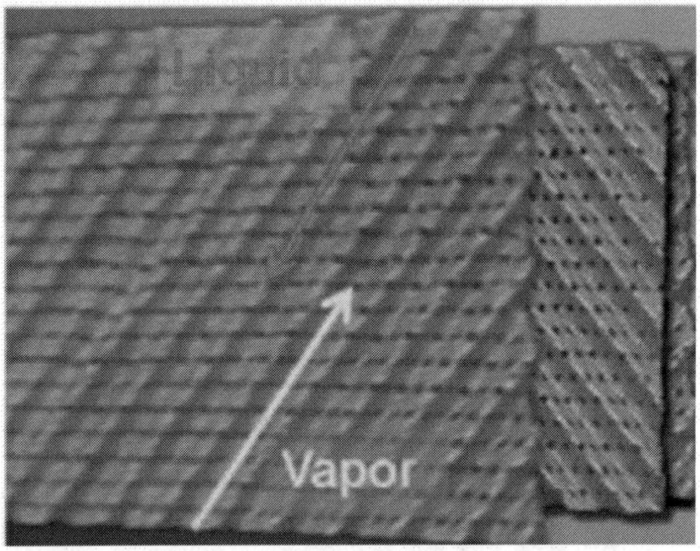

Figure 12. The Distillation Group Criteria for Structured Packings in High Pressure Distillation

Figure 13. Illustration of Vapor Back-Mixing in Structured Packing Corrugated Channels

Liquid Drops Back-mixing Theory

At high liquid rates, structured packings with surface areas approximately 250 m^2/m^3 have insufficient installed surface area to handle easily detachable low-surface-tension hydrocarbon fluid. Under such conditions the partially 'free' liquid chokes a proportion of the corrugation channels thus forcing vapor to ascend through less loaded channels. It is in these channels where liquid drop back-mixing is likely to occur. To support this hypothesis, the work of Andreas Kerst[12] was adopted in which the behavior of free jets and falling films at different pressures was studied. A customized photographic technique was employed to observe the flow regime behavior at different pressures and liquid rates. In general, for a constant liquid mass rate, a smooth liquid film undergoes a flow regime transition to fine drops on progressing from low to high pressures that approach a critical value. This is illustrated in Figure 14.

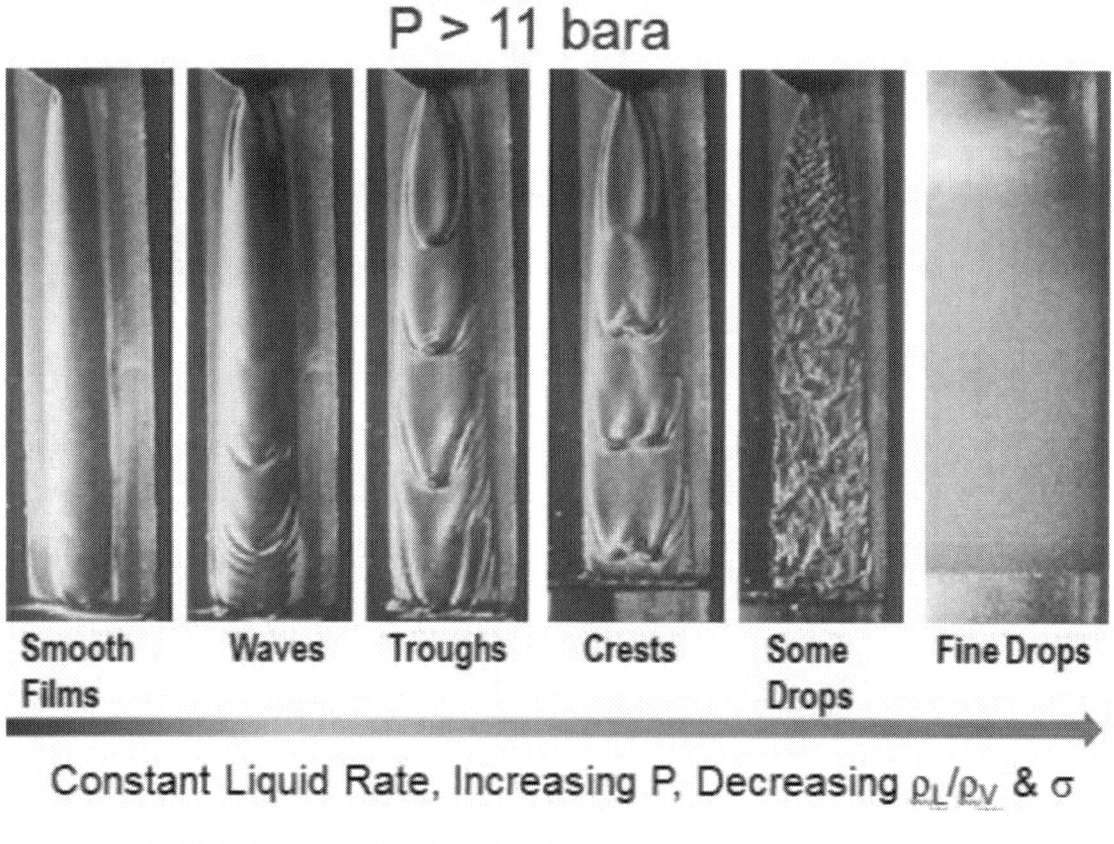

Figure 14. Digital Image of Falling Films Flow Transition from Low to High Pressures Reproduced from Andreas W. Kerst: Flow regimes of free jets and falling films at high ambient temperature, Chem. Eng. Science 55 (2000)

Since structured packing can be considered as a falling film device, the preferred driving force for mass transfer is the total 'occupation' of the packing surface with thin liquid films to enhance interfacial area for mass transfer kinetics. Based on the work of Kerst, the smooth liquid films undergo transitions from waves, troughs, crests, coarse drops to fine drops as pressure is increased. When applied to structured packing, liquid is sheared off from the packing surface into drops as the pressure and physical properties (e.g. surface tension, liquid/vapor densities) approach critical values. In effect this reduces significantly the influence of packing surface in promoting interfacial area that ultimately leads to a decline in bed separation efficiency. The net result is back-mixing of liquid drops that potentially extend the height of the structured packing element. *Figure 15* is an illustration of liquid drop back-mixing.

Figure 15. Illustration of Liquid Drop Back-Mixing in Structured Packing Corrugated Channels

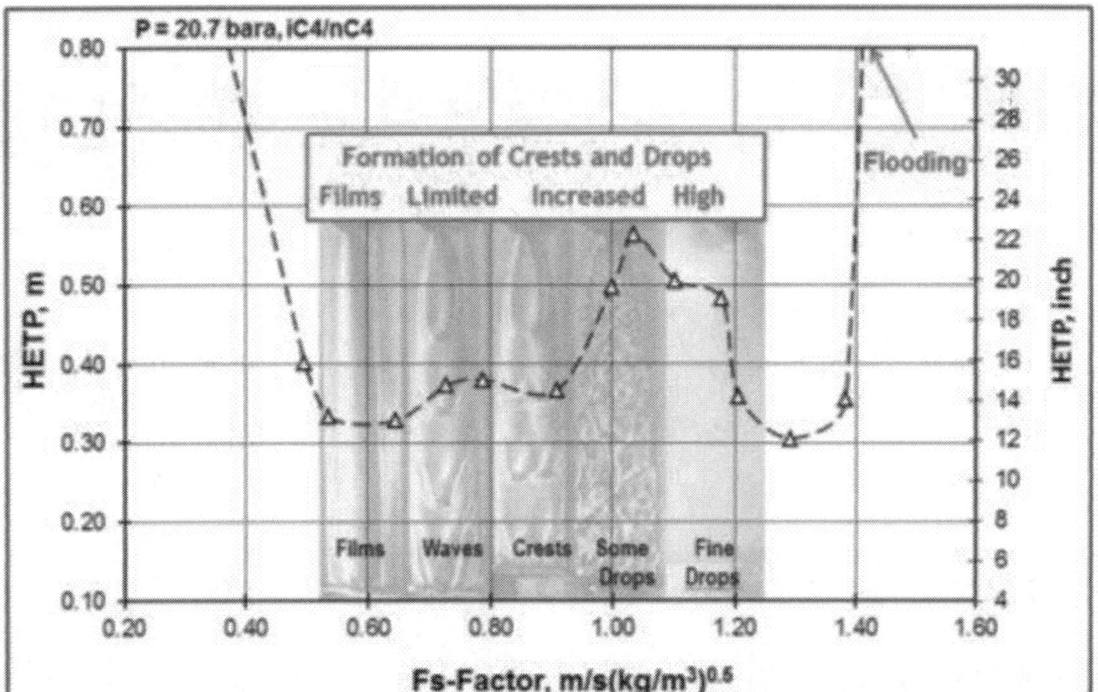

Figure 16. Impact of Flow Regime Transitions from Films to Drops on Mellapak 250.Y HETP's

If flow regime transitions from smooth films to crests to drops are applied to FRI Mellapak 250.Y test data at high pressure (greater than 11 bara), *Figure 16* is an example of corroborating those flow regimes with bed separation performance. That is, in the preloading regime, smooth liquid film flows result in low HETP's (high efficiency). On progressing into the loading regime, the formation of crests and drops increases. The formation of a complete liquid drops regime and the resultant liquid drop back-mixing lead to the HETP hump at approx.. 60 to 90% of the flooding vapor rate.

Observation of Flow Regimes from FRI Windows

One of FRI's greatest assets is the ability to film flow behavior of trays, packings and liquid distributors with hydrocarbon systems in a commercial scale column. *Figure 17 (a)* and *(b)* are screen shots from FRI Film No. 509[13] that show change in liquid flow behavior with the Tubed Drip Pan (TDP) liquid distributor from low to high pressure during the Mellapak 250.Y tests. In *Figure 17 (a)* the liquid flow from the distributor at low pressure is in the form of clear discrete streams. At higher pressure, *Figure 17 (b)*, liquid discharges as cones of mist and fine drops. Since no filming was done of the structured packing itself, 50 mm Pall Rings were filmed in 1964 (FRI Film No. 10)[14] at high pressures up to 34.5 bara. Figure 17 (c) shows swarms of tiny drops in and around the Pall Rings at this pressure. One can speculate that similar flow behavior occurred within the Mellapak 250.Y when tested at high pressure.

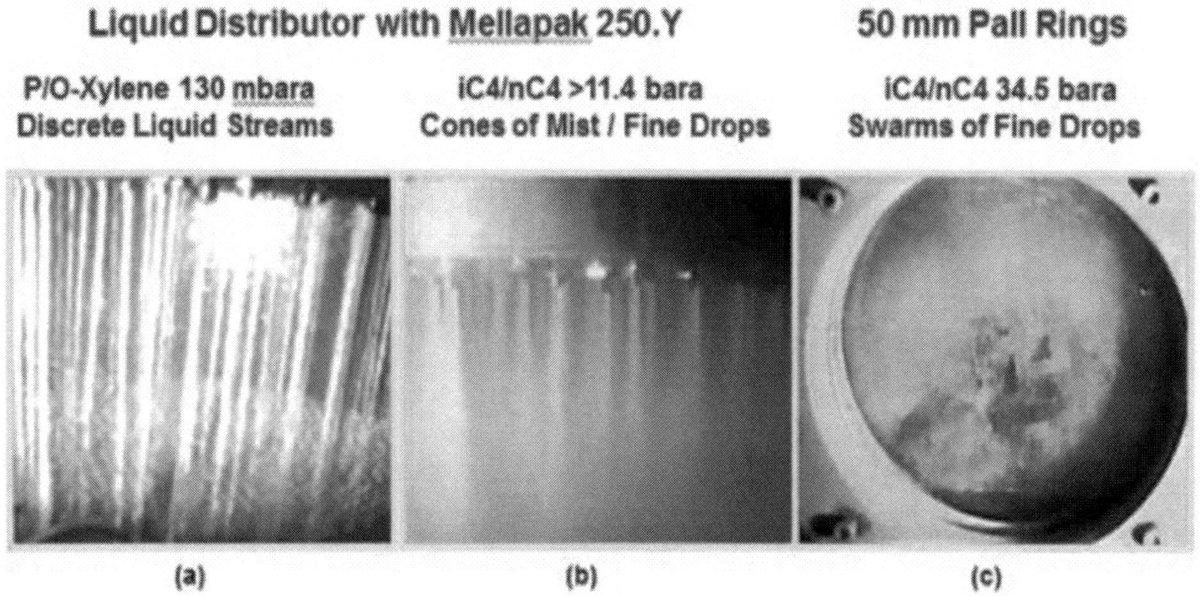

Figure 17 Flow Regime Behavior from Low to High Pressure from FRI Films
Parts (a) and (b) are from FRI Film 509 and Part (c) is from FRI Film 10

Concluding Remarks

In the 1980's and 90's many light end hydrocarbon columns were revamped from conventional trays to structured packing owing to the latter having significant pressure drop and capacity advantages. However spectacular failures were reported owing to capacities and efficiencies falling far below expectations. Dedicated high-pressure distillation testing by FRI with a 250 m^2/m^3 corrugated sheet metal structured packing revealed the existence of a strong deterioration in separation efficiency in the loading region known as the efficiency hump. The findings in this paper showed that a combination of high flow parameters (high pressure), high liquid flux, and physical properties such liquid density, vapor density and surface tension approaching their critical values all contributed to the HETP hump and hence less efficiency.

Vapor back-mixing or liquid drop back-mixing or a combination of both lead to the packing becoming less effective in radial spreading of both the vapor and liquid. Back-mixing leads to a decrease in the two-phase effective contacting area that ultimately results in an appreciable loss of bed separation efficiency.

While high capacity trays have enjoyed much success with light hydrocarbon distillations over the last 20 years, it remains unclear whether high capacity structured packings (Mellapakplus™ or Flexipac® HC® for instance) can solve the problems with light ends hydrocarbon high pressure distillation. Since the efficiency hump is still not fully understood, and there those in the industry who still wish to employ structured packings (e.g. light ends distillation off-shore) further experimental work or CFD would be necessary.

Nomenclature

C_6/C_7	Cyclohexane/n-Heptane	[-]
FP	Flow Parameter	[-]
HETP	Height Equivalent to a Theoretical Plate	[m]
iC_4/nC_4	iso/normal butane	[-]
ln	logarithm	
L	Liquid	[kg/hr]
L_{Flux}	Liquid Flux	[$m^3/(h.m^2)$]
p/o-Xylene	para/ortho Xylene	
TDP	Tubed Drip Pan liquid distributor	
V	Vapor	[kg/hr]
x	Liquid composition	[-]
α	Relative volatility	[-]
ρ_L	Liquid density	[kg/m^3]
ρ_V	Vapor density	[kg/m^3]
σ	Surface Tension	[N/m]

References

(1) Spiegel, L. and Meier, W., Correlations of the Performance Characteristics of the Various Mellapak Types, Inst. Chem. Eng. Symp. Der. **1987**, 104, A203

(2) Billet, R. *Packed Towers in Processing and Environmental Technology*, VCH Verlagsgesellschaft: Veinheim, Germany, 1995

(3) Fitz, C. W., Kunesh, J. G. and Shariat, A., Performance of Structured Packings in a Commercial-Scale Column at Pressure 0.02-27.6 bar, Ind. Eng. Chem. Res. 38 (**1999**), 512-518

(4) Shariat, A. and Kunesh, J. G., Packing Efficiency Testing on a Commercial Scale with Good (and Not So Good) Reflux Distribution, Ind. Eng. Chem. Res. (**1995**), 34, 1273.

(5) Fitz, C. W., Shariat, A. and Spiegel, L., *Performance of Structured Packing at High Pressure*, Distillation Column Design and Operation – I, AIChE Spring Meeting, Paper 25e, Houston, TX, March **1995**.

(6) Fitz, C. W. and Kunesh, J. G., *Performance of Structured Packing at Low Liquid Rates*, Modeling of Distillation Column Hydraulics and Mass Transfer – I, AIChE Spring Meeting, Paper 187f, Chicago IL, November **1996**.

(7) Olujic, Z., *Standardization of Structured Packing Efficiency Measurements - A Work in Progress Version 1*, Delft University of Technology, 2628 CA Delft, The Netherlands, Encouraged and Supported by BASF SE, March **2008**.

(8) Zuiderweg, F. J., Olujic, Z.; Kunesh, J. G., *Liquid Backmixing in Structured Packing in High-Pressure Distillation*, Inst. Chem. Eng. Symp. Der. **1997**, 142, 865.

(9) Nooijen, J. L., Kusters, K. A.; Pek, J. J. B., *The Performance of Packing in High-Pressure Distillation Applications*, Inst. Chem. Eng. Symp. Der. **1997**, 142, 885.

(10) Sloley, A., Structured packing applications: Structured packing in high pressure services, The Distillation Group, January **2001.**

(11) Kister, H.Z., *Distillation Design: Packing Design and Operation*, New York: McGraw-Hill, **1992**, Page 459

(12) Kerst, A. W., Judat, B.: Schlünder, E. U., *Flow Regimes of Free Jets and Falling Films at High Ambient Temperature*, Chem. Eng. Science 55 (**2000**), 4189-4208.

(13) FRI Film No. 509, *FRI Tests of Structured Packing*, Film Line Productions, Hollywood, CA, MCMLXXXIX, FRI, Inc.

(14) FRI Film No. 10, *Vapor-Liquid Interaction of iC4-nC4 System at 165 and 500 psia (11.4 and 34.5 bara) with 2 inch (50 mm) Diameter Carbon Steel Pall Rings*, Progress Report for December 1964, FRI, Inc.

A STUDY OF AXIAL MIXING INFLUENCE ON THE MASS TRANSFER TRANSFERABILITY FROM ABSORPTION TO DISTILLATION CONDITIONS

Lukas Valenz, Frantisek J. Rejl[1], Jan Haidl
University of Chemistry and Technology, Prague, Czech Republic

Introduction

Distillation and absorption are treated as analogical mass-transfer processes between liquid and gas phase on the packing. The analogy between processes assumes that the same set of models can be used for description of both processes, while there exists a clear relation between their parameters. While description of both processes by complete set of the models is theoretically possible, practically it faces insurmountable difficulties due to lack of the model parameters and computational complexity. Therefore deep simplification is often introduced and reflects the specifics of the individual process. Among these simplifications belongs description of the phase hydrodynamics by a plug flow model which doesn't need any parameter for its utilization.

In our previous contribution at AIChE Spring meeting 2014 we have presented the HETP data of RSP250Y packing recalculated from the k_La and k_Ga data obtained by standard absorption experiments using Billet-Schultes model. The *HETPs* calculated out of the simulations were barely half of the experimental ones when the plug-flow hydrodynamic model was used. On the other hand, when axial dispersion in the liquid phase was taken into account, the out of simulation calculated *HETP*s matched the experimental ones with only about 16 % and 9 % error (for C_6/C_7 and MeOH/PrOH system, respectively). The acquisition of the absorption mass-transfer data and their recalculation from absorption to the distillation conditions seemed to be therefore successful in this particular case, but incorporation of the more sophisticated hydrodynamic model was clearly inevitable.

In this paper we present result of similar procedure for larger spectra of packings and systems and also present axial mixing data of the packings evaluated from the distillation experiments. Analysis of the distillation data revealed also relative impact of the individual phenomena (mass-transfer resistance in the liquid/gas phase, axial mixing of the liquid phase) on the separation efficiency.

Methods

Mass-transfer coefficients and effective interfacial area of the packings under absorption conditions has been measured using standard aqueous systems (O_2/water – k_La, SO_2/aq. NaOH - k_Ga, CO_2/aq. NaOH - a_{EFF}) for various flows of the gas and liquid phase. These data has been used for calculation of the HETP; the Billet-Schultes correlation was used for recalculation of the data to the distillation conditions. The axial mixing in the liquid phase has been measured using dynamic experiments involving NaCl tracer with the air/water system. The axial mixing in the gas phase has been measured

[1] presenting and corresponding author

using helium tracer with the same system and was found negligible with respect to that in the liquid phase. Experiments have been performed in the i.d. 0.295m column.

Distillation experiments have been performed with MeOH/EtOH and C_6/C_7 systems in the atmospheric distillation column of i.d. 0.15m under total reflux. *HETP* has been evaluated out of the composition in the head and in the bottom. Molar fraction profiles in the liquid and in the gas phase have been measured and were used for simulation of the column behavior out of which k_La, k_Ga and axial mixing (in terms of Bo_L) have been evaluated.

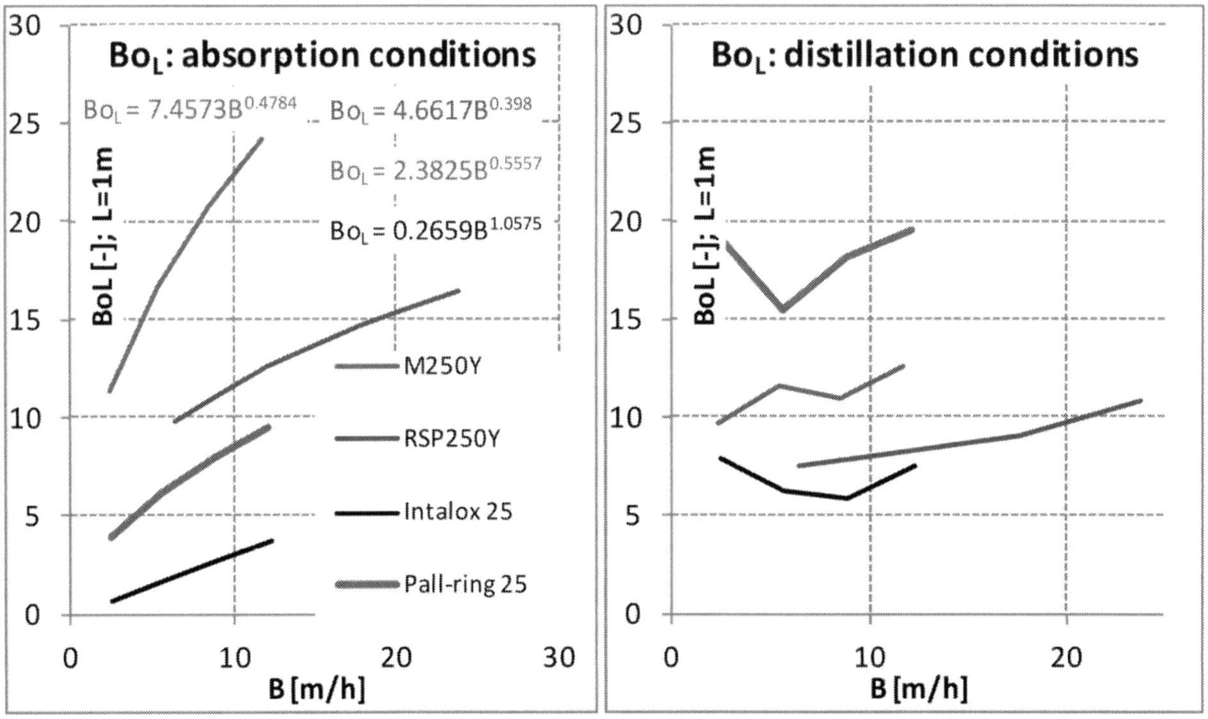

Fig. 1A,1B: axial mixing AIR/WATER system (1A), axial mixing DISTILLATION (1B)

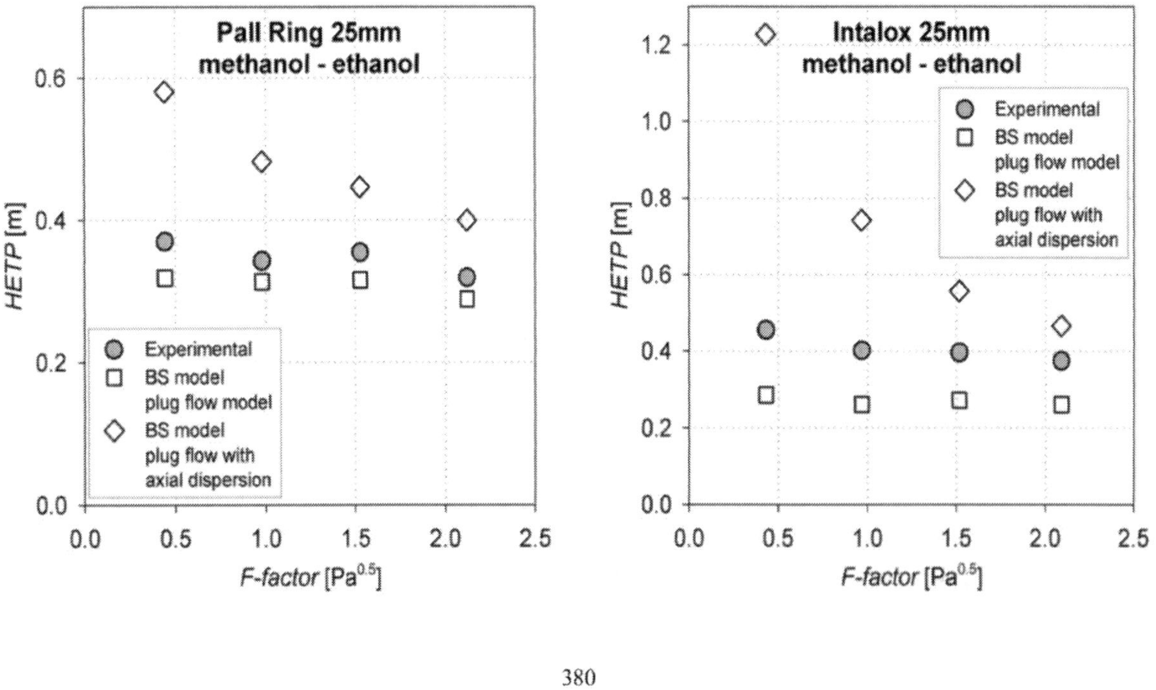

Fig. 2A

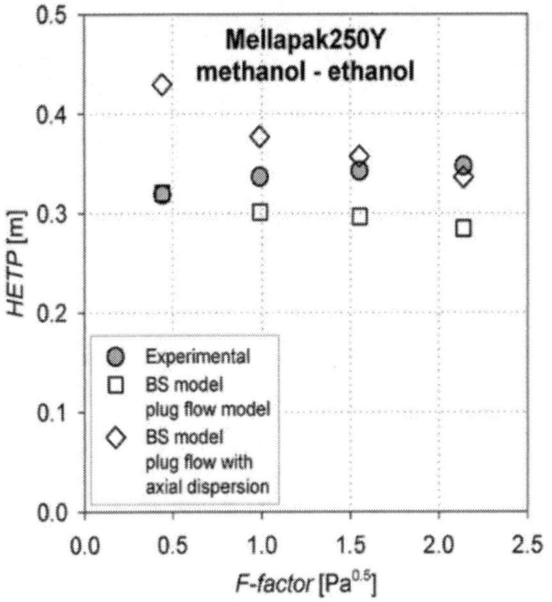

Fig. 2B

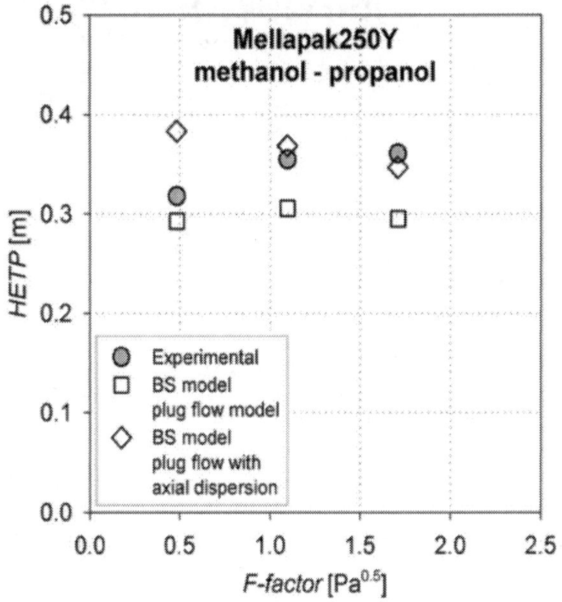

Fig. 2C

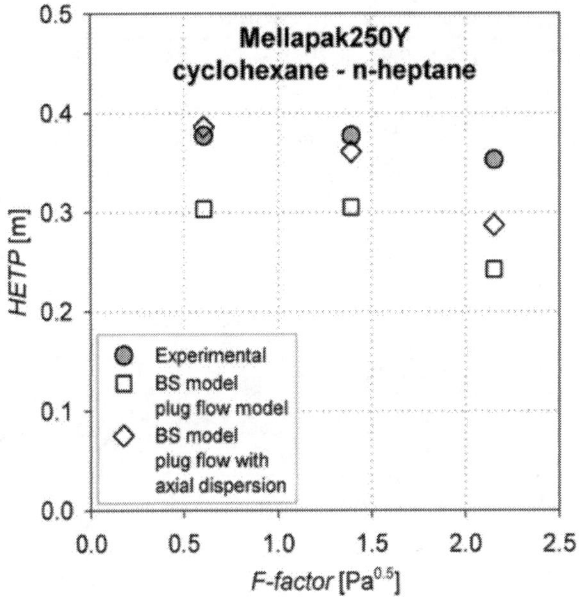

Fig. 2D

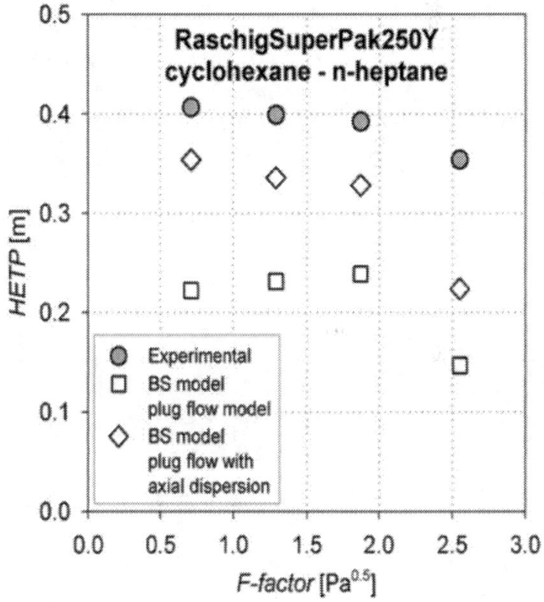

Fig. 2E

Fig. 2F

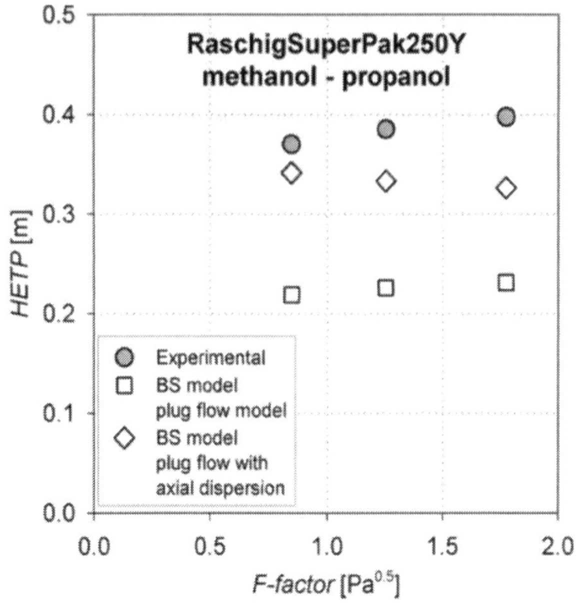

Fig. 2G

			RELATIVE EFFECT OF INDIVIDUAL PHENOMENA ON SEPARATION EFFICIENCY (UNDER DISTILLATION CONDITIONS)		
PACKING	SYSTEM	LOAD	MASS TRANSFER IN THE LIQUID PHASE	MASS TRANSFER IN THE GAS PHASE	AXIAL MIXING OF THE LIQUID PHASE
M250Y	MeOH/EtOH	10kW	18%	50%	32%
M250Y	MeOH/EtOH	40kW	16%	63%	21%
M250Y	C_6/C_7	8kW	18%	67%	15%
M250Y	C_6/C_7	24kW	26%	62%	13%
Pall 25	MeOH/EtOH	10kW	15%	72%	13%
Pall 25	MeOH/EtOH	40kW	23%	62%	15%
Intalox-25	MeOH/EtOH	10kW	13%	64%	23%
Intalox-25	MeOH/EtOH	40kW	8%	64%	28%
RSP250Y	C_6/C_7	9kW	18%	58%	24%
RSP250Y	C_6/C_7	28kW	24%	53%	24%

Tab. 1: Relative effect of individual phenomena on separation efficiency (distillation conditions)

Results

On the basis of the experiments performed in the absorption column the axial mixing is found to occur in the largest scale for the Intalox-25 packing (low Bo_L values), to be of medium scale for Pall-ring 25 and Raschig SuperPak 250Y and the smallest was found for Mellapak 250Y (high Bo_L values) (see Fig 1A). Generally the axial mixing was found to be higher for random packings than for structured packing and the values of Bo_L were found to increase with the flow of the liquid phase (with the power of 0,4-1,06).

HETP calculated out of the absorption k_La, k_Ga is for all packings and systems too optimistic (too low) with respect to the experimental ones (see Fig 2A-2G). The HETP calculated using air/water axial mixing data (recalculated to the distillation conditions with Fairs correlation) overestimates the distillation *HETP* data significantly for random packings while for the RSP250Y introduction of the axial mixing improves the agreement. For the M250Y packing the results are inconclusive (see Fig 2A-2G).

The axial mixing data evaluated from the distillation experiments (see Fig 1B) confirm the highest degree of axial mixing for the Intalox-25 packing. However, severe mixing is also found for Raschig SuperPak 250Y and axial mixing on the Mellapak 250Y seems to be much more intense under distillation conditions than for the air/water system. On the other hand the mixing on Pall Ring 25 is found lower than for the air/water system and even lower than that for Mellapak 250Y. The strong dependence of the Bo_L on the liquid phase flow vanishes under distillation conditions.

Assessment of the relative importance of the individual phenomena not surprisingly attributes major influence to the gas-side mass-transfer resistance (see tab. 1), which forms at least 50% of the "resistance against the separation" for all packings and systems. However, while for the Pall rings and for MeOH/PrOH distillation on M250Y the second most important factor is the liquid-side mass-transfer resistance, for other cases it is the axial mixing of the liquid phase.

Conclusions

Present modeling of the absorption and distillation process often assumes plug flow of the phases in the column. This assumption was found unjust, as the axial mixing forms 13-32% of the overall resistance against separation and is in fact one of the most important phenomenon which occurs in these processes. Relatively poor *HETP* results of the Intalox-25 are not caused by some extremely low values of k_La and k_Ga, but by severe axial mixing in the liquid phase on this packing. This finding should be considered also from the point of view of a new packing design.

Usage of the plug-flow model for evaluation of the k_La, k_Ga data out of standard absorption experiments leads to involution of the mixing effect in their values. The same holds for the distillation experiments, but the effect unfortunately doesn't cancel. The effect of axial mixing on the evaluated k_La, k_Ga values is different for both processes as: 1) the axial mixing in terms of the Bo_L number might differ considerably under absorption and distillation conditions 2) the effect of axial mixing on the overall separation efficiency is strongly dependent on the number of transfer units – in which the standard absorption and distillation experiments usually differ also.

The axial mixing in the liquid phase doesn't have to be only phenomenon whose neglect causes difficulties in common description of the absorption and distillation process. It was shown, however, that without taking it into account the successful transfer of the data from one process to another is achieved only by chance.

Acknowledgement

Authors are grateful to Raschig the Czech Science Foundation for supporting the project through Grant 13-01251S, and to the financial support from specific university research (MSMT No 20/2015).

References

1. Billet, R. and Schultes, M. (1999), „Prediction of Mass Transfer Columns with Dumped and Arranged Packings. Updates Summary of the Calculation Method of Billet and Schultes.", in *Trans IChemE*, 77 (Part A), p. 498-504.

Symbols

a_{EFF}	effective interfacial area of the column, $m^2 \cdot m^{-3}$
k_Ga	volumetric gas phase mass-transfer coefficient, s^{-1}

$k_L a$	volumetric liquid phase mass-transfer coefficient, s^{-1}
HETP	height equivalent to the theoretical plate, m

Subscripts

G	gas phase
L	liquid phase

Tracerco
Providing Insight Onsite

Monitor Fouling in Distillation Towers with Gamma Scanning

A.I.Ch.E. 2015 Spring Meeting Distillation Troubleshooting and Lessons Learned April 29, 2015

FOULING

Tracerco
Providing Insight Onsite

- "Nine hundred (900) case histories of malfunctioning towers reported over the last 50 years were surveyed and analyzed."
- "Plugging, especially of tray active areas, packing and distributors tops the malfunctions list."
- "The number of plugging/coking incidents reported over the last decade is higher than that reported over the previous four decades. Plugging/coking is here to stay, and most likely will continue to top the tower malfunction list."

Reference: **Kister, H.Z., "What Caused Tower Malfunctions in the Last 50 Years?"** ***Trans. IChemE*****, Vol.81, Part A, Jan 2003**

A.I.Ch.E. 2015 Spring Meeting Distillation Troubleshooting and Lessons Learned April 29, 2015 Process Diagnostics

CAUSES OF FOULING

Tracerco
Providing Insight Onsite

- Coking (Refineries)
- Scale, corrosion products
- Precipitation, salting
- Solids in feed
- Polymer formation

Reference: **Kister, H.Z., "What Caused Tower Malfunctions in the Last 50 Years?"** ***Trans. IChemE*****, Vol.81, Part A, Jan 2003**

A.I.Ch.E. 2015 Spring Meeting Distillation Troubleshooting and Lessons Learned April 29, 2015 Process Diagnostics

FOULING, WHAT TO DO?

Tracerco
Providing Insight Onsite

- "The number of plugging/coking incidents reported over the last decade is higher than that reported over the previous four decades. Plugging/coking is here to stay, and most likely will continue to top the tower malfunction list."

Reference: **Kister, H.Z., "What Caused Tower Malfunctions in the Last 50 Years?"** ***Trans. IChemE*****, Vol.81, Part A, Jan 2003**

A.I.Ch.E. 2015 Spring Meeting Distillation Troubleshooting and Lessons Learned April 29, 2015 Process Diagnostics

FOULING, WHAT TO DO?

Tracerco
Providing Insight Onsite

Two Choices:

- "Go Along" or Passive – continue to operate and react when a fouling problem causes an operational problem.
- Proactive – monitor for presence of fouling and monitor the progress of fouling in order to anticipate and plan ahead for operational problems.

MONITORING FOR FOULING

Tracerco
Providing Insight Onsite

- Monitor Process Variables - ΔP and others
- Product Composition
- Operating History
- Periodic Gamma Scanning

A.I.Ch.E. 2015 Spring Meeting Distillation Troubleshooting and Lessons Learned April 29, 2015 Process Diagnostics

Trayed Column Scanning Procedure

Tracerco
Providing Insight Onsite

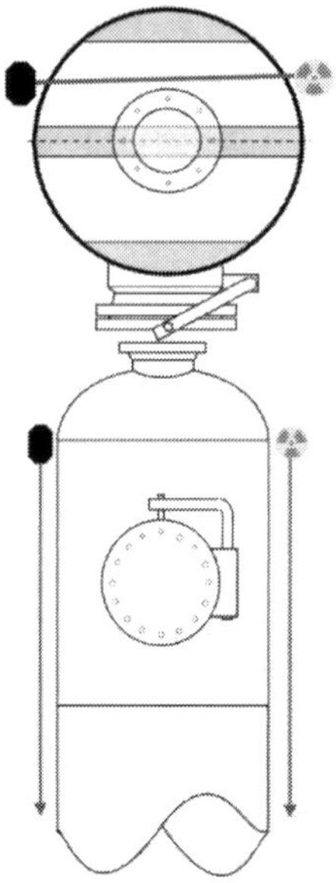

A.I.Ch.E. 2015 Spring Meeting Distillation Troubleshooting and Lessons Learned April 29, 2015

Process Diagnostics

BASELINE

6 months after baseline, Trays 2 and 3 were flooding

Another 7 months later Trays 2 – 7 now flooding

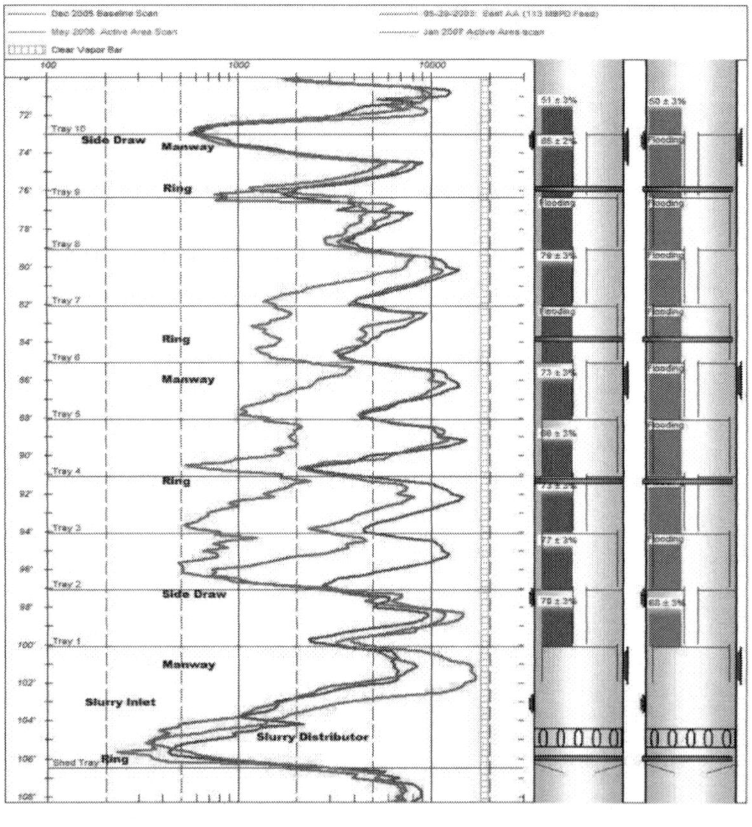

Packed Column Scanning Procedure

Tracerco
Providing Insight Onsite

$$I = I_0 \, e^{-\rho\mu\chi}$$

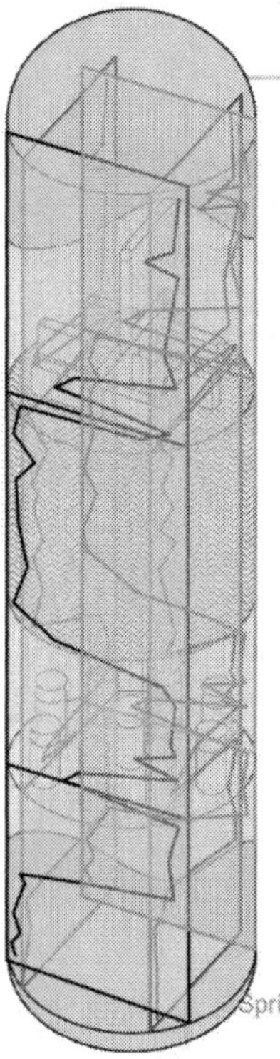

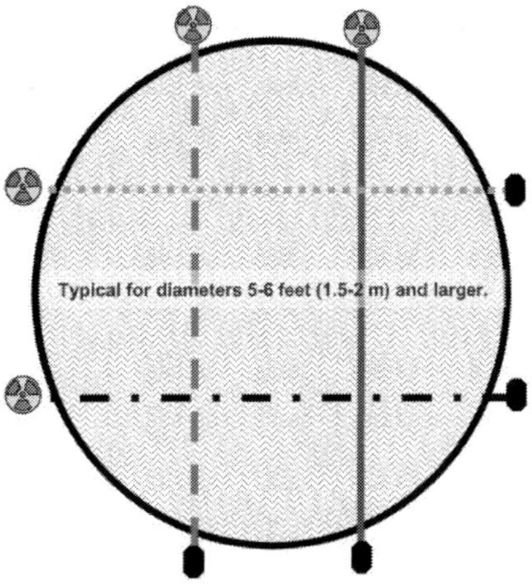

Process Diagnostics

Scan Dates:

Baseline Scan Jul 2009

Jan 2010

May 2011

Jan 2012

Oct 2012

Jan 2014

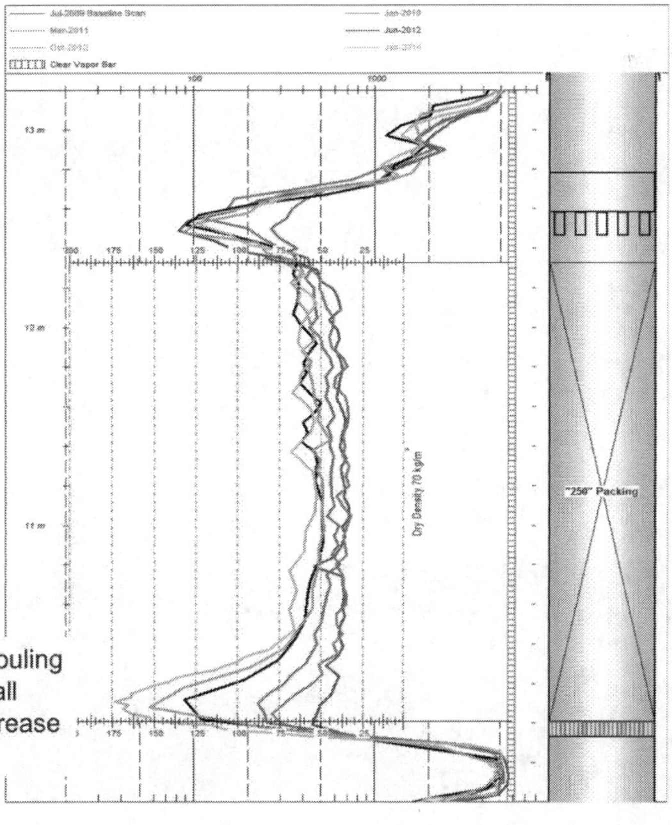

Scans showed progress of fouling in bottom of bed, while overall bed density continued to increase due to more liquid retention.

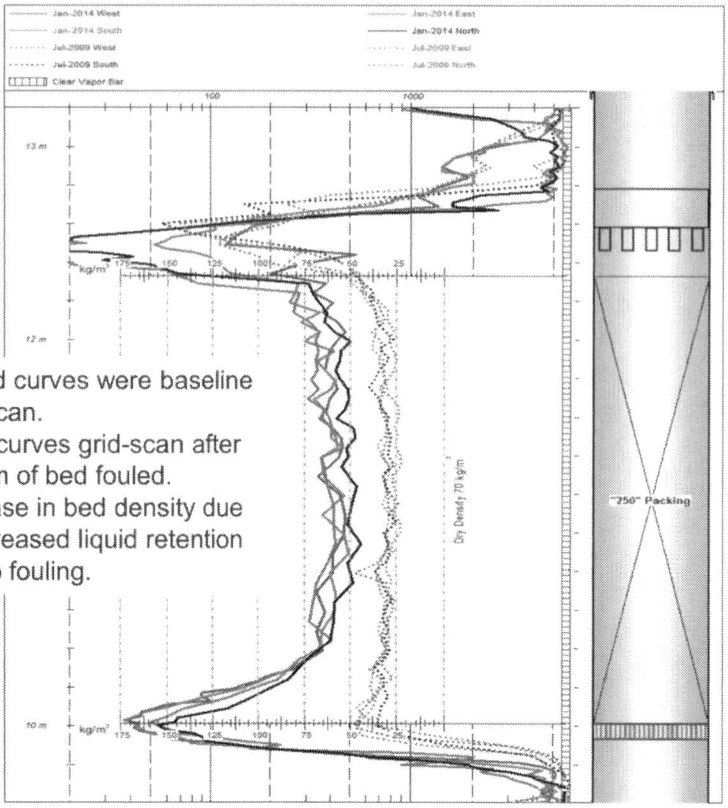

Tracerco
Providing Insight Onsite

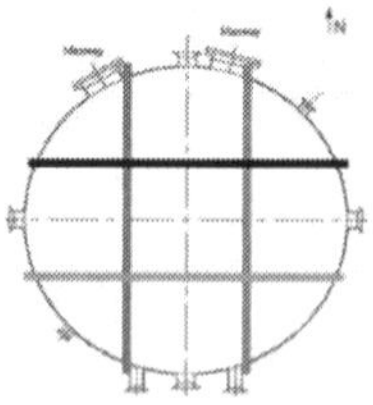

Process Diagnostics

ThruVision™ Tomography Scan

Tracerco
Providing Insight Onsite

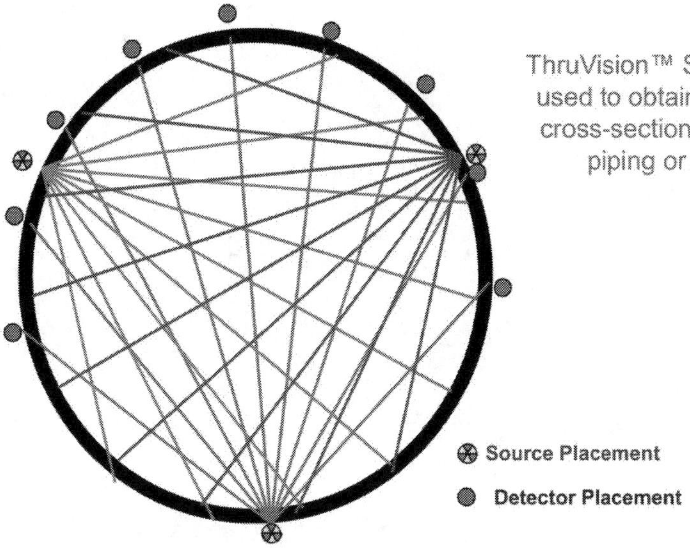

ThruVision™ Scanning technology is used to obtain a density profile of a cross-section of a column, reactor, piping or other equipment.

A.I.Ch.E. 2015 Spring Meeting Distillation Troubleshooting and Lessons Learned April 29, 2015 Process Diagnostics

Tru-Grid™ vs. ThruVision™

Tracerco
Providing Insight Onsite

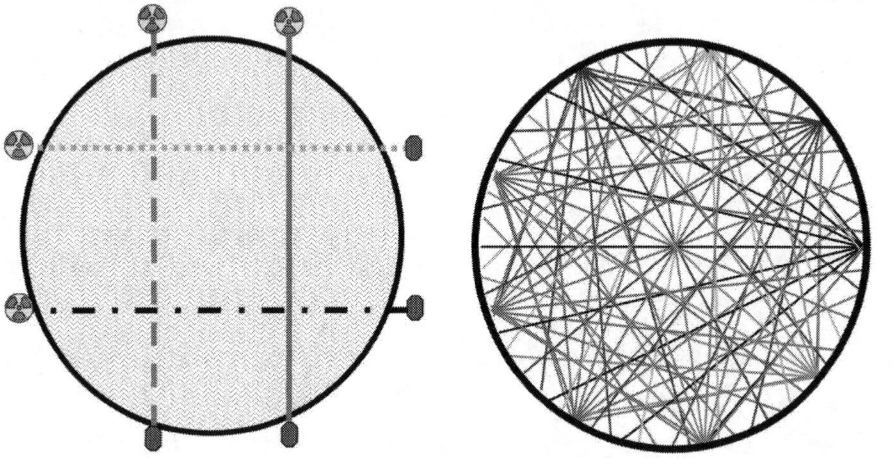

35 ft. (10.7 m) Diameter:

2% Area Coverage **40%** Area Coverage

A.I.Ch.E. 2015 Spring Meeting Distillation Troubleshooting and Lessons Learned April 29, 2015 Process Diagnostics

MONITORING BED AVERAGE DENSITY

Tracerco
Providing Insight Onsite

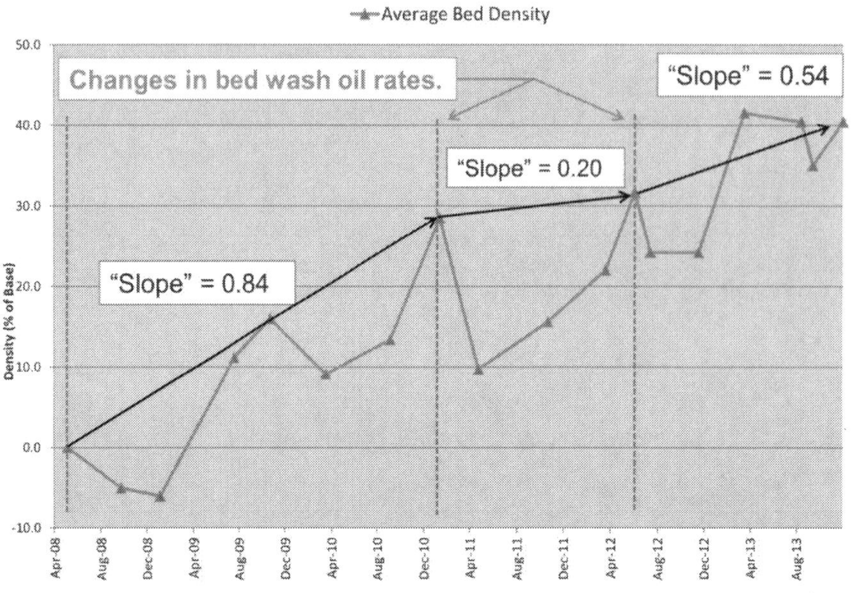

VACUUM TOWER WASH BED ThruVision™ SCANS

Tracerco
Providing Insight Onsite

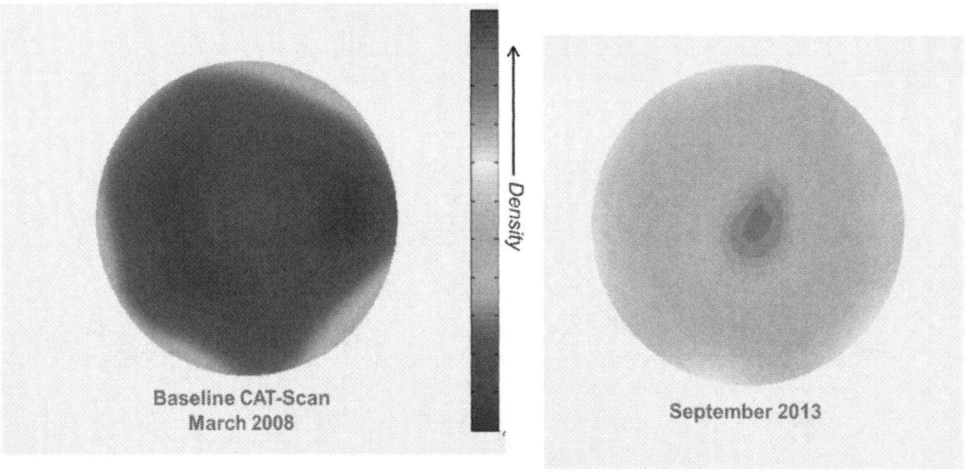

VACUUM TOWER WASH BED

Tracerco
Providing Insight Onsite

Coke in packing at end of cycle

A.I.Ch.E. 2015 Spring Meeting   Distillation Troubleshooting and Lessons Learned   April 29, 2015   Process Diagnostics

CONCLUSIONS

Tracerco
Providing Insight Onsite

- Fouling (plugging) is very prevalent in the processing industry.
- You want 2 questions answered:
  - Where is the fouling / plugging?
  - What is the progress or extent of the fouling?
- Gamma scanning has proven to be an essential measurement tool to supplement process variable and operating history data analysis.

A.I.Ch.E. 2015 Spring Meeting   Distillation Troubleshooting and Lessons Learned   April 29, 2015   Process Diagnostics

15AIChE
Spring Meeting, Austin, TX

Quick and Easy Troubleshooting of a Packed Tower: Thermal Imaging as a Novel Method

J. H. Barnard
(Corresponding Author)
Sasol Ltd, Secunda Synfuels Operations
PdP Kruger Street, Secunda, 2302
hendri.barnard@sasol.com

E. P. Du Toit
Sasol Ltd, Secunda Synfuels Operations
PdP Kruger Street, Secunda, 2302

Prepared for Presentation at
American Institute of Chemical Engineers
2015 Spring Meeting
11th Global Congress on Process Safety
Austin, Texas
April 27-30, 2015

UNPUBLISHED

Quick and Easy Troubleshooting of a Packed Tower: Thermal Imaging as a Novel Method

J. H. Barnard
(Corresponding Author)
Sasol Ltd, Secunda Synfuels Operations
PdP Kruger Street, Secunda, 2302
hendri.barnard@sasol.com

E. P. Du Toit
Sasol Ltd, Secunda Synfuels Operations
PdP Kruger Street, Secunda, 2302

Keywords: thermal imaging, packed distillation tower, maldistribution, troubleshooting

Abstract

This paper presents thermal imaging as a successful troubleshooting method for packed distillation towers. A distillation tower showed poor performance in the top section, indicated by an unstable top temperature. The methodology of investigation combined surveys drawn from the equipment's available instrumentation on temperature, pressure and flow, together with taking thermal scans to identify possible root causes of the poor performance. The thermal scans explicitly uncovered severe maldistribution throughout the tower, ranging from the top packed bed and pump around system all the way down the bottom of the tower. The thermal scans clearly show the vapour channeling up one side of the tower with the liquid channeling down the other side; there were no indications of flooding.

A damaged bed limiter, pump around liquid distributor and random packing were identified as the main possible root causes of the poor performance prior to entering the tower. The tower internals were inspected and it was found that the random packing in the top bed was fouled, conglomerated and severely crushed. A support grid was installed as a bed limiter and some of the random packing conglomerates were stuck in the grid, contributing to the maldistribution.

The maldistribution was completely rectified with the installation of a new bed limiter together with new random packing. This study shows that thermal imaging can effectively be used to troubleshoot packed towers on an industrial scale.

1. Introduction

The effects of maldistribution in packed towers are well understood in industry. It is however difficult to determine the extent of the maldistribution within a packed bed while the tower is still in operation. Even though traditional troubleshooting methods like pressure surveys and studying the temperature profiles can partly reveal where the problem originates, these methods require a high level of experience and is often time consuming. The interpretation of gathered data by various analysts will also differ, and any theory can only be confirmed by decommissioning the tower, leading to downtime and production loss.

This paper presents an alternative method of troubleshooting industrial scale packed towers.
An acid gas scrubber, operating with an unstable top section temperature profile, was the test subject of this novel method. Traditional troubleshooting methods were applied with no conclusive results. The tower was decommissioned; the top section was inspected and no major defects were found. After commissioning the top temperature remained unstable.

For the investigation presented in this paper, an infra-red (IR) camera was used for further troubleshooting and the images clearly revealed fluid channeling in the packed section of the tower. If this was known beforehand, it would have been clear that an inspection of the top tray section would be unnecessary.

Thermal imaging offers the opportunity to visually inspect the inside of an uninsulated packed tower while it is still in operation. This eliminates unnecessary downtime and production losses. It also allows for detecting the origin of the problem and a limited level of skill is involved for capturing images for troubleshooting and interpretation.

2. Process Description

The purpose of the acid gas scrubber is to remove all the water and carbon dioxide from ammonia, yielding ammonia of a higher concentration. In the tower the ammonium carbonation reaction is used to ensure that all the carbon dioxide is removed from the ammonia before sending the ammonia for further purification. The ammonia would be in excess and this causes the carbon dioxide to reacts till completion. The excess ammonia is removed at the top of the tower.

Two sections of packing are used to enlarge the contact area for the reaction to take place. The reaction is an exothermic reaction where excess heat must be removed and this is done by middle and top pump around systems. Figure 1 shows a simplified process flow diagram of the acid gas scrubber and supporting equipment.

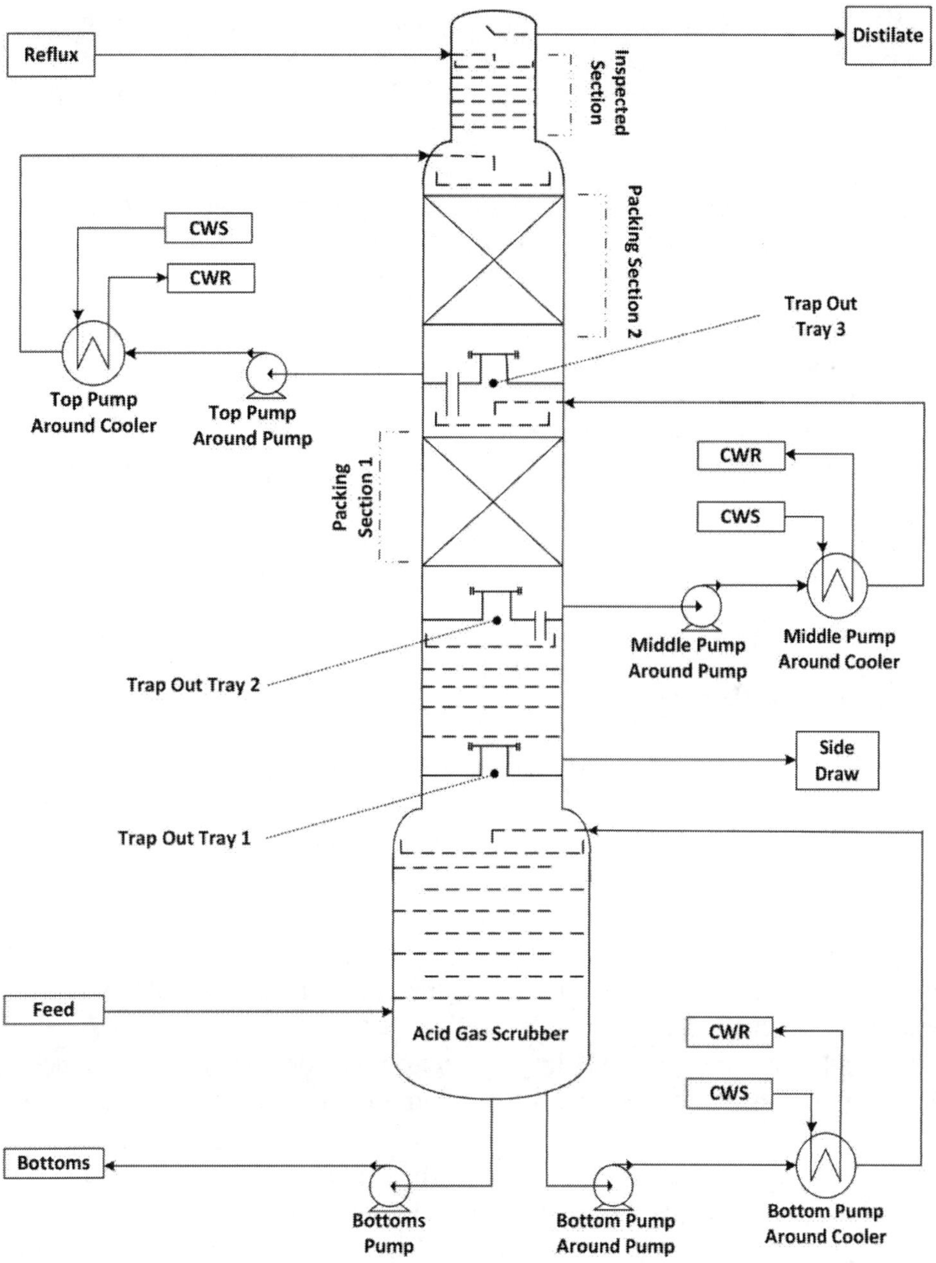

Figure 1: Simplified process flow diagram of the acid gas scrubber tower.

The overheads product from a preceding tower enters the tower below the first tray. Steam in the feed to the tower needs to be condensed which is done by a bottom pump around system. The bottoms pump around and cooler configuration is used for temperature control of the bottom section.

The middle and top pump around systems are utilized to address the temperature dependency of the formation reaction for ammonium carbonate; temperature is controlled to favor the formation reaction. A stream from the Trap Out Tray 2 in the middle section is pumped through the middle pump around cooler to attain the required temperature after which it is then reintroduced above Packing Section 1.

The vapour leaving the middle pump around passes through a chimney tray into the upper middle section of the tower. The top pump around system also aids in the formation of ammonium carbonate. A stream from Trap Out Tray 3 is drawn off and pumped through the top pump around cooler and reintroduced above the second section of random packing in the upper middle section of the tower. A stream with a high purity of ammonia is used as a reflux and is introduced right at the top of the tower. This stream also controls the temperature at the top of the tower and washes down any the ammonium carbonate present. The ammonium carbonate is removed from the tower at Trap Out Tray 1 by the side draw.

3. Problem Definition

The factory consists of two phases that operate in parallel. One of the acid gas scrubber towers was experiencing an unstable top temperature. The top temperature of this tower was swinging 10 °C higher and 10 °C lower than the set point and this posed a risk contamination of the overheads product to the downstream units. The top temperature is usually very stable and any upset in this temperature can cause the overhead stream to be contaminated with the unwanted acid gases. The tower was decommissioned, and no major defects were found.

4. Conventional Troubleshooting

The following steps were taken to try and alleviate the problem:

- It was found that the temperature transmitter used for the control of the top temperature was operated outside its specified range. The range was corrected but this did not lead to any improvement in the temperature control;
- The top temperature control loop was investigated. The control loop was found to be functioning as intended. It had the same tuning parameters as for the acid gas scrubber towers on the other phases;
- The temperature at the bottom of the top section was determined to be much higher than on the other phases, indicating that the problem might actually be in the middle section of the tower;
- The top reflux, fed to the top of the tower which plays an integral part in controlling the top temperature, was investigated and it was found to be different when compared to the other phases. This could also be detected by visual inspection as this was the only reflux line that indicated ice formation on the outside of the line. The tower operation was adjusted to allow for similar compositions of the reflux when compared to the other phases. This resolved the ice formation on the line and did influence the temperature, however the temperature at the bottom of the top section remained higher than normal;
- The performance of the tower was investigated, all pump around temperatures and tower pressures were within normal operating limits; and
- Finally thermal scans of the tower were taken to determine if the packing inside the middle section of the tower was still intact.

5. Thermal Imaging Results and Discussion

The following thermal scans were captured of two acid gas scrubber towers. Pictures of the unstable tower (Tower 1) were compared to a tower showing stable operation (Tower 2).

Figure 2 shows the IR scans of the top section of the acid gas scrubber (Western side of the tower), while Figure 3 is a simplified general arrangement drawing of the top section of the acid gas scrubber.

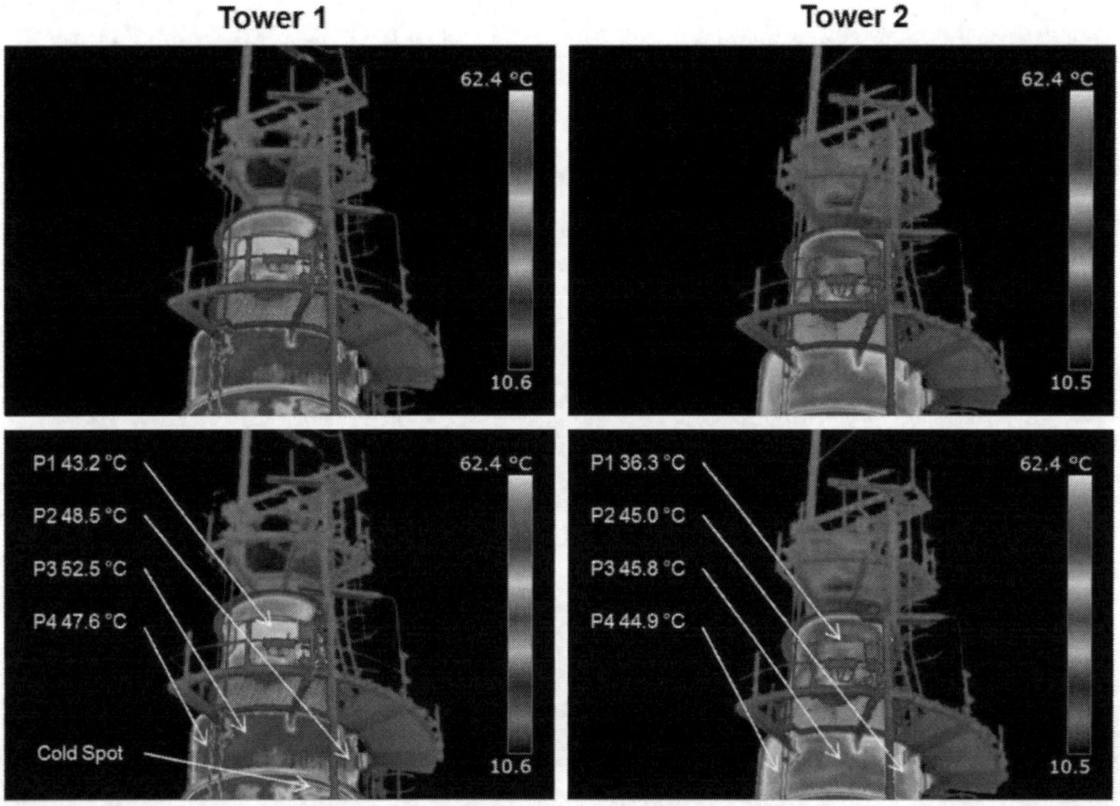

Figure 2: Thermal scans taken of the top section of the acid gas scrubber towers. Tower 1 (on the Left) is the tower showing the deviations from normal operations. Tower 2 (on the right) shows stable operation.

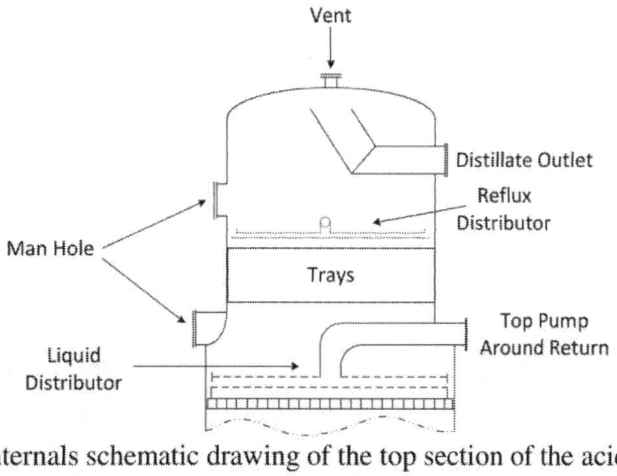

Figure 3: A simplified internals schematic drawing of the top section of the acid gas scrubber tower.

From Figure 2 it can be seen that there is a different temperature distribution in Tower 1 when compared to Tower 2. The average operating temperature of the section just above the liquid distributor is 45 °C (the temperature seen in Tower 2), Tower 1 shows a maximum temperature of 52.5 °C. Towards the top of the tower the same temperature differences can be seen with Tower 1 operating 6.9 °C higher than Tower 2. The bottom right side of Tower 1 in Figure 2 also shows a cold spot which is the first sign of liquid/vapour maldistribution.

Figure 4 below shows the IR scans of the top pump around return section of the acid gas scrubber (Western side of the tower), while Figure 5 is a simplified general arrangement drawing of the middle pump around return section of the acid gas scrubber.

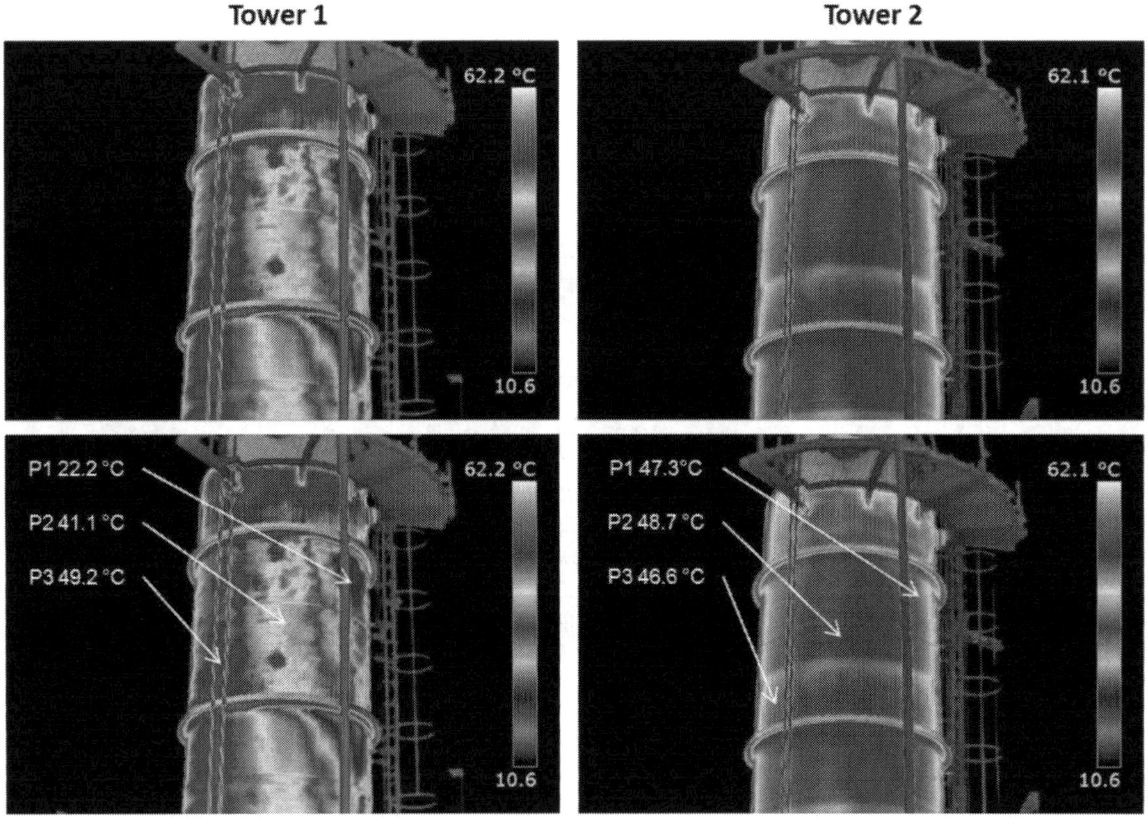

Figure 4: Thermal scans taken of the top pump around return section of the acid gas scrubber towers. Tower 1 (on the Left) is the tower showing the deviations from normal operations. Tower 2 (on the Right) shows stable operation.

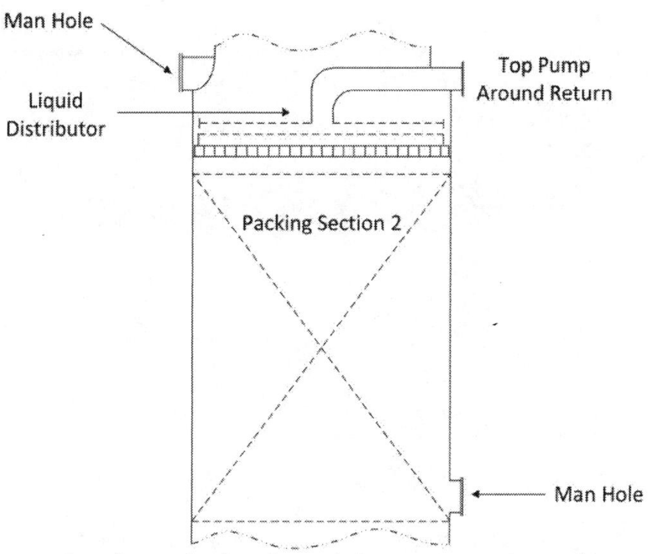

Figure 5: A simplified internals schematic drawing of the top pump around return section of the acid gas scrubber tower.

From Figure 4 it can be seen that there is a large temperature gradient in the cross-section view of Tower 1 with a maximum cross-sectional temperature difference of 27 °C. When this is compared to the same section of Tower 2 the maximum cross-sectional temperature is only 2.1 °C indicating a more uniform radial temperature distribution. Visual inspection of the thermal scans clearly indicates that the liquid and vapour is unevenly spread through the packing in Tower 1.

The position of the cold spot when compared to the general arrangement drawing of this section (Figure 5) shows that the liquid/vapour maldistribution originates from the middle pump around return line. The preliminary conclusion was made that this was due to the liquid distributor being damaged or fouled, causing the liquid/vapour maldistribution starting from this section or channeling of liquid and vapour through the random packing section. Packed towers are very sensitive to distribution and maldistribution is detrimental to packing efficiency (Kister; 1990: 35).

In order to investigate the temperature effect the maldistribution has on the entire tower, thermal scans were taken from a distance away, allowing for visual inspection of both the top section of the tower as well as the entire middle pump around. Figure 6 shows the IR scans of the top section as well as the entire middle pump around section of the acid gas scrubber (South-Eastern side of the tower). Figure 7 is a simplified general arrangement drawing of the top and middle pump around section of the acid gas scrubbers.

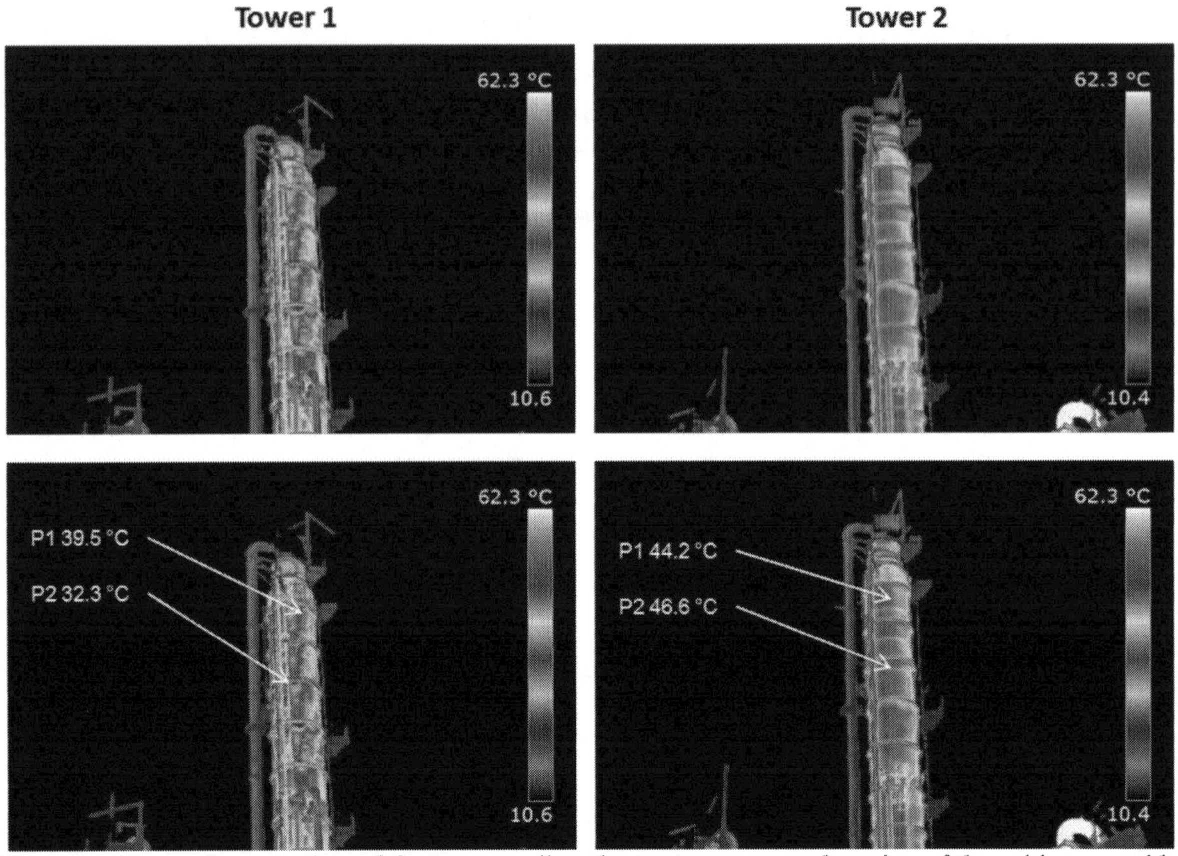

Figure 6: Thermal scans taken of the top as well as the top pump around section of the acid gas scrubber towers. Tower 1 (on the Left) is the tower showing the deviations from normal operations. Tower 2 (on the Right) shows stable operation.

From Figure 6 it can be seen that there is a definite maldistribution when comparing Tower 1 with Tower 2 can be seen. The South-Eastern side of Tower 1 is much colder than the same side of Tower 2. This maldistribution propagates through the entire middle pump around of Tower 1. Tower 2 also shows a much better temperature distribution through-out tower.

This indicates that there is a maldistribution of the liquid and vapour throughout the entire top packing (Packing Section 2) of the tower – the cause of the temperature instabilities seen in the top section.

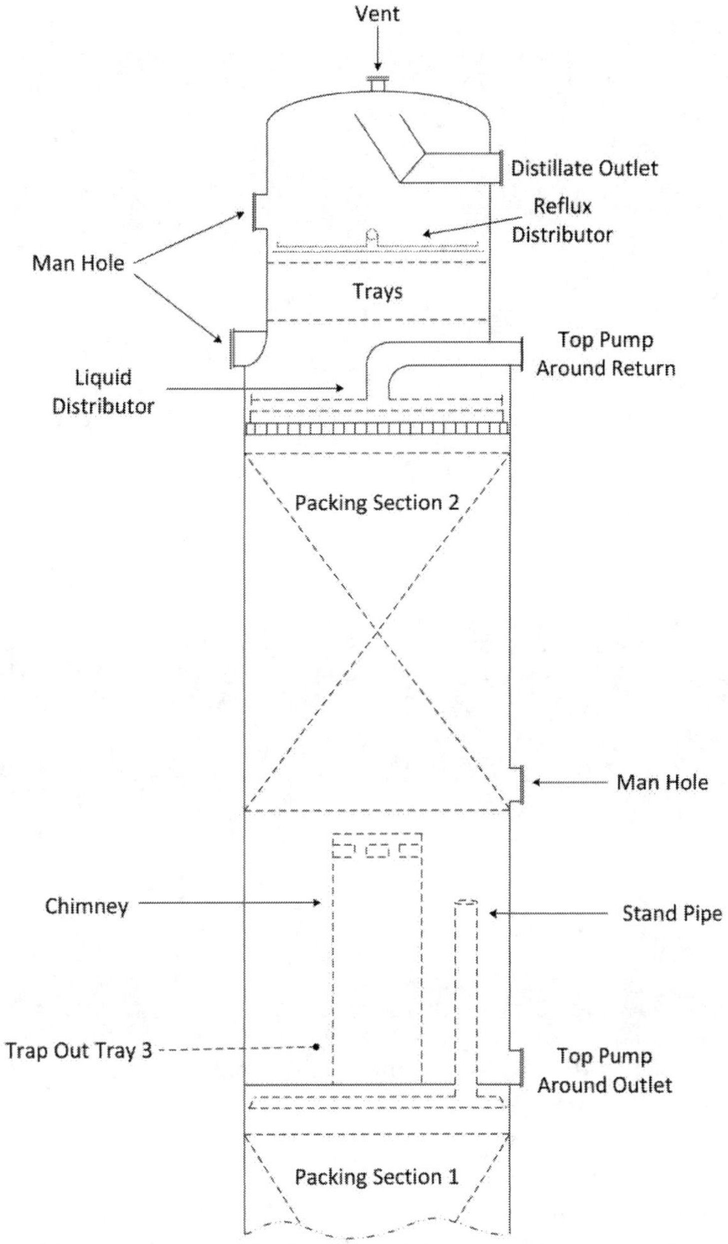

Figure 7: A simplified internals schematic drawing of the top and top pump around section of the acid gas scrubber tower.

It was decided to take more scans from different angles around the towers to try and match what was seen on Figure 6. Figure 8 and Figure 9 were taken from different angles and the maldistribution can be seen on all of these scans.

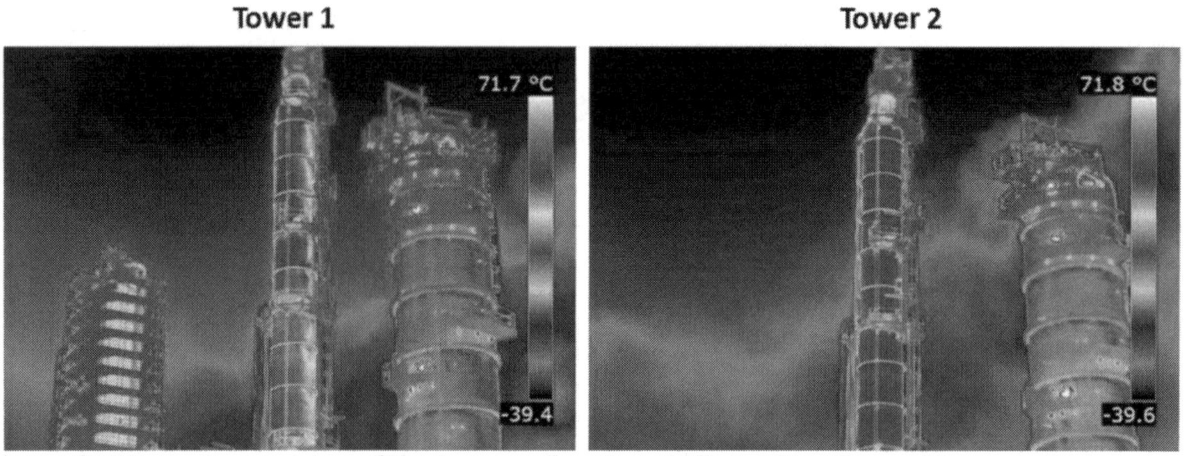

Figure 8: Thermal scans taken of the top as well as the middle section of the acid gas scrubber towers. Tower 1 (on the right) is the tower showing the deviations from normal operations. Tower 2 (on the left) shows stable operation.

Figure 9: Thermal scans taken of the top as well as the middle pump around section of the acid gas scrubber towers. Tower 1 (on the right) is the tower showing the deviations from normal operations. Tower 2 (on the left) shows stable operation.

The vapour liquid maldistribution can be seen on both Figure 8 and Figure 9. Figure 8 shows the hotter side of Tower 1 while Figure 9 shows the colder side of Tower 1. In both these pictures Tower 2 shows an evenly distributed temperature through-out the entire tower.

From the thermal scans the three main deviations possibly contributing to the observed vapour liquid maldistribution were identified as:

- Uneven or damaged top pump around liquid distributor.
- Blocked or damaged random packing.
- Blocked or damaged bed limiter.

6. Internal Inspections and Findings

Tower 1 was opened during the routine statutory maintenance turnaround. The top pump around liquid distributer was inspected and no deviations were found.

Figure 10 and Figure 11 below show findings from the inspection of the packing as well as the hold down grid.

Figure 10: Damaged random packing removed from the packing sections.

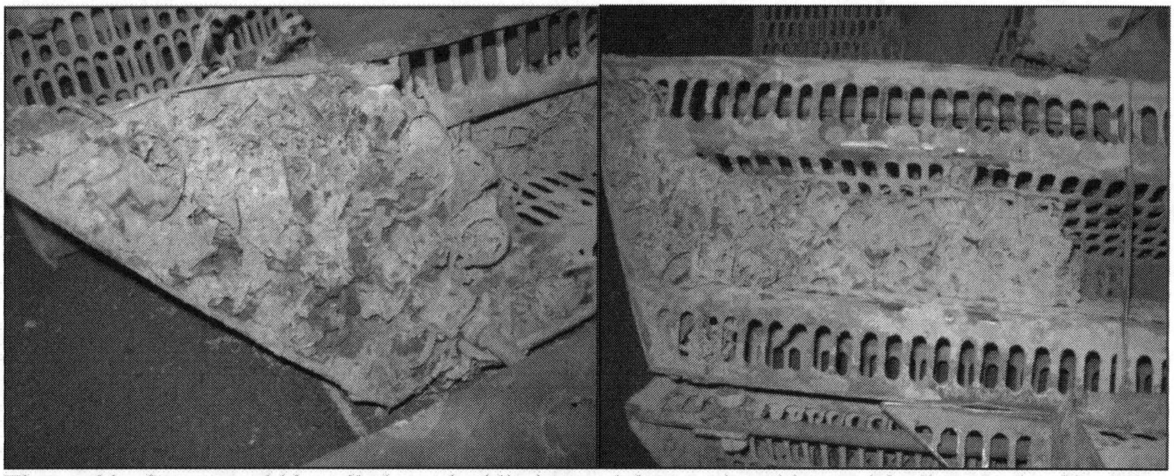

Figure 11: Support grid installed as a bed limiter and damaged packing and fouling causing blockages in the support grid.

Figure 10 shows the damaged and fouled random packing which was found in packing section 2. The maintenance team struggled to remove the packing due to the packing being crushed and sticking together. Fouled and damaged random packing leads to liquid and vapour maldistribution through packing sections and can cause the severe temperature gradient observed in the packing section. Liquid distribution plays a vital role in the effective operation of a packed tower. Poor distribution reduces the effective wetted packing and promotes liquid channeling (Xu, 2000). The random packing was replaced with new packing.

Figure 11 shows the random packing being stuck in the support grid, installed as a bed limiter, contributing to the maldistribution seen on the IR images. Upon further investigation it was found that support grid was installed when the original hold down grid was unavailable. A hold down grid should not be used with plastic packing. Since the hold down grid rests directly on packing it can lead to the plastic packing being compressed (Kister; 1990: 223). The support grid was removed and replaced with a new bed limiter seen in Figure 12.

Figure 12: New bed limiter before it was installed.

7. Conclusion

From the case study presented it is clear that thermal imaging proves to be a viable option in packed tower troubleshooting. The benefit of being able to visually inspect the flow pattern inside a packed tower while it is in operation cannot be overstated. Regular visual inspections of this kind even allows for identifying problems even before it has any major effect on production.

These thermal scans can be used to identify problems with equipment before they are decommissioned for statutory maintenance. Thus proving to be a significant aid in planning ahead and procurement of spares to ensure that downtime is minimized.

Furthermore, troubleshooting packed towers in the traditional way, using pressure and temperature surveys, are time consuming and more regular than not requires a great deal of expertise. However, thermal imaging offers the opportunity to perform a temperature-based visual inspection on the flow patterns inside the tower. Very little expertise is required for capturing the images and defects can easily be identified by anyone that regularly inspects the thermal scans for a specific tower. The thermal scans can also aid in improving the understanding of flow patterns within a tower.

One major drawback for the use of thermal imaging is that the tower's lagging/insulation would need to be removed before the thermal image can be taken. However, depending on the type of lagging/insulation and the size of the column, the costs of removal and replacement of lagging/insulation to understand what the current operation in the tower is can be insignificant when compared to product loss or increased utility usage.

8. Acknowledgements

We gratefully acknowledge the support and input of our supervisor Carla Cronje, without which this study could not have been completed. We thank Jakes van der Walt for his help and reliance on us regarding the use and availability of the infrared camera. Furthermore we would also like to thank the various anonymous reviewers for their feedback.

9. References

[1] Kister, HZ (1990) *Distillation Operation*, McGraw-Hill, New York

[2] Xu, SX (2000) "Quantitatively Measure and Assess Maldistribution in Industrial Packed Towers"

Meeting: Spring AIChE Meeting
Austin, TX – April 29th 2015
Distillation 2015 Paper 179d

Title: *"Steps for Troubleshooting Amine Sweetening Plants"*

Authors: Glenn Shiveler, Sulzer Chemtech USA Inc. – Tulsa, OK
Harold Wandke, Sulzer Chemtech USA Inc. – Tulsa, OK

Abstract:

Amine gas treating (also called "sweetening") is a well-proven method for removing acid gases from a gas stream, and is commonly used in natural gas plants, petrochemical plants, and refineries. An amine unit typically consists of two towers—a contactor to allow the selected amine to remove the acid gases, followed by another column that regenerates the amine using a reboiler or direct steam injection, plus the associated heat exchangers.

The proper design and operation of the plant is critical to meet product quality and environmental standards while minimizing operating costs to the plant. There are multiple potential problems that can result in an underperforming amine sweetening plant, including inadequate hydraulic capacity of the equipment, damaged equipment, non-optimal operating conditions, heat stable salts, and poor amine quality. Determining the cause of a malfunctioning sweetening plant can be difficult because the investigating engineer can either misdiagnose the root cause of the problem, or fail to identify all of the contributing factors to the primary problem. Deductive reasoning is a critical element in the review of the potential causes of the problem.

This paper presents a list of recommended steps for performing a thorough investigation of an underperforming amine sweetening plant. The steps include collecting operating plant data to eliminate unlikely causes and then focusing in on the probable causes based on the plant data. A discussion of the role of thermal scan and gamma-scan data to diagnose hydraulic problems with the mass transfer equipment is included. Finally, guidelines for key operational parameters for the safe operation of tray hydraulics and packing capacity will be presented. This paper covers only gas/liquid contacting.

Introduction

The sweetening of natural gas has been performed for many years by contacting the sour gas with an aqueous amine solvent. The traditional gas sweetening process utilized by the natural gas processing industry is depicted in Figure 1. The first step is to contact the sour gas with the lean amine solvent in the amine absorber, where hydrogen sulfide and carbon dioxide are picked up by the solvent by a reversible chemical reaction with the amine. The sweetened gas leaves the top of the absorber and rich amine containing the absorbed hydrogen sulfide and carbon dioxide leaves the bottom. Rich amine from the absorber passes through a flash separator, then through a cross exchanger economizer in its path to the regenerator. Stripping of the hydrogen sulfide and carbon dioxide is performed in the regenerator unit using a reboiler to provide enthalpy to reverse the absorption reaction. The overhead gas is cooled by the condenser to return water reflux to the regenerator. The remaining off-gas can be processed by a sulfur recovery unit or sent to a flare. The lean amine stream produced at the bottom of the regenerator is passed through the cross exchanger, filters and a trim cooler before the lean solvent is returned to the absorber. The two columns provide a closed loop system that requires only an occasional addition of fresh amine to keep the amine strength at an acceptable level.

Figure 1 – Natural Gas Sweetening Plant Process Flow Diagram

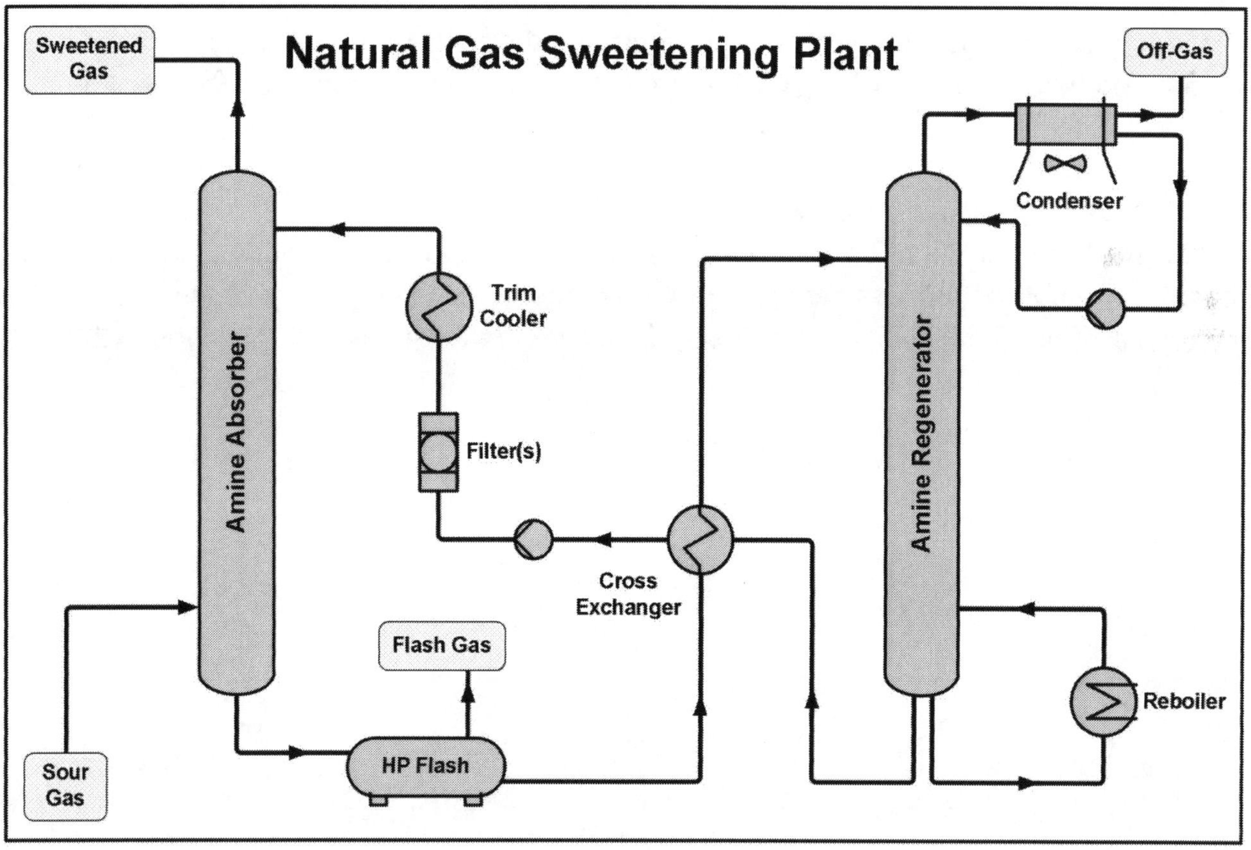

The performance of amine sweetening plants can be adversely affected by various problems. Some problems relate to the amine solvent quality, which can be susceptible to foaming and contaminants. Gas inlet composition and operating conditions of some plants drift from the design basis over the course of time. As a result, it might not be obvious that a change in the reactant ratio that has occurred. Some plant problems are due to operating conditions that exceeds the hydraulic capacity of the mass transfer internals. Others are due to equipment damage. Plant operators often contact process experts and equipment vendors to perform troubleshooting of plant operating problems. In the troubleshooting process, there are several steps to the process of investigation for amine sweetening plants required to understand the root cause for the performance problem.

First Step: Gathering Evidence

The first step in any troubleshooting investigation is to gather data and information from the operating control room. Data is necessary to understand the facts about the problem, and for disclosure to consultants. It is recommended to start the investigation with a review of the plant design basis by comparing the current operating conditions to the original design basis or to a previous period of acceptable plant operation. Prepare a data spreadsheet table that lists all of the available process instrument readings for the current operation of the entire sweetening plant for side-by-side comparison to the readings for acceptable operation of the process. Stream flow rates, compositions, temperature and pressure are especially important. Providing the data for the original design basis and previous good operation helps third party consultants to gain perspective and understand the performance problem in its context.

Analysis of the data table will show some readings that differ slightly from expectation, while some process variables might deviate significantly from expectation. An example of a data table is shown in Figure 2 below. Usually, the process variables that show the greatest deviation from the expectation from the design simulation provide the suspects for the focal point for the investigation.

Figure 2 – Example of Plant Operating Data Table

Plant Data Table

Process Variable	Instrument Tag	Units	Original Design Basis	Last Good Operation	Current Problem Operation
Inlet Gas Pressure	PI-101	kPag			
Inlet Gas Temperature	TI-103	°C			
Inlet Amine Temperature	TI-121	°C			
Absorber Pressure Drop	DPI-110	mbar			
HP Separator Pressure	PI-131	kPag			
Regenerator Top Pressure	PI-141	kPag			
Regenerator Pressure Drop	DPI-144	mbar			
Regenerator Top Temperature	TI-142	°C			
Regenerator Bottom Temperature	TI-143	°C			
RA Temperature from Cross-X	TI-152	°C			
LA Temperature from Cross-X	TI-153	°C			
Inlet Gas Flow Rate	FI-100	Nm3/hr			
Inlet Gas - H2S Composition		ppm-wt			
Inlet Gas - CO2 Composition		ppm-wt			
Outlet Gas Flow Rate	FI-109	Nm3/hr			
Outlet Gas - H2S Composition		ppm-wt			
Outlet Gas - CO2 Composition		ppm-wt			
Lean Amine Circulation Rate	FIC-102	LPM			
Amine Strength		wt%			
Residual H2S in LA Stream		mol/mol			
Residual CO2 in LA Stream		mol/mol			
Regenerator Steam Consumption	FI-153	kg/hr			

At times, the plant operators begin the investigation knowing the symptoms of the problem before they fully understand the root cause. As such, there is risk that the investigating engineer might jump to a premature conclusion about the suspected cause for the problem. Study of the operating data table coupled with deductive reasoning is necessary to focus the investigation towards understanding the root cause of the plant operating problem.

Second Step: Review Plant Operation with a Simulation

There are many process variables that can affect the performance of an amine sweetening plant. Changes in the inlet gas or solvent composition could have occurred during the course of plant operation since its original design. The reaction between the amine and the acid gases (H_2S/CO_2) can affect the mass and energy balance for the amine absorber. It can be difficult to extrapolate the differences observed in the process variables from those of acceptable plant operation based only on the stream table from the original design basis. It is recommended that a process simulation study be conducted to ascertain the expected flow rate, composition, temperature and pressure for the process streams and equipment. It is necessary to have mass flow rates and physical properties from a simulation for performing hydraulic calculations over the mass transfer units in order to properly review the equipment capacity and pressure drop.

One problem is that plant operators might not have access to a process simulation package suitable for review of an amine sweetening plant. Therefore, plant operators will contact third party consultants and equipment vendors for technical support. Having the data spreadsheet from the data gathering step will facilitate the communication process of the plant operating data with technical consultants. Review whether the number of actual trays or packing height is suitable for meeting the outlet hydrogen sulfide and carbon dioxide gas composition objectives is best achieved with a rate-based simulation.

Third Step: Review Pressure Drop

The next step is to rate the mass transfer internals using the equipment vendor's proprietary rating program. The existing internals should be evaluated using the results of the simulations that were performed in step two. Sulzer has its hydraulic calculations available on-line for its trays and packings in a computer program called SulCOL.

A common symptom of a sweetening plant problem is high pressure drop over the mass transfer vessel. Sometimes the amine absorber is observed to show higher pressure drop than predicted by the hydraulic rating program. It is necessary to review the expected pressure drop for the mass transfer equipment at the operating conditions using the hydraulic calculations of an equipment supplier to diagnose hydraulic flooding. Amine absorbers and regenerators experience a normal amount of foaming that is associated with absorption of hydrogen sulfide and carbon dioxide gas and the reaction of these with the amine. Pressure drop is can be 10% to 20% higher than the calculated pressure drop due to normal foaming. If the actual pressure data exceeds the prediction by more than 50%, it is a possible sign of hydraulic flooding in the mass transfer equipment.

It is necessary to calculate the pressure drop as a baseline for comparison to the plant operating data and to review the plant operating data for any time-related trends for pressure drop. Sometimes there are repeated cyclic changes in pressure drop, and operating room data can show cyclic waves or saw-tooth patterns over regular time periods. If there is cyclic behavior for the absorber pressure drop, it is necessary to observe the local maximum and minimum pressure drop during a cycle, and the time period for the cycles. Snap shots of the control room pressure drop data should be provided to process consultants.

Fourth Step: Review Equipment Capacity

Although it is common to evaluate the conditions at the inlet and outlet of each column, it is important to review the column profile from the simulation. For the absorber, there is usually a stage or segment that has peak temperature in the column profile. Review the mass transfer equipment hydraulic calculations for the top and bottom inlet/out conditions, as well as review the hydraulic loads for the peak temperature location in the column profile. Often times, the hydraulic conditions for the flow rates and properties associated with the peak temperature has the highest pressure drop, and the mass transfer devices could have the highest risk for hydraulic flooding.

In general, the dry pressure drop for trays should be below 0.6 mbar (1.5 inches H_2O) for safe operation. There is low to moderate risk for dry pressure drop between 0.6 to 0.8 mbar (1.5 to 2 inchesH_2O), and a higher risk for dry pressure drop that exceeds 0.8 mbar (2 inches H_2O) in amine absorbers. Tray hydraulic calculations should consider a system factor of 0.75 for absorbers as a baseline. The downcomer inlet velocity should below 0.11 m/s (0.35 ft/sec) for safe tray operation as there is a risk for a downcomer inlet velocity limitation if it exceeds 0.12 m/s (0.4 ft/sec). Review the downcomer head loss and downcomer stack-up to ensure that the values are within the equipment vendor's acceptable limits.

For packed absorbers, the packing pressure drop should not exceed 6 mbar/m (0.75 inches H_2O/ft). Review the capacity of the liquid distributor as shown on the distributor assembly drawings to confirm it is suitable for the operating conditions. Review the velocity of the amine inlet pipe and the gas inlet device to confirm whether it is within the limits for safe design. If there is reason to suspect liquid maldistribution, then it is necessary review the installation and levelness of the liquid distributor troughs, as well as the drip point density of the distributor.

Fifth Step: Review Amine Quality

Changes in the sweetening plant performance are sometimes due to changes in the amine quality over the course of time. High pressure drop can be caused by poor solvent hygiene, and it is necessary to review the plant operating log for amine strength, lean solvent quality and heat stable salts (HSS). Amine strength is usually expressed in weight percent. Solvent quality is the residual H_2S/CO_2 gas in the lean solvent produced by the regenerator. Local laboratories can perform tests for amine strength and quality with reasonable accuracy. Testing of HSS should be performed by qualified laboratories operated by the solvent vendors, who have experience in diagnosing various amine plant problems due to specific contaminates.

There are various kinds of contaminants that can affect the amine solvent quality. Some contaminants enter the plant through the gas pipeline and breakthrough the knock out drum. Gas feed contaminants can include methanol, glycol, caustic, compressor lubrication oils, corrosion inhibitors, drilling mud and other contaminants. Sometimes, pressure drop spikes observed in the amine absorber occur after the scouring pig is used in the pipeline. A vital clue to a gas feed contaminant can be proven by review of the pressure drop over the knock out drum prior to the pressure drop observed in the amine absorber.

Sometimes HSS metal ions affect the performance of the amine solvent. Metal ions can include sodium, calcium, chlorides and potassium. Other solvent contaminants can include chlorides, sulfate, bicine, formate and acetate. Some plants can experience changing concentration levels for HSS contaminants over the course of operation. In the case of proprietary mixed amine solvents, it is necessary to review solvent contaminants with the solvent vendor to diagnose this issue.

Sixth Step: Thermal Imaging

Most amine absorbers are equipped with few temperature instruments. Having temperature instruments only for the inlet/outlet streams is insufficient to gain perspective of the absorber temperature profile.

Many amine absorber vessels are uninsulated. It is possible to use a thermal imaging pyrometer to review the skin temperature of the absorber. The cost to rent or purchase a pyrometer is low compared to the cost of a gamma-scan, so it is possible for plants to screen for potential temperature related problems in the mid-section of the absorber by thermal imaging.

Data from the thermal imaging pyrometer can provide insight of the absorber temperature profile for comparison to the simulation prediction, and also to locate the peak temperature position in the absorber. Modern pyrometers can record the temperature data along a straight line, so it is possible to obtain the operating temperature profile for the column to compare versus the simulation prediction. Thermal images can reveal hot spots that can reveal the location in the absorber where the reaction takes place between the amine and hydrogen sulfide and carbon dioxide. The hot spots can reveal hydraulic disturbances or maldistribution involving the column internals.

Seventh Step: Gamma-Scan

If all else fails, hydraulic problems involving the mass transfer internals can be evaluated with a gamma-scan. This test is similar to performing an X-ray through the vessel. The scan data shows the location of metal internals and the presence of liquids. Often times the gamma-scan test is conducted at two flow conditions for comparison: one test at stable operating conditions and one test at suspected flooding or unstable conditions. For vessels equipped with trays, the gamma-scan report can show which trays are highly loaded and whether any trays are hydraulically flooded or damaged. For vessels equipped with packing, a grid gamma-scan report can show the liquid flow profile, and whether the liquid is properly distributed or maldistributed through the packing. Conducting the gamma-scan at two operating conditions provides a comparison that can show the location where hydraulic flooding occurs in the column internals.

A gamma-scan test is a costly and time consuming consideration for the gas plant and should be used as a last resort to investigate the column. It is recommended that the plant conduct a thermal imaging study, and perform a reasonable study of the plant operation and amine quality study prior to conducting the gamma-scan.

Eighth Step: Internal Inspection

It is possible that the gamma-scan in step seven might not adequately reveal a cause for the operating problem. In the case of random or structured packing, it is possible that minor maldistribution from the liquid distributor might not be noticed by a gamma-scan. Common causes for liquid maldistribution can be an out-of-level liquid distributor, plugged distribution holes, and damaged inlet pre-distributor pipes. In the case of trays, it is possible that a hydraulic anomaly for one tray might not be adequately explained by the gamma-scan. Common causes for tray problems include missing valves, corroded deck or downcomer panels, blockage inside of downcomers and mechanical damage to tray parts. In some cases, there is no substitute for performing visual inspection of the column internals to determine the cause of an absorber or regenerator operational problem that is not conclusively identified by any of the previous investigation steps.

Notes About Foaming

There is a normal amount of foaming in amine absorbers that is due to the absorption of H_2S/CO_2 gas into the solvent and the exothermic reaction with amine. Normal foaming increases the gas-liquid mass transfer area. There are absorbers that can exhibit signs of more severe foaming compared to well behaved units. Severe foaming can have several causes:

1. High concentration of hydrocarbon components that forms a separate liquid phase.
2. Presence of lubrication oils in the solvent.
3. High concentration of HSS and amine degradation products.
4. Overly high dry tray pressure drop.
5. Particulates present in the solvent (drilling mud, corrosion products, and suspended solids).

There are several aspects about the nature of foaming. The solvent absorbs hydrogen sulfide and carbon dioxide gas into the liquid followed by exothermic reaction with the amine that can produce foaming. Operation of the absorber with a high peak temperature can be accompanied by severe foaming.

Many plants conduct regularly scheduled foam tests on the lean amine solvent by bubbling nitrogen gas in a flask, then record the time for foam bubbles to dissipate. Based on this test, sometimes plant engineers make the claim that their solvent is non-foaming. This is a misleading conclusion, as the nitrogen foaming test only measures the relative capability for the solvent to foam for one day with inert gas compared to other times. However, this test can diagnose whether the presence of heavy hydrocarbons or chemical contaminants are affecting the foaming capability of the solvent. At the same time, this test might be unreliable in diagnosing foaming caused by HSS or amine degradation components.

Many modern sweetening plants utilize anti-foam to control foaming in the amine absorber, although some plants do not. A program for using anti-foam is developed through consultation with the anti-foam vendor. There are two types of anti-foam programs: constant anti-foam injection via a metering pump, and anti-foam addition in batches. In the case of batch addition of anti-foam, there is usually a pressure drop threshold criteria that triggers the operators to inject the next anti-foam spike before the absorber reaches it limit for pressure drop or foam over.

Anti-foam has a limited life-time in the recirculated amine solvent in the sweetening plant. In the case of spike addition, the benefit for the anti-foam spike diminishes over the course of time because it is slowly removed when the solvent is circulated through the activated carbon filter.

If there is a pattern of cyclic pressure drop observed over the absorber, there is a process dynamic issue responsible for the cycles. The anti-foam program should be reviewed to control the changing foaming conditions in the absorber.

Notes on the Absorber Energy Balance

The interaction between gas and liquid in the absorber results in the reaction between amine and H_2S and CO_2. Modern rate-based simulation models for amine units consider a stepwise mass transfer model that has a mass and energy balance over each segment as shown in Figure 3. Mass transfer segments review the mass and energy inputs, and reviews the heat of reaction that develops a temperature profile throughout the absorber. If there is sufficient liquid flow rate in the absorber, the enthalpy from the reaction will be dissipated by the enthalpy of the liquid stream. Operation at a lower liquid flow rate will elevate the peak temperature or the range of peak temperature in the mass transfer section. If the peak temperature segment should exceed a critical threshold, this will cause the water vapor pressure to increase to the point that it becomes a significant contributor to the vapor volumetric flow rate. The peak temperature point in the stepwise model is usually accompanied by the highest vapor volumetric flow rate in the absorber profile. It is possible that flooding can occur in the middle of the absorber. Absorber operation at high peak temperatures in the region of 85° to 90°C (185° to 195°F) can exhibit abnormal foaming behavior.

For absorbers designed with physical solvents, the process designer considers the minimum L/G ratio, and specifies the solvent circulation rate with a recommendation that includes a margin of design safety factor. The text book method for determining the minimum L/G ratio is to plot the solute composition for the equilibrium line and operating line on a stage diagram, and then adjust the operating line slope until it intersects the equilibrium line. This text book graphical analysis works for isothermal absorbers without considering any enthalpy associated with the solute. For amine absorbers, however, it is necessary to consider the effect of the exothermic reaction of the solute with the solvent and to review the resulting temperature profile in order to evaluate the minimum L/G ratio for achieving the outlet gas objective. The review of the minimum L/G ratio for an amine plant requires a sensitivity study using a qualified rate-based simulation package for the solvent.

Figure 3 – Mass and Energy Balance for Absorber Segment

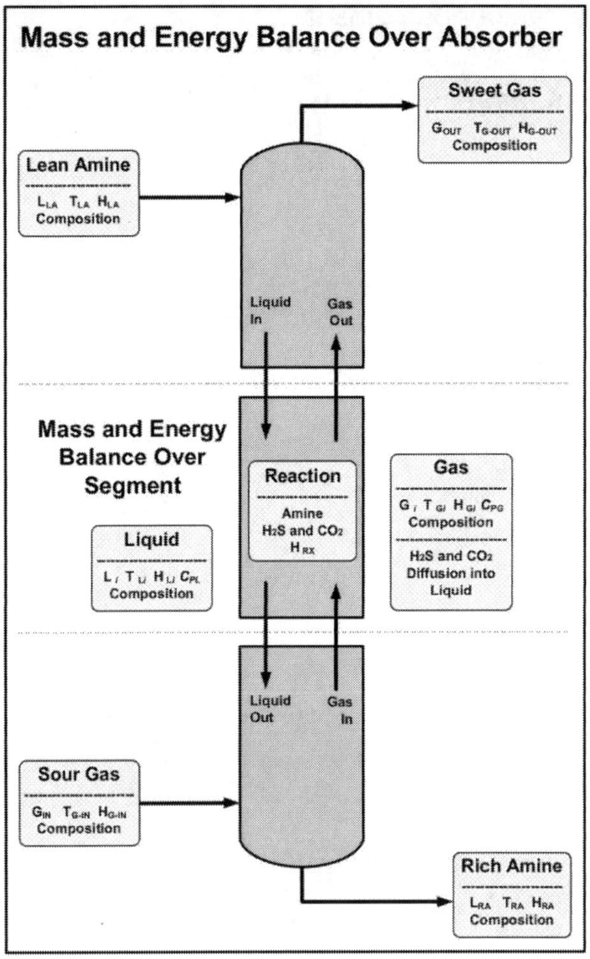

In view of the risk for a hydraulic capacity limitation in the interior of the amine absorber, it is necessary to evaluate the absorber with an accurate rate-based simulation program followed by review with the equipment vendor's hydraulic calculations. Most amine absorbers are equipped with temperature instruments for the inlet and outlet steams, and these vessels usually lack temperature instruments in the mid-section of the absorber where the peak temperature could occur. Temperature data from a pyrometer can provide insight into the temperature profile over the absorber to review the peak temperature and can confirm whether the actual temperature profile over the absorber matches the prediction from the simulation model.

The location and width of the peak temperature zone in the absorber can be an indicator of the adequacy of the solvent flow rate. Most well designed plants have a narrow peak temperature that occupies up to two or three actual trays or up to 10% of the packing height. An absorber temperature profile that extends over most of the mass transfer equipment exchange length is an indication of inadequate amine circulation rate or inadequate contacting of the solvent with the gas. Figure 4 shows the temperature profiles from a rate-based simulation for different liquid/gas mass flow ratios. The temperature profiles for high L/G ratios have a sharp peak temperature point and the peak temperature is located near the lower 20% of the mass transfer equipment height. In contrast, the temperature profiles for lower L/G ratios have a wider peak temperature zone and the peak temperature point is located at higher elevations in the mass transfer equipment.

Figure 4 – Temperature Profile as Function of Liquid / Gas Mass Flow Rate

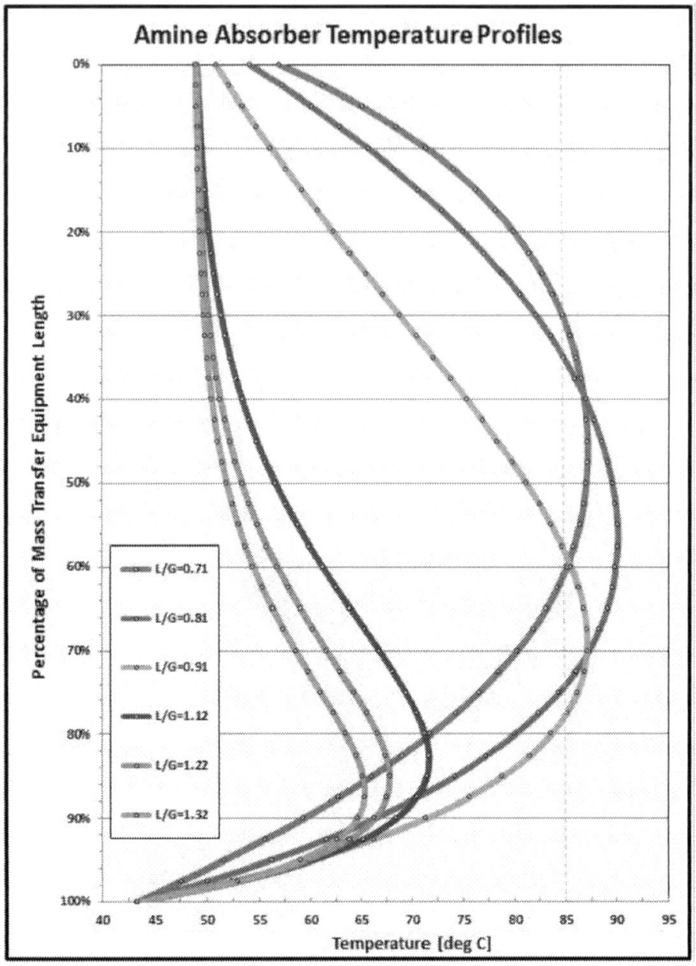

Notes on the Amine Regeneration Shortfall

In a healthy amine sweetening plant, it is possible to observe an incremental improvement in lean amine quality when the reboiler duty is increased in the regenerator. Some plants observe a limitation in the amount of residual hydrogen sulfide and carbon dioxide loading in the lean amine quality when they operate at the maximum possible reboiler duty. There are four potential causes for this shortfall of solvent regeneration:

1. Capacity limit for the regenerator mass transfer equipment.
2. Damaged trays or packing in the regenerator.
3. Leaky cross exchanger.
4. HSS problem with the amine solvent.

There are several avenues of investigation that can establish the root cause for solvent regeneration problems when coupled with deductive reasoning.

A first step is to perform a process simulation over the regenerator; then perform the hydraulic calculations. Next, review the pressure drop over the regenerator. If the pressure drop over the regenerator trays is below expectation, it is evidence for damaged or missing trays. If there are fluctuations observed over the regenerator with a higher pressure drop than expected, it is evidence for a possible hydraulic capacity limit for the mass transfer equipment. Performing a gamma-scan over the regenerator might confirm these two theories.

A leaky cross exchanger can be reviewed by taking lean amine samples upstream and downstream of the cross exchanger. Also, a leaky cross exchanger problem might be observed following a maintenance turnaround if cleaning was performed on the cross exchanger.

Amine regeneration shortfall due to HSS can be confirmed by sending lean amine samples to a qualified laboratory to review the various HSS contaminants. Diagnosing HSS as the root cause is usually confirmed when the three other causes for amine regeneration shortfall can be ruled out during the investigation along with confirmation of HSS from the laboratory report.

Final Step: Establishing the Root Cause

Some suspected causes can be eliminated by use of deductive reasoning in the investigation. Review of a comprehensive plant data table will identify which plant instruments and operating variables deviate most when compared to good operation. Use of the investigative process using plant data to review the mass transfer equipment hydraulics will focus on at least one clearly identified cause for the operating problem.

It is possible that the investigation might reveal more than one contributing factor to the plant performance problem. This is sometimes possible for old plants that are revamped to handle significantly different new operating conditions. This can also be possible for plants that suffer from HSS contamination of their solvent. Multiple causes for amine plant problems can be more challenging to diagnose and sometimes require operating data at several operating conditions to clearly identify the problem by thorough data analysis.

Troubleshooting Example 1

An amine absorber was designed for an off-shore platform based on an assumption for the inlet CO_2 composition from the exploratory gas well. The sweetening plant was designed by the engineering firm using a proprietary formulated amine solvent. Three years later when the platform was started-up at the well site, the operator found that the absorber was not meeting the CO_2

specification. The initial investigation revealed that the inlet CO_2 composition was approximately 5% higher compared to the original design basis. Initially, the height of the structured packing was the suspected cause for the platform's absorber performance problem.

When the higher CO_2 composition was reviewed in the sweetening plant simulations, the reported outlet CO_2 composition was still significantly higher compared to the simulation prediction performed both by the process designer and by Sulzer. A more thorough investigation included the completion of the plant operating data table similar to Figure-2. Afterwards, it was observed that the lean amine temperature was approximately 10°C higher than the original design basis. When the simulation considered the actual inlet amine temperature to the absorber, there was a reasonably good match between the simulation and the platform reported operating data. This focused the investigation on the performance of the lean amine trim cooler. It was determined that the root cause for the problem was insufficient trim cooler duty to deliver lean amine at the design flow rate and inlet temperature to the absorber.

Troubleshooting Example 2

An amine absorber located in Mexico was revamped from valve trays to MellapakPlus™ structured packing in order to improve the selective absorption of H2S using MDEA solvent [1]. The first retrofit was unsuccessful due to an undersized rich amine outlet nozzle. The absorber was observed to have "hot spots" when the outside vessel temperature was reviewed with thermal images from a pyrometer. The unit was also suspected to suffer gas maldistribution based on the thermal images, so the second revamp included the addition of a vapor distributor chimney tray. The second revamp was successful in providing good hydraulic operation of the packing, and the temperature profile was a closer match to the prediction from the rate-based process simulation.

The plant had consulted an additive vendor, who recommended a program for batch addition of anti-foam to control a cyclical pattern of foaming and pressure drop observed in the amine absorber. However, the plant operators complained that there were some days when the unit achieved the 4ppm outlet H_2S specification with a predictable amine circulation rate, and some days when the unit had difficulty meeting the H_2S specification with the maximum available amine circulation rate.

The performance of the regenerator was reviewed with a simulation sensitivity study for the heat duty that showed erratic performance for the lean amine quality delivered at the maximum available heat duty for the reboiler. There was a shift in the lean amine quality data from the simulation prediction based on the regenerator duty. Review of the temperature profile from the thermal scans of the absorber sometimes showed a reasonable match to the simulation prediction. But sometimes, the absorber showed a broad and flat temperature profile above the 85°C maximum temperature recommended by the solvent vendor. Review of the daily foam cell tests and review of the inlet knock-out drum performance pressure drop data did not reveal any breakthrough of solvent contaminants into the absorber. Review of the amine quality by at the solvent vendor's laboratory revealed the solvent experienced fluctuations in HSS that reduced the number of amine molecules available for reaction with the H_2S solute. The unit had historically operated with moderate to high Na^+ salt contaminations as the root cause for diminished performance of the solvent. The unit also suffered from variable bicene, formate, acetate and chloride contaminant problems, which explained the occasional daily problems in meeting the H_2S specification as a contributing factor to the HSS problem.

Troubleshooting Example 3

An amine absorber located in Russia was placed into service, and start-up was limited to a gas flow rate ceiling that was approximately 60% of the design capacity due to severe foaming reported by the plant operators [2]. Both Sulzer and the process designer performed a thorough review at the original design basis at the plant reported operating conditions. Multiple reviews of the design basis and operating conditions using different simulation packages failed to provide any explanation for the reported problem. Review of the tray hydraulic calculations also failed to provide any explanation for the reported heavy foaming problem. The plant consulted an additive vendor and initiated an anti-foam injection program, which also failed to alleviate the observed heavy foaming problem.

The operating company was forced to take the sweetening plant off-line. Sulzer recommended that inspection be performed on the absorber trays to confirm the tray weir and downcomer clearance settings, and review of the mechanical condition of the trays. During inspection, the inspectors observed different colors of pasty coatings at different tray elevations in the absorber. Samples were submitted to a laboratory, which revealed the unit suffered from micron sized particulates that matched the drilling mud used in one of the upstream production wells. The plant's inlet knock-out separator was not suitable for removal of the fine particulates found in the absorber. The sweetening plant used a mixture of gas from several wells to increase the plant capacity, and they later installed an improved filter particulate. After several months of operation, the troublesome gas well had sufficiently expelled its drilling mud and the sweetening plant was able to operate at its full design capacity. In this example, the heavy foaming problem could not be adequately addressed by changing of any of the plant's process control variables. The root cause could not be resolved until the fine particulates were eventually eliminated from the gas well with the fine particulate drilling mud.

Troubleshooting Example 4

An amine absorber unit was revamped with high capacity tray in order to increase the gas handling capacity of the sweetening plant. The maintenance turnaround included cleaning of all process equipment in the entire plant. Following start-up, the absorber failed to provide the same removal of H_2S that was achieved prior to the turnaround. The plant checked operation at the same gas flow rate achieved prior to the turnaround, again with the same shortfall in meeting the H_2S specification. The initial investigation focused on the new high capacity trays. Thorough review of the process simulation and tray hydraulic calculations did not reveal any explanation for the shortfall in the absorber performance.

The plant took lean amine samples from a sample port located immediately downstream from the regenerator. The plant's laboratory confirmed the regenerator was delivering the expected amine quality. The solvent vendor's laboratory confirmed the solvent was not affected by any HSS issue. One operator took a lean amine sample from the surge tank, and noticed it had a slight different color. Analysis of amine samples from the surge tank revealed different residual amine quality compared to the samples taken downstream from the regenerator.

The turnaround had also included cleaning of the cross exchanger tubes using scrubbing brushes. Those tubes were carbon steel, and it was suspected that some tubes might have been punctured during the cleaning procedure coupled with corrosion. The plant ordered a replacement cross exchanger. Inspection of the corroded cross exchanger confirmed the root cause was pin sized holes in several of tubes that provided leakage from the rich amine stream that contaminated the lean amine stream.

Troubleshooting Example 5

An amine absorber was equipped with valve trays. Several months after start-up, the absorber removal efficiency suffered and the unit had difficulty meeting the CO_2 and H_2S removal objectives. The plant responded by increasing the amine circulation rate to its maximum flow capacity. The unit operated for several months, and then the plant operators noticed a gradual reduction in the absorber removal efficiency. A gamma-scan was taken over the vessel that revealed that the bottom two trays were displaced, and there was a difference in the scan lines for the five trays located above the gas inlet compared to the other trays at higher elevation.

The plant decided to perform a revamp to improve the reliability of the absorber by installing random packing. The plant did not remove the tray support rings or downcomer clamping bars located in the interior of the packed section. The revamp was only partially successful in improving the absorber performance.

During the turnaround, the plant inspected the trays prior to their demolition. The trays immediately above the inlet gas had lost most of their valves. Trays at higher elevations were observed to have fewer lost valves. Closer inspection of the tray deck revealed notches in the tray deck where valve action of the valve legs fretted away at the carbon steel tray deck. The valves were 410S ferritic stainless steel material. There is the potential risk for galvanic corrosion for contact of 410S with carbon steel in the presence of aqueous electrolytes. There was a pattern for valve loss on the second tray above gas inlet that strongly suggested that liquid cross flow from the downcomer to the outer periphery of the deck was partially responsible for eroding the carbon steel tray decks.

The root cause for the problem was damage to the two bottom trays during a gas flow rate excursion. The subsequent problem was continued damage to the trays that caused the loss of valves due to the combination of gas flow direction and valve wear against the carbon steel tray decks. In this case, the plant made the decision for a potentially risky revamp from trays to random packing. The absorber could have been retrofitted with heavy-duty trays for the bottom two trays, and with austenitic stainless steel valves for the upper section trays for a lower cost compared to the packing revamp.

Conclusion

Writing of this paper is motivated by investigation problems when reported by plant operators. Sometimes the plant does not adequately provide sufficient operating history for the trouble-shooting analysis to conclusively determine the cause. Sometimes, there is a previous process simulation performed on the plant when it operated under different operating conditions and plant personnel do not have a rate-based simulation available to perform a simulation study for the current operating conditions. There can be significant issues that affect the solvent reaction kinetics for the absorber and it is difficult to extrapolate predictions for deviations of plant performance from a previous simulation.

Revamps of existing amine sweetening plants can be complicated by pre-existing process problems. The investigation of a problem with a revamped sweetening plant should include a thorough review of all potential issues, including the operating solvent L/G ratio and solvent quality. A revamp that considers the installation with higher capacity or higher efficiency mass transfer equipment could potentially be compromised by contributing problems associated with HSS or solvent quality that go undiagnosed during the revamp.

This paper has presented important steps for an engineering investigation to understand the root cause of amine sweetening plant performance problems. The process requires a comprehensive plant data table to place the operating problem in its context and to compare to conditions for acceptable operation. Plant operators who can provide a plant operating data table will facilitate the trouble-shooting analysis by third-party investigators. The investigation should include conducting a process simulation study and hydraulic capacity and pressure drop calculations. The use of deductive reasoning can remove unlikely suspects and focus the investigation on the factors behind the root cause of the problem.

Literature References:

[1] *"Retrofit of a H2S Selective Amine Absorber Using MellapakPlus Structured Packing"*; G.H. Shiveler, G.S. Solis, L.H.P. Gonzalez, M.L. Bueno; Y2005 Spring AIChE Meeting Atlanta; April 2005.

[2] *"DEA Treater Revamp Targets Foaming"*; Waldo de Villiers, William Green, Glenn Shiveler; Y2009 AIChE Spring Meeting Tampa; paper 26f; April Y2009.

Biographical Information for Authors:

Glenn Shiveler has worked for nearly 30 years in the gas processing industry designing mass transfer equipment. He now works as an applications specialist in the Sulzer office located in Tulsa, Oklahoma. He has co-authored two previous papers presented at AIChE on trouble-shooting amine plant operational problems. He is a chemical engineer who attended Rutgers University, and holds membership in AIChE, ACS and ISA.

Harold Wandke holds B.S. and M.S. degrees in chemical engineering from Oklahoma State University. His first 6 years of professional employment were spent designing, building, and operating petroleum refining and petrochemical pilot plants. He has spent the last 21 years with Sulzer Chemtech, where his duties have included mass transfer equipment and design (both process and mechanical); project management; and plant commissioning and start-up. He is a registered professional engineer in the state of Oklahoma.

Flooded Condenser Controls: Principles and Troubleshooting

By

Henry Z. Kister, Fluor, Aliso Viejo, CA, USA

Presented at the Distillation Topical Conference, AIChE Spring Meeting, Austin, Texas, April 26 - April 30, 2015

Flooded Condensers Controls: Principles and Troubleshooting

Henry Z. Kister, Fluor Corp., Aliso Viejo, CA

Flooded condenser control is most popular with total condensers generating liquid product only. The principles of these methods had been described in the literature more than 60 years ago, yet some of these methods, like the hot vapor bypass and the flooded drum methods, continue to be among the most troublesome distillation controls. Chin's classic paper on distillation pressure control[1] describes the principles, good practices and how many of the potential problems can be avoided. Most of the reported problems could have been avoided with a better understanding of the principles and learning from past experiences[2].

A lot of experience has been accumulated with these methods over the four and a half decades since Chin's paper came out. This paper provides an updated detailed description of the principles and applies them to address many of the most common traps that had caused problems in the application of flooded condenser control methods.

BACKGROUND

Flooded condenser control is most popular with total condensers generating liquid products only. The condenser area is partially flooded by condensate. The flooded tubes do not contact the vapor and perform little condensation. Column pressure is controlled by manipulation of the flooded area. Raising the liquid level in the condenser floods additional tubes, which reduces condensation area, thereby raising tower pressure. Conversely, lowering the liquid level in the condenser exposes more tubes, which increases the condensation area, and subsequently lowers the column pressure.

Although the flooded area performs little condensation, it serves the vital purpose of subcooling the condensate before it leaves the exchanger. The subcooling is beneficial when pumping volatile liquids[3]. Although the subcooling requires some heat transfer area, this area is not always added in exchanger design. Some designers[3] are comfortable to have the subcooling area come from the overplus allowed for in the exchanger design, while others[4] oversize the condensers by as much as 25% to ensure subcooling, especially when the subcooling is critical, as in the hot vapor bypass scheme.

Figure 1 shows common flooded condenser control arrangements, all with the condenser mounted above the reflux drum. Figure 1a has the control valve in the condenser liquid outlet. The control valve required is small, and should be located as close to the reflux drum as possible to maximize static head[1] when the condensate enters at the top of the drum. For condensate entrance at the bottom of the reflux drum (like in Figure 2a), the valve should be located at the lowest horizontal leg. This method is simple, linear, and maintains the same pressure in the column and the drum. It is therefore often favored[1, 5].

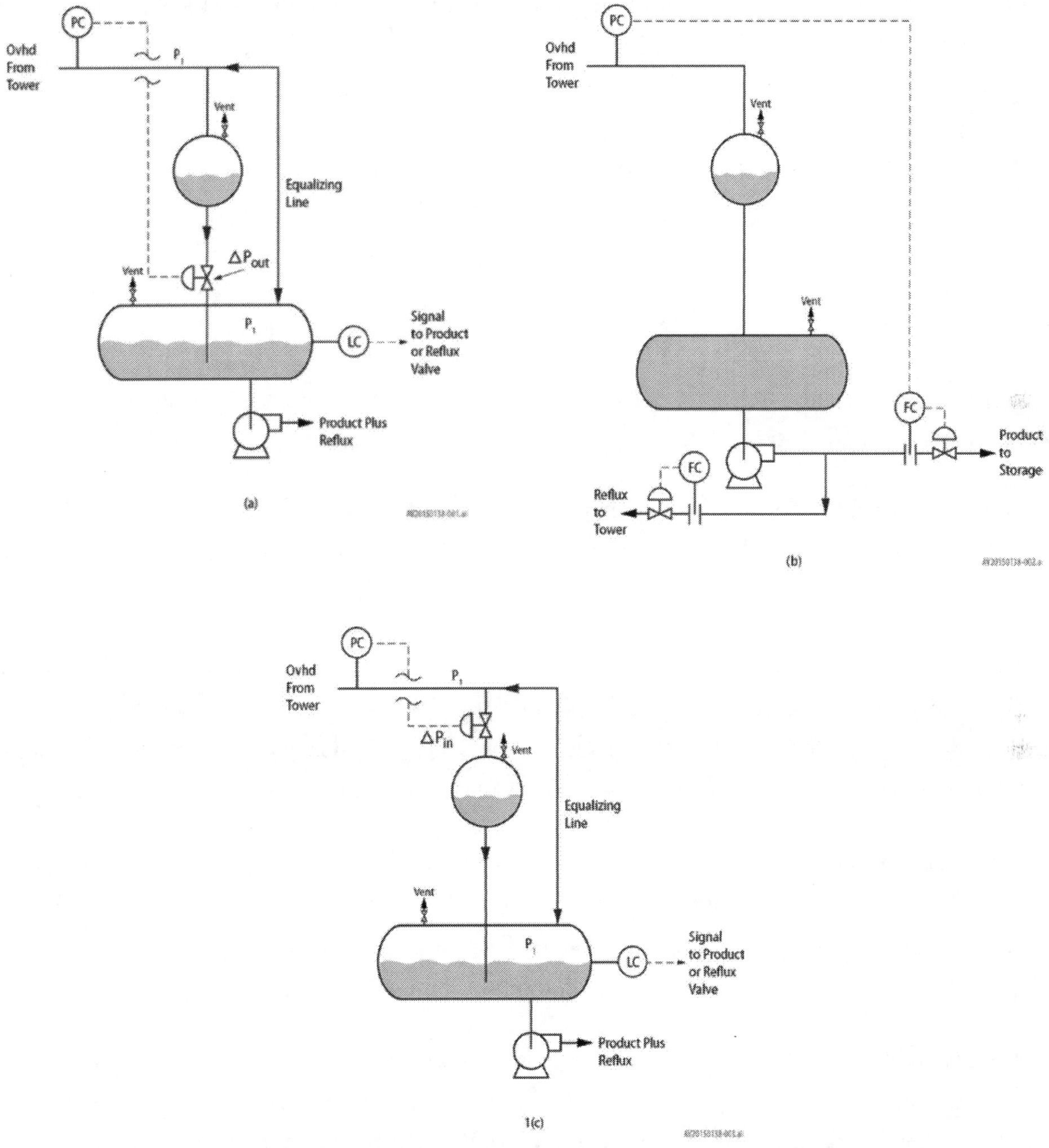

Figure 1. Common flooded condenser controls arrangements. (a) Control valve in the condensate liquid line; (b) Flooded drum; (c) Control valve in the vapor line.

This control method requires that a pressure-equalizing line is included[1, 5, 6, 7]. Without this line, pressure in the reflux accumulator will be unsteady. A smaller equalizing line is required when the subcooled liquid is introduced near the bottom of the drum as shown in Figure 1a.

Figure 1b illustrates the flooded reflux drum method. Here the drum runs full of liquid, and the level control of the reflux drum is eliminated. Sometimes the reflux drum itself can be omitted. The pressure controller directly controls distillate flow. Due to the tight pressure control usually required, distillate flow rate controlled by this method is likely to fluctuate. These fluctuations may destabilize downstream units. This method should therefore only be used when the product goes to storage[1, 5, 8, 9], and should be avoided when the product goes directly to another unit.

This method is sometimes used to control reflux flow. This practice is not recommended[5]. Here reflux flow rather than product flow is likely to fluctuate, and this can destabilize the tower.

The flooded drum method suffers from many other potential issues, which are separately described later in this paper.

Figure 1c shows a flooded condenser scheme similar to that of Fig. 1a, but with the control valve located at the condenser vapor inlet. Similar to method in Figure 1a, the condenser liquid outlet line must enter near the bottom of the reflux drum, and a pressure-equalizing line is required.

When the control valve is in the condenser inlet, the condensation pressure is lower than in the earlier schemes, which requires additional condenser surface area. If no additional area is provided, tower pressure needs to be raised, which increases energy consumption. The control valve required is large and may be impractical with large overhead lines[1].

Figure 2 shows additional flooded condenser control schemes, all containing a control valve in the condenser vapor bypass. Figures 2a and 2b are analogous to Figures 1a and 1c, the difference being the addition of the bypass control valve. This control valve helps overcome the pressure drop in the condenser, especially when the liquid head is small.

Figure 2c shows the hot vapor bypass pressure control method. Here the control valve location is switched from the condenser route in Figures 1a or 1c to the hot vapor bypass. The condenser is at ground level, typically 10-20 ft below the reflux drum (which may be horizontal or vertical). The condensate must be subcooled so that liquid surface in the drum is colder than in the condenser. This generates a difference in vapor pressure large enough to lift condensate from the condenser into the drum. To reduce column pressure, the valve is throttled. This reduces hot vapor supply to the drum liquid surface, lowering the drum surface temperature, and, therefore, the drum vapor pressure. This increases the vapor pressure difference between the condenser and the drum, which in turn sucks liquid from the condenser into the drum. This exposes additional tubes in the condenser, and increases condensation rate, which lowers column pressure.

The hot vapor bypass arrangement permits the condensers to be mounted at ground level instead of on a platform above the reflux drum. Locating large cooling water condensers at ground level eliminates the need for massive condenser support structure and for piping cooling water to high elevations, and provides easy access for maintenance. The piping is simple, the control valve is small, and the response is fast[3, 10]. These advantages can translate into handsome savings in steelwork, platforms, trolleys, and maintenance. These savings can be major in large installations, especially where a battery of condensers rather than a single exchanger is used. However, this method suffers from many potential issues, which are described in detail elsewhere[11].

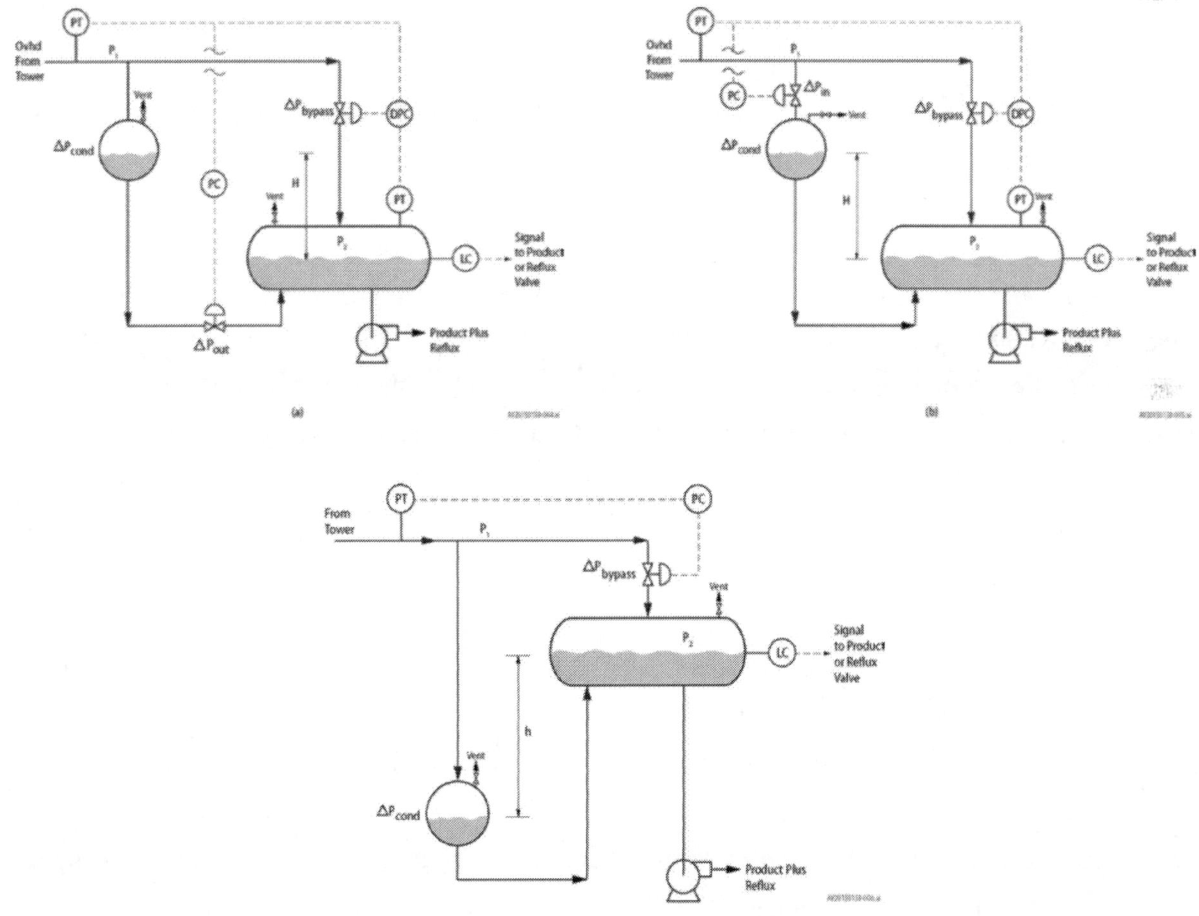

Figure 2 Flooded condenser controls with valves in the vapor bypass (a) Pressure control valve in the condensate liquid line; (b) Pressure control valve in the vapor line.; (c) Hot vapor bypass.

HYDRAULICS OF FLOODED CONDENSERS

Figure 2a shows a common arrangement. The condenser is elevated above the drum, with the condensate descending into the drum by gravity. The column pressure control valve is in the condensate line from the condenser to the drum, which enters the drum below the liquid level. Up to this point this scheme is the same as that in Figure 1a. The difference is that in Figure 2a there is also a control valve in the condenser bypass. The presence of this control valve renders the pressure in the drum lower than at the condenser, which introduces a vapor pressure effect. Assuming negligible line pressure losses, a pressure balance on the condenser gives

$$P_1 - P_2 = -H + \Delta P_{cond} + \Delta P_{out} \qquad (1)$$

Where

P_1 is the pressure at the junction between the vapor line to the condenser and the condenser bypass, psia:
P_2 is the pressure inside the reflux drum, psia;
H is the head differential between the condenser liquid level and the reflux drum liquid level, psi
ΔP_{cond} is the condenser pressure drop, psi;
ΔP_{out} is the pressure drop across the condensate outlet control valve, psi.

The density used to calculate the head differential H is the difference between the liquid and vapor densities, due to the presence of a static leg of vapor in the bypass line. The vapor density is based on the drum pressure and drum vapor space temperature. The liquid density is best approximated as the density of the subcooled liquid leaving the condenser.

If there were no control valve in the bypass, then $P_1 = P_2$, and Eq. 1 becomes

$$H = \Delta P_{cond} + \Delta P_{out} \qquad (2)$$

Eq. 2 describes the configuration in Figure 1a (the bypass acting as a pressure equalizing line). Eq. 2 states that for the Figure 1a configuration, the head required to drain the condenser must be high enough to overcome the condenser pressure drop plus the pressure drop of the condenser outlet valve. This condition may not be satisfied when the condenser pressure drop ΔP_{cond} is high and/or the available liquid head H is low. Adding the valve in the bypass, i.e. converting the Figure 1a scheme to Figure 2a, changes the difference between P_1 and P_2 to ΔP_{bypass} giving

$$P_1 - P_2 = \Delta P_{bypass} \qquad (3)$$

Combining Eq. 1 and Eq. 3 gives for the scheme in Figure 2a

$$H + \Delta P_{bypass} = \Delta P_{cond} + \Delta P_{out} \quad (4)$$

So the pressure drop across the bypass valve helps the gravity head push the liquid from the condenser into the drum.

For the arrangement in Figure 2b, the pressure drop at the condenser inlet ΔP_{in} (psi) replaces the pressure drop at the condenser outlet ΔP_{out}, and Eq. 4 becomes

$$H + \Delta P_{bypass} = \Delta P_{cond} + \Delta P_{in} \quad (5)$$

For the arrangement in Figure 2c, there are no control valves at the condenser inlet nor outlet, and the liquid head h (psi) is the difference between the reflux drum liquid level and the liquid level in the condenser, i.e. $h = -H$. The value of H in Eq. 4 is negative as the liquid level in the reflux drum is at a higher elevation than that of the condenser. Eq. 4 then becomes

$$\Delta P_{bypass} = \Delta P_{cond} + h \quad (6)$$

VAPOR PRESSURE DIFFERENTIALS

For as long as there is no hydraulic restriction in the condenser bypass line (Figure 1a), and the pressure drop of the condenser inlet line and at the condenser entry is low, the pressure is the same at the condenser as at the reflux drum. In this case the vapor pressure differential between the condenser surface and the drum liquid surface is zero. In flooded condensers (Figure 1b) there is no vapor space in the drum so the vapor pressure differential between the condenser and drum liquid has no impact on the controls. In all other situations, the vapor pressure differential plays a role, often a major one.

The pressure difference between the condenser and drum in schemes 2a through 2c introduces vapor pressure effects. In Figures 2a and 2c, $P_1 - \Delta P_{cond}$ is the vapor pressure of the liquid surface in the condenser, assuming most of the condenser pressure drop is near the condenser inlet (usually a reasonable assumption for total condensers). In Figure 2b, the vapor pressure of the liquid surface in the condenser is $P_1 - \Delta P_{in} - \Delta P_{cond}$. In all three schemes (Figure 2a through 2c), P_2 is the vapor pressure at the reflux drum surface.

For ($P_1 - \Delta P_{in} - \Delta P_{cond}$) to be higher than P_2 , the temperature at the drum surface needs to be lower than that at the condenser surface. The hot vapor provided by the bypass condenses onto the drum liquid surface, which keeps the surface hot. At steady state, the valve induces sufficient vapor to maintain the drum liquid surface temperature at the value corresponding to the desired vapor pressure P_2. There is a heat flow from the hot liquid surface to the subcooled liquid underneath and also atmospheric heat losses from the vapor space of the drum. These heat flows need to be matched by condensing the hot vapor from the bypass.

As long as the drum surface remains steady, most of the heat flow from the surface to the subcooled bulk liquid is by conduction. Since process liquids are good thermal insulators the conduction heat transfer from the hot surface to the subcooled liquid is small. In reality, convection and some bulk movement raise the heat transfer from the surface to the subcooled bulk liquid, but even accounting for these, the bypass vapor flow rate can easily match the heat demand at the drum liquid surface.

The heat balance for the drum was discussed in detail in our work on hot vapor bypass controls[11], the scheme in Figure 2c. The heat balance discussion in that paper fully applies to other flooded condenser control schemes (1a, 1c, 2a, and 2b).

UNIQUE VAPOR PRESSURE CONSIDERATIONS FOR HOT VAPOR BYPASS

While the vapor pressure effects are important for understanding all flooded condenser schemes, they become extremely important with hot vapor bypass schemes with submerged condensers (Figure 2c). This is because in this scheme, the vapor pressure differences directly determine the flooded height in the condenser. In the elevated condenser schemes (Figures 2a and 2b) the condenser inlet or outlet valve directly determine the flooded height, with the bypass mainly used to provide sufficient pressure drop per equations 4 and 5.

With the hot vapor bypass scheme (Figure 2c), the vapor pressure difference provides the driving head that pumps the condensate liquid from the condenser into the reflux drum (Eq. 6). Eq. 6 can be combined with Eq. 3 to give

$$P_1 - P_2 - \Delta P_{cond} = h \tag{7}$$

Eq. 7 shows that the vapor pressure difference is balanced by the liquid head lift. If higher pressure is required in the tower, there is a need to flood more area in the condenser, ie, to reduce the liquid head lift h. To achieve this, the vapor pressure differential $P_1 - P_2 - \Delta P_{cond}$ needs to be reduced. With P_1 and ΔP_{cond} constant, this is achieved by opening the hot vapor bypass to raise P_2. Opening the hot vapor bypass heats up the liquid surface in the drum, which raises the vapor pressure P_2. Conversely, to reduce column pressure there is a need to lower the liquid level in the condenser, i.e., to raise the liquid head lift. This is achieved by closing the bypass valve. This cools the surface in the drum, lowers P_2, and the larger $P_1 - P_2$ difference sucks liquid from the condenser into the drum, thus exposing more condenser area for condensation. These mechanisms are described in detail in Ref. 11.

AVOIDING OPERATING PROBLEMS

As stated earlier, our survey of tower malfunctions[2] found the flooded condenser controls are among the most troublesome distillation pressure control methods. It is therefore critical that engineers involved with flooded condensers develop a good understanding for its principles and a good appreciation of the common operating problems.

Vapor to vapor, liquid to liquid lines hook-up – Correct piping is mandatory for the success of all flooded condenser schemes. Bypass vapor must enter the vapor space of the reflux drum. The bypass should be free of pockets where liquid can accumulate; any horizontal runs should drain into the reflux drum. We are familiar with cases in which a pocket in the hot vapor bypass in Figure 2c led to severe oscillations and column pressure swings. Most important, liquid from the condenser must enter the reflux drum near the bottom of the drum, well below the liquid surface. This is imperative with the hot vapor bypass scheme (Figure 2c), and highly recommended with the others. The rule is "vapor to vapor, liquid to liquid".

If the liquid enters at the bottom of the drum with an upward momentum (as shown in Figures 2a – c), a horizontal baffle should be added above the inlet to spread the momentum of the incoming liquid jet. As recently reported[11, 12], liquid jets at velocities of a few feet per second can easily penetrate through quite a few feet of drum liquid, bringing a variable amount of subcooled liquid to the drum liquid surface, and disturbing the surface. In some cases[11,12], such disturbances caused a massive amount of liquid to be suddenly sucked from the condenser into the drum.

Fig. 3 depicts one case[14] in which violation of this practice led to severe pressure fluctuations, inability to keep column pressure, and a capacity bottleneck. In the original scheme (upper configuration), subcooled liquid mixed with vapor at its dewpoint, and vapor collapse occurred at the site of mixing. The rate of vapor collapse varied with changes in subcooling, overhead temperature and condensation rate. Variation of this collapse rate induced pressure fluctuations and control valve hammering.

In the red piping in the bottom configuration shows the fix that eliminated the problem. The liquid and vapor lines were separated, and the vapor line was modified so that it introduced vapor into the top of the reflux drum. After these changes were made, the tower pressure no longer fluctuated, and the problem was completely solved.

This case is one example of more than two dozen cases that the author is familiar with. This variation violates the "vapor to vapor, liquid to liquid" rule above and is the most common cause of poor performance of hot vapor bypass schemes. A number of these cases were reported in the literature[5, 6, 13, 14].

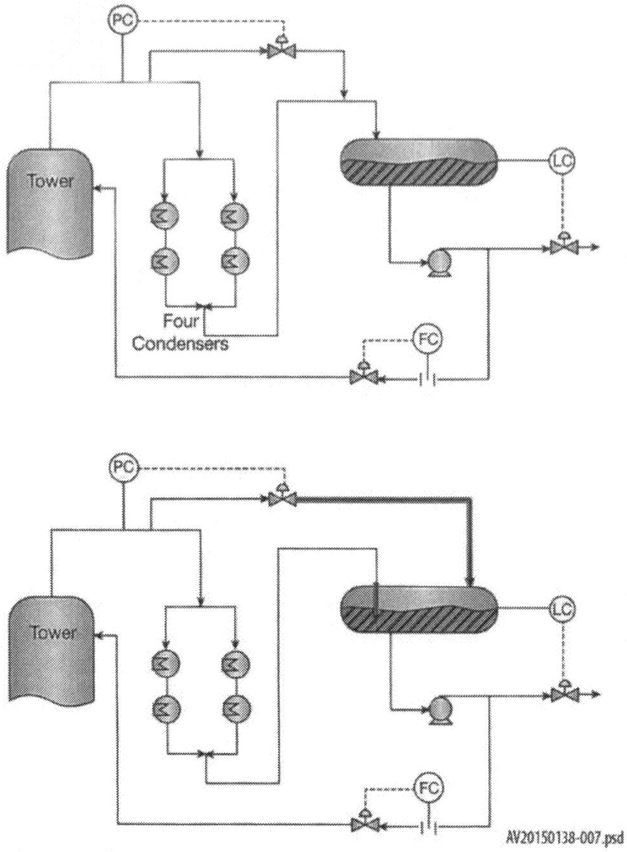

Figure 3: Hot Vapor Bypass Hookups: upper: incorrect, led to pressure fluctuations; lower: modified, good pressure control. (from Ref. 11. Reprinted courtesy of AIChE).

Any subcooled liquid streams entering the drum must also enter at or near the bottom of the drum. In one case[6] subcooled liquid entered the drum vapor space (presumably due to unflooding of the liquid inlet). The vapor space was 100°F hotter; rapid condensation sucked the liquid leg between the drum and condenser into the drum in seconds.

With the systems in Figures 1a, 2a, and 2b there is some debate in the literature whether the liquid should be introduced into the vapor space or into the liquid at (or at least near) the bottom of the drum. One source[1] prefers liquid entry above the liquid level so that drum level does not affect condenser level.

The author strongly prefers liquid entry below the liquid level so that cold liquid is introduced at the bottom of the drum. This practice is shared by others[10, 15]. Introducing subcooled liquid above the liquid level is likely to cause collapse of vapor when the cold liquid meets the surface liquid; this in turn may result in pressure fluctuations and possible hammering. Further, introducing the subcooled liquid near the surface largely drops the vapor pressure in the drum, tremendously raising the demand of hot vapor from the bypass, and often overwhelming the capacity of the bypass. If a much larger bypass is not available, the pressure inside the drum can fall as far down as the vapor pressure of the liquid at the subcooled temperature. The author is familiar with cases in which this pulled vacuum inside the reflux drum. In other cases, where the bypass was large, the liquid almost entirely lost its subcool, causing cavitation of the reflux pump. Splashing subcooled liquid onto the drum surface can also lead to generation of static electricity. The higher the difference between bubble point and subcooled liquid, the more aggravated these issues become. Having the liquid line entering the drum liquid also has the advantage of providing a better seal to the condenser and avoiding vapor from blowing through the condenser[3].

A common design practice is to introduce the liquid from the top of the drum via a slotted or perforated pipe, so that most of the liquid is introduced below the drum liquid level, but some is splashed on to the liquid surface. This is better than introducing all the liquid into the vapor space but not as good as introducing the liquid near the bottom of the drum, and has been troublesome. The larger the opening that discharges liquid into the vapor space the more troublesome this method is likely to be, especially with a high degree of subcooling like cold winter nights or low-rate operation. The issues are identical to those described in the previous paragraph. The author had experiences of severe hammering when liquid was introduced from the top of the drum with perforated or slotted pipes. The hammering ceased after the slots in the vapor space were blocked.

Decanting Water – If the reflux drum is used to decant small quantities of free water from condensed hydrocarbons or other water-insoluble organics, the entry point of the condensate liquid (and other subcooled liquid streams that may contain free water) should be at the opposite end of the drum to the end at which the liquid product and reflux are withdrawn. The water removal boot should be just upstream of the point where the reflux and product streams are withdrawn[16], as illustrated in Figure 4 for the control system in Figure 1a. In many cases, a short standpipe (about 6 to 12 inches tall) is used for the reflux and product draw to positively avoid water in them[16], but this technique may lead to water accumulation in the drum and possibly corrosion when the interface level controller in the boot malfunctions and water containing acidic components is not adequately removed from the drum. Reference 17 describes a related experience.

Non-condensables – Flooded condenser methods are suitable only for total condensers, although some less satisfactory variations are also available for partial condensers[5]. The methods in Figures 1 and 2 can handle small amounts non-condensables such as those introduced during startups or upstream upsets. To handle these non-condensables, vents are required on the condenser and drum. The condenser vents can be directed to the vapor space of the drum, to an upstream unit, or elsewhere. The drum vents, and if frequent venting is anticipated also the condenser vents, should be board-operated. A colleague of the author[18] experienced frequent high pressure, flaring and instability in a debutanizer flooded condenser system due to breakthrough of non-condensables from an upstream tower that experienced control issues. Adding a manual board-operated condenser vent to an upstream system that opened upon high pressure mitigated the problem.

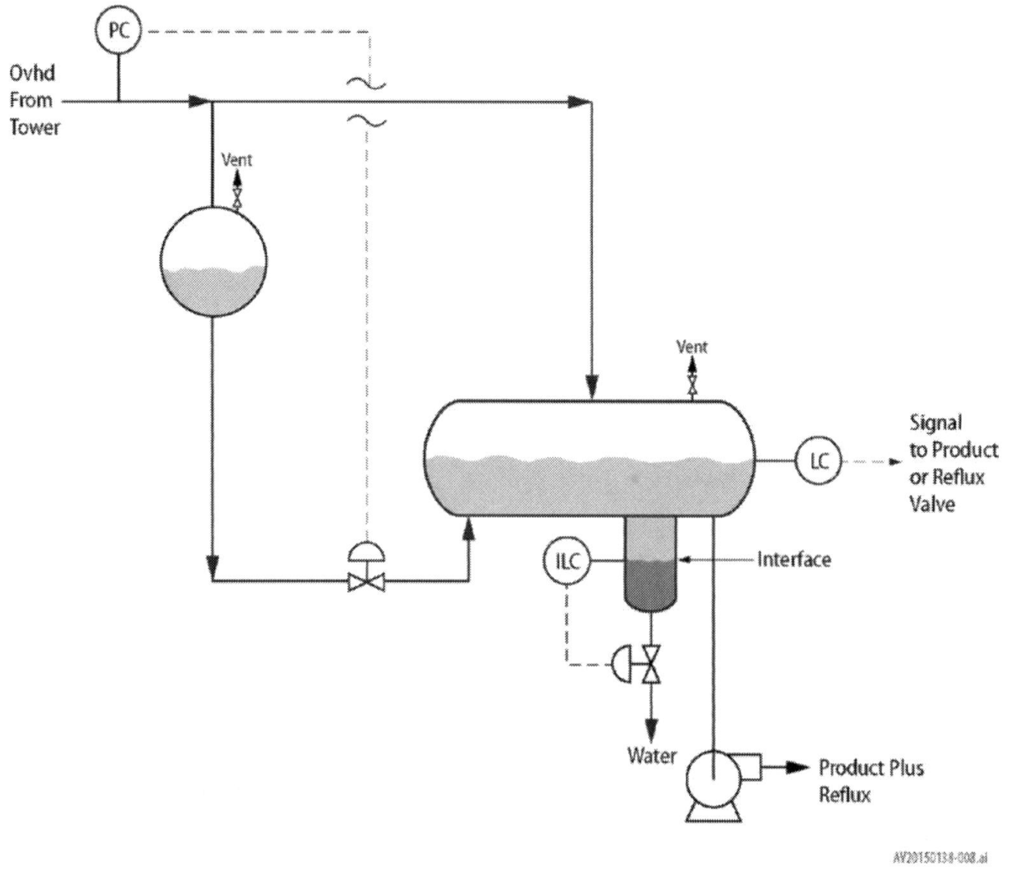

Figure 4. Flooded condenser control of Figure 1a with water decanting

Insufficient Subcooling – Flooded condenser control methods produce subcooled reflux and product. This subcooling is beneficial in avoiding Net Positive Suction Head (NPSH) problems in the pump and/or flashing problems in the reflux and/or product line. Such flashing can lead to instability, poor reflux distribution at tower inlet, slug flow, even hammering[21]. The instability may be particularly severe with the flooded drum method (Figure 1b) due to its potential interaction with the column pressure control, and is further discussed below in connection to this method. However, other flooded condenser methods are not immune.

Subcooling is diminished when the condenser nears its maximum capacity. This may be the natural maximum limit, or caused by fouling, non-condensable accumulation, condenser drainage, or other issues. Subcooling is also diminished when liquid is splashed onto the surface of the drum as discussed earlier. Finally, many advanced controls use pressure minimization strategies such as Shinskey's floating pressure control[8, 19]. These strategies reduce tower pressure during periods of favorable ambient temperatures to minimize energy consumption and in high-pressure towers (>150 psia) also maximize capacity. The pressure reduction brings the condenser closer to its limit, and by doing so, minimizes subcooling.

Issues with insufficient subcooling can be avoided for as long as the condenser is not near its capacity limit. The key is to avoid splashing the condensate liquid onto the drum surface, minimize condenser fouling, properly vent non-condensables from the condenser, and to monitor the subcooling. Some override control or simply an operator advisory to limit the pressure minimization upon low subcooling may also be beneficial.

Air Condensers – Air condensers are elevated above the reflux drum, so the only compatible flooded condenser schemes are those for condensers mounted above the drum (Figure 1, Figure 2a and 2b). The hot vapor bypass method (Figure 2c) requires mounting the condenser below the drum and is incompatible with air condensers. We have seen poor pressure control with the Figure 2c scheme with an elevated air condenser. In one case[11] condensation was by an elevated air condenser followed by a ground-level cooling water condenser. The hot vapor bypass control of Figure 2c worked well when the liquid level was in the water condenser, but became unstable in cold days when the air condenser supplied the total condensation duty and the liquid level climbed to the air condenser. The solution was to reduce the air condenser duty by shutting fans and closing louvers so the liquid level remains in the water condenser.

Air condensers have small ratios of height to width. A change in liquid level of one inch may result in an entire tube row being covered or uncovered. Typically an air condenser will have very few rows, often well below 10, so covering or uncovering one gives a bump in heat transfer.

It is common to slightly slope several bottom tube rows, or all rows, towards the outlet so that the movement of liquid level up or down the tubes is less bumpy[4].

Flooded-drum controls (Figure 1b): unique issues As stated earlier, a major issue is that the product flow rate using this method tends to fluctuate, often severely. This method is therefore only suitable when the product goes to storage. Other issues include:

- Besides the condenser venting issue above, the flooded drum method has an additional venting issue. Accumulation of noncondensables in the drum may unflood the drum and interrupt the control action. These noncondensables need to be vented from near the top of a liquid-full drum, so they need to be vented to a facility that can handle liquid like a knock-out drum. If noncondensables accumulation is infrequent, manual venting from the top of the drum is often sufficient to maintain satisfactory operation. If noncondensables accumulate frequently, or the column is run unattended, automatic venting is required.
- Figure 5 illustrates an automatic vent system that worked well[5, 13, 20]. A second pressure controller (PC No. 2), a level controller, and a control valve in the vent line are added. The set point of PC No. 2 is lower than that of the normal pressure controller (PC No. 1). When the drum is full, the level controller keeps PC No. 2 tripped off, and the vent valve is closed. Drum unflooding (due to noncondensable accumulation) is sensed by a drop in drum level. The lower level activates PC No. 2. Since the set point of PC No. 2 is lower than PC No. 1, it opens the vent valve. As the pressure falls, PC No. 1 closes, helping to build up the drum level. As soon as the drum refills, the level controller trips PC No. 2, and the vent valve closes.

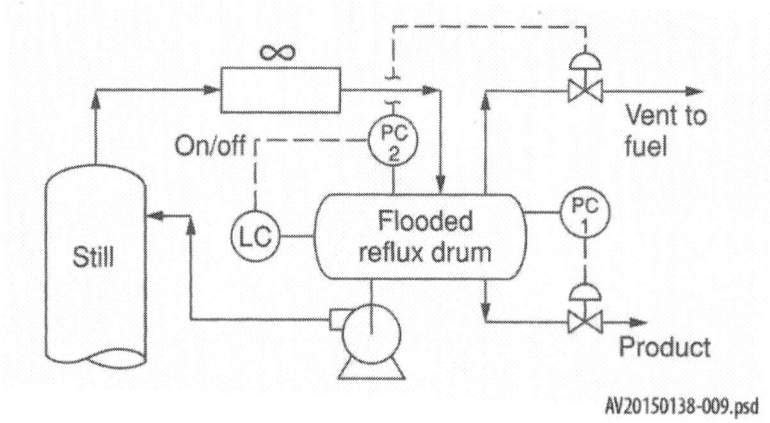

Figure 5 Flooded drum automatic vent system that worked well (from Ref 20. Reprinted courtesy of AIChE)

- Since the venting required above is from a liquid-full drum, leaks from the vent line are likely to be liquid, giving a product loss much greater than a leaking vapor valve. Flashing of liquid across the valve can chill the vent line, sometimes causing icing up or "weeping" on the outside of the pipe due to atmospheric moisture condensation, with possible overchilling or corrosion of the vent pipe. Lieberman[16] describes one case of icing up due to such a leak from a debutanizer flooded drum.

- Unless the product is subcooled and at a significantly higher pressure than storage, it is best to take the product to storage from downstream of the reflux pump (as shown in Figure 1b). If the product is taken directly from the drum, flashing may occur downstream of the control valve, or it may be difficult to get product into storage when storage pressure is high. Either may cause instability or back excessive liquid into the condenser, thereby reducing its capacity and possibly leading to a relief situation. In one depropanizer[19], pressure variations in elevated propane storage bullets downstream induced intermittent flashing and slug flow in the product line even though the product was pumped. Collapse of vapor due to elevation and pressure changes is believed to have caused transient shock waves, hammering, and chattering of the relief valves in the pump discharge circuit. The chattering was eliminated by adding a back pressure controller that prevented the flashing.

- With the flooded drum method, a failure of the reflux pump often produces a relief situation. The condensate has nowhere to go, quickly floods the condenser, condensation ceases in the flooded region, and the tower pressure rises until the relief valve lifts. In other flooded condenser schemes (e.g., Figure 1a) where the reflux drum is not flooded, the vapor space in the drum provides operators with a few minutes to take action before the liquid fills the condenser causing the relief valve to lift. The author is familiar with one plant that replaced a flooded drum by a new non-flooded drum that was over twice the volume (to accommodate a vapor space) just to prevent recurrence of such a relief.

- Any relief valve mounted on the flooded drum (Figure 6) is most likely to discharge liquid, which may not provide adequate relief and may cause problems in the flare system. Further, when the drum relief valve is elevated above the liquid level in the condenser, a vapor pocket may form in the valve inlet line during warm weather. Upon cooling (e.g. at night), the vapor pressure of the liquid in the small pocket largely drops. The vapor pocket may collapse, forcing a liquid rush that will hit the relief valve and caused chattering, as reported in two cases[19]. The need for a relief valve on the drum should be critically reviewed, as usually the relief valve on the tower should be able to relieve the drum.

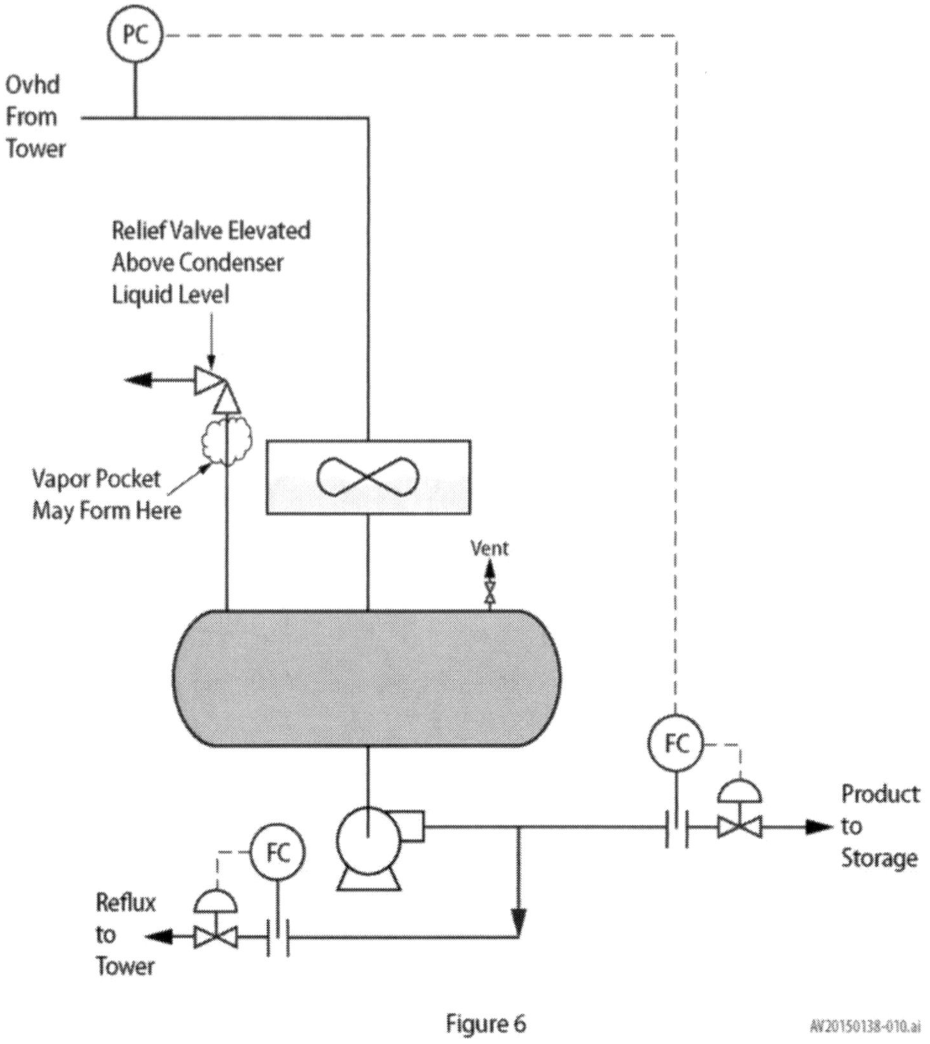

Figure 6. Relief valve mounted on a flooded drum

Valve in the condenser vapor inlet (Figure 1c, 2b): unique issues As mentioned earlier, this method places a backpressure valve in the overhead vapor line, thereby reduces condenser temperature difference and capacity.

To minimize pressure drop, the overhead valve is often designed for a small pressure drop when fully open. This often leads to valve oversizing. When oversized, the valve operates barely open during winter and cold spells. In this situation, very small valve movements cause large fluctuations to tower pressure. In one case[16], it was necessary to throttle a manual valve upstream in the line to force the control valve to operate close to its half-open position.

This method is prone to liquid hammering if the valve excessively closes. In one case[21], the valve closed fully under some startup conditions. Vapor downstream of the valve rapidly condensed, causing liquid to be rapidly drawn from the reflux drum, which in turn generated a liquid hammer that shook the whole unit. The problem was solved by changing the valve so that it would not fully close. The author is familiar with other similar unpublished cases.

Valves in both the condenser and bypass lines (Figures 2a, 2b): unique issues The addition of the second valve generates potential interaction between the pressure and differential pressure control loops. There is also the question of which variable should be used to control the bypass. An article by Zak Friedman[22] specifically addresses this question.

Friedman advocates controlling the bypass using a separate drum pressure controller (Figure 7a) rather than the differential pressure controller in Figures 2a and 2b. Chin[1] reports one successful case with the Figure 7a system in a situation where liquid head was small and the valve in the bypass was needed. We are familiar with experiences of troublesome interaction between the pressure controllers. This interaction was also discussed by Friedman[22] and by Smith[10]. Friedman believes that in this interaction the two control loops help each other. However, both Friedman and Smith also report that the key for success for this scheme is to tune the drum pressure fast and the column pressure slow, much like a level controller. Tuning one loop fast and the other slow is effective for decoupling control loop interactions, as Friedman points out, but here it is achieved at the expense of a loose control on the tower pressure. In our experience, tower pressure is the most important column control variable, as pressure affects everything in the tower, namely temperatures, vaporization, condensation, flows. It therefore needs to be tuned fast so it does not wander. Level does not affect much as long as it stays within limits, so it can move slower and be allowed to wander. But not pressure. So we would not recommend this scheme.

Another issue with using a separate drum pressure controller is that every time the set point is changed on the tower pressure controller the same change needs to be made on the reflux drum pressure controller[10]. The reflux drum pressure set point needs to be lower than the tower pressure set point, making the scheme prone to major upsets upon operator error. To overcome this issue, Smith[10] proposes an advanced control that subtracts an appropriate bias from the tower set point to provide the set point for the reflux drum pressure controller.

A common alternative to the two pressures control scheme in Figure 7a is the differential pressure (dP) control scheme is Figure 2a and 2b. Friedman notes that with this scheme the two controllers tend to fight each other: Upon an increase in column pressure the condenser outlet valve (Figure 2a) opens, lowering the liquid level in the condenser, while the dP controller opens the bypass valve, which raises the liquid level in the condenser. Friedman states that he saw the

Figure 2a scheme at least 100 times, and none of them worked, due to the negative interaction between the tower pressure control and the differential pressure control. The author has seen this scheme a few times (far less than 100), and in most of these times it seemed to work. Further, the author had polled delegates that attended his "Practical Distillation Technology" seminars and have this control scheme, and about 80% of them reported that it does work. We concur with Friedman that if both controllers are tuned fast, there will be an interaction that may not be favorable. It may well be that this system is successful when tower pressure fluctuations are small, but runs into trouble when the fluctuations are larger.

Friedman shows a way out in Figure 7b. Instead of controlling the actual dP, he sets the dP as the difference between the set point on the pressure controller and the drum pressure. "Have your cake and eat it too" as he puts it. This scheme has the strength of Figure 2a, which is keeping a set dP, without getting into the negative interaction that concerns Friedman. Also, it permits fast tuning of the pressure controller, which is key. We concur with Friedman's in liking Figure 7b. The key for its success (as with other dP controls on the bypass) is to keep a low dP across the bypass (typically 5 psi). Just enough to overcome the pressure drop in the exchanger and the piping downstream of the bypass takeoff including the control valve, keeping the condensate valve comfortably on control. Here the pressure should be tuned fast and the dP a little slower. The dP is only there to provide enough pressure drop across the condensate valve.

Another simple way out is to place the dP valve on manual. We have seen this implemented successfully to overcome the interaction.

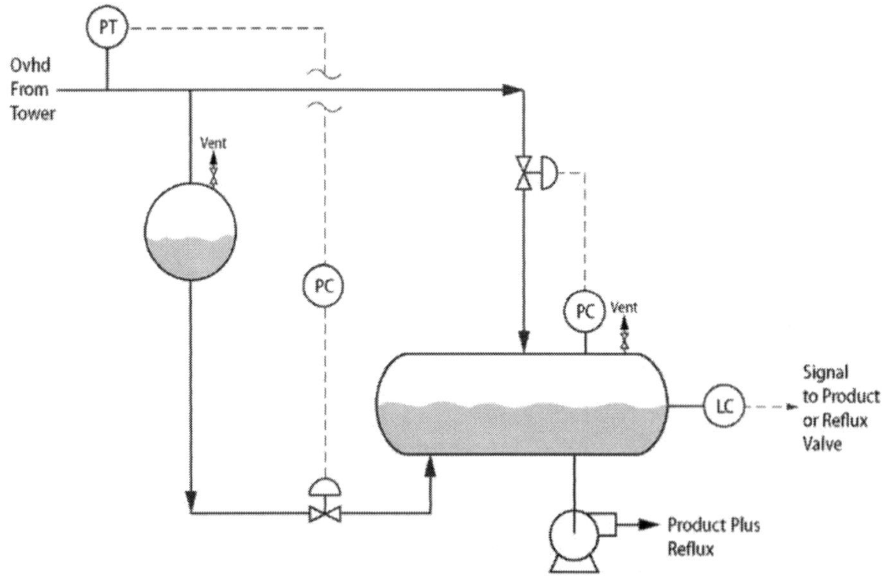

7(a)

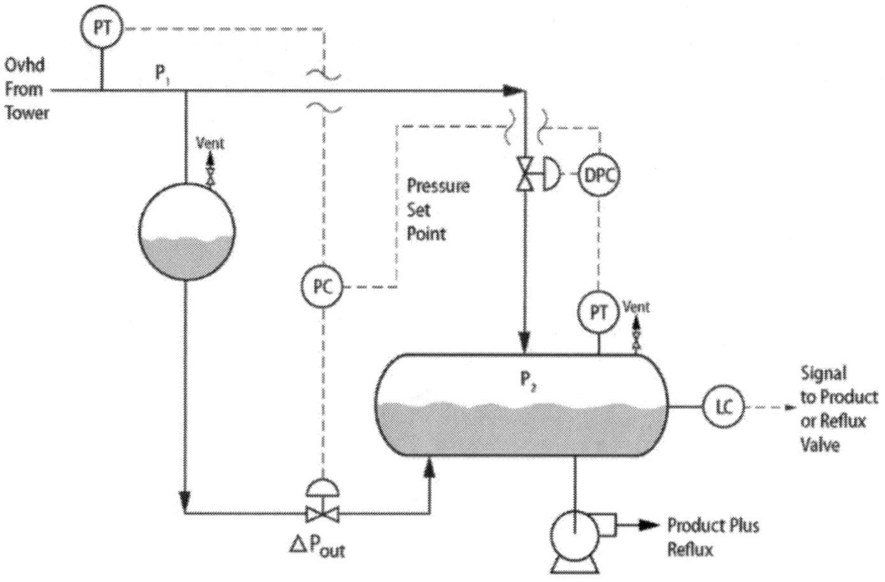

Figure 7. Alternatives to differential pressure control on the vapor bypass. (a) Separate pressure control on the reflux drum, not favored by the author; (b) Using the pressure set point in the differential pressure controller.

Hot vapor bypass controls (Figure 2c): unique issues With the submerged condensers hot vapor bypass, the issues are so numerous that it took a full article to do them justice. The reader is referred to our recent article on the subject[11] for complete coverage. A summary is below:

- Correct piping is mandatory for the success of the hot vapor bypass control method. As described earlier, the bypass vapor must enter the vapor space of the reflux drum (Figure 2c or 3b). The bypass should be free of pockets where liquid can accumulate; any horizontal runs should drain into the reflux drum. Most important, liquid from the condenser, as well as any other subcooled liquid streams like the reflux pump minimum flow recycle, must enter the reflux drum near the bottom of the drum. References (4-6, 13, 14, 23) report experiences in which incorrect piping led to instability, poor control, and hammering.

- Operation may be troublesome if the drum liquid surface is agitated. Cases of instability due to surface agitation have been reported[1, 6, 11, 12]. Agitation may be due to high-velocity impingement of the hot vapor jet on the liquid surface, due to upward-directed subcooled liquid jet reaching the liquid surface, as well as other causes. Agitation of the liquid surface can be avoided by judicious baffling[11].

- Because of the liquid leg between the condenser and the drum, sudden reduction in drum pressure can rapidly suck the liquid right out of the condenser, causing a major upset[11,12]. There is also the possibility of U-tube oscillations[5, 9, 24]. Both issues can be mitigated by adding a throttling valve in the liquid leg between the condenser and the drum.

- This technique can suffer from interaction between the drum and the condenser liquid levels[5, 8, 9, 11, 12, 24]. To minimize the interaction, the pressure controller should be tuned much tighter than the drum level controller[8, 9]. This can be an issue if the reflux drum is small, and the level controller needs to be tuned fast to avoid overflow or loss of level. This scenario is uncommon; we have encountered this situation only once, but another case was reported in the literature[25].

- Because of the liquid leg between the condenser and the drum, non-condensables accumulate in the condenser and need venting from their accumulation points. If a vent line is absent, instability and capacity bottlenecks may result, as described in one published case history[26].

- Leakage of vapor through the bypass valve at the closed position can substantially reduce condenser capacity. In one case[27], closing a manual valve in the hot vapor bypass increased condenser capacity by 50%.

- Undersizing of the bypass control valve may lead to inability to keep the tower pressure up in cold winter days when the drum is not insulated. In one case[28], poor pressure control due to undersizing was improved by installing a throttling valve in the liquid line from the condenser to the drum.

- In some cases, the hot vapor bypass control valve is manipulated by the drum pressure instead of the tower pressure, as in a case described in Ref. 28. Dynamically, this control is inferior because the vapor volume in the drum is much smaller than in the tower and more variable in response to ambient changes.

- The reflux drum vapor space may need to be insulated to minimize interference from rain and snow storms[1, 5, 6, 11]. This issue was reported to be more pronounced with narrow boiling-range mixtures [1, 6], and at high pressure[6], where small temperature changes have a large effect on the split of overhead flow between the condenser the bypass. At the other extreme, the incidence of Rayleigh fractionation with wide-boiling mixtures (heavy components condensed out without mixing with the remaining mixture), can also interfere with this control system[3].

- The amount of subcooling and vapor bypass rates can only be determined empirically. Simplified sizing procedures, however, are available[6, 29], but these are based on heating all the subcooled liquid to its bubble point, and therefore are grossly conservative. More reasonable sizing criteria can be inferred from the principles discussed in Reference 11.
- There is an inverse response problem[15]; when pressure rises, it closes the bypass valve, which initially causes a further increase in pressure until the condenser level begins to fall.

In summary, this method has some major advantages, but its operation can be troublesome.

CONCLUDING REMARKS

Flooded condensers are, and will continue to be, the prime tower pressure control methods for total condensers. They have been some of the most troublesome distillation controls that have been reported, yet our experience is that they can be quite trouble-free when correctly designed and applied. It is hoped that the principles and experiences described in this paper will pave the way for their successful, trouble-free application for tower pressure controls.

REFERENCES

1. Chin, T. G., "Guide to Distillation Pressure Control Methods," Hydrocarbon Proc. 58(10), 1979, p. 145.
2. Kister, H. Z., "What Caused Tower Malfunctions in the Last 50 Years?", Trans. IChemE, Vol 81, Part A, p. 5, January 2003.
3. Whistler, A. M., "Locate Condensers at Ground Level," Pet. Ref. 33(3), 1954, p. 173.
4. Schneider, D. F., and M. C. Hoover, "Practical Process Hydraulic Considerations", Hydroc. Proc., p. 47, August 1999.
5. Kister, H. Z., "Distillation Operation", McGraw-Hill, NY, 1990.
6. Hollander, L. "Pressure Control of Light-Ends Fractionators," ISA J. 4(5), 1957, p. 185.
7. Sloley, A. W., "Effectively Control Column Pressure", Chem. Eng. Progr., p. 38, January 2001.
8. Shinskey, F. G., "Distillation Control for Productivity and Energy Conservation", 2d ed., McGraw-Hill, New York, 1984.
9. Nisenfeld, A. E., and R. C. Seemann, "Distillation Columns", Instrument Society of America, Research Triangle Park, North Carolina, 1981.
10. Smith, C. L., "Distillation Control – An Engineering Perspective", Wiley, NJ, 2012.

11. Kister, H. Z., and Hanson, D. W., "Control Column Pressure via Hot-Vapor Bypass", Chem. Eng. Progr., p. 35, February, 2015.

12. Kister, H. Z., and Hanson, D. W., "Troubleshooting a Hot Vapor Bypass Control Instability", 10th International Conference on Distillation and Absorption, Friedrichshafen, Germany, September 14 – 17, 2014.

13. Kister, H. Z. "Distillation Troubleshooting", Wiley – Interscience, NJ, 2006.

14. Kister, H. Z., and J. F. Litchfield, "Distillation: Diagnosing Instabilities in the Column Overhead", Chem. Eng., p.56, September 2004.

15. Buckley, P. S., W. L. Luyben, and J. P. Shunta, "Design of Distillation Column Control Systems", Instrument Society of America, Research Triangle Park, NC, 1985.

16. Lieberman, N. P. "Process Equipment Malfunctions", McGraw-Hill, NY, 2011.

17. Cardoso, R. , and H. Z. Kister, "Refinery tower Inspections: Discovering Problems and Preventing Malfunctions", Proceedings of the Distillation Topical Conference, AIChE Spring Meeting, San Antonio, TX, April, 2013.

18. Parsons, K., Private communication, March, 2015.

19. Mani, G., "Avoid Condensation-Induced Transient Pressure Waves", Hydrocarbon Processing, p.41, January 2011.

20. Kister, H. Z., and T. C. Hower, Jr., "Unusual Case Histories of Gas Processing and Olefins Plant Columns," Plant/Operations Prog. 6(3), 1987, p. 151.

21. Steinmeyer, D. E., and A. C. Mueller, "Why Condensers Don't Operate as They Are Supposed To," Panel Discussion, Chem. Eng. Prog. 70(7), 1974, p. 78.

22. Friedman, Y. Z., "Condenser Hot Vapor Bypass Control", Hydroc. Proc., p.150, June 2008.

23. Duguid, I., "Take this Safety Database to Heart", Case MS16, Chem. Eng., p. 80, July 2001 (includes database at www.che.com/CEEXTRA).

24. Lupfer, D. E., "Distillation Control for Utility Economy", Proceeding of the 53rd Annual Convention, Gas Processors Association, March 25-27, Denver, Colorado, p.159, 1974.

25. Laird, D., and J. Cornelisen, "Control-System Improvements Expand Refinery Processes", Oil & Gas J., p.71, September 25, 2000.

26. Sloley, A. W., "Simple Methods Solve Exchanger Problems", Oil & Gas Journal, p. 85, April 20, 1998.

27. Lieberman, N. P., "Troubleshooting Process Operations", 4th Ed., PennWell Books, Tulsa, OK, 2009

28. Hartman, E. L., and T. Barletta, "Reboiler and Condenser Operating Problems", PTQ, p. 47, Summer 2003.

29. Durand, A. A., "Sizing hot Vapor Bypass Valve", Chem. Eng., August 25, p.111, 1980.

Fractionation Research, Inc.
RESEARCH - KEY TO BETTER DESIGN

CFD Simulation of Thick Sieve Tray Dry Pressure Drops

X.P Li (ZJUT)
J. R. Whiteley (Oklahoma State University)
M.R. Resetarits (Independent Consultant)
Tony Cai (FRI)

Kister Distillation Symposium

2015 AIChE Spring Meeting
Austin, TX

29 April 2015

Background

FRI

- Dry pressure drops of perforated plates were studied in 50's and 60's
- Most studies employed small size pipes or cylinders with perforated plates
- Very few dry tray pressure drop data were measured with actual tray configurations
- Few data available on the dry tray pressure drops of sieve trays with **thick decks**
- The deck thickness was limited by the punching capability

FRI Work

- Measured the thick sieve tray dry pressure drops with a large column having the same configuration as commercial trays
- Simulated the dry pressure drops were simulated using the CFD
- Compared to the experimental data
- Obtained the dry pressure drop data for the tray deck that could not be tested

3 2015 AIChE Spring Meeting Sieve Tray Dry Pressure Drops

Tray Panels

4 2015 AIChE Spring Meeting Sieve Tray Dry Pressure Drops

Test Set-up

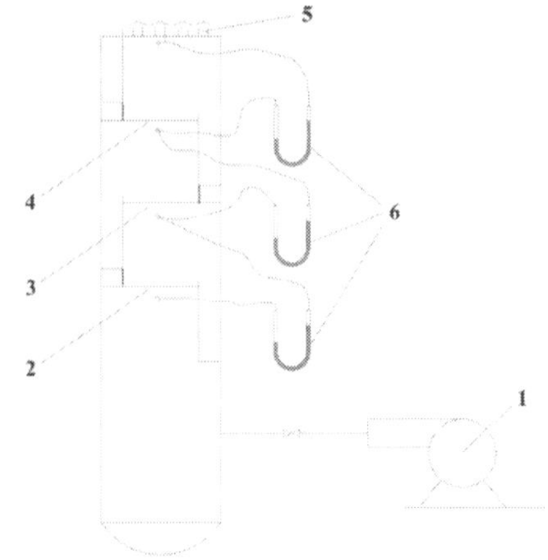

1 - Blower   2 - Bottom Tray   3 - Middle Tray   4 - Top tray
5 - Chimney Tray   6 - U-tube Water Manometer

5   2015 AIChE Spring Meeting   Sieve Tray Dry Pressure Drops

Parameters of Trays Tested

Tray Design No.	Material	Hole Diameter, d	Plate Thickness, t	t/d
		mm (inch)	mm (inch)	
1	Carbon Steel	13.2 (0.520)	2.0 (0.079)	0.152
2	Carbon Steel	13.2 (0.520)	3.8 (0.150)	0.288
3	Carbon Steel	13.6 (0.535)	5.8 (0.228)	0.426
4	Carbon Steel	6.2 (0.244)	2.0 (0.079)	0.323
5	Carbon Steel	6.3 (0.248)	3.8 (0.150)	0.603

CFD Simulation

- All holes were punched
- Punch exit (shape side) facing vapor/air flow (i.e., burrs down)
- Nominal open area 14%

6   2015 AIChE Spring Meeting   Sieve Tray Dry Pressure Drops

CFD Simulation Domain

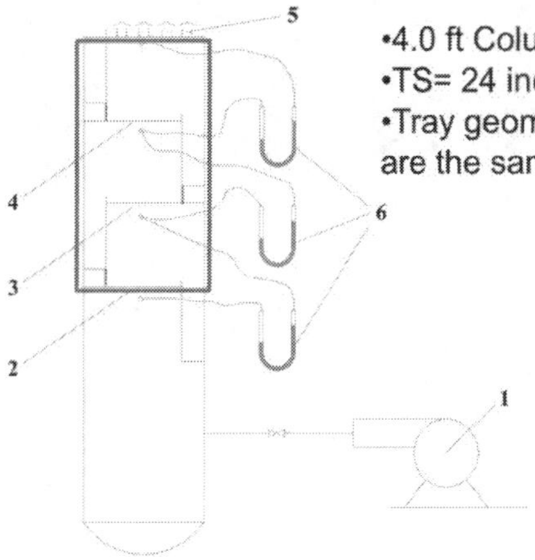

- 4.0 ft Column
- TS= 24 inches
- Tray geometries and hole layout are the same as what tested

1 - Blower 2 - Bottom Tray 3 - Middle Tray 4 - Top tray
5 - Chimney Tray 6 - U-tube Water Manometer

Sieve Deck Simulation Cases

Cases	Hole Diamter, d	Deck Thickness, t	t/d	ΔP Measured
	inch (mm)	inch (mm)		
1	0.520 (13.2)	0.150 (3.8)	0.288	YES
2	0.535 (13.6)	0.228 (5.8)	0.426	YES
3	0.535 (13.6)	0.535 (13.6)	1.000	NO
4	0.535 (13.6)	0.803 (20.4)	1.500	NO

CFD model

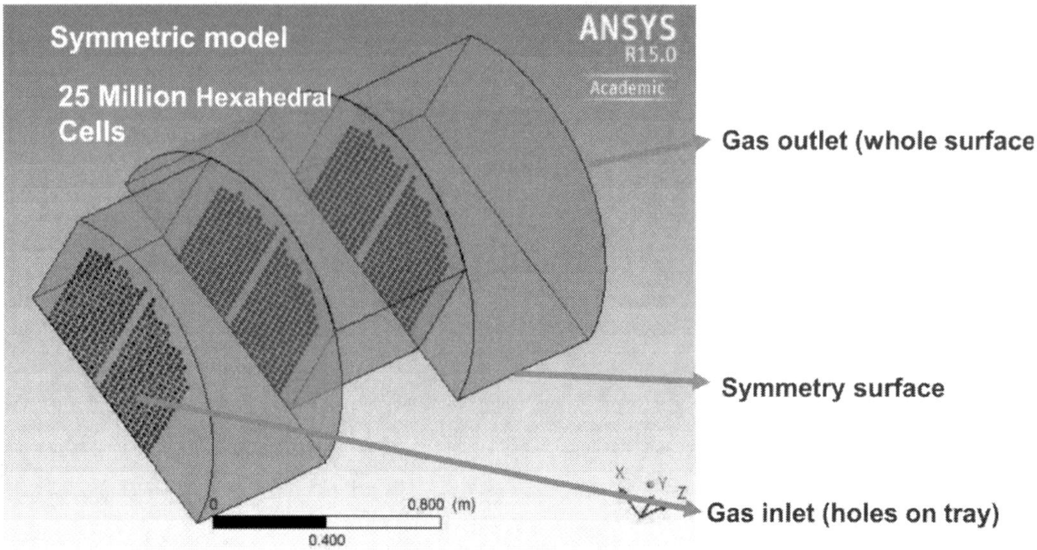

The upper two tray decks are modeled.

Meshes are refined near tray deck.

CFD model

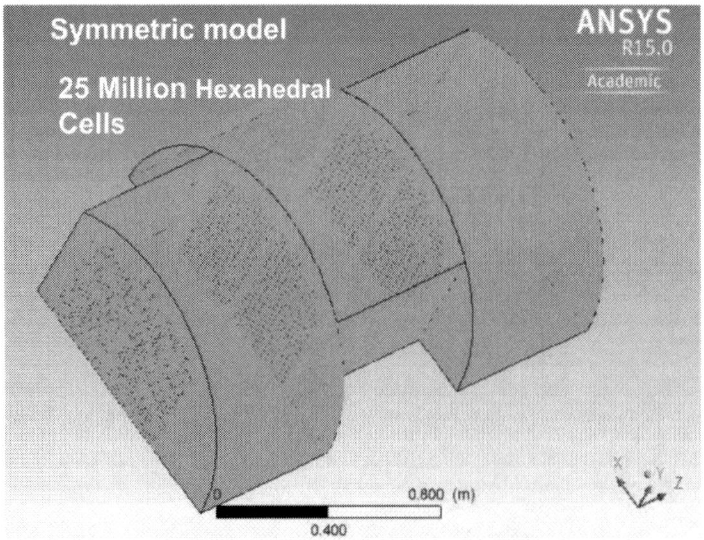

The upper two tray decks are modeled.

Meshes are refined near tray deck.

Meshing Schemes

11 2015 AIChE Spring Meeting Sieve Tray Dry Pressure Drops

Simulation method

- **Steady state simulation**
- **Shear-stress transport k-ω turbulence model**
- **Boundary conditions:**
  - **Symmetric**
  - **Velocity inlet-uniform hole velocity**
  - **Pressure outlet-atmospheric pressure**
- **Fluid air:**
  - **ρ=1.165 kg/m^3**
  - **μ=1.86x10^{-5} kg.s/m or Pa.s**

12 2015 AIChE Spring Meeting Sieve Tray Dry Pressure Drops

Simulation Platform

- **All simulations were conducted on the OSU High Performance Computing Center (HPCC)**
- **Total 24 cores**
- **Each case (flow rate) needed 7000 or more iterations, took 36 hours or more to complete**

Pressure Drop vs Iterations

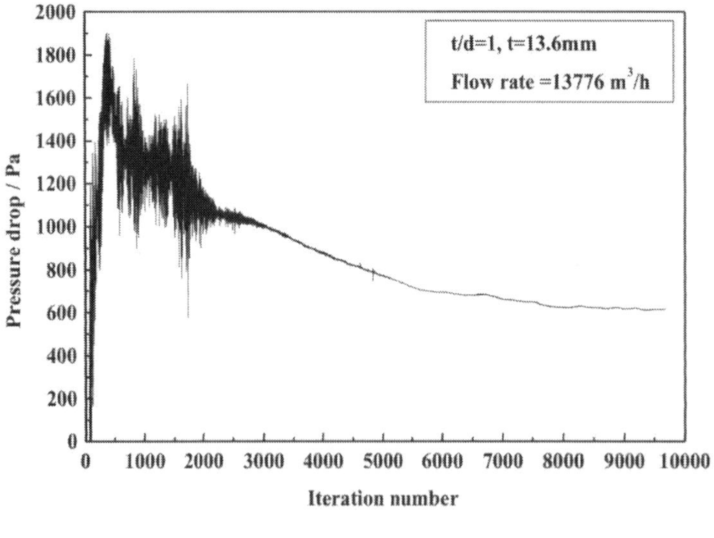

Simulation is stopped when the curve of pressure drop becomes steady

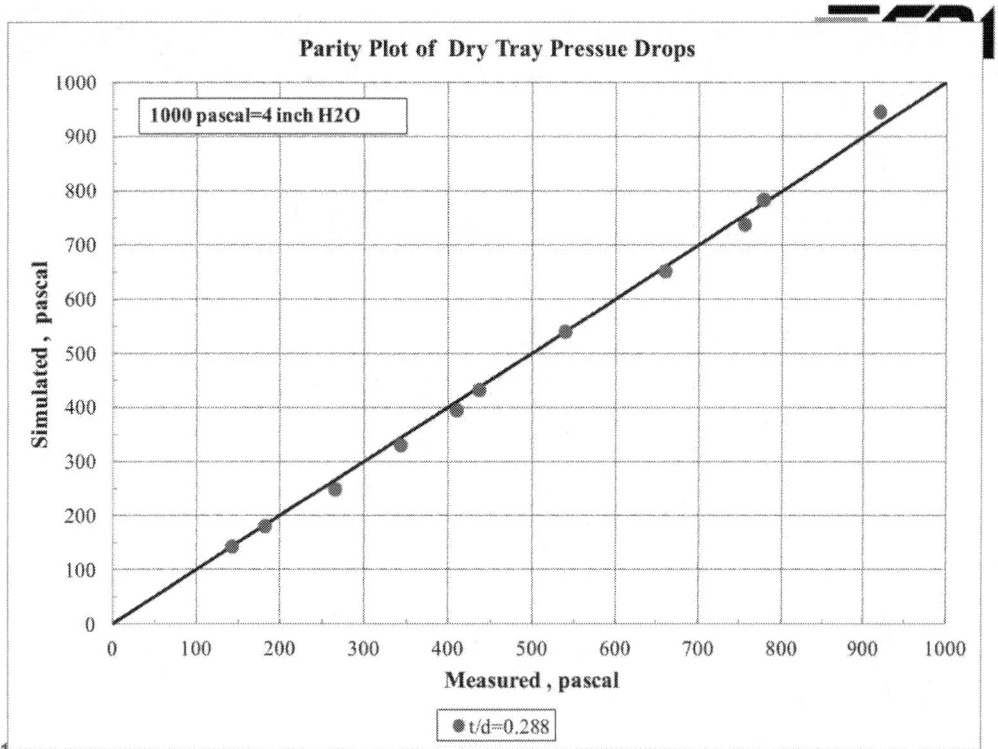
Parity Plot of Dry Tray Pressue Drops
1000 pascal=4 inch H2O
Simulated , pascal
Measured , pascal
0
100
200
300
400
500
600
700
800
900
1000
t/d=0.288

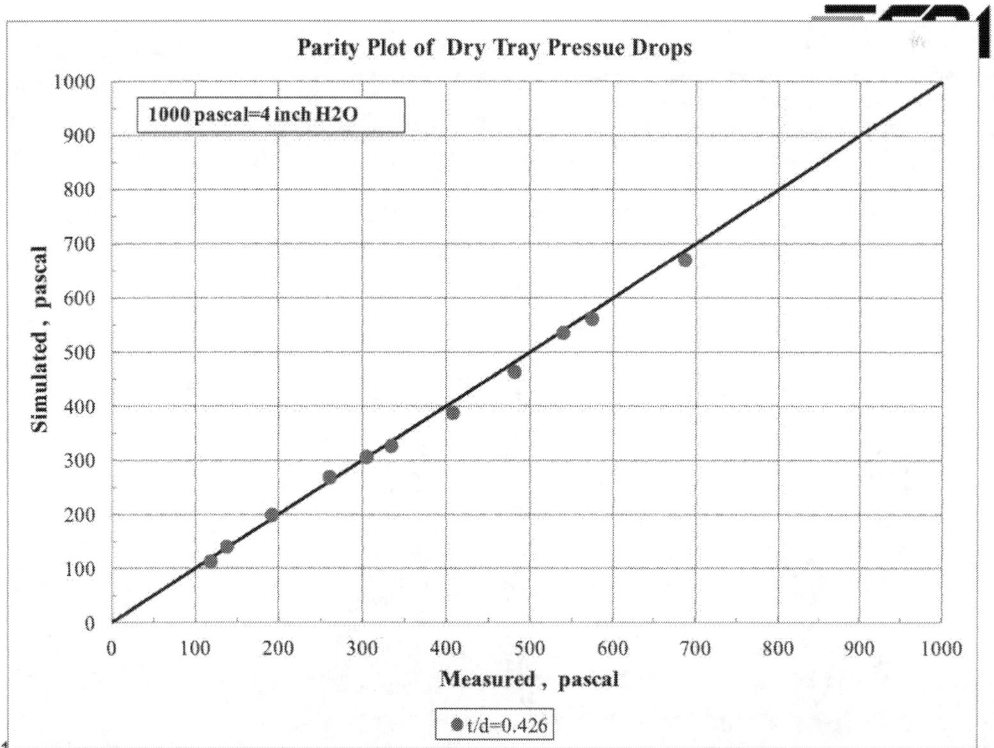
Parity Plot of Dry Tray Pressue Drops
1000 pascal=4 inch H2O
Simulated , pascal
Measured , pascal
0
100
200
300
400
500
600
700
800
900
1000
t/d=0.426

Simulation results (t/d=0.288)

Velocity m/s

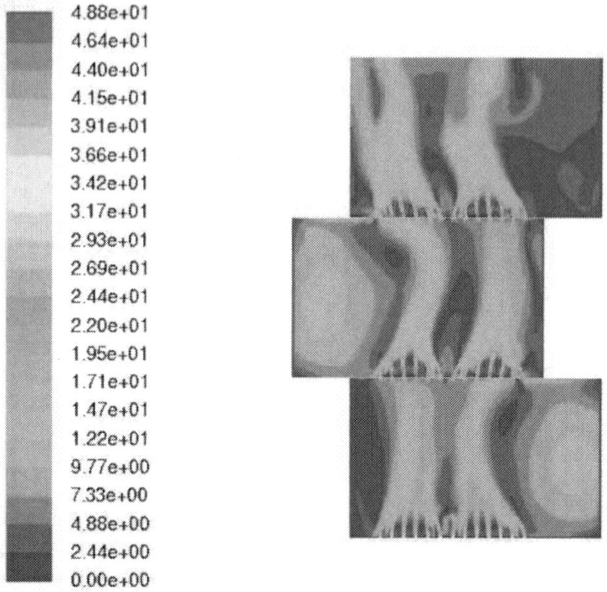

V_{hole}=30.7 m/s

Simulation results (t/d=0.288)

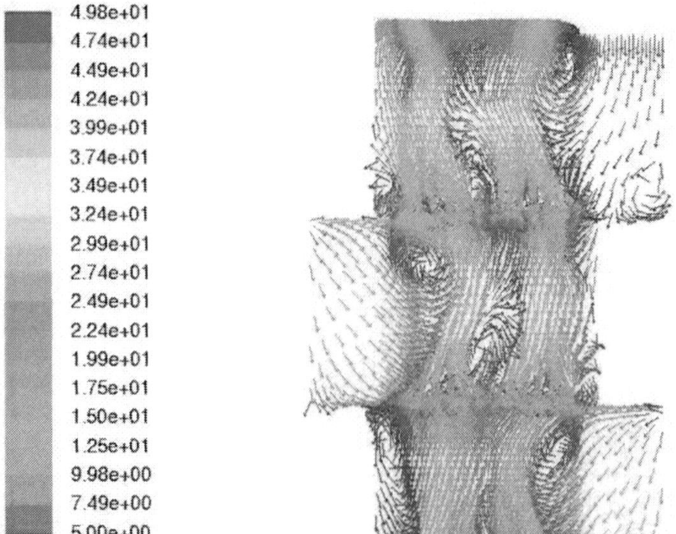

V_{hole}=30.7 m/s

Velocity vectors at the symmetric plane

Simulation results (t/d=0.288)

Stream lines

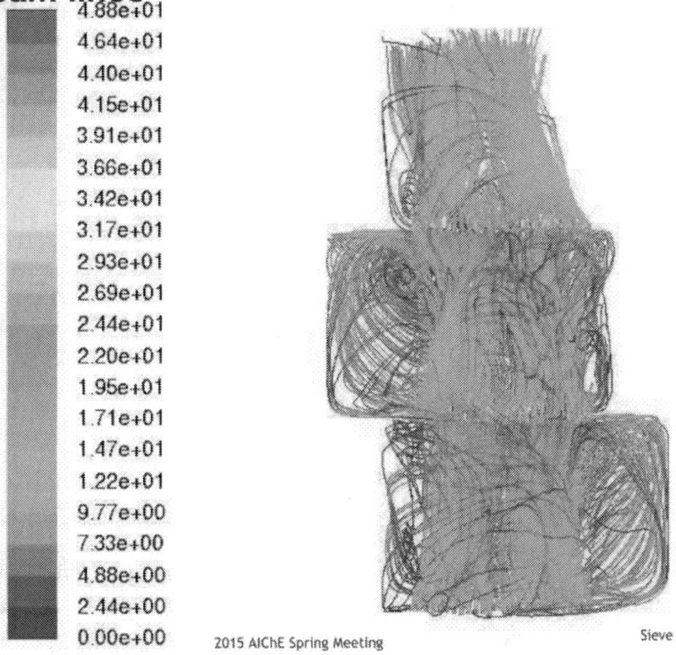

V_{hole}=30.7 m/s

Simulation (t/d=0.288)

Velocity, m/s

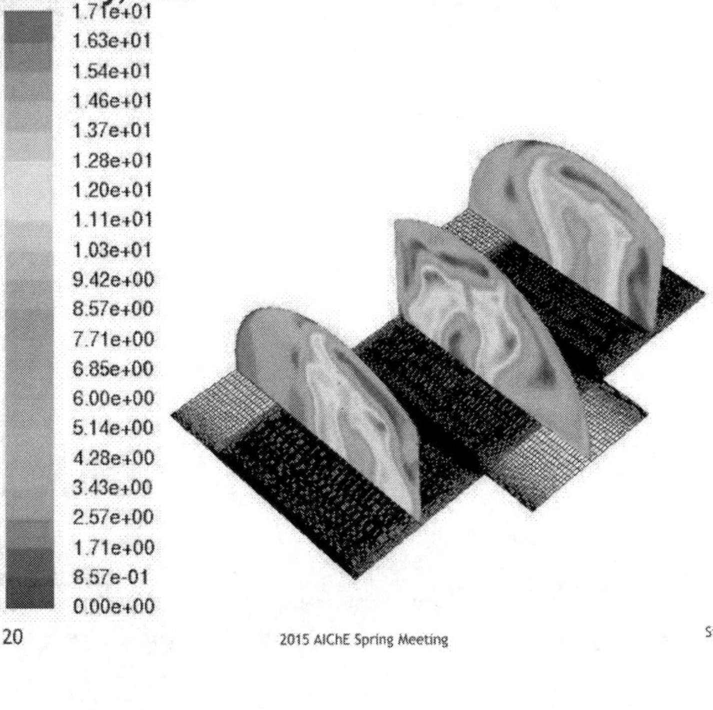

V_{hole}=30.7 m/s

Simulation (t/d=0.288)

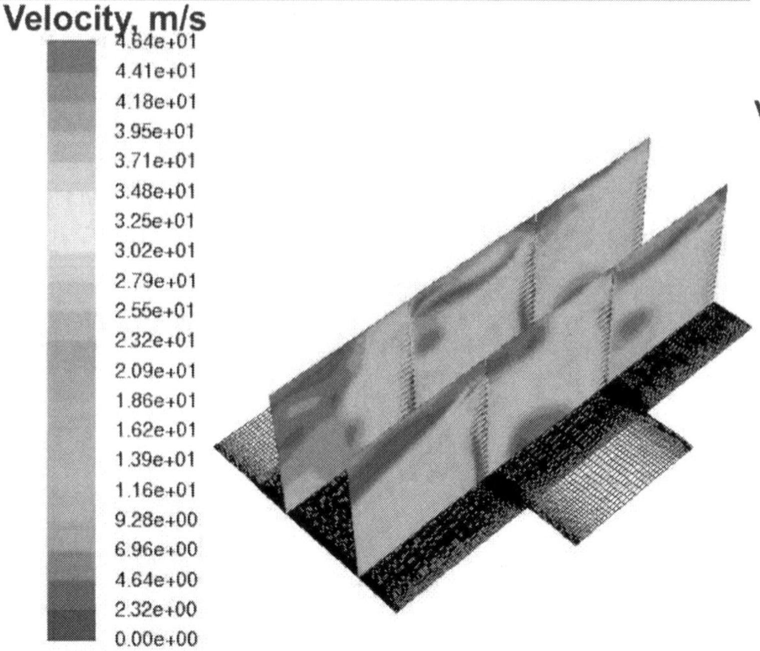

V_{hole}=30.7 m/s

Simulated Dry Tray Pressue Drops vs Hole Velocity

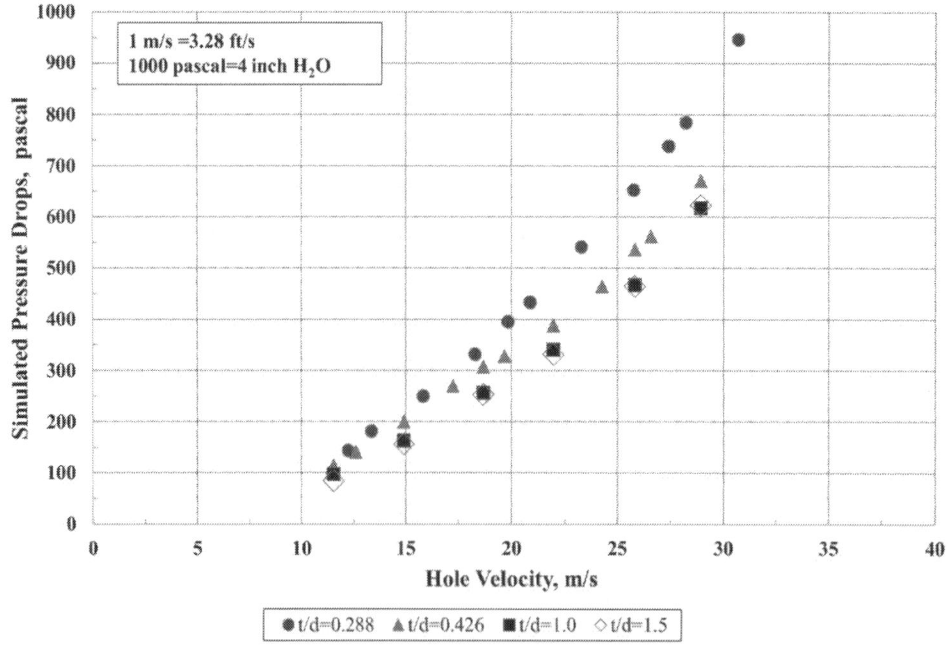

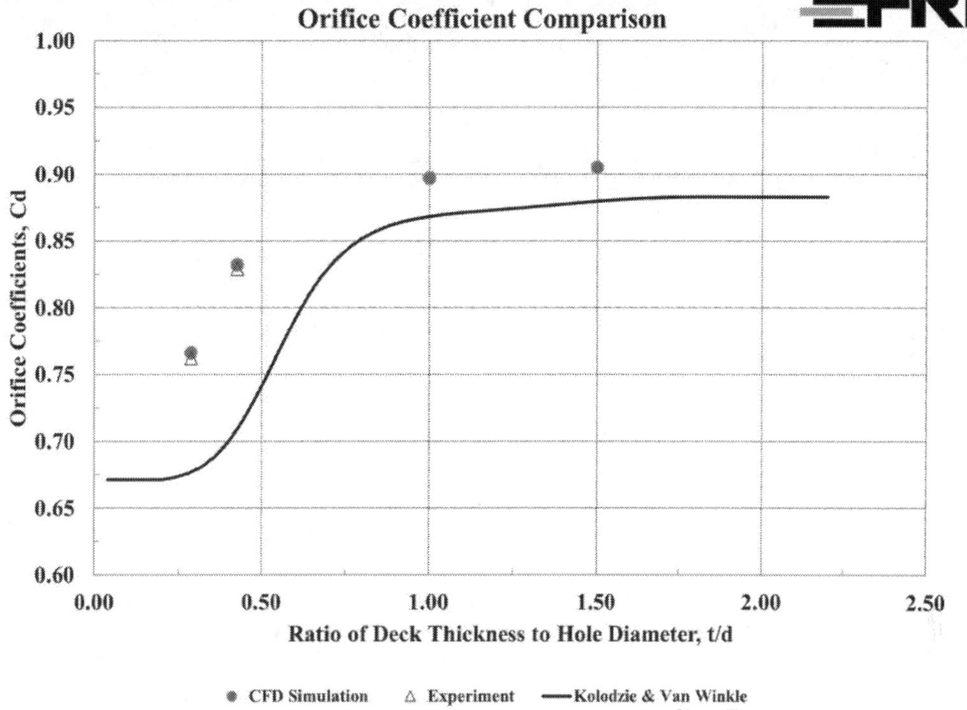

Summary

- The dry pressure drops of the commercial size thick sieve trays were measured and simulated using CFD
- The CFD simulation results matched the experimental data well
- The dry pressure drop for the thick deck that could not be tested were simulated
- Flow velocities were NOT uniformly distributed due to the effect of jetting from the deck orifices
- The simulated Cd's matched the experimental data, but are higher than those predicted by literature models

Fractionation Research, Inc.

RESEARCH - KEY TO BETTER DESIGN

CFD Simulation of Thick Sieve Tray Dry Pressure Drops

Thank You for Your Attentions!

Kister Distillation Symposium

2015 AIChE Spring Meeting
Austin, TX

29 April 2015

Ripple® Trays for Fouling Applications

AIChE 2015 Spring Meeting
Paper # 398200

Steven G. Schon, P.E.
Arkema Inc.

15 AIChE
Spring Meeting, Austin, TX

Introduction

- Tray Fouling
- Approaches to Mitigate Fouling
- Ripple® Trays
- Plant Experience
- Conclusions

15 AIChE
Spring Meeting, Austin, TX

2

Tray Fouling

- **Mechanisms**
  - Thermal Decomposition / Coking
  - Reaction / Polymerization
  - Precipitation
  - Settling
- **Locations**
  - Tray Deck
  - Tray Holes / Valves
  - Underside
  - Downcomers

15 AIChE Spring Meeting, Austin, TX

3

Tray-Deck Fouling

Photo – courtesy of Technip Stone & Webster

15 AIChE Spring Meeting, Austin, TX

4

Tray Orifice (Valve) Fouling

Photo – courtesy of Technip Stone & Webster

15 AIChE Spring Meeting, Austin, TX

5

Under-side Fouling

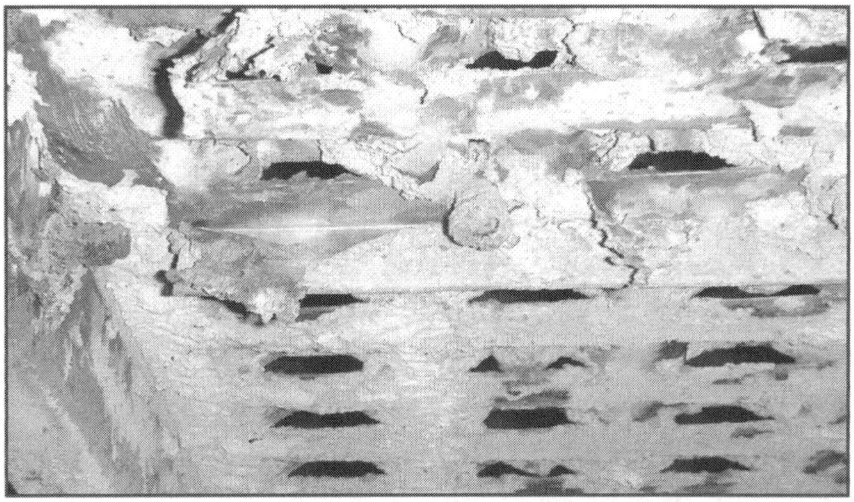

Photo – courtesy of Technip Stone & Webster

15 AIChE Spring Meeting, Austin, TX

6

Fouled Trays With Downcomers

Photo – courtesy of Technip Stone & Webster

15 AIChE Spring Meeting, Austin, TX

7

Approaches to Mitigate Fouling

- **Hydraulics**
  - Increase liquid velocity ("washing" effect)
  - Decrease vapor velocity / increase spacing (entrainment)
- **Additives**
  - Polymerization inhibitors
  - Surfactants
  - Dispersants
- **Materials of Construction**
  - Polished finishes
  - Non-stick coatings (expensive; questionable durability)
- **Fouling-tolerant Geometries**
  - Large orifices / No moving parts (valves)
  - Counter-current flow through holes / eliminate downcomers

15 AIChE Spring Meeting, Austin, TX

8

Ripple® Trays

Photo – courtesy of Technip Stone & Webster

15 AIChE
Spring Meeting, Austin, TX

9

Ripple® Trays

- **Counter-Current Contact**
- **No Downcomers**
- **Constant Wetting-Washing Action**
  - Wets underside with downcoming liquid
  - Additives reach underside via liquid
- **Wave Action Promotes Agitation**
  - No stagnant areas on tray deck
  - Reduced Solids Accumulation
- **Polished Finish**
  - Slippery surface to minimize adhesion
    - Punched/drilled holes need to be cleaned to smooth edges

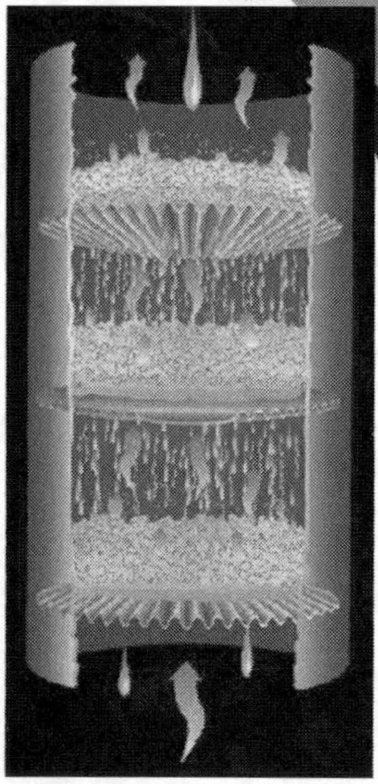

Image – courtesy of Technip Stone & Webster

15 AIChE
Spring Meeting, Austin, TX

10

Flow Characteristics
Ripple® vs. Dual-Flow Trays

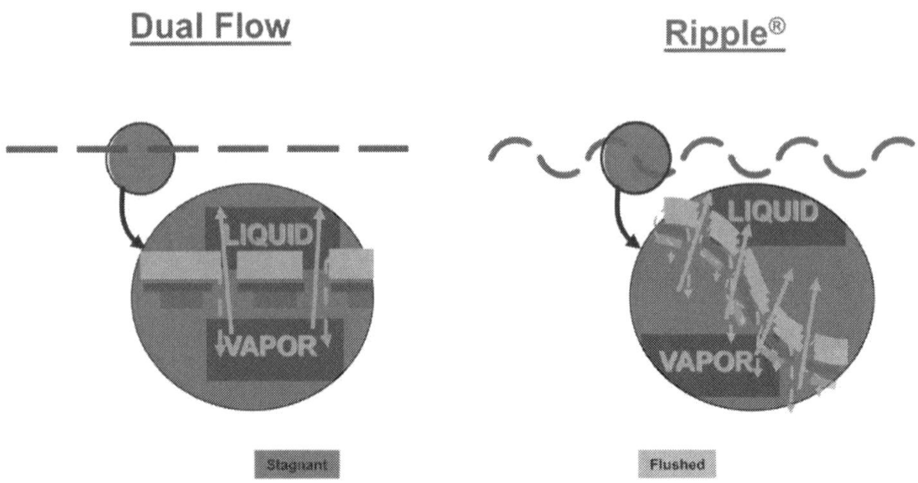

15 AIChE Spring Meeting, Austin, TX

11

Ripple® Tray Loading Schematic

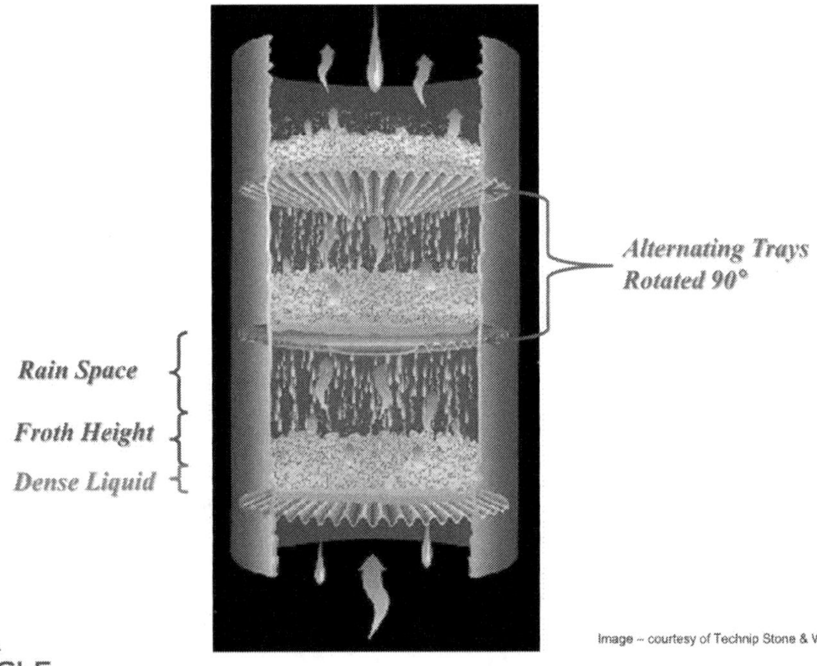

Image – courtesy of Technip Stone & Webster

15 AIChE Spring Meeting, Austin, TX

12

Plant Experience – Monomer Absorber

- **Hot Monomer-laden Gas**
- **Polymerization Fouling, Especially Under-side**
  - Quenching / absorption of vapor near feed (warm zone)
  - Condensation of un-inhibited hot vapors
- **Original: Sieve Trays**
  - 3-month cleaning cycle
- **Replacement: Ripple® Trays**
  - Bottom section only (sieve trays above)
  - 1½+ years without cleaning

15 AIChE Spring Meeting, Austin, TX

13

Monomer Absorber After 1 Year With RT

Pan Above Ripple® Tray section

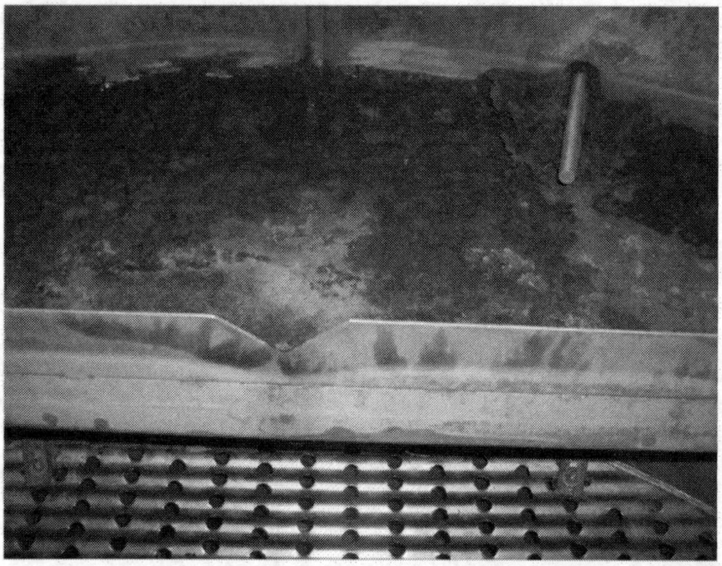

15 AIChE Spring Meeting, Austin, TX

14

Monomer Absorber After 1 Year With RT

Ripple® Tray – Top Side

15 AIChE Spring Meeting, Austin, TX

15

Fouling in Absorber After 1 Year With RT

Ripple® Tray – Under Side

15 AIChE Spring Meeting, Austin, TX

16

Monomer Absorber After 1 Year With RT

Sump

- Solids accumulated in bottom rather than on trays

15 AIChE Spring Meeting, Austin, TX

17

Other Considerations

- **Tray Efficiency**
  - Low to Moderate for Ripple® Trays
    - **Fractionation**: Better than DualFlow; worse than sieve/valve
    - **Absorption**: Same as DualFlow; better than sieve/valve
- **Capacity**
  - TS&W claims higher capacity than sieve/valve/DualFlow
    - Full active area (no downcomer)
    - Allows closer spacing
- **Turn-down**
  - Limited (~ 2/3 range of sieve & valve trays)
- **Pressure Drop**
  - Higher tray dP than DualFlow or sieve/valve
  - But, if fewer trays (more efficient vs. DualFlow) → lower column dP
- **Cost**
  - Proprietary → relatively expensive
    - Technip Stone & Webster
    - Off-patent; other vendors, but no hydraulic design capability
  - Potential to reduce tower diameter → reduce investment for new unit

15 AIChE Spring Meeting, Austin, TX

18

Ripple® Tray Capacity & Turn-down

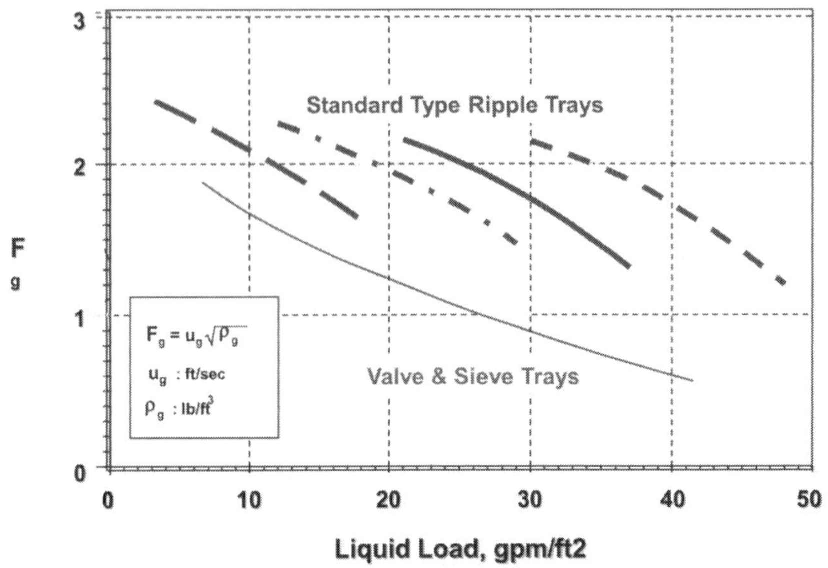

Image – courtesy of Technip Stone & Webster

15 AIChE Spring Meeting, Austin, TX

19

Ripple® Trays - Conclusions

- **Effective for (Polymerization) Fouling Service**
  - > 6-times longer cleaning interval vs. sieve trays
  - Works best in conjunction with additives / inhibitors
    - Washes underside with additives from downcoming liquid
- **Well-suited for Direct-contact Heat & Mass Txfer**
  - Quench contacting
  - Vapor absorbers
- **Consider for De-bottlenecking in Other Services**
  - Eliminate downcomers (increase active area)
  - Reduce tray spacing / more trays (offset lower efficiency)
  - Vapor-rate limited columns
  - Columns with limited turn-down
- **Relatively Expensive, But Can Have Attractive Payback**
  - **Fouling**: Replace only trays in troublesome section
    - Transition chimney between tray types (redistribute V & L)
  - **Debottleneck**: Full re-tray may avoid need for new larger-ø tower
- **Mature Technology, Though Not Well-known**
  - Over 500 Ripple® Tray towers

15 AIChE Spring Meeting, Austin, TX

20

Questions?

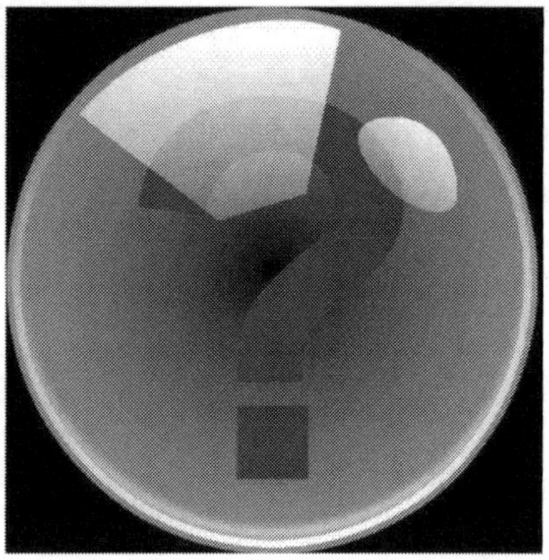

15 AIChE
Spring Meeting, Austin, TX

21

NGL Unit Debottlenecked with UOP ECMD™ Trays and High Flux™ Tubing for Significant Production Expansion and Energy Optimization

Shanda G. Fry
Senior Design Engineer
UOP LLC, A Honeywell Company

Joshua M. Moll
Design Engineer
UOP LLC, A Honeywell Company

Patrick Ronan
Process Optimization Engineer
Chevron Phillips Chemical Company LP

Lee Scanlon
Olefins & NGL Technical Services Engineer
Chevron Phillips Chemical Company LP

Prepared for Presentation during the
New Frontiers in Tray Design and Operation Session
2015 AIChE Spring Meeting
Austin, Texas, USA, April 29, 2015

Abstract

UOP high performance distillation and heat transfer equipment have been used extensively in a wide variety of services in the petrochemical and refining industry for more than 40 years. This paper will illustrate the optimization and expansion of Chevron Phillips Chemical's natural gas liquids (NGL) Unit 21 in Sweeny, Texas by coupling the design of new tower internals with new tubing in the associated stab-in reboiler and overhead condenser. Revamping the bottlenecked deethanizer tower with UOP Enhanced Capacity Multiple Downcomer (ECMD™) trays, along with revamping the condenser and stab-in reboiler with UOP High Flux™ tubes, increased unit production rate more than 65% within the existing limitations of the available heating and cooling mediums and plant footprint.

Project Objectives

Capacity at Chevron Phillips Chemical Company LP's Sweeny NGL fractionation complex was increased by approximately 19% in 2013. The project was driven by the increased production of NGLs from shale formations in the region and an opportunity to secure additional long-term NGL supplies into the facility. The NGL fractionation complex consists of multiple units. For economic as well as logistical reasons, the full capacity increase of the expansion project would be achieved at Unit 21, representing a much greater and challenging expansion for that single train. Evaluation of Unit 21 showed that the existing deethanizer would need to be debottlenecked to achieve the targeted capacity increase, while project economics dictated that the existing column and exchanger shells would need to be reused.

In April 2010, Chevron Phillips Chemical contacted UOP to identify an expansion solution for increasing tray capacity and handling the corresponding increase in reboiler and condenser duties. For the expansion project, the following design objectives were given for the modifications to the existing equipment:

- Change operating conditions in tower to maximize capacity (>65% increase) and optimize utility consumption
- Reuse the existing tower with minimal modifications
- Maintain external dimensions of replacement exchangers
- Reuse existing piping and foundation
- Accomplish equipment modifications within the turnaround schedule
- Guarantee performance of the tower system

Post-Revamp Results

The revamp of the deethanizer tower and exchangers with UOP's high performance mass and heat transfer equipment allowed for more than a 65% increase in the feed rate to the column (compared to the maximum pre-expansion feed rate), while maintaining the existing footprint. Table 1 highlights the capacity increase of the expansion project. The data in the table has been normalized using the highest pre-expansion feed rate as the basis (i.e. pre-expansion feed rate is 100%). Additional tower and exchanger capacity (approximately 18%) is expected before flooding occurs.

Table 1. Normalized Operating Data – Before Revamp vs. After Revamp

	Pre-Expansion Operating Data[2]	July 2014 Operating Data[3]
Feed Rate[1]	100	167
Feed Ethane Content, vol%	mid 40s	low 40s
Feed Propane Content, vol%	< 30	> 30
Overhead Product Rate[1]	36	52
Overhead Propane Content, %	2.5	2.5
Bottoms Product Rate[1]	64	115
Observed Reflux Ratio	1.6	2.0
Observed Tray Efficiency, %	-	69
Number of Trays	35	35
Tray Type	1-/2-pass Valve	ECMD Trays
Percent of Flood, %	~100	84

[1] Values have been normalized based on pre-expansion feed rate = 100%.
[2] Normalized values for a 6 hr. period of stable operation on January 13, 2011, representing the highest pre-expansion feed rate while maintaining good cut points.
[3] Normalized values for a 4 hr. period of stable operation from July 14-15, 2014 (post-revamp).

The deethanizer tower was revamped with UOP ECMD trays on a tray-for-tray basis with the column pressure optimized to achieve 90% of flood at the design rates (see Appendix A for more information on UOP ECMD trays). The overhead condenser and bottoms stab-in reboiler were retrofitted with OD coated / ID bare UOP High Flux tubing (see Appendix B for more information on UOP High Flux tubing). This revamp marked the first commercialization of OD coated High Flux tubing with a tube diameter of 15.875 mm.

UOP provided a process guarantee, in addition to a mechanical guarantee, for the new trays and tubes in the deethanizer tower and exchangers. The process guarantee covered the desired feed rate and desired product purities at the design conditions that coincide with operation at 90% of flood.

Modifications to the tower and exchangers were determined to be the most cost effective way to meet the project objectives.

All of the modifications were accomplished within the scheduled project outage of 26 days.

Revamp Design Approach

The key to truly maximizing the achievable production from the deethanizer was to approach the design of the tower and exchangers as an overall unit, and not evaluate each piece of equipment independently. This approach allowed the designers to adjust the process conditions – in this case, operating pressure – to gain the most benefit from all three pieces of equipment until the highest capacity was achieved within the heating and cooling constraints. Many companies have benefited from this approach, which can only be done with the combined use of UOP's ECMD trays and High Flux tubes (McGuire and Sobczyk 2008).

The deethanizer tower originally contained 35 conventional valve trays in a dual-diameter shell. The condenser uses propylene refrigerant to condense the overhead vapor to generate reflux, and the reboiler uses LP / MP steam to partially vaporize the column bottom stream. The pre-expansion deethanizer configuration and equipment details are provided in Figure 1.

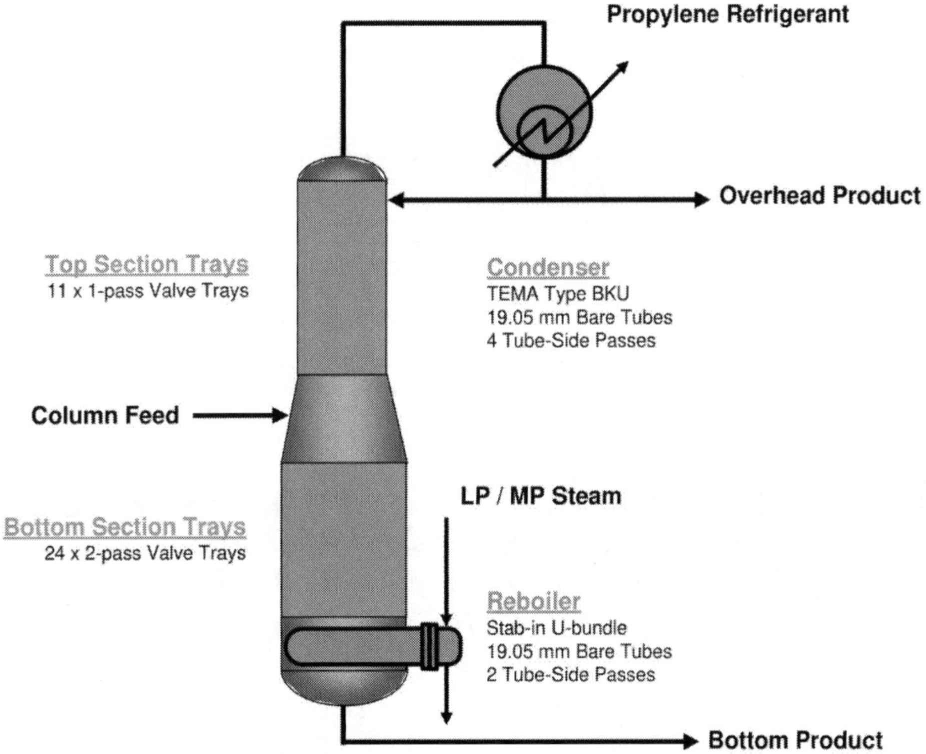

Figure 1. Pre-Expansion Configuration of Deethanizer Unit

Early on in the project, the towers, exchangers, and propylene refrigeration system were identified as bottlenecks for the expansion. To eliminate the need to modify the steam turbine compressor driver, Chevron Phillips Chemical determined they needed to operate the propylene refrigerant 2.32 kg/cm^2 higher than the original design conditions. The increase in the propylene refrigerant pressure would require an increase in the column overhead pressure to obtain a sufficient temperature driving force between the condensing overheads and boiling propylene refrigerant. Unfortunately, the increased operating pressure reduces the capacities of the tower and the reboiler due to smaller delta between liquid and vapor densities, higher reflux ratios required for the given separation at elevated pressures, lower surface tensions, and lower mean temperature difference (MTD) in the reboiler. As a result, when setting the design pressure for the column overhead, there is a trade-off between unloading the condenser system and unloading the trays and reboiler system. To maximize the overall capacity of the deethanizer unit, these competing benefits had to be balanced.

Revamping the deethanizer condenser with High Flux tubes resulted in a minimum MTD for the column overhead vapor and the propylene refrigerant, thus minimizing the column overhead pressure as much as possible. Once the overhead pressure was determined, UOP simulated the column at the desired conditions and confirmed that a tray-for-tray revamp with ECMD trays would not only provide a sufficient number of theoretical stages for the given separation, but would also maximize the column capacity. The original tray count could be maintained considering the tray efficiency of MD and ECMD trays is similar to that of conventional trays in this service and the separation is not very sensitive to changes in tray efficiency (i.e. a 10% loss in tray efficiency only results in a 4% increase in the reflux rate). UOP applied a design efficiency of 65.7%, relative to commercially-available simulation models, which is conservative for ECMD trays in this service (McGuire et al. 1997). Figure 2 shows the efficiency vs. reflux rate curve for the design separation. The x-axis (reflux rate) has been normalized using the design reflux rate as the basis (i.e. design reflux rate is 100%).

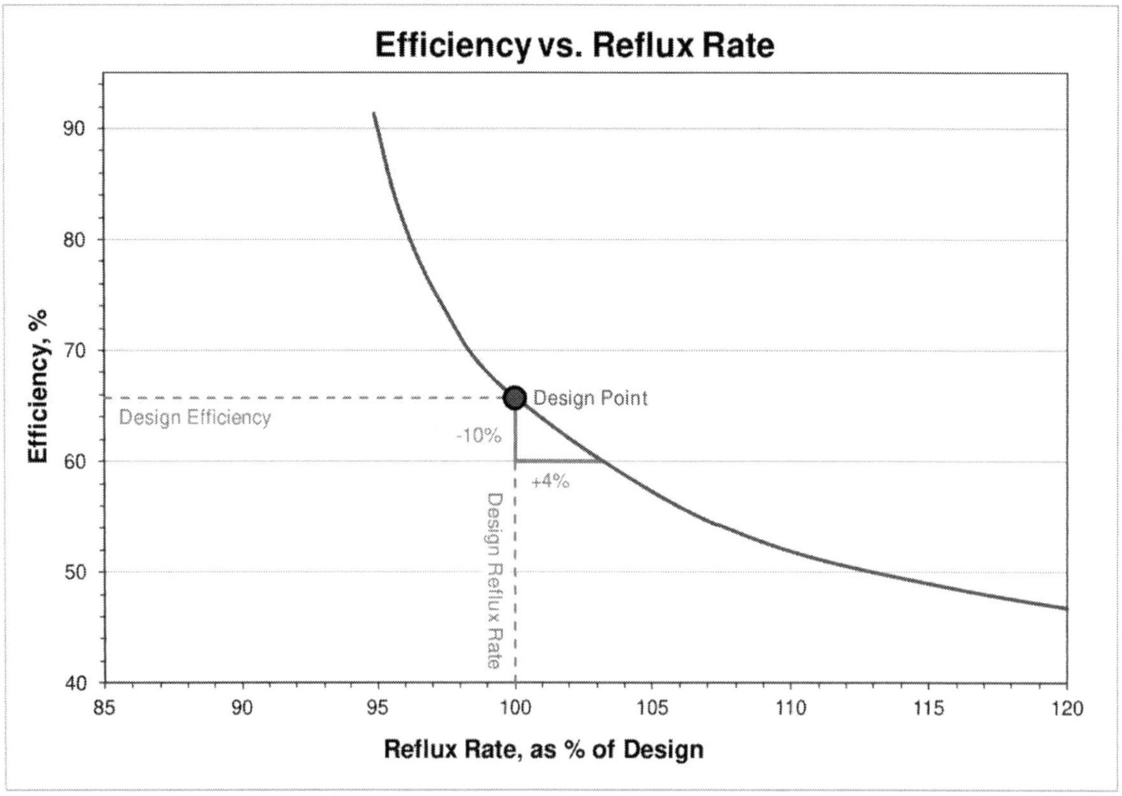

Figure 2. Efficiency vs. Reflux Rate Curve for Design Separation

A sketch of the deethanizer column revamp is shown in Figure 3. A tray-for-tray revamp with 35 ECMD trays allowed for the column capacity to be maximized, relative to a multi-tray revamp configuration.

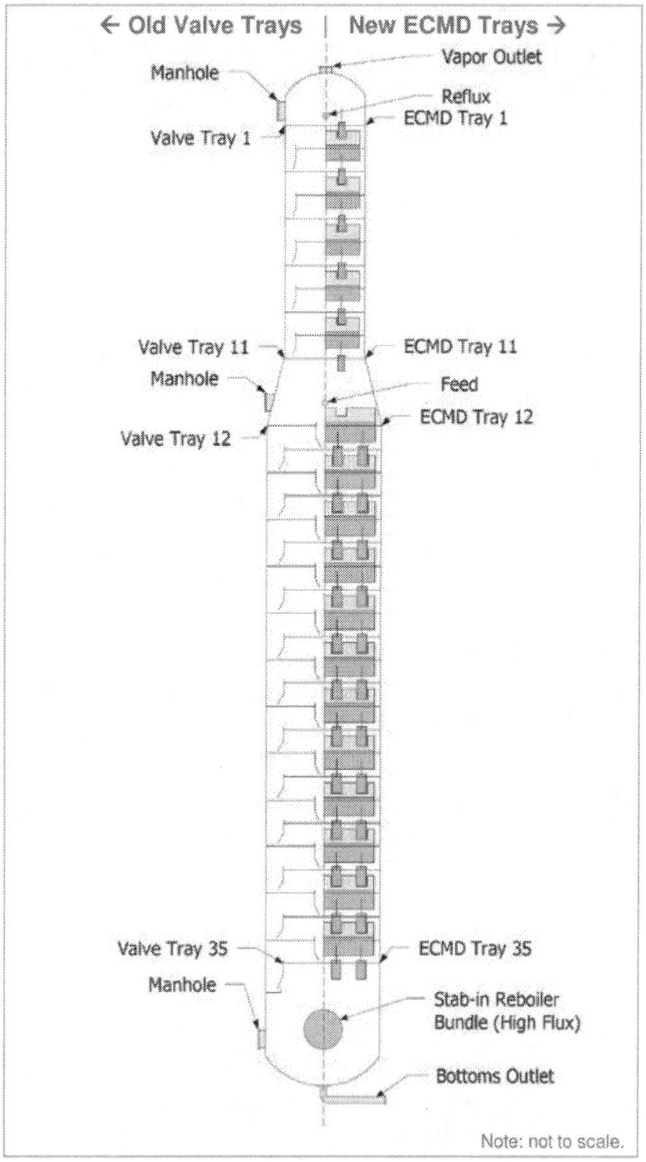

Figure 3. Deethanizer Revamp Sketch

After the tray revamp scheme was determined, UOP evaluated the reboiler and condenser for the new conditions.

The reboiler duty requirement increased by 50% relative to the original bare tube design. The boil-up duty could not be met with the existing bare tube bundle or a 1-for-1 High Flux tube bundle replacement using the existing low pressure steam.

The optimum revamp solution for the deethanizer reboiler involved a maximum area bundle replacement with 15.875 mm diameter OD coated / ID bare High Flux U-tubes oriented on a 20.64 mm square tube pitch. Reducing both the tube diameter and pitch ratio (1.5 to 1.3) allowed a heat transfer area increase of more than 20% without modification to the existing column port diameter. This design allowed the reboiler to stay on low pressure steam, rather than more costly medium pressure steam. Table 2 shows the normalized performance comparison between the original bare tube design and the proposed High Flux tube design.

Table 2. Normalized Reboiler Design Comparison

	Original Bare Tube Design	High Flux Revamp Design
Duty[1]	100%	150%
Area[1]	100%	123%
U-value[1]	100%	146%
MTD, °C	38	32
Percent of Max Bundle Q/A, %	64	97

[1] Values have been normalized based on original 1966 bare tube design = 100%.

The high overall heat transfer coefficient and MTD of this steam driven reboiler warranted a close evaluation of the bundle critical heat flux. In addition, the combination of high reduced pressure, low surface tension, and low latent heat of the boiling fluid contributed to lowering the bundle critical heat flux. Compared with the original bare tube design, the higher duty and lower MTD resulted in a 22% increase in the bundle heat flux. The new calculated bundle heat flux was 97% of the Heat Transfer Research Inc. (HTRI) maximum bundle heat flux and well above HTRI's 70% guideline. This high heat flux, and the substantially larger amount of vapor being generated within the bundle, has the potential to cause liquid starvation and vapor blanketing of the upper tube rows. This condition can reduce both the heat transfer performance and capacity of the exchanger. Therefore, the revamp design incorporated a 110 mm wide vertical bypass lane within the tube layout. This open channel allowed the rapidly boiling vapor to more easily exit the tube field, alleviating bundle critical heat flux concerns. Figure 4 depicts the High Flux bundle containing the vapor by-pass lane.

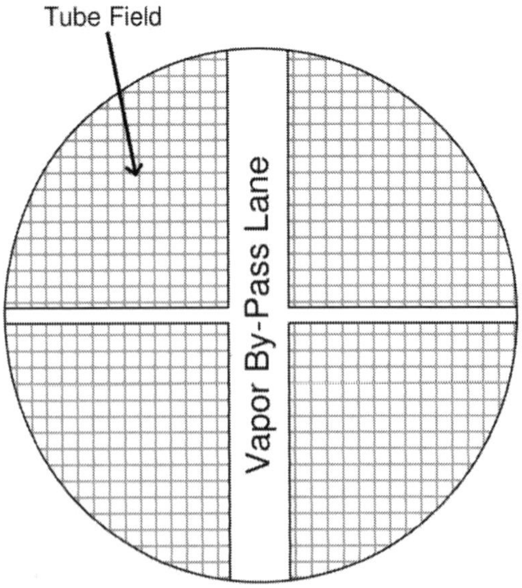

Figure 4. Deethanizer Reboiler Tube Arrangement

The expected condenser duty at 100% of expansion project rates was 70% higher than the original bare tube duty (88% higher at the expected flood point of the ECMD trays), so there was no way the bare tube bundle could accommodate the duty at the target propylene pressure.

UOP determined that a bundle replacement with the same number of 19.05 mm OD coated / ID bare High Flux tubes could accommodate the target duty at the minimum propylene pressure limit (7.24 kg/cm^{2}g).

After the initial exchanger evaluations, there was some iteration regarding allowable product purities (initially tighter specifications were considered to allow for some margin) – they had to be reconsidered based on handling the desired feed rate within the existing column constraints (i.e. column diameter and tray spacing). Once Chevron Phillips Chemical determined the proposed purities would not negatively impact the overall plant production, the system pressure was optimized to prevent the column, condenser, and reboiler from becoming a bottleneck. The proposed tray and reboiler stab-in bundle designs were checked and confirmed acceptable, while improvements were made to the proposed condenser design.

In order to maximize the benefits of the condenser revamp, 15.875 mm diameter tubes were evaluated. Smaller diameter tubes arranged on a tighter tube pitch permitted more heat transfer area to be packed within the existing kettle port diameter. There was some discussion about the allowable tube side pressure drop, since changing to smaller tubes would result in a higher pressure drop for the overhead stream. But since the effect on the reflux temperature was minimal, it was decided that the benefit of increased refrigerant pressure outweighed tube side pressure drop concerns and a pressure drop of 0.7 kg/cm^2 was allowed. The increase in heat transfer area allowed the MTD to be minimized and the propylene boiling pressure to be increased by 0.39 kg/cm^2. The tighter approach temperature was still more than adequate to sustain stable nucleate boiling with UOP's High Flux tubing. The increase in propylene pressure would also further unload the refrigerant compressor by reducing the compression ratio, and lowering the utility costs associated with the propylene refrigeration system.

The optimum condenser revamp involved 15.875 mm OD coated High Flux tubes, arranged on a 19.84 mm triangular pitch. Four tube-side passes and the original tube pitch ratio of 1.25 were maintained. Table 3 details the normalized performance comparison between the original bare tube design and the proposed High Flux tube design.

Table 3. Normalized Condenser Design Comparison

	Original Bare Tube Design	High Flux Revamp Design
Duty[1]	100%	188%[2]
Area[1]	100%	102%
U-value[1]	100%	338%
MTD, °C	7.3	4.0

[1] Values have been normalized based on original 1966 bare tube design = 100%.

[2] The High Flux revamp design duty was 171% of the original bare tube design duty plus 10% additional design margin.

Considering the new duty and rates were nearly 90% higher than the original bare tube design, shell-side liquid entrainment was evaluated for potential impact on the downstream propylene refrigerant compressor. Entrainment calculations showed that maintaining a liquid level at least

70% of the bundle height ensured adequate wetting of the top tube rows with a negligible impact on capacity or performance.

Performance Results

Equipment modifications for the expansion project were completed in early 2013 during the scheduled project outage. The plant successfully started up in the middle of February 2013.

Operating data from July 2014 confirmed that the target feed rate increase had been achieved, with tighter purity specifications than design. Table 4 compares the normalized operating data with the normalized revamp design conditions. Evaluation of the data suggests the efficiency of the ECMD trays is approximately 69%, relative to commercially-available process simulators, which is slightly higher than that of the conventional trays that were replaced. The efficiency vs. reflux rate curve of the operating data is shown in Figure 5. The x-axis (reflux rate) has been normalized using the design reflux rate as the basis (i.e. design reflux rate is 100%).

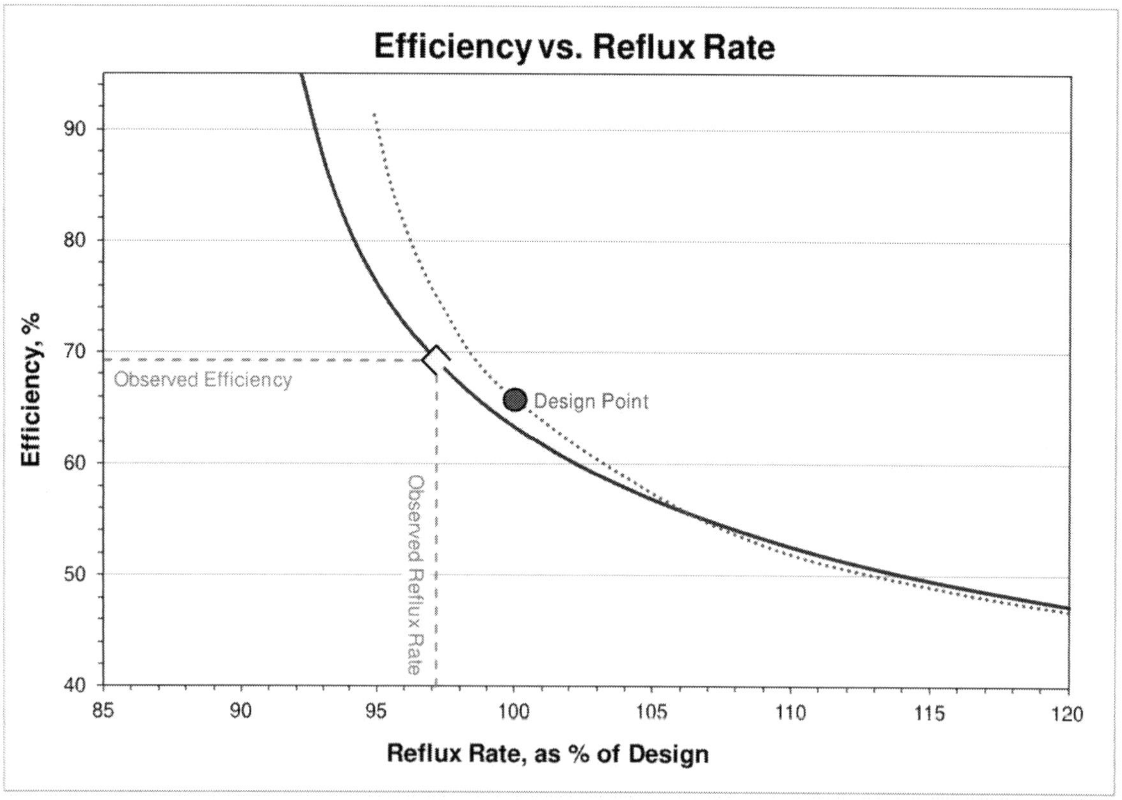

Figure 5. Efficiency vs. Reflux Rate Curves – Design vs. Observed

The higher tray efficiency and changes in the expected feed composition maintained a lower operating pressure. Chevron Phillips Chemical confirmed their ability to push the capacity in excess of 104% of design rates without running at the higher column pressure considered at the design stage. Based on UOP's simulation of the July 2014 operating data, the ECMD trays were operating at about 84% of flood, which meant the trays were expected to process an additional increase of 18% in the feed rate before flooding.

Table 4. Normalized Operating Data vs. Revamp Design

	Pre-Expansion Operating Data[2]	ECMD Tray Revamp Design	July 2014 Operating Data[3]
Feed Rate[1]	100	165	167
Feed Ethane Content, vol%	mid 40s	mid 40s	low 40s
Feed Propane Content, vol%	< 30	< 30	> 30
Overhead Product Rate[1]	36	59	52
Overhead Propane Content, %	2.5	4	2.5
Bottoms Product Rate[1]	64	106	115
Tray Efficiency, %	-	65.7	69.2
Reflux Ratio	1.6	1.8	2.0
Condenser Duty[4]	-	171%	179%
Reboiler Duty[4]	-	150%	146%
Column Pressure Drop[1]	1.0	1.7	1.3
Number of Trays	35	35	35
Tray Type	1-/2-pass Valve	ECMD Trays	ECMD Trays
Percent of Flood, %	~100	90	84

[1] Values have been normalized based on pre-expansion feed rate = 100%.

[2] Normalized values for a 6 hr. period of stable operation on January 13, 2011, representing the highest pre-expansion feed rate while maintaining good cut points.

[3] Normalized values for a 4 hr. period of stable operation from July 14-15, 2014 (post-revamp).

[4] Condenser and reboiler duty ratios based on original 1966 bare tube design duties = 100%.

Figure 6a and Figure 6b show the capacity graphs for the ECMD trays in the top and bottom sections, respectively. The graphs compare the July 2014 operating data with the design point. The x-axis (liquid load) and y-axis (v-load) have been normalized using the design rates as the basis (i.e. design rates are 100%). As previously stated, the lower column operating pressure improved the separation and resulted in increased capacity. The improved separation is a product of a larger delta between liquid and vapor densities, resulting in lower reflux ratios required for the given separation. The increased capacity is a product of increased surface tensions, which allows for higher vapor loads. These phenomena are represented by the lower vapor and liquid loads compared to design conditions and by the different "Vapor Max" and "DC Backup" lines for the design and July 2014 data points.

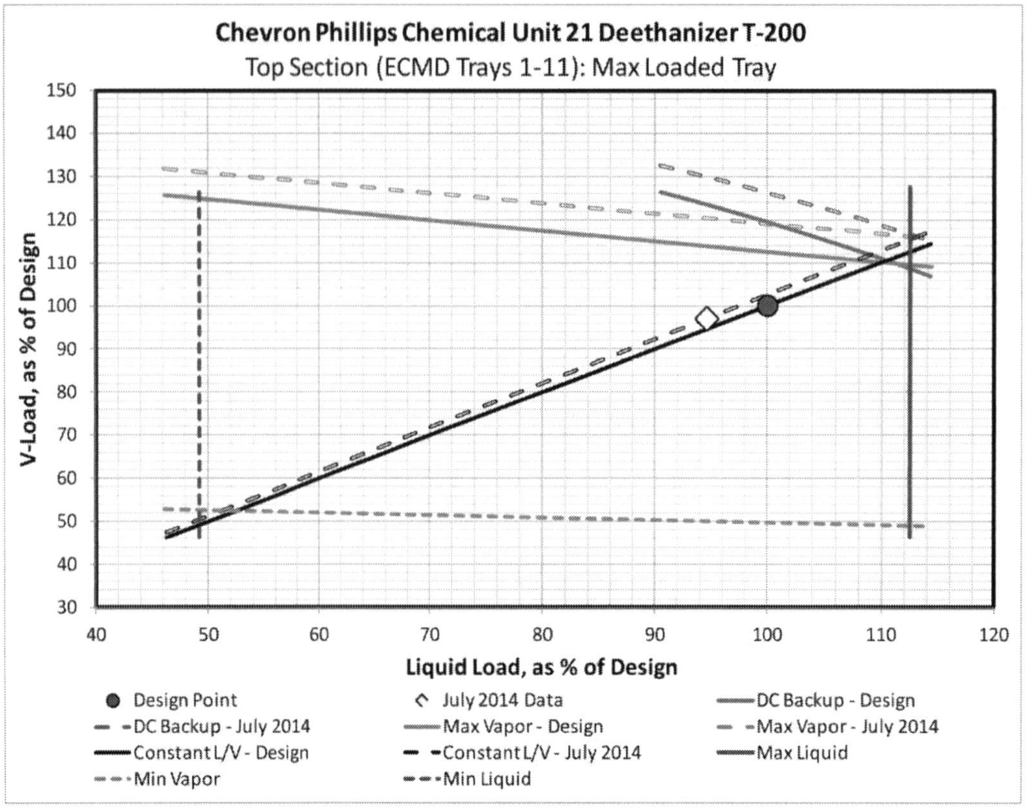

Figure 6a. Capacity Graph – Top Section

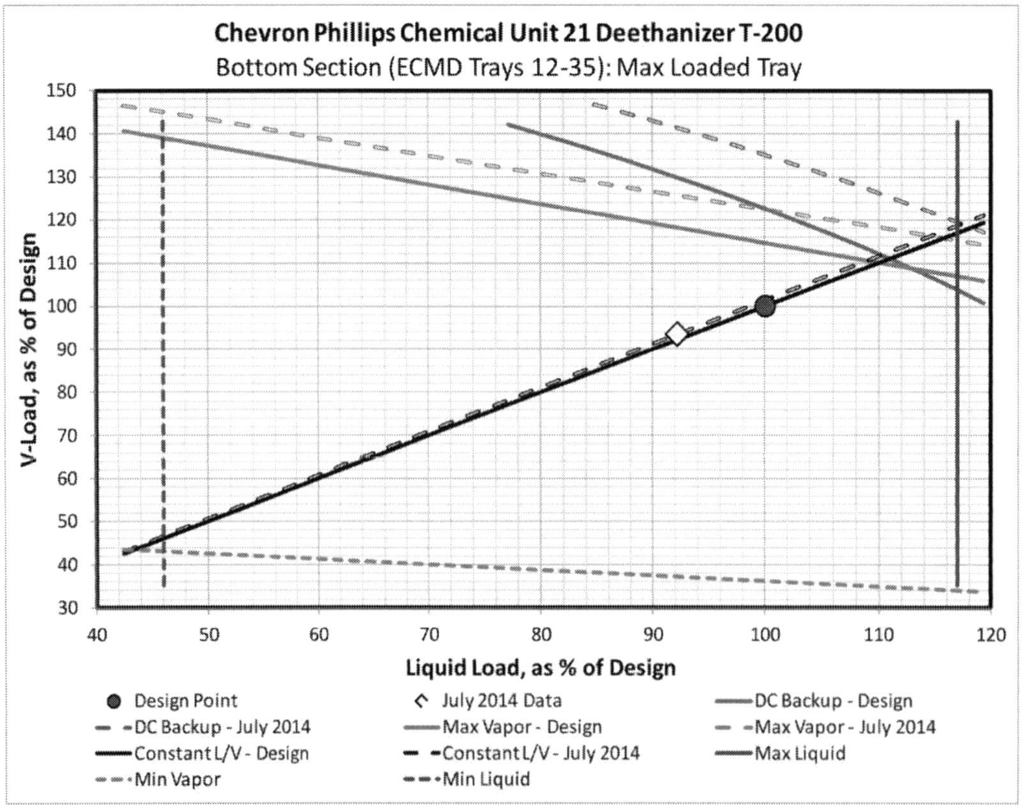

Figure 6b. Capacity Graph – Bottom Section

At one point during the design of the ECMD trays, Chevron Phillips Chemical expressed concerns about the capacity that was offered by UOP regarding the downcomer velocity and jet flood values. In systems that operate at elevated pressures, like the deethanizer, the delta between the liquid and vapor densities is relatively low, making it more difficult for liquid and vapor separation. If not taken into account, this oversight can result in trays that flood earlier than predicted by the different flood correlations.

Based on FRI information, Chevron Phillips Chemical developed maximum criteria for downcomer velocity based on vapor density. For the deethanizer top section with a vapor density of 69.0 kg/m^3, this criteria would have restricted the design downcomer velocity to 220 m^3/(m^2-hr) in order to avoid potential downcomer choking or liquid and vapor separation issues. For the deethanizer bottom section with a vapor density of 87.8 kg/m^3, the system would be limited to an even lower downcomer velocity.

Based on the design rates, the downcomer velocities in the top and bottom sections were close to 375 m^3/(m^2-hr) for the proposed ECMD trays, which greatly exceeded Chevron Phillips Chemical's design criteria for the system conditions, and led them to question the proposed design. UOP stated that their design guideline is around 367 m^3/(m^2-hr) for MD and ECMD trays.

For Chevron Phillips Chemical to be comfortable with the proposed tray design, they required UOP to provide examples of successful operating experience at high downcomer velocities, in a system similar to the deethanizer tower. UOP provided Chevron Phillips Chemical with a plot similar to the one shown in Figure 7 containing the downcomer velocity vs. vapor density for UOP's MD and ECMD tray experience in deethanizer and demethanizer services (i.e. typical services with high vapor densities).

Although the proposed design downcomer velocities were slightly higher than what UOP normally designs for with ECMD trays (considering the vapor density), UOP determined that the higher downcomer velocity was a reasonable trade-off in order to limit the percent of jet flooding to 90%. Based on UOP's in-house data, directionally, the high percent of jet flood constituted more design risk than the high downcomer velocity. Also, UOP had anticipated that there would be a slight reduction in the downcomer velocity when the final design was done, as a result of optimizing the downcomer configuration (i.e. top and bottom downcomer areas).

For the jet flood capacity concerns, Chevron Phillips Chemical was able to confirm that they had past experience with operating ECMD trays at 90% flood and higher at their plants. Consequently, Chevron Phillips Chemical felt that operating the proposed retrofitted deethanizer tower at 90% of flood should be routinely achievable.

During the final design, UOP was able to optimize the tray design to balance the downcomer area and active area in order to lower the downcomer velocities below UOP's design guideline, while keeping the percent of flood less than 90%. The resulting design downcomer velocities in the top and bottom sections were 366 m^3/(m^2-hr) and 367 m^3/(m^2-hr), respectively. These design points are included in Figure 7 for reference.

UOP also plotted the observed downcomer velocities and vapor densities for the ECMD trays based on UOP's simulation of the July 2014 operating data. These values prove that the ECMD trays can handle much higher downcomer velocities than anticipated by conventional design criteria.

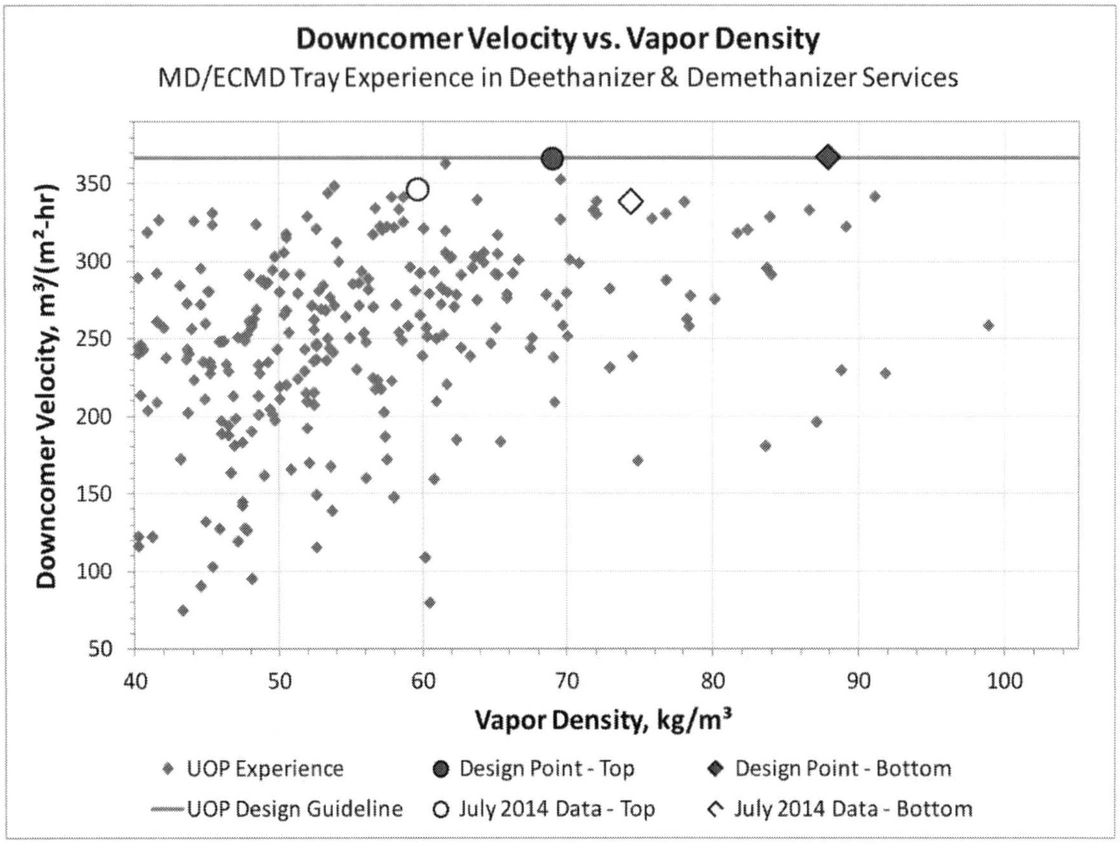

Figure 7. Downcomer Velocity vs. Vapor Density – UOP Experience

The July 2014 operating data also shows that the High Flux bundles are satisfying the expansion requirements. The measured cold-end approach temperature between boiling and condensing streams in the condenser was less than 1°C, and simulations of the operating data indicate the design U-value had been exceeded. Table 5 summarizes the observed condenser performance compared to the High Flux design.

Table 5. Normalized Condenser Performance

	Original Bare Tube Design	High Flux Revamp Design	July 2014 Operating Data[2]
Duty[1]	100%	188%	179%
Area[1]	100%	102%	102%
U-value[1]	100%	338%	394%
MTD, °C	7.3	4.0	3.2

[1] Values have been normalized based on original 1966 bare tube design = 100%.

[2] Normalized values for a 4 hr. period of stable operation from July 14-15, 2014 (post-revamp).

The July 2014 operating data shows that the reboiler is also meeting the goals of the revamp. The LP steam is being throttled to minimize the condensing pressure. The condensate pressure is ~1 kg/cm² less than design. The steam enters superheated and is subcooled to a pinch. Table 6 summarizes the observed reboiler performance compared to the High Flux design. The revamp design was based on the assumption that the steam and condensate would be saturated. The average operating heat transfer coefficient appears lower than design because the steam sensible heat transfer accounts for approximately 15% of the total operating duty.

Table 6. Normalized Reboiler Performance

	Original Bare Tube Design	High Flux Revamp Design	July 2014 Operating Data[2]
Duty[1]	100%	150%	146%
Area[1]	100%	123%	123%
U-value[1]	100%	146%	141%
MTD, °C	38.3	31.9	32.2

[1] Values have been normalized based on original 1966 bare tube design = 100%.
[2] Normalized values for a 4 hr. period of stable operation from July 14-15, 2014 (post-revamp).

Summary

The deethanizer unit at Chevron Phillips Chemical's Sweeny plant was bottlenecked with conventional separation and heat transfer equipment. By revamping with UOP's ECMD trays and High Flux tubing, Chevron Phillips Chemical achieved a capacity increase greater than 65%. At the same time, the implemented changes achieved all of Chevron Phillips Chemical's design objectives. Operating data since the 2013 startup demonstrates that the tower and exchangers are operating better than design.

As a result of the revamp, both Chevron Phillips Chemical and UOP are pleased with the performance of the Unit 21 deethanizer system. The revamp of the deethanizer unit adds to the long history of working with Chevron Phillips Chemical to maximize the benefits from operating their various distillation units located in different plants throughout the country. Chevron Phillips Chemical has also used UOP heat and mass transfer equipment in other areas of the Sweeny complex (Shakur et al. 1999).

References

1. P. J. McGuire and M. R. Sobczyk, "Maximizing Return on Assets and Process Optimization of New Units with UOP Equipment Technologies" 2008.

2. P. J. McGuire, M. S. M. Shakur, A. G. Cantrell, J. Valverde, and D. Webb, "Replacing MD™ Trays in Amoco Deethanizer Yields 20% Capacity Increase" American Institute of Chemical Engineers Spring National Meeting, Houston, Texas, March 12, 1997.

3. M. S. M. Shakur, R. E. Tucker, K. J. Richardson, M. R. Sobczyk, R. D. Prickett, C. Polito, and S. E. Harper, "Increase C2 Splitter Capacity with ECMD Trays and High Flux Tubing" American Institute of Chemical Engineers Spring National Meeting, Houston, Texas, March 18, 1999.

Appendix A – MD and ECMD Tray Product Description

For more than forty years, UOP has been a leader in supplying high performance distillation technology, including MD and ECMD trays. These trays have been used in over 1000 applications, and the performance and benefits of these trays are discussed below.

MD and ECMD trays are designed for medium to high pressure service, where the liquid to vapor ratio is relatively high. Whereas conventional trays must go to two-pass, four-pass, and even six-pass trays to accommodate these high liquid rates, the unique design of the MD and ECMD trays make them ideally suited for these applications.

An MD tray consists of a series of long, trough like downcomers, separated by sections of perforated plate (see Figure A-1). These downcomers are terminated in the vapor space of the tray below, and are spouted to allow liquid to flow directly onto the panel of the tray below eliminating the need for receiving pans. This configuration allows for larger active area and downcomer area, thus increasing the tray capacity.

Figure A-1. MD Tray Configuration

The most unique feature of the MD tray is the large number of downcomers, resulting in a very long outlet weir length. This extended weir length reduces the liquid height on the tray, and as vapor rises through this thinner liquid layer, lower froth heights are generated compared to conventional trays. This lowers the percent of jet flood and further increases the capacity. The lower froth height also allows these trays to be designed for a lower dry tray pressure drop, as lower vapor velocities are required to prevent the liquid from weeping through the decks.

The combination of no receiving pans, lower froth heights, and lower dry tray pressure drops all serve to significantly reduce the pressure drop of the MD tray. In fact, the pressure drop of an MD tray is only about 60 percent of that of a comparable conventional tray. As the pressure drop across the tray makes up the largest component of the downcomer backup, MD trays have much lower downcomer backups than conventional trays. The combination of low froth heights and low downcomer back-ups allows the trays to be installed at much smaller tray spacings than conventional trays. In the case of revamps, installing a larger number of trays at smaller tray spacings can significantly increase the number of theoretical trays generated within the column, allowing for improved product purities or reduced energy requirements for a given separation. In new units, installing the trays at smaller tray spacings can result in a significant reduction in overall column height.

Figure A-2 shows a side view of an MD tray. The box-like structure of the downcomers provides all of the mechanical support, and no additional bolting bars or beams are required. This photo also shows another unique feature of the MD tray, that consecutive trays are rotated ninety degrees. As shown in Figure A-3, the crossing of the downcomers on consecutive trays forms a "unit cell", and all of these cells on the tray perform exactly the same as the others. As the diameter of the column increases, the number of downcomers also increases, such that the unit cell dimensions remain relatively constant. This allows the MD tray to be scaled up to very large diameters, with no fall-off in efficiency or capacity, which is often experienced with conventional trays at large diameters.

Figure A-2. 90 Degree Tray Rotation

Figure A-3. Unit Cell Geometry

The crossing of the downcomers also results in the MD tray being self-balancing, i.e., the trays themselves are able to correct a maldistribution problem. If there is a problem in the column that results in all of the liquid being distributed to one side of the column, the downcomers will act very quickly to redistribute the liquid evenly across the tray, resulting in maximum capacity and efficiency.

For 30 years, the MD tray was the industry standard for high capacity trays. However, in the late 1980's, a need developed for an even higher capacity tray, in order to revamp columns that already contained MD trays, as well as for future expandability for customers who wanted to use MD trays in new units. To address this need, UOP developed the ECMD tray.

Figure A-4. ECMD Tray Configuration

An ECMD tray is very similar to an MD tray in configuration, but differs in two main aspects (see Figure A-4). An ECMD tray uses slotted decks to introduce some of the vapor horizontally into the liquid to reduce the froth height further, providing for increased vapor capacity. Disengagement plates are used over the downcomers to redirect the slot-propelled froth into the downcomer, and to enhance vapor disengagement. These trays provide a 15-20% capacity increase over MD trays, while maintaining the same separation efficiency, and have been installed in hundreds of columns since 1990.

Other UOP Mass Transfer Products

UOP offers a full range of mass transfer products for various high capacity and high efficiency applications (see below). This paper is focused on the benefits provided by one of our high capacity tray products. The other products also provide many benefits to customers looking to increase / improve production of existing columns, or looking for capital savings on the installation of grassroots columns.

- Family of MD Trays
  - **Multiple Downcomer (MD) Trays** – High Capacity
    [+30-50% more capacity than conventional trays]
  - **Enhanced Capacity (ECMD) Trays** – Higher Capacity
    [+45-70% more capacity than conventional trays]
  - **Parallel Flow (PFMD) Trays** – High Efficiency & High Capacity
    [+45-70% more capacity than conventional trays with similar / better efficiency]
- **SimulFlow Devices** – Highest Capacity
  [+105%-130% more capacity than conventional trays]
- **Slotted Sieve Trays** – High Efficiency & Low Pressure Drop
- **Multiple Upcomer (MU) Trays** – Liquid-Liquid Extraction

ECMD Trays

MD Trays

PFMD Trays

Slotted Sieve Trays

SimulFlow Devices

MU Trays

Appendix B – High Flux Tube Product Description

High Flux tubing features a nucleate boiling enhancement that significantly increases the heat transfer performance of shell and tube heat exchangers. The surface consists of a porous metal coating metallurgically bonded to the inside or outside surface of a bare tube. With conventional tubing, boiling begins randomly and usually at pits and scratches in the surface. With High Flux tubing, capillary action draws liquid into the porous layer and vaporizes from the large number of nucleation sites that are integral to the structure. These sites allow vapor bubbles to be generated at a wall superheat of less than 1°C. This is considerably lower than that required to initiate nucleate boiling on the surface of a plain tube. Vigorous nucleate boiling, at low wall superheat, results from a combination of good matrix thermal conductivity, highly extended microsurface area, and a large number of interconnecting channels within the porous surface. As a result, the boiling heat transfer coefficient improves by a factor of ten or more. This produces overall heat transfer coefficients that are up to two to five times higher than conventional tubing in boiling applications. The relative performance improvement will be a function of the shell side and tube side fluids, exchanger geometry, and design fouling factors for the specific application.

See Figure B-1 for a visual boiling comparison for High Flux, bare, and low fin tubes.

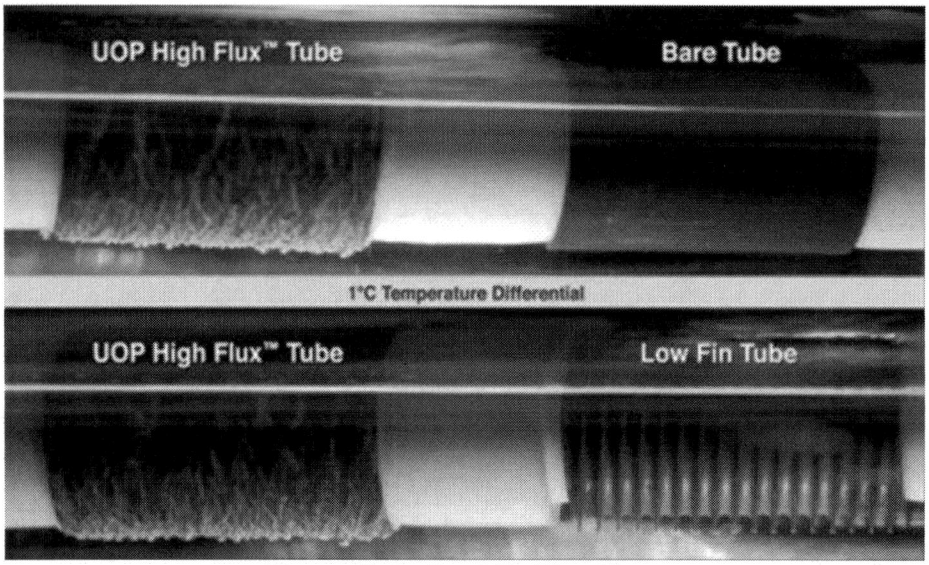

Figure B-1. Pool Boiling Test

High Flux tubing is available in most common materials such as carbon steel, stainless steel, copper alloys and aluminum, with the coating being the same material as the tubular substrate. The manufacturing process ensures a mechanically strong surface highly resistant to abrasion or erosion. The porous coating on one tube surface can also be combined with sensible- and condensing-side enhancements (finning or fluting) on the other surface to produce a doubly-enhanced High Flux tube to meet performance requirements. Horizontal thermosyphons, kettles and stab-in reboilers, with boiling taking place on the shell side of the exchanger, use OD coated tubing with lengths manufactured up to 15240 mm. Vertical thermosyphon reboilers use ID coated tubes with lengths up to 6100 mm.

UOP's experience in refinery and petrochemical applications dates back more than 40 years. Ideal High Flux exchanger candidates typically use condensing hydrocarbons or steam to vaporize steam or hydrocarbon streams. Reduction in capital investment, reduction in the number of exchanger shells, and efficient operation at small temperature differences are some reasons cited by operating companies or contractors for selection of High Flux heat exchangers.

Figure B-2 shows a photograph of the High Flux tube with surface magnification.

Figure B-2. High Flux Boiling Surface

Other UOP Heat Transfer Products

UOP offers a full range of enhanced heat transfer products for boiling and condensing applications (see photo below). This paper was focused on the benefits provided by one of our tube products. Similar benefits can be offered by the other products, as needed to satisfy a specific application.

UOP Heat Transfer Tube Products
High Flux and High Cond Tubes

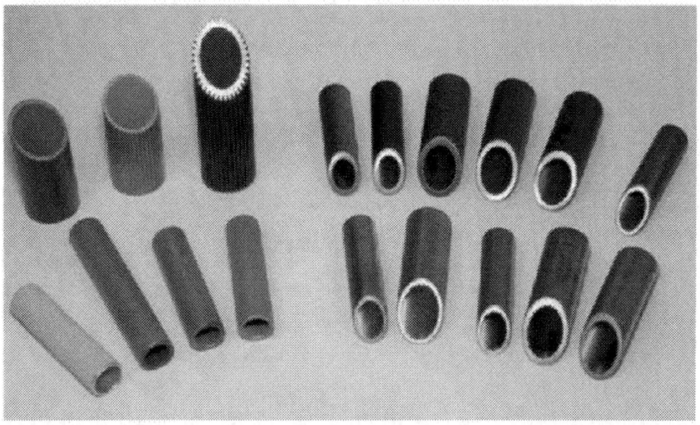

CHANNEL BAFFLES ON TRAYS

By

Daniel R. Summers, P.E.
SULZER CHEMTECH USA, Inc.

and

Dr. Simon Chambers
Formerly of Fractionation Research Inc.

Presented at the AIChE Spring National Meeting

Topical 8:

Kister Distillation Symposium

Paper 206c

April 29, 2015
Austin, Texas

Background

Frequently, distillation tray designers are posed with the problem of handling moderate to high volumes of liquid with small amounts of vapor. Side Strippers, Sour Water Strippers and Wastewater Strippers are typical examples of such applications where this problem arises. Tray performance is critical in these services and a shortfall in performance could have severe detrimental effects downstream. Conventional wisdom is to provide blanking or "dead zones" on the tray decks where no contacting occurs. A better practice is to provide Channel Baffles which contract the active area of the tray without reducing the flow path length of the device. The impact of channel baffles on tray performance was measured with a water cooling technique in Aston University's 2.44 meter diameter air-water simulator. Temperature isotherms (analogous to concentration profiles) of the developed flow patterns on a channel baffled sieve tray are compared to the same circular sieve tray. This paper will discuss the issues shown above and describe good Channel Baffle design practices that one can employ to ensure that a good, low vapor loaded, design is achieved.

Design Practice

When a trayed tower is repurposed or the operational range is severely reduced, the tower diameter can be many times larger than it needs to be. In these situations, there are several options that a designer can employ to optimize the new operating range. The first option is to simply add blanking plates (or strips) to the tray decks such that the open area, and sometimes the active area as well, is reduced. For movable valve trays instead of adding blanking plates, one can "crimp" the valve legs on numerous valves to force them to stay closed or even replace numerous valves with fully closed flush seating valve units. The problem with this simple approach is that flow path length and vapor/liquid contact time is reduced. In a critical service, the loss of tray efficiency cannot be tolerated or will result in heightened energy consumption per unit mass of product.

A second option is to redesign the trays altogether with larger downcomers. This option is definitely more costly than adding blanking or crimping valves and will also result in shorter flow path lengths similar to the blanking option above. However, the attractiveness of this option is typically the speed with which a revamp can be achieved. Adding blanking or crimping valves will typically take more time in the tower than replacing the trays completely. Frequently, when the tower revamp is the critical path during a turnaround, the total cost of installing completely new replacement trays can be justified by enabling the unit to come back online just one day sooner. There have been several occasions when an oversized 4-pass trayed tower has been successfully revamped by Sulzer with 2-pass trays having larger envelope downcomers. In this situation, a total tray revamp would actually result in longer flow path lengths and improved tray performance. This idea is not part of this discussion but it did need to be mentioned.

A third option is to "gut the tower" and add a smaller diameter "sleeve" that would contain the new trays. The sleeve is non pressure bearing and has been frequently used to reduce the effective diameter in packed towers. To the best knowledge of the authors, no one has attempted this idea with a trayed tower. This idea should, in reality, be cost prohibitive especially since the sleeve will have to be assembled and welded together inside the tower with all the parts fitting through the vessel manholes. The cost should be higher for trayed towers than packed towers because tray support rings and bars will additionally need to be welded into the "sleeve."

Finally, a cost effective high performance option is the employment of channel baffles on trays, hence, the subject of this discussion.

Channel Baffles

In 1988 there was a US Patent[1] issued that described the use of parallel baffles on trays to solely improve tray performance, but not turndown. In addition, a brief research into the archives of Nutter Engineering revealed that Channel Baffles have been employed as far back as the early 1980's. However, the popular method (even today) of reducing a trayed tower's operating range is through the employment of blanking [2,4]. Channel Baffles actually are "blanking" except it is different than the traditional method in that the blanking is actually in the direction of liquid flow on the tray. Traditional blanking involves blanking strips (or valve crimping) placed on the tray decks perpendicular to flow. The reason for blanking perpendicular to flow is to ensure that there is no opportunity for liquid to find a way to flow across a tray without contacting vapor. What Channel Baffles do is prevent liquid from finding that "non-contacting" path across a tray while blanking the tray parallel to liquid flow.

The preferred way to incorporate Channel Baffles is to blank the area nearest the vessel wall and place sufficiently tall baffles that will keep liquid (and froth) from getting into the blanked area, see *Figure 1*. This was described first by Sloley and Fleming in 1994[3]. The active area is reduced by the amount of blanked area and the baffles prevent liquid from entering that same blanked area. With the reduced active area, the tray deck openings can now be placed closer together and provide for good mixing between the vapor and liquid phases. The resulting flow path length is thus entirely dedicated to contacting between the liquid and vapor thus ensuring maximum mass transfer potential.

Figure 1 – Typical 1-Pass Channel Baffle Application

Figure 2 shows two different designs for a 1,350 mm (54 in.) ID tray, that have exactly the same open area. In *Figure 2b*, two large blanked areas, perpendicular to flow, are employed. The resulting flow path length is only 264 mm (10.4 in.) and if the outlet weirs are not picketed, the weir loading is very low at 6.62 $m^3/(h.m)$. *Appendix A* shows the detailed hydraulics of this potential design. Please note that with the low weir loading and the large flow path width (larger than the weir length) the potential exists for 17.7% of the active area to be stagnant (outside the dashed lines on the figure). A Channel Baffle design, shown in *Figure 2a*, has a reasonable flow path length of 787 mm (31 in.), and a 457 mm (18 in.) wide flow width (large enough for a tray manway) and a weir loading of 16.2 $m^3/(h.m)$ (1.8 gpm/inch) with no need for weir picketing. In addition, there is zero risk for stagnation on the trays. Therefore, tray efficiency with a channel baffle design should be far superior to an application with blanking. *Appendix B* shows the detailed hydraulics of the superior Channel Baffle design.

Figure 2a

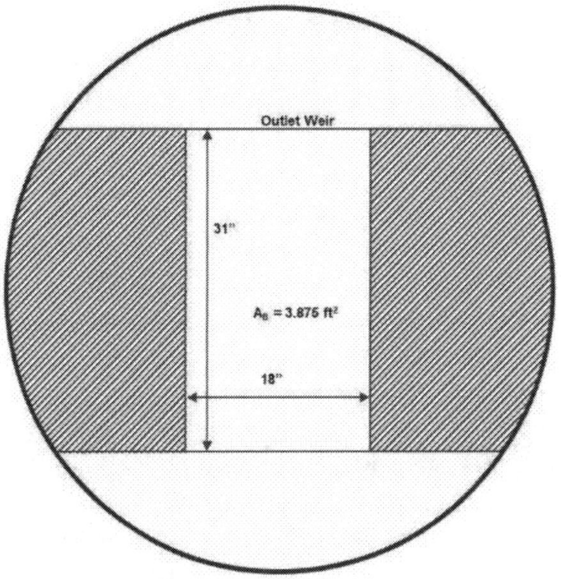

Diameter = 54"
Weir Loading = 1.81 gpm/inch
Pressure Drop = 1.88 mmHg
Deck Holdup = 7.46 lbs
Stagnation = None

Figure 2b

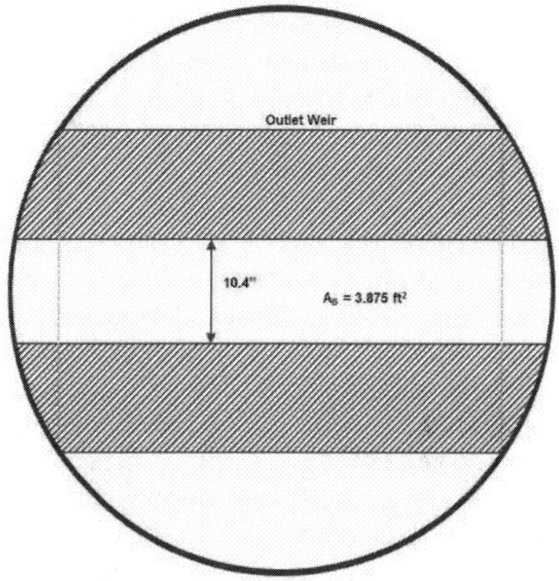

Diameter = 54"
Weir Loading = 0.74 gpm/inch
Pressure Drop = 1.73 mmHg
Deck Holdup = 5.83 lbs
Stagnation = 17.7% of Active Area

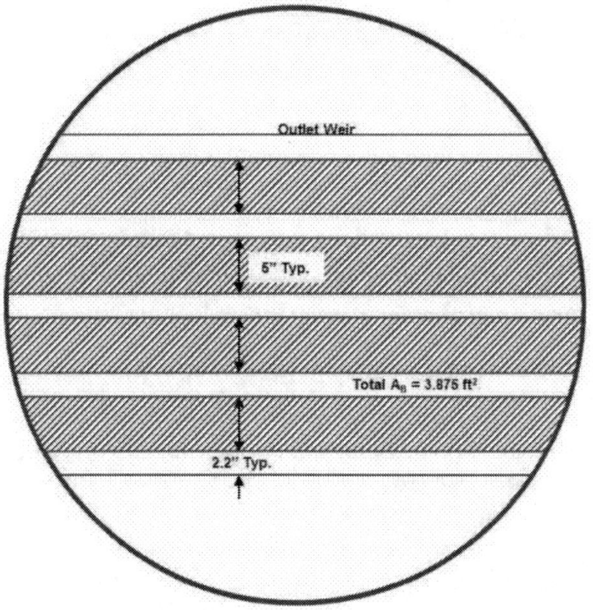

Figure 3 – Blanking Strip Alternative

One can argue that instead of two large blanked areas, as depicted in *Figure 2b*, one should instead incorporate numerous small blanked areas as shown in *Figure 3*. Standard practice for blanking strips are to make them no wider than 100 to 140 mm (4 to 5.5 inches)[4,5]. In this example, one needs to employ four blanking strips that are 127 mm (5 in.) wide each to get the effective active area (and open area) to the same level as shown in *Figure 2b*.

However, having five active sections of 56 mm (2.2 in.) in length each is a little absurd because only a single row of valves (or sieve openings) could be placed in each of the five active sections and there would be innumerable opportunities for liquid to get across this tray without contacting any of the vapor. Again, a channel baffle design as shown in *Figure 2a* would be superior.

Design Considerations

The above leads to two issues; 1.) When does one employ Channel Baffles and 2.) How does one best employ them? Before these questions can be answered, the authors assume that the desired open area has already been determined based on pressure drop, turndown and tray stability considerations[6].

Channel Baffles should be employed when the spacing between valves for the given open area gets larger than 133 mm (5.25 in.) and for sieve trays when the spacings between the 12.7 mm (½ in.) holes gets greater than 76 mm (3 in.), see *Figures 4* and *5*. These values come from a number of sources but the most recently documented source is from page 43 of Sloley and Fleming[3]. Figure 4 was generated based on a typical minimum valve density of 65 valves/m^2 (6 per ft^2). Round valves could be substituted for the rectangular BDH™ valves shown in *Figure 4*. If the valves were placed on a perfect equilateral triangle, the row and tier spacing would be 232 mm (9.12 in.) and 67 mm (2.65 in.) respectively and the percent open area would be 8.4% based on a 12.7 mm (½ in.) valve lift. For movable valves this is a fairly small open area (keeping in mind that the open area when the valves are fully closed is about 9.5 times less than when fully open). Therefore, it is the author's opinion that channel baffles should be considered to be employed when the valve density falls below 65 valves/m^2 (6 per ft^2).

The relationship between open area and pitch for sieve openings is;

$$Pitch = Dp\left(\frac{\pi}{fp\sqrt{12}}\right)^{0.5}$$

Based on this equation, for a hole diameter (Dp) of 12.7 mm (1/2 in.) and a pitch of 76 mm (3 in.) the minimum fraction perforation (fp) is 0.025 or 2.5% of the active area, see Figure 5. This also is a very small open area. Therefore, channel baffles should be considered to be employed when the sieve tray open area falls below 2.5%.

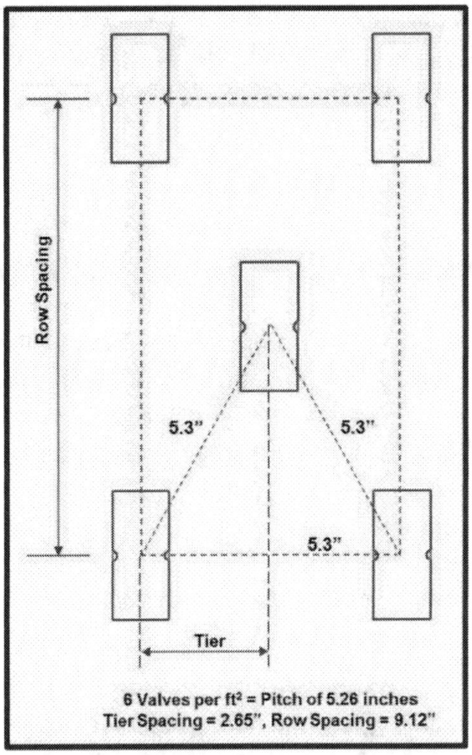

Figure 4 – BDH Valves with Perfect Triangular Pitch at min Valve Density

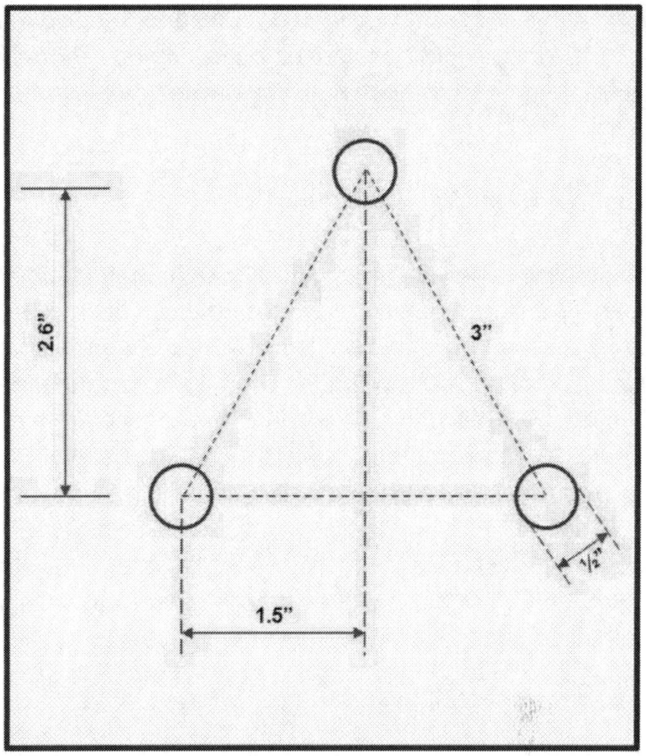

Figure 5 – Typical maximum hole spacing for 12.7 mm hole diameter sieve trays

Channel baffles help with the spacing of openings on trays at low vapor loads. However, if the liquid loads are very high, such as a weir loading in excess of 107 $m^3/(h.m)$ (12 gpm/inch), prior to channel baffle placement, then adding channel baffles will obviously make the weir loading higher still. Under these conditions, one needs to seriously consider what mass transfer can realistically be achieved. With a "speeding" liquid flowing across a tray and a very low open area, will the vapor trying to pass through the tray have sufficient liquid-vapor contact time to achieve satisfactory tray efficiency? High weir loadings in excess of 116 $m^3/(h.m)$ (13 gpm/inch) can be tolerated when the jet flood on the trays is sufficiently low. This subject was addressed by Resetarits and Ogundeji[7] to show that hydraulically, higher weir loadings are tolerable at low vapor loadings such as those where channel baffles would be employed.

When channel baffles are employed, the baffles need to be of sufficient height to minimize carryover to the "blanked" portion of the tray. Those baffles will typically be as tall as the tray spacing minus a few inches to avoid interference with the integral trusses of the tray panels above. This elevation will ensure that the majority of the flowing liquid will remain within the confines of the channel baffles especially under conditions that have extremely low weir loading (< 4.47 $m^3/(h.m)$ or 0.5 gpm/inch). Baffles that only extend to 50% of the tray spacing are ineffective and highly discouraged. If any liquid does find its way over the baffles, the "blanked" portions of the tray decks need to have provisions to enable liquid escape into the downcomer. Typically, no outlet weir is apportioned to the blanked sections of the tray. Gasketing of the "blanked" sections is typically not necessary, but it must be realized that any miscellaneous open area will enable some portion of the vapor to totally bypass the liquid. Therefore it is highly recommended that any miscellaneous openings in the blanked area be limited to the sum total of 1% of the tray open area. As an example, the tray in *Figure 2a* has an open area of 325 cm^2 (0.35 ft^2). One percent of this area is 3.23 cm^2 (0.5 in^2). This means that two individual openings in the blanked sections of ½" x ½" each could be tolerated but no more. Therefore, lip-slot tray panel connections (because there are numerous openings at the slots) should not be used with channel baffle trays.

Channel Baffle Tray Research

As started earlier in this paper, although channel baffled trays are not new, experimental data to quantify their performance on a commercial scale is scarce. To address the issue, the two-phase flow patterns of channel baffled sieve trays were studied as part of a comprehensive investigation into the impact of flow patterns on tray performance involving different hole size sieve trays. Studies were performed in Aston University's 2.44 m diameter air-water simulator as shown in *Figure 6*. The rectangular tray is the circular sieve deck fitted with vertical walls (channel baffles) that reduce active area to a rectangle. Two perforation diameters were utilized in the experiments: 6.35 and 12.7 mm. The work of Porter et al 1992[8], Chambers 1993[9] and Fenwick 1996[10] provide complete descriptions of the air-water simulator setup, experimental techniques and data collections. *Table 1* provides details of two tray designs employed, the channel baffle design as well as the air and liquid flow rates employed.

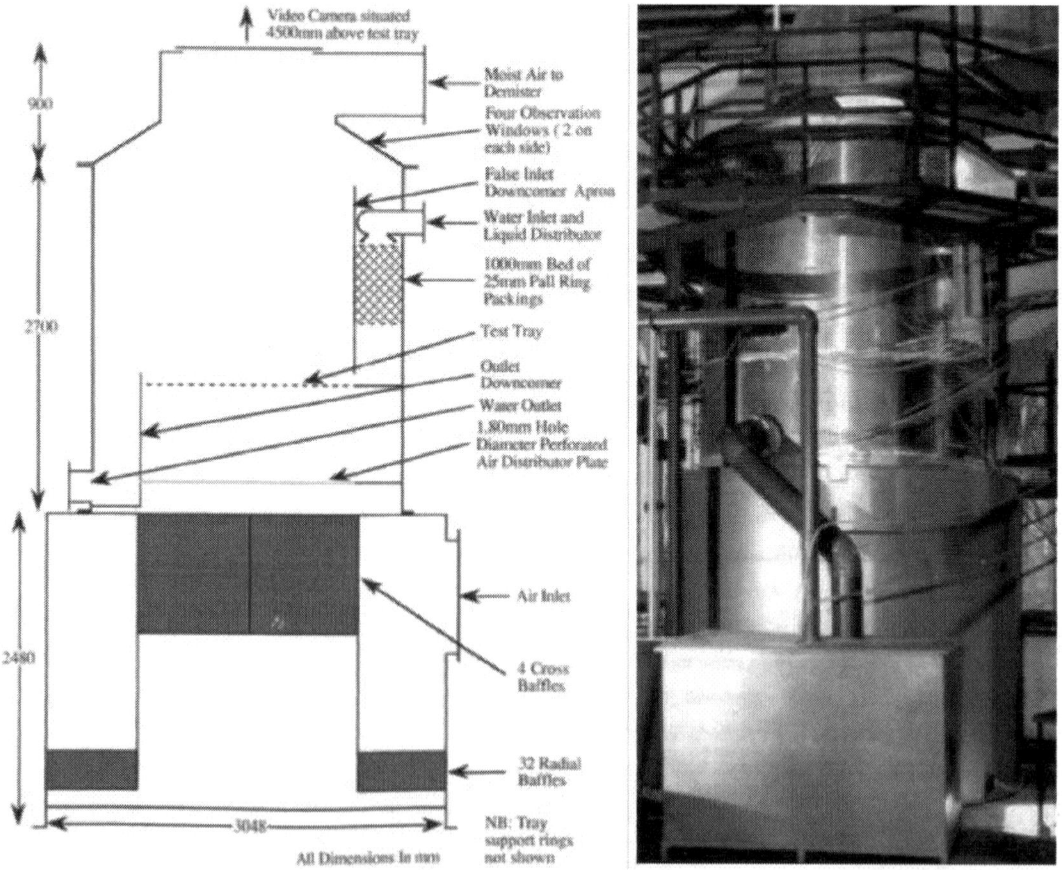

Figure 6 – Setup of the Aston University 2.44 Diameter Air-Water Simulator

Specifications			Study	Hole Diameter mm	DC_{CL} mm	h_{OW} mm	Air Velocity m/s	10^4 Water Weir Load $m^3/(s.m)$
Test Column	Diameter	2440 mm	Water Only	0.0	20	20	0.0	25
	Material	Aluminium		0.0	50	50		50
	Flow Path Length	1925 mm		0.0				100
	Chordal Weir Length / Column Dia. (W/D)	0.615		0.0				150
				0.0				200
								250
Circular Trays	Weir Length	1500 mm	Air / Water	6.35 & 12.7	20	20	1.00	25
	Hole Area % A_B	10.0%		6.35 & 12.7	50	50	1.25	50
	Deck Thickness	1.5 mm		6.35 & 12.7			1.50	100
	Edge of Hole	Sharp		6.35 & 12.7			2.00	150
	Active Area	4.189 m^2		6.35 & 12.7			2.50	200
	Downcomer Area	0.243 m^2		6.35 & 12.7				250
Channel Baffled Trays	Weir Length	1260 mm	Air / Water	6.35 & 12.7	20	20	1.00	50
	Flow Path Length	1925 mm		6.35 & 12.7	50	50	1.25	100
	Active Area	2.426 m^2		6.35 & 12.7			1.50	150
	Baffle Height	450 mm		6.35 & 12.7			2.00	200
	Downcomer Area	0.243 m^2		6.35 & 12.7			2.50	250

Table 1 – Tray Specifications, Configurations and Flow Rates Employed in Experiments

Direct Observations

With the circular trays only, flow patterns were determined qualitatively by comparing water only flow on a non-perforated tray[11] with air-water two-phase flow on two sieve trays as shown in *Figure 7*. On the sieve trays directional flow pointers were employed. Those observations compliment temperature profile measurements from the water-cooling technique when comparing the flow patterns developed on the circular sieve trays to that on the rectangular trays[9, 10].

With water only flow and with most air-water flow rates, liquid circulation up to 30% of the tray active area at the sides was observed. The effect of air flow and sieve hole size delayed the onset of circulation which occurred at much higher weir loads compared to the water only experiments. Operations charts in *Appendix C* summarize the extent of liquid circulation for various air and water flow rates.

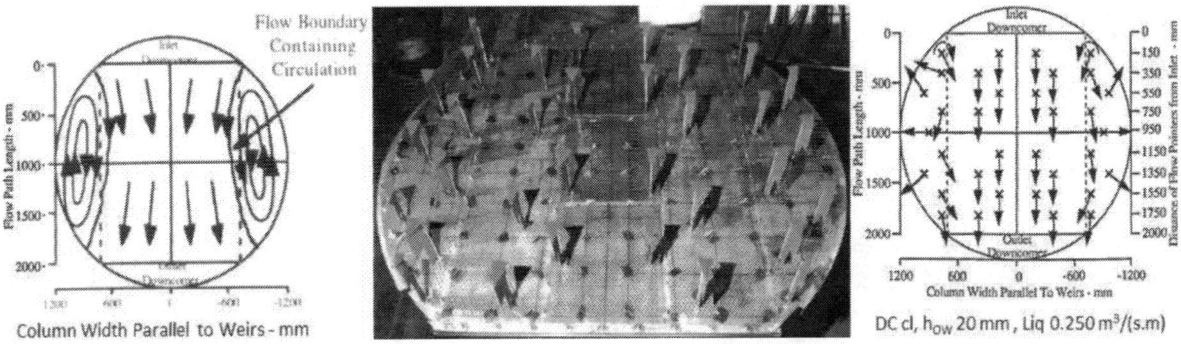

Figure 7 – Layout of Thirty Four Flow Pointers Employed on the Circular Sieve Trays

Water Cooling

Water cooling involves the cooling of hot water as it crosses the test tray by evaporating small fractions of water with unsaturated air at ambient temperature. Since water temperature is analogous to liquid concentration and heat transfer analogous to mass transfer, the heat transfer involves measurement of water temperature as it is cooled by air. Temperature measurements on the circular trays employed 108 Platinum Resistance Thermometers (PRT) and 70 PRT's on the channel baffled rectangular tray as shown in *Figure 8*. A complete description of the water cooling technique including interpretation of results can be found in Porter et al 1992[8], Chambers 1993[9] and Fenwick 1996[10].

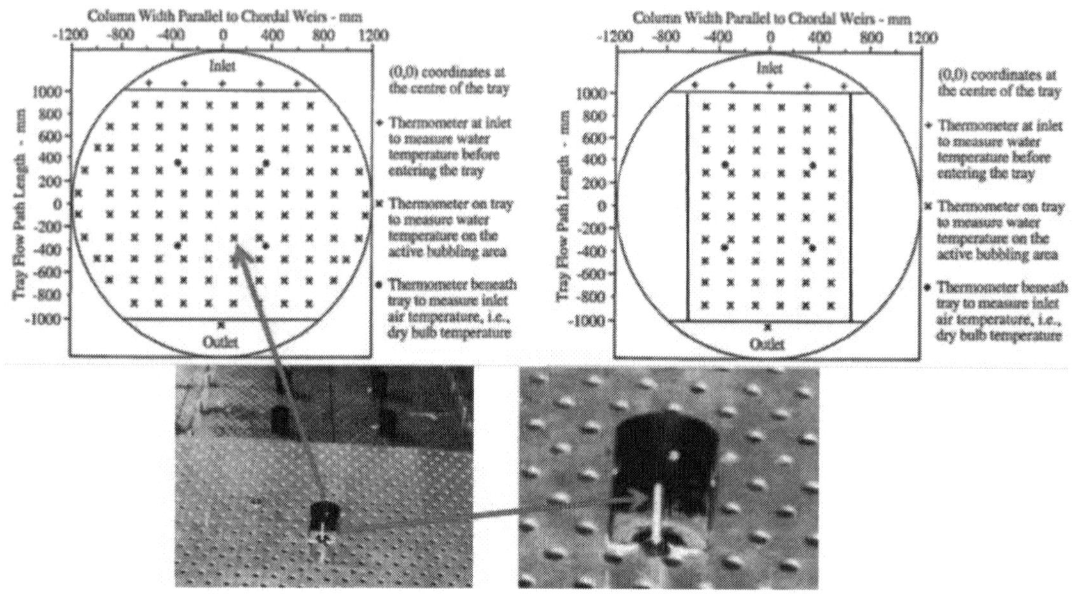

Figure 8 – Temperature Measurements on Circular and Rectangular Tray Designs with Platinum Resistance Thermometers

Graphical Representation of Temperature Profiles

Temperature values on conversion into interpolated temperature profiles are used to generate lines of constant temperature in the form of isothermal contour maps. Isotherms show lines of equal temperature increments and are identified with reduced temperature values between 1.0 and 0.0.

The shape of the isotherms provide guidance on how the air-water flow patterns developed on the circular and rectangular tray decks would potentially impact tray performance. If the liquid flow pattern yields uniform liquid residence time (plug flow) on the tray and the air flow distribution is uniform, straight lines of constant liquid temperature (isotherms) parallel to the downcomers is the expected result.

Deviations from non-idealized flow are quantified with variation in temperature measurements. The longer the liquid residence time on the tray, the cooler it becomes. Consequently the stagnant zones at the side of the tray will have lower temperatures compared to the active zone. The latter results in "U-shaped" isotherms indicative of preferential liquid flow along the active region between downcomers with stagnant zones of slow moving, stationary or circulating liquid (coldest liquid temperature) at the side of the tray. Inverted "U-shaped" isotherms are another example in which liquid is travelling faster along the sides of the tray or there is disproportionate gas flow through the middle of the tray compared to the sides.

With the rectangular tray, where the channel baffle walls eliminate stagnant zones, straight parallel isotherms to the downcomers would be expected. *Figures 9a to c* show simplified two-dimensional isotherm contours of ideal and non-idealized flows. Figure 9d shows the anticipated ideal isotherms profile for the channel baffled rectangular tray.

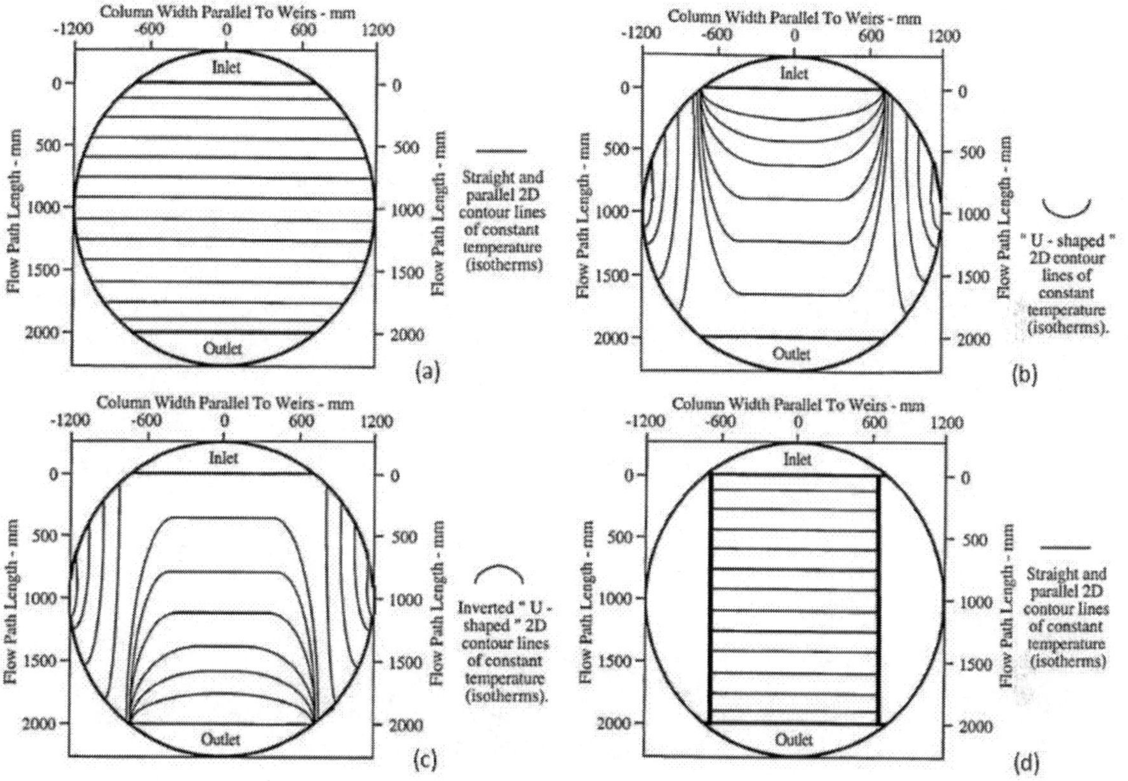

Figure 9 – Simplified 2-Dimensional Isotherms of Ideal and Non-idealized Flows on Circular and Rectangular Sieve Trays

Results and Discussion

Owing to the large amount of data collected, remarks shall be restricted to a fixed air velocity and low, medium and high weir liquid loads. *Figure 10* is an example of the isotherm contours of the flow patterns developed with the 6.35 mm (0.25 inch) diameter perforated channel baffle sieve tray compared to the circular tray at a fixed air velocity of 1.5 m/s and 0.005 to 0.020 $m^3/(s.m)$ (2 to 8 gpm/inch) weir loadings. The downcomer clearance and outlet weir height were both 20 mm. *Figure 11* presents similar comparisons between the rectangular and circular trays with 12.7 mm (0.5 inch) diameter sieve holes. Further isotherm profiles are in *Figures D1* and *D2* in Appendix D.

Overall, the measured temperature isotherms on the rectangular tray deviated somewhat from the simplified profiles shown in Figure 9 owing to the impact of frictional drag and sieve hole size on the air flow when forced into the liquid cross flow.

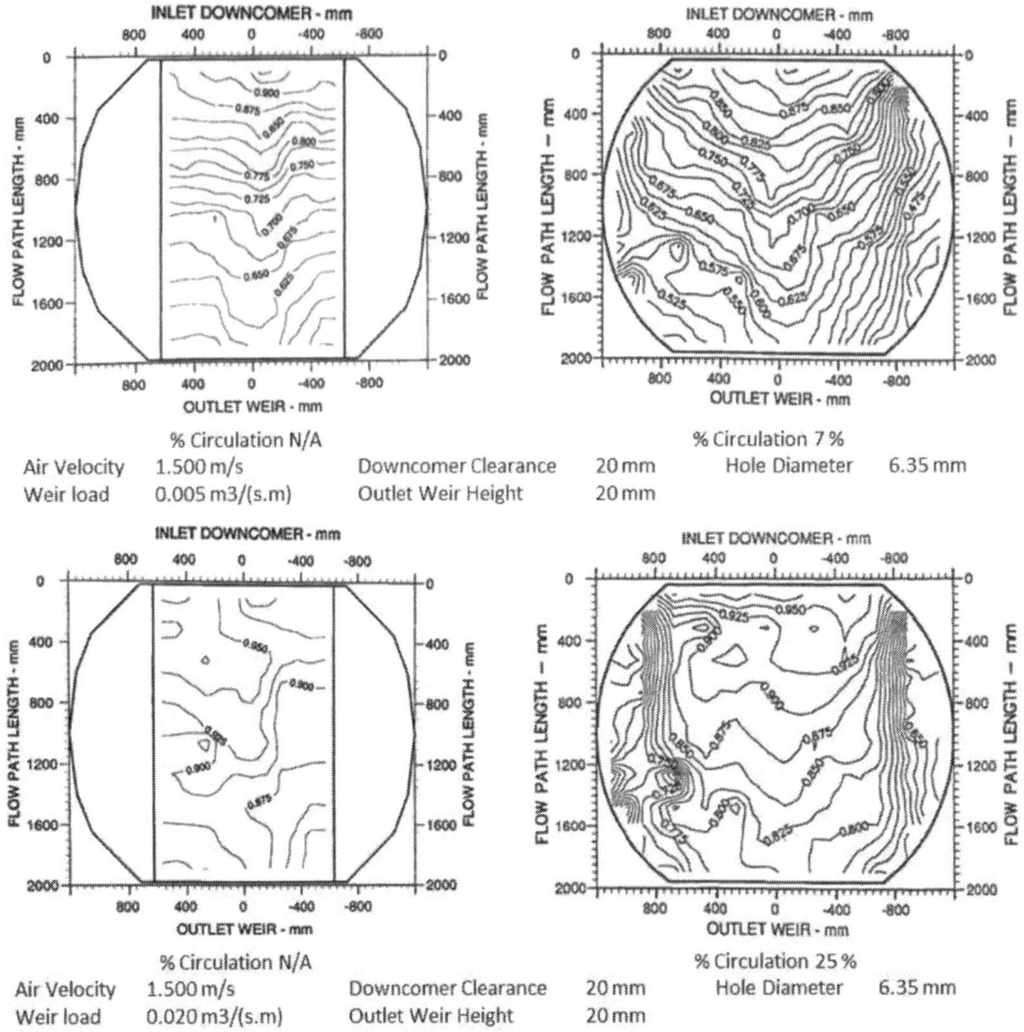

Figure 10 – Isotherm Contours Comparison of Rectangular and Circular Trays (ϕ = 6.35 mm) at Low and High Weir Loads

For both the 6.35 and 12.7 mm hole size rectangular trays, *Figures 10* and *11* show approximate parallel isotherms to the downcomers at a weir loading of 0.005 $m^3/(s.m)$. With the same weir loading, the circular tray exhibits more of a U-shaped isotherm profile along the centerline of the active zone between downcomers. On increasing the weir loading from 0.010 to 0.025 $m^3/(s.m)$, the U-shaped isotherms on the circular tray became more severe on the active zone, with tightly spaced isotherms that are almost perpendicular to the downcomers. The latter is indicative of colder liquid and longer residence times due to stagnancies at the tray sides. In the case of the rectangular tray, the increasing weir load distorted the isotherms such that they were spaced further apart as well as producing a mixed shaped profile. Additionally the reduced temperature change along the tray flow path on the rectangular tray was less than that on the circular tray, indicative of faster moving liquid within the narrow channels than on the wider circular tray. For example in Figures 10 and 11 with air velocity set at 1.50 m/s and weir load 0.020 $m^3/(s.m)$, we see in Table 2 the following:

Tray	Hole Dia. mm	Reduced Temp, Tr	Distance from Inlet Downcomer, mm	Hole Dia. mm	Reduced Temp, Tr	Distance from Inlet Downcomer, mm
Rectangular	6.35	0.950	0 – 700	12.7	0.925	450 - 800
Rectangular	6.35	0.875	1200 – 1900	12.7	0.850	1200 - 1900
Circular	6.35	0.950	150 – 250	12.7	0.950	100 - 350
Circular	6.35	0.875	900 - 1575	12.7	0.850	1450 - 1650
Circular	6.35	0.800	1500 - 1900	12.7	0.825	1700 – 1900

Table 2 – Reduced Temperatures of Isotherms and Profile Distance from Inlet on the Rectangular and Circular Trays

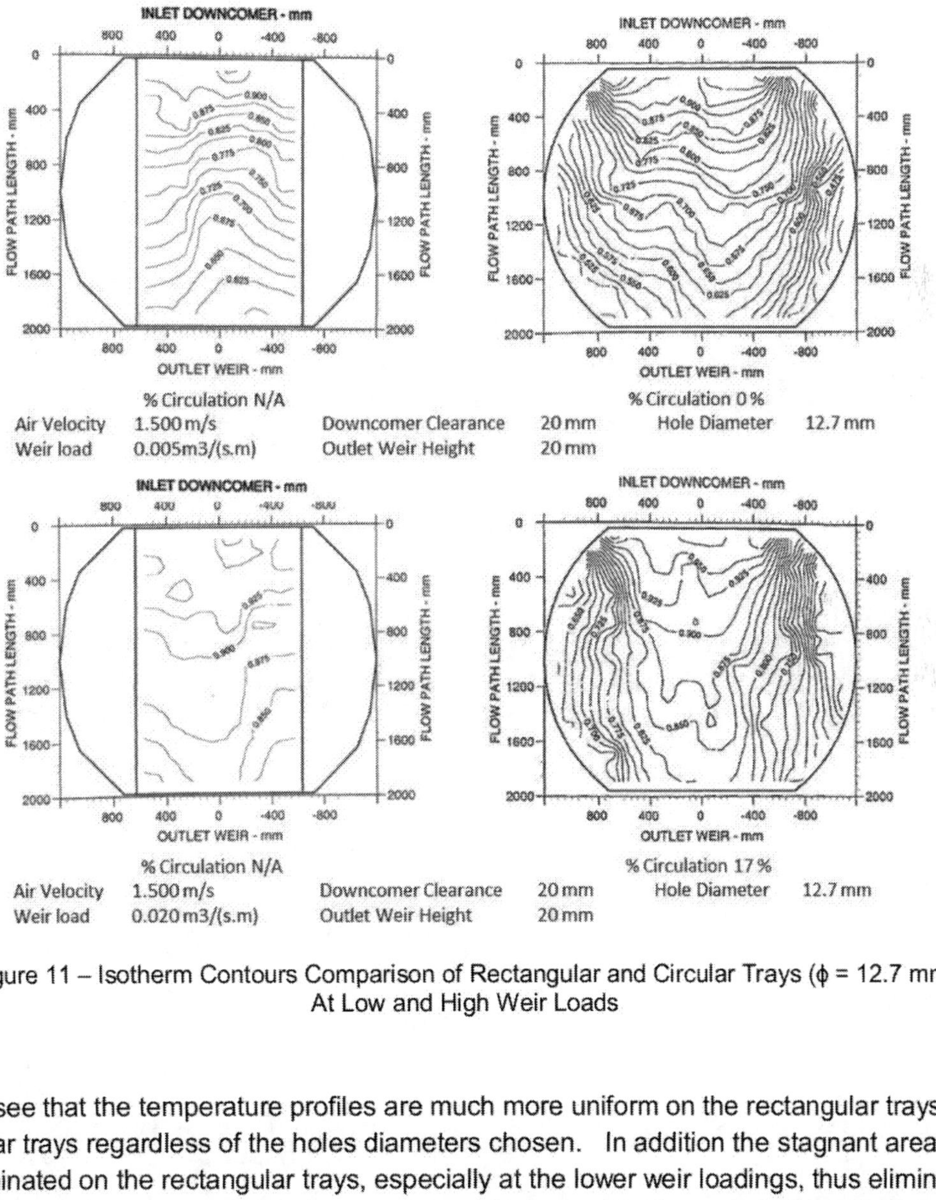

Figure 11 – Isotherm Contours Comparison of Rectangular and Circular Trays (ϕ = 12.7 mm) At Low and High Weir Loads

One can see that the temperature profiles are much more uniform on the rectangular trays than on the circular trays regardless of the holes diameters chosen. In addition the stagnant areas have been eliminated on the rectangular trays, especially at the lower weir loadings, thus eliminating the potential for remixing and reducing their mass transfer potential. As a result, it is inferred that the mass transfer efficiency would be superior on the rectangular channel baffle trays in comparison to the circular trays.

Overall further investigations would be of high interest by measuring tray efficiencies based on composition profiles for both a circular and channel baffled tray. Such tests s could be carried out at a world renowned research facility such as Fractionation Research Inc.'s 2.44 m low pressure column section (pending of course on the approval of its membership). FRI is the only experimental facility capable of handling such tests.

Nomenclature

D	Column Diameter
DC_{CL}	Inlet Downcomer Clearance
h_{OW}	Outlet Weir Height
Tr	Reduced Temperature (displayed on Isotherm contour graphs)

References

(1) Parker, B.M. and Parker, T.J, US patent 04762651, 9 August 1988

(2) Perry, R.H. and Green, D., *Perry's Chemical Engineering Handbook*, 8th Edition, New York: McGraw-Hill, 2008 p.14-47

(3) Sloley, A.W. and Fleming, B., *Successfully Downsize Trayed Columns*, CEP, March 1994 pp 39-47

(4) Kister, H.Z., *Distillation Operation*, New York: McGraw-Hill, 1990, p150

(5) Summers, D.R., Kister, H.Z., Clancy-Jundt, B., and Miller, R., *Troubleshooting and Rectification of a Giant C3 Splitter Tower Problem, Part 2 – Rectification*, AIChE Spring Meeting, Paper 119b, New Orleans, LA, April 2014

(6) Summers, D. R., *Tray Stability at Low Vapor Loads*, Conference Proceedings of "Distillation and Absorption 2010", p.611, Eindhoven, The Netherlands, September 12-15, 2010.

(7) Resetarits, M.R. and Ogundeji, A.Y., *On Distillation Tray Weir Loadings*, AIChE Spring Meeting, Paper 48a, Tampa, FL, April 2009.

(8) Porter, K.E., Yu, K.T., Chambers, S. and Zhang, M.Q,. 1992, "Flow Patterns and Temperature Profiles on a 2.44 m Diameter Sieve Tray", Trans. IChemE **Vol. 70, Part A,** September 1992.

(9) Chambers, S., 1993, "Flow Patterns on Sieve Trays", Ph.D. Thesis, University of Aston in Birmingham, Birmingham, United Kingdom

(10) Fenwick, K.S., 1996, "Flow Patterns on Distillation Trays", Ph.D. Thesis, University of Aston in Birmingham, Birmingham, United Kingdom.

(11) Hine, C. J., 1990, "Effect of Liquid Flow Patterns on Distillation Trays", Ph.D. Thesis, University of Aston in Birmingham, Birmingham, United Kingdom.

Appendix A
Example Tray Hydraulics Representing Figure 2b

SULTECH 8.3

SULZER

Reference:	0091265
Customer:	Unknown
Service:	*******
Item:	D-1266
Date Run:	13-Mar-2015

FINAL DESIGN

TRAY DESIGN Section 1 **Trays 1-6 with Inlet and Outlet Blanking**

Geometry: Tray4

Deck Type:	BDH	Tray Thickness [in]:	0.0787	Valve Density [1/ft²]:	6.5
Number of Passes:	1	Material:	304	Valve Lift [in]:	0.5000
Inner Diameter [in]:	54.00	Tower Area [ft²]:	15.90	Valves:	25
Tray Spacing [in]:	18.00	Active Area [ft²]:	3.88	Open Area [ft²]:	0.35
		Active Area [%]:	24.36	Open Area [%]:	8.99

Downcomer Data ---- BLANKING EMPLOYED ----

Type: CUSTOM DESIGN	Side	
Top Width [in]:	11.50	
Bottom Width [in]:	11.50	
Clearance Height [in]:	1.00	
Outlet Weir Height [in]:	1.25	
Outlet Weir Length [in]:	44.22	
Inlet Weir Height [in]:	0.00	
Rec. Pan Depth [in]:	0.00	
Pan Width [in]:	N/A	
Radius Tip:	No	
Downcomer Top Area [ft²]:	2.477	Downcomer Area [ft²]: 2.48
Downcomer Bottom Area [ft²]:	2.477	Downcomer Area [%]: 15.574
Clearance Length [in]:	44.22	

Fluid Data: Design Flow Rates

	Top #1	Tray #6	#1- Case R	#6 Case R
Gas:				
Flow Multiplier [%]:	100	100	100	100
Mult. Gas Rate [lb/h]:	6964.6	6858.3	5320.1	5248.3
Density [lb/ft³]:	3.0847	2.9232	3.0584	2.8500
Viscosity [cP]:	0.0086	0.0084	0.0086	0.0083
QV [cfs]:	0.63	0.65	0.48	0.51
Liquid:				
Flow Multiplier [%]:	100	100	100	100
Mult. Liquid Rate [lb/h]:	5698.8	5592.5	5690.5	5618.8
Density [lb/ft³]:	23.03	21.40	23.17	21.10
Surface Tension [dyne/cm]:	3.01	2.62	3.05	2.57
Viscosity [cP]:	0.046	0.041	0.046	0.040
QL [gpm]:	30.85	32.59	30.62	33.19
System Factor:	1.00	1.00	1.00	1.00

Hydraulic Data

Jet Flood [%]:	20	21	15	17
Downcomer Flood [%]:	7	7	6	7
Downcomer Froth Backup [%]:	25	26	22	24
Downcomer Clear Liquid [in]:	4.22	4.42	3.97	4.26
Weir Loading [gpm/in]:	0.70	0.74	0.69	0.75
DryDrop [in. H2O]:	0.63	0.63	0.63	0.63
Pressure Drop [mmHg]:	1.76	1.73	1.75	1.71

Number of trays: 6
Design Notes:
Engineer Name: File: F:\DRS\Channel Baffles.sult

 Page 1 of 2

Appendix B
Example Tray Hydraulics Representing Figure 2a

SULZER

SULTECH 8.3

Reference:	0091265
Customer:	Unknown
Service:	*******
Item:	D-1266
Date Run:	13-Mar-2015

--

FINAL DESIGN

--

TRAY DESIGN Section 1 **Trays 1-6 with 18" Channel Baffles**

--

Geometry: Tray2

Deck Type:	BDH	Tray Thickness [in]:	0.0787	Valve Density [1/ft²]:	6.5
Number of Passes:	1	Material:	304	Valve Lift [in]:	0.5000
Inner Diameter [in]:	54.00	Tower Area [ft²]:	15.90	Valves:	25
Tray Spacing [in]:	18.00	Active Area [ft²]:	3.88	Open Area [ft²]:	0.35
		Active Area [%]:	24.36	Open Area [%]:	8.99

--

Downcomer Data ---- CHANNEL BAFFFLES EMPLOYED ----

Type: CUSTOM DESIGN	Side		
Top Width [in]:	11.50		
Bottom Width [in]:	11.50		
Clearance Height [in]:	1.00		
Outlet Weir Height [in]:	1.25		
Outlet Weir Length [in]:	18.00		
Inlet Weir Height [in]:	0.00		
Rec. Pan Depth [in]:	0.00		
Pan Width [in]:	N/A		
Radius Tip:	No		
Downcomer Top Area [ft²]:	2.477	Downcomer Area [ft²]:	2.477
Downcomer Bottom Area [ft²]:	2.477	Downcomer Area [%]:	15.574
Clearance Length [in]:	18.00		

Fluid Data: Design Flow Rates

	Top #1	Tray #6	#1- Case R	#6 Case R
Gas:				
Flow Multiplier [%]:	100	100	100	100
Mult. Gas Rate [lb/h]:	6964.6	6858.3	5320.1	5248.3
Density [lb/ft³]:	3.0847	2.9232	3.0584	2.8500
Viscosity [cP]:	0.0086	0.0084	0.0086	0.0083
QV [cfs]:	0.63	0.65	0.48	0.51
Liquid:				
Flow Multiplier [%]:	100	100	100	100
Mult. Liquid Rate [lb/h]:	5698.8	5592.5	5690.5	5618.8
Density [lb/ft³]:	23.03	21.40	23.17	21.10
Surface Tension [dyne/cm]:	3.01	2.62	3.05	2.57
Viscosity [cP]:	0.046	0.041	0.046	0.040
QL [gpm]:	30.85	32.59	30.62	33.19
System Factor:	1.00	1.00	1.00	1.00

Hydraulic Data

Jet Flood [%]:	19	20	15	16
Downcomer Flood [%]:	7	7	6	7
Downcomer Froth Backup [%]:	28	29	25	27
Downcomer Clear Liquid [in]:	4.82	5.09	4.51	4.85
Weir Loading [gpm/in]:	1.71	1.81	1.70	1.84
DryDrop [in. H2O]:	0.63	0.63	0.63	0.63
Pressure Drop [mmHg]:	1.92	1.88	1.91	1.86

Number of trays: 6
Design Notes:
Engineer Name: File: F:\DRS\Channel Baffles.sult

 Page 2 of 2

Appendix C
Operations Charts with Percent Circulating Flow for 6.35 and 12.7 mm Hole Sieve Trays

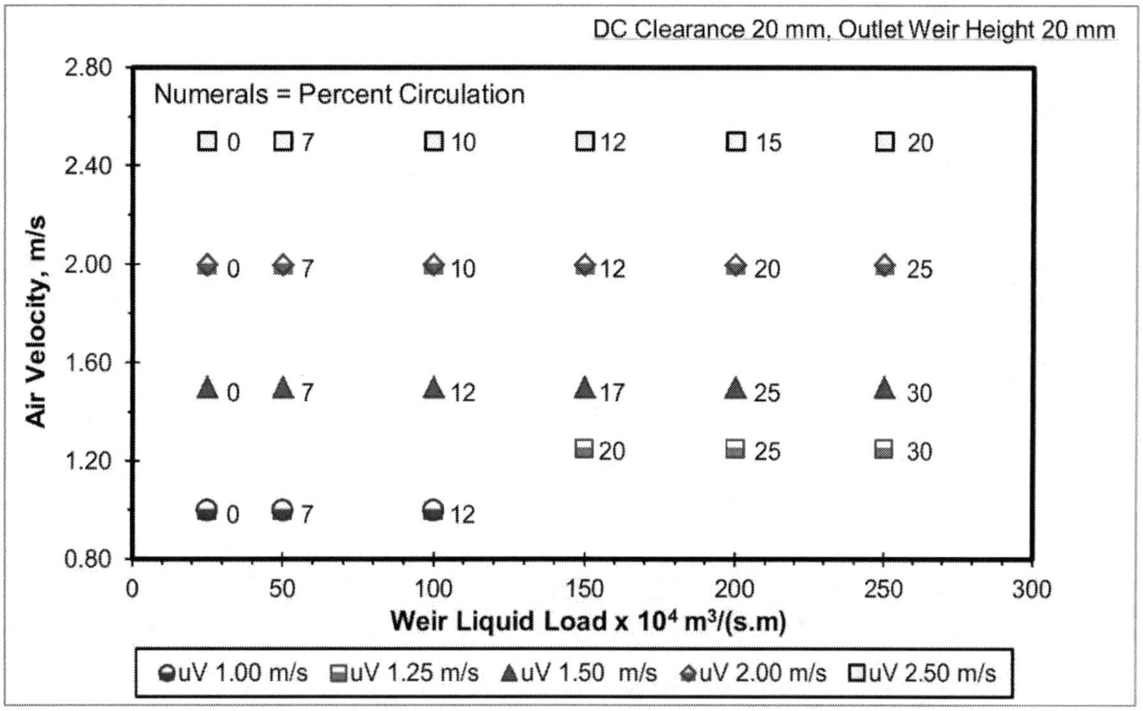

Figure C1 Operations Chart with Percent Circulation for 6.35 mm Hole Sieve Tray

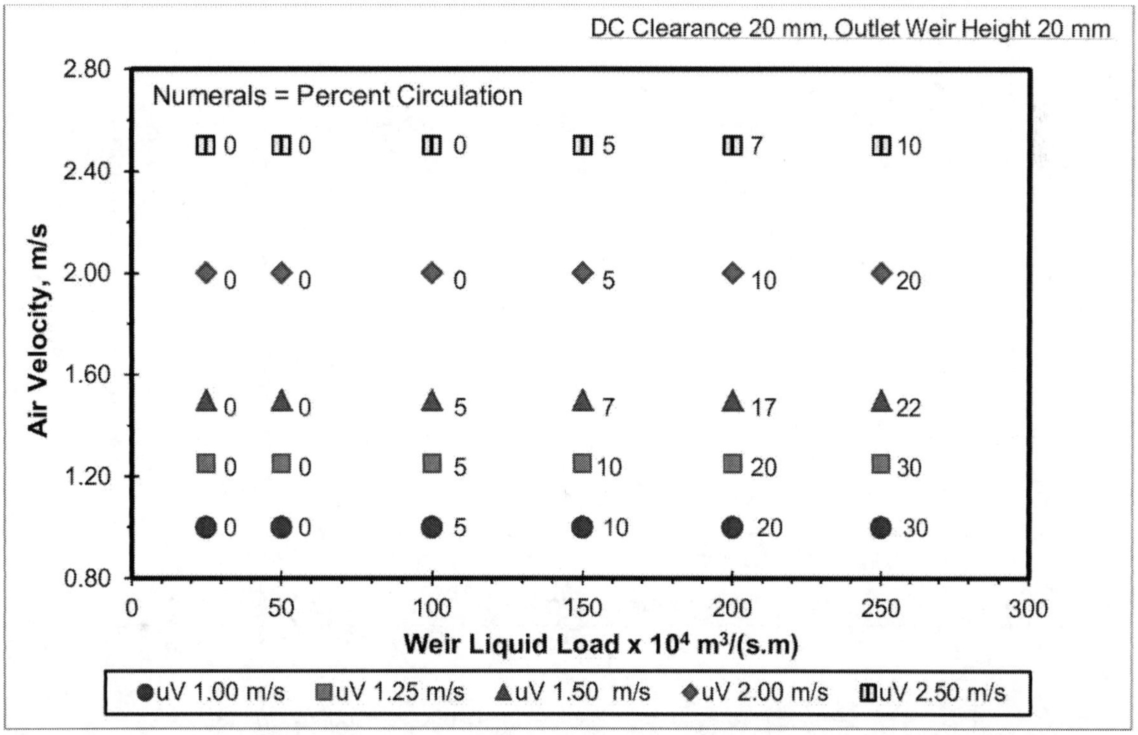

Figure C2 Operations Chart with Percent Circulation for 12.7 mm Hole Sieve Tray

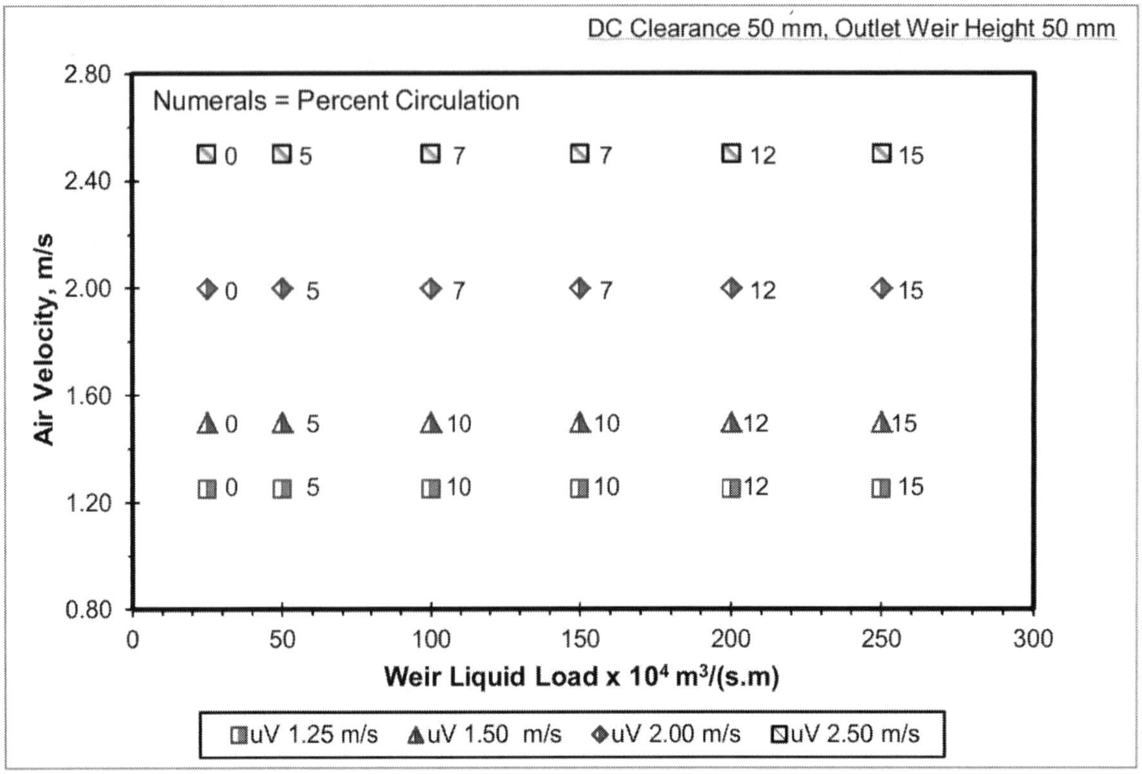

Figure C3 Operations Chart with Percent Circulation for 6.35 mm Hole Sieve Tray

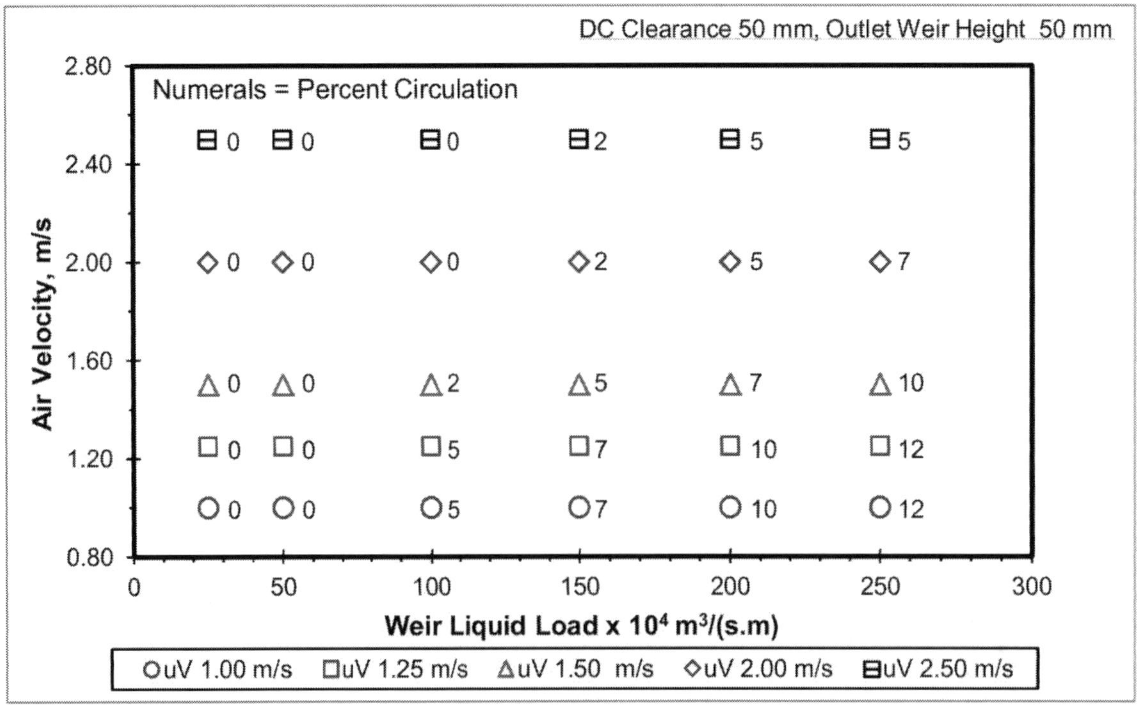

Figure C4 Operations Chart with Percent Circulation for 12.7 mm Hole Sieve Tray

Appendix D
Two-Dimensional Isotherms Comparisons of Rectangular and Circular Sieve Trays

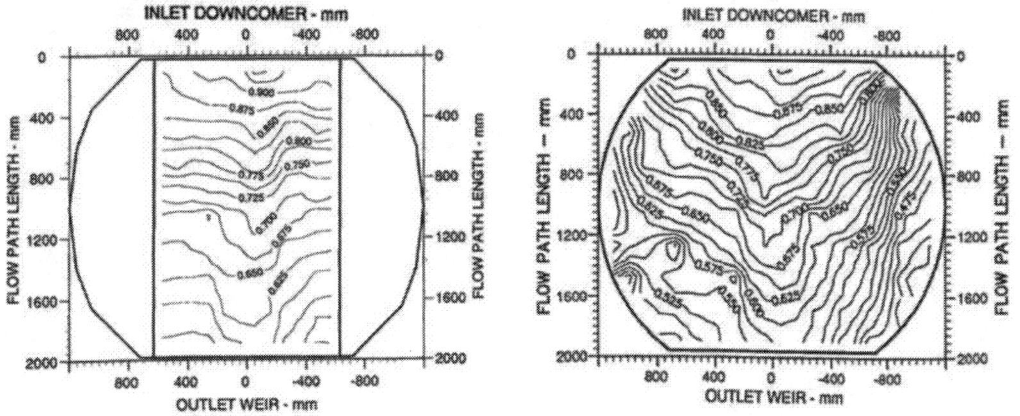

% Circulation N/A   % Circulation 7 %

Air Velocity 1.500 m/s   Downcomer Clearance 20 mm   Hole Diameter 6.35 mm

Weir load 0.005 m3/(s.m)   Outlet Weir Height 20 mm

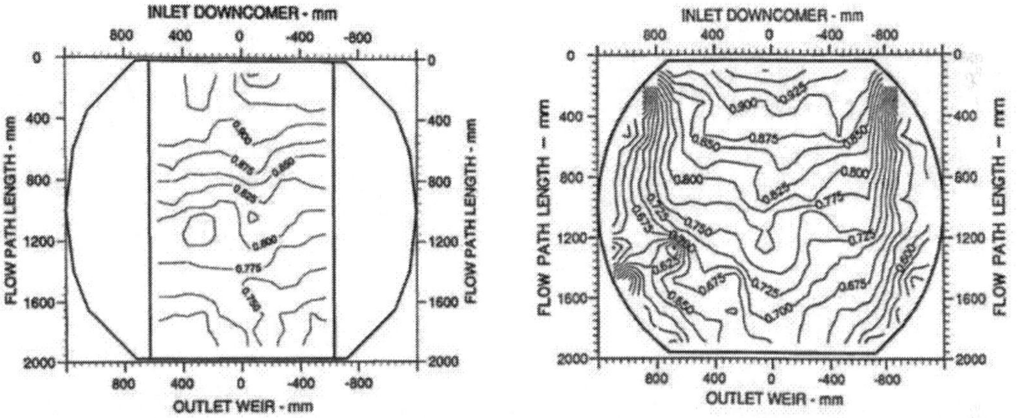

% Circulation N/A   % Circulation 12 %

Air Velocity 1.500 m/s   Downcomer Clearance 20 mm   Hole Diameter 6.35 mm

Weir load 0.010 m3/(s.m)   Outlet Weir Height 20 mm

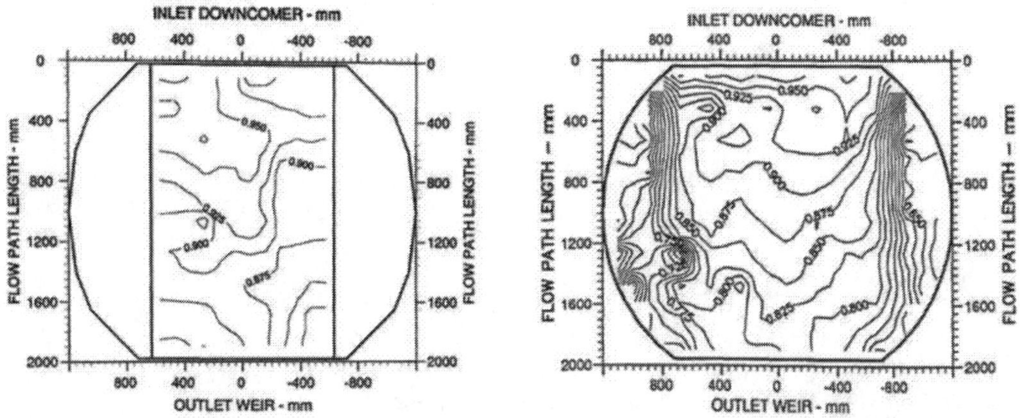

% Circulation N/A   % Circulation 25 %

Air Velocity 1.500 m/s   Downcomer Clearance 20 mm   Hole Diameter 6.35 mm

Weir load 0.020 m3/(s.m)   Outlet Weir Height 20 mm

Figure D1 – Isotherm Contours Comparison of Rectangular and Circular Trays (ϕ = 6.35 mm)

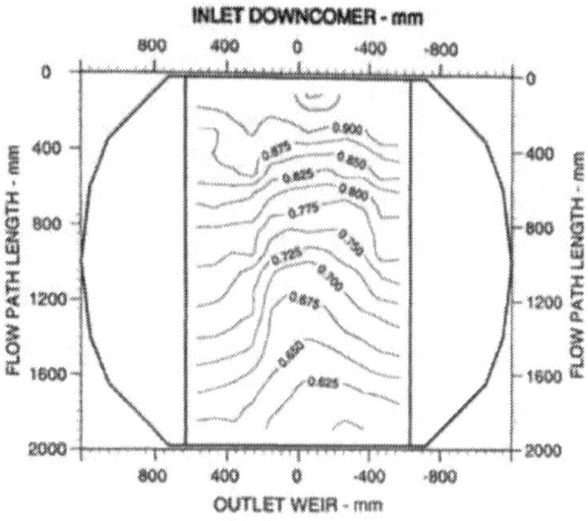

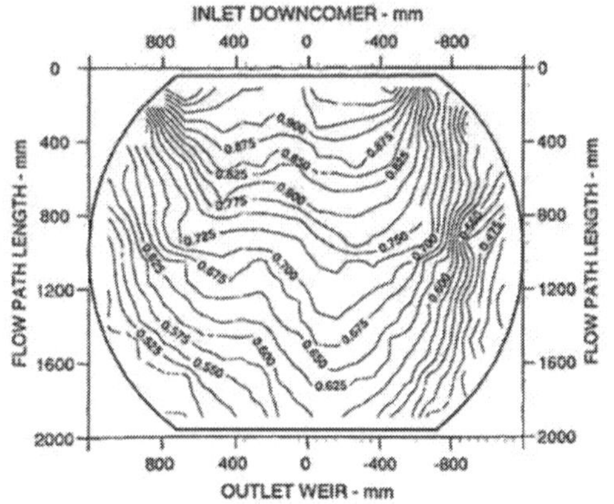

% Circulation N/A		% Circulation 0 %	
Air Velocity 1.500 m/s	Downcomer Clearance	20 mm	Hole Diameter 12.7 mm
Weir load 0.005 m3/(s.m)	Outlet Weir Height	20 mm	

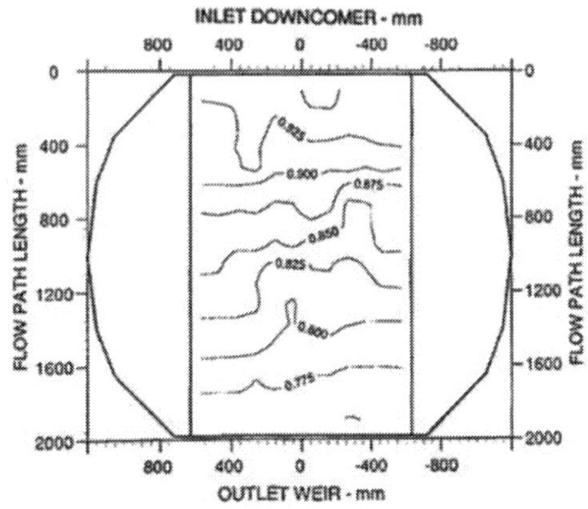

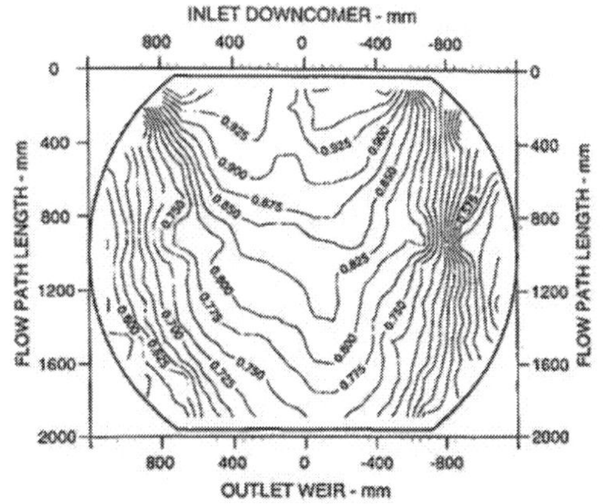

% Circulation N/A		% Circulation 5 %	
Air Velocity 1.500 m/s	Downcomer Clearance	20 mm	Hole Diameter 12.7 mm
Weir load 0.010 m3/(s.m)	Outlet Weir Height	20 mm	

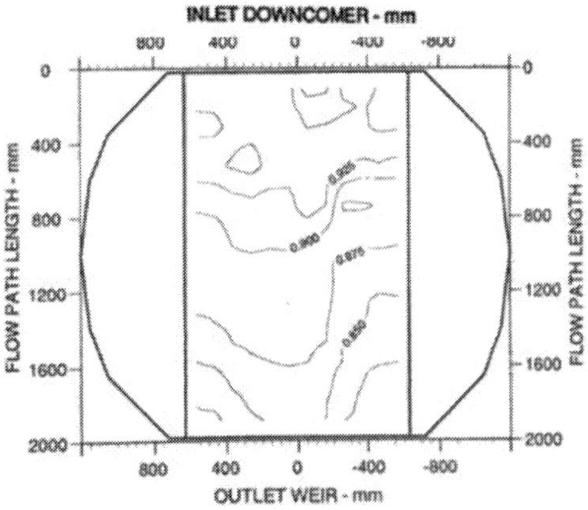

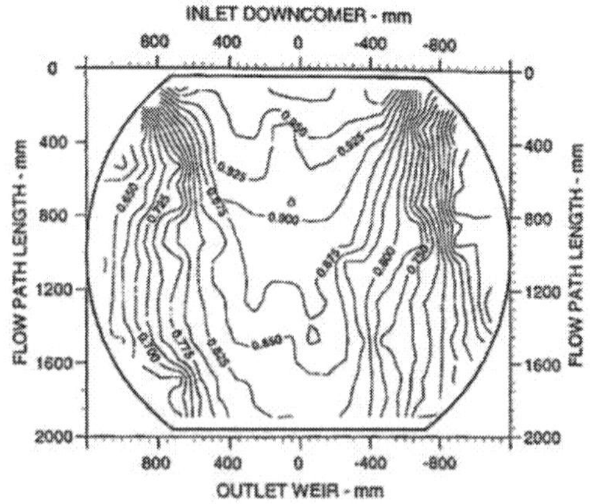

% Circulation N/A		% Circulation 17 %	
Air Velocity 1.500 m/s	Downcomer Clearance	20 mm	Hole Diameter 12.7 mm
Weir load 0.020 m3/(s.m)	Outlet Weir Height	20 mm	

Figure D2 – Isotherm Contours Comparison of Rectangular and Circular Trays (ϕ = 12.7 mm)

Systematic Field Tests Turn Loss into Gain in High-Capacity Trays

Part 2: Reverse Vapor Cross Flow Channeling in High-Capacity Trays: Diagnosis and Fixup

By

Henry Z. Kister, Fluor, Aliso Viejo, CA, USA

and

Neasan O'Shea (retired) and Dan Cronin, Phillips 66
Whitegate Refinery, County Cork. Ireland

Presented at the Distillation Topical Conference, AIChE Spring Meeting, Austin, Texas, April, 2015

Summary

This paper originates from a systematic troubleshooting effort that successfully diagnosed two complex independent problems which severely limited the capacity of an atmospheric crude distillation tower. The crude tower had been retrayed (by others) with modern high-capacity trays, but upon start-up achieved less than 80% of its previous capacity.

The troubleshooting identified two bottlenecks: premature flooding on some of the heavy naphtha-diesel fractionation trays, and a phenomenon which we termed "excursion," in which sudden temperature reductions around the diesel draw, accompanied by distillate yield loss, occurred upon rate increase.

Our previous paper[1] focused on diagnosing and fixup of the excursion bottleneck. This paper focuses on identifying and fixup of the high-capacity trays bottleneck.

Multichordal gamma scans with quantitative analysis[5] provided detailed mapping of the froth heights, entrainment, froth densities, clear liquid heights, and hydraulic gradients on the trays. These profiles showed conclusively that the trays were bottlenecked by a previously unknown mechanism which we termed "Reverse Vapor Cross Flow Channeling (RVCFC)". This phenomenon is closely related to vapor cross flow channeling (VCFC), but is induced by the forward push on high capacity trays rather than by the hydraulic gradient on the trays. Like VCFC, this phenomenon sets in when there is a significant channeling initiator, when the friction pressure drop of the trays is too low to counter the channeling, and when the ratio of flowpath length to tray spacing is high.

Based on this diagnosis, the tower was retrayed with the same type of tray but with a design that gave significantly larger friction pressure drop. The new trays also had two passes which halved the ratio of flowpath length to tray spacing. The modifications completely eliminated the bottleneck and the RVCFC, permitting higher-than-design capacities to be achieved.

Systematic Field Tests Turn Loss Into Gain in High-Capacity Trays

Part 2: Reverse Vapor Cross Flow Channeling in High-Capacity Trays: Diagnosis and Fixup

Henry Z. Kister, Fluor, Aliso Viejo, CA, USA and Neasan O'Shea (retired) and Dan Cronin, Phillips 66, Whitegate Refinery, County Cork, Ireland

Introduction

The last 2-3 decades saw tremendous improvements in high-capacity trays and packings. Modern trays and packings have permitted towers to be debottlenecked to throughputs and separations not previously envisioned. Yet, occasionally a high-capacity tray or packing debottleneck falls short of achieving its design targets.

A common reaction is to blame the trays or packing, and to seek a sledge-hammer solution, such as abandoning the new trays and replacing them by structured packing, or even building a new tower. Sledge-hammer solutions like this are not only costly, but are also highly counter-productive and may have a negative environmental impact[3]. Often they may obscure simple solutions that can be implemented immediately (rather than at the next turnaround), and minimize losses, down time, and are environmentally-friendly..

In most situations where high-capacity trays or packings fail to achieve their design expectations, it is hydraulic details associated with the trays or other internals that cause the premature bottlenecks. The correct strategy to address these is not tossing the trays out; rather, it is good troubleshooting that seeks the root cause and eliminates the bottleneck.

Good troubleshooting is systematic and thorough. It combines interacting with the operating staff to collect operating experience and obtain a good problem definition, conducting well-targeted field tests, applying process diagnostics (such as gamma scans and surface temperature surveys), simulation, and hydraulic analysis. A vital element is good testing of each theory and hypothesis with valid field tests. One good field test can humble a thousand expert opinions. This approach was adopted by our team.

This paper describes a case history where a high-capacity retray of a crude fractionator (by others) led to a capacity loss. A thorough and systematic investigation utilizing extensive field tests identified and diagnosed two completely independent bottlenecks that caused the revamp to fall short of its expectations. The diagnosis led to a very simple on-line solution that overcame both bottlenecks, involved no hardware changes, cost little to implement, and recovered most of the losses, circumventing any need to shut the crude unit down. Longer term, the diagnosis led to correct, low-cost internals modifications that produced capacities well-above the design.

Background

In the 1990's, Irish Refining Company operated a 52,500 BPSD (7,070 t/d) crude unit processing North Sea crudes (this refinery is now owned by Phillips 66). Products from the unit included light and heavy naphthas, diesel, and atmospheric residue. The diesel was produced to a cloud point spec of 0°C while the heavy naphtha (HN) was produced to an ASTM D86 95% point spec of 200°C.

Initially, the 4 meter ID atmospheric crude tower contained 34 single-pass jet tab trays. At a later time, the tower was revamped (Figure 1) to increase fractionation efficiency and to permit the refinery to intermittently take a kerosene side-cut while maintaining the same throughput of 52,500 BPSD (7,070 t/d). To achieve this, trays 11 – 20, i.e., the trays in the HN-diesel fractionation section, were replaced by single-pass high-capacity Nye trays. Structured packings were installed in the wash and stripping sections. A new kerosene side draw was installed at the inlet to Tray 16, approximately mid-way between the HN and diesel draws. A side stripper was added to strip the kerosene. Initially, the kerosene rundown system was incomplete so the kerosene side draw was not operated.

Figure 1 shows the unit with the mass balance data following the revamp, with the unit operating at a reduced crude rate of 38,700 BPSD (5,200 t/d). This material balance and operating conditions are shown in black print. Also shown on the diagram, in red print, are the mass balance and operating conditions following the operational debottleneck to 49,100 BPSD achieved by taking a kerosene draw that will be discussed later in this paper.

The atmospheric tower contained two pumparounds (BPA and TPA) and two active side draws. The upper side draw was heavy naphtha (HN), the lower was diesel. The diesel draw was steam-stripped in a side stripper.

The naphtha end point was maintained using a temperature control (TRC) just below Tray 23. This TRC adjusted the TPA duty. The TPA was drawn from a sump just under the downcomer from Tray 21 to Tray 20, so that any liquid not drawn overflowed onto Tray 20 and became reflux to the naphtha-diesel fractionation section.

The diesel cloud point was maintained at another temperature control (TRC) just above Tray 10. This TRC adjusted the BPA duty. The BPA was drawn from a total draw chimney tray just below Tray 9. The level control on the chimney tray manipulated the wash-back which was the reflux to the packed wash bed below.

After the retray, the tower was not able to achieve the 52,500 BPSD (7,070 t/d) it had previously achieved. At throughputs exceeding 38,000 BPSD (about 5200 t/d) the tower would start to flood with a rise in pressure drop and the HN product going off-specification with heavy components. Initial gamma scans showed the tower to be flooding with the flood initiating near the, as yet unused, kerosene draw (around Trays 14-16) and building up till Tray 23. The pumparound Trays 21-23 were jet tab trays and were not changed in the revamp.

In addition, a severe instability that we termed an "*Excursion*" would set in at about the same rates as the flooding initiated. A relatively small increase in BPA duty or reduction in coil outlet temperature (COT) would trigger a massive reduction in tray 10 temperature, usually

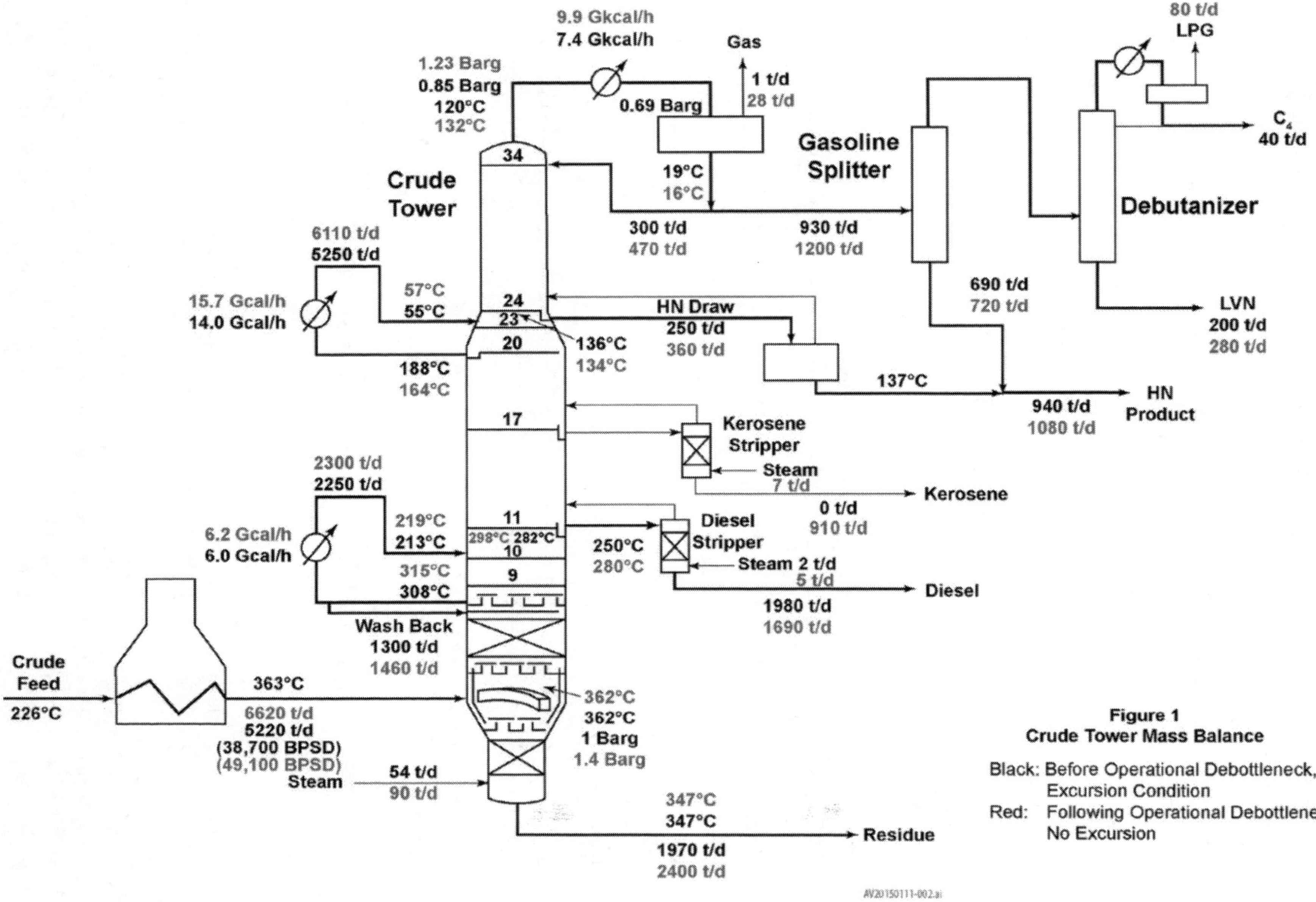

Figure 1
Crude Tower Mass Balance

Black: Before Operational Debottleneck, Excursion Condition
Red: Following Operational Debottleneck, No Excursion

accompanied by a HN yield loss, with the lost yield ending up in the atmospheric residue. Part 1 of our paper[1] focused on identifying the root cause of the excursion. This Part 2 of our paper focuses on troubleshooting the cause of the premature flood.

Focus on Channeling to Explain the Premature Flood

Trays 11-20 are high-capacity Nye trays (Figure 2). These trays differ from conventional trays by having a "Nye unit" at the tray floor, right under the downcomer. The Nye unit is closed at the top, perforated at the side and underneath. Vapor entering the unit from the tray below via the bottom holes is turned horizontally towards the tray and exits via the side holes. This horizontal wind imparts a horizontal "push", or horizontal momentum, to liquid droplets in the intertray space, blowing them towards the outlet downcomer, and preventing their entrainment into the tray above. By reducing this entrainment, the horizontal push is the capacity-enhancement mechanism on the Nye tray. The same horizontal push mechanism, produced by a variety of devices and geometries, is employed by most of today's high-capacity trays.

Figure 3 (drawn roughly to scale) shows the tray dimensions. The active areas of trays 11 to 16 contained long-legged venturi (smooth orifice) uncaged moving valves with 16.2% open area (minimum open area, expressed as a percent of active area), while trays 17 to 19 contained small-diameter moving valves with 14.6% open area. There were also small differences in downcomer and active area designs, as shown in Figure 3.

Hydraulic calculations both by the vendor and the troubleshooting team (Table 1, "Before Operational Debottleneck") showed that at 38,700 BPSD crude charge, the most highly loaded trays above the diesel draw and below the top pumparound (trays 11 to 20) should have been operating comfortably below any capacity limit. The calculations showed that at these loads even conventional trays should have operated without flood.

Eye-catchers in Figure 3 and Table 1 are the very long flow paths, giving a huge ratio (3.6 to 4) of flow path length to tray spacing, the very large weir loads (100 – 116 m^3/h m of outlet weir), the venturi valves, and the large open areas (14.6% to 16.2%). Previous work showed that venturi valves, large open areas (>13-14% with sharp-orifice moving valve trays), large ratios of flow path length to tray spacing (> 2:1), and high weir loads (>50 -60 m^3/h m of outlet weir) renders trays prone to vapor cross flow channeling (VCFC)[5-7]. VCFC is encountered when under the influence of a hydraulic gradient, vapor preferentially channels through the tray outlet and middle, generating a high-vapor-velocity region with high entrainment and premature flood. At the same time, the tray inlet region remains vapor-deficient, which promotes excessive weeping and poor tray efficiency.

VCFC is not the only form of channeling experienced on distillation trays. There are reports[(2,8)] of other forms of channeling, such as due to vapor maldistribution or multi-pass tray maldistribution. One thing they have in common is that they have only been experienced at large tray open areas, high ratios of flow path lengths to tray spacing, and high weir loads, conditions that apply also to the trays involved in this case.

An argument against VCFC is that forward push was used as cure to VCFC. In VCFC, the channeling is caused by an excessive hydraulic gradient[5,6,9]. Using "push valves", i.e. slots or valves that enter some of the vapor onto the tray with a horizontal velocity in the direction of the

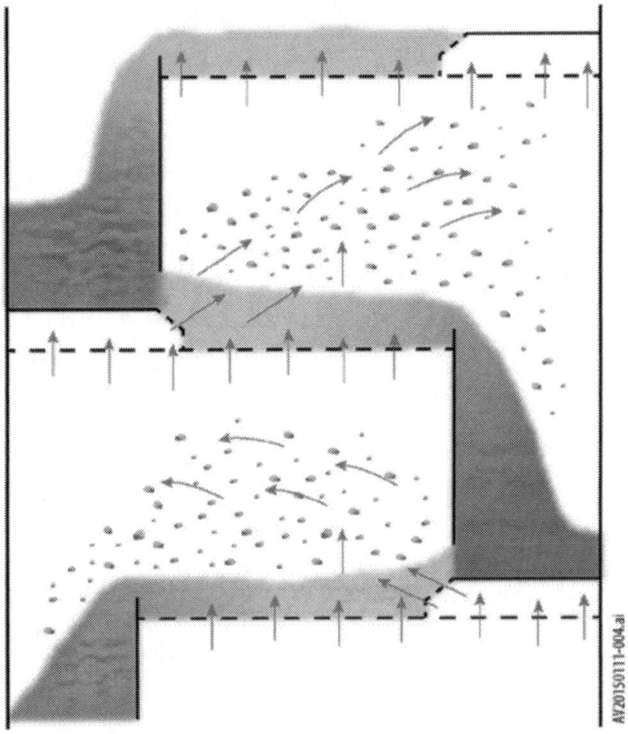

Figure 2. Schematic of a Nye™ Tray

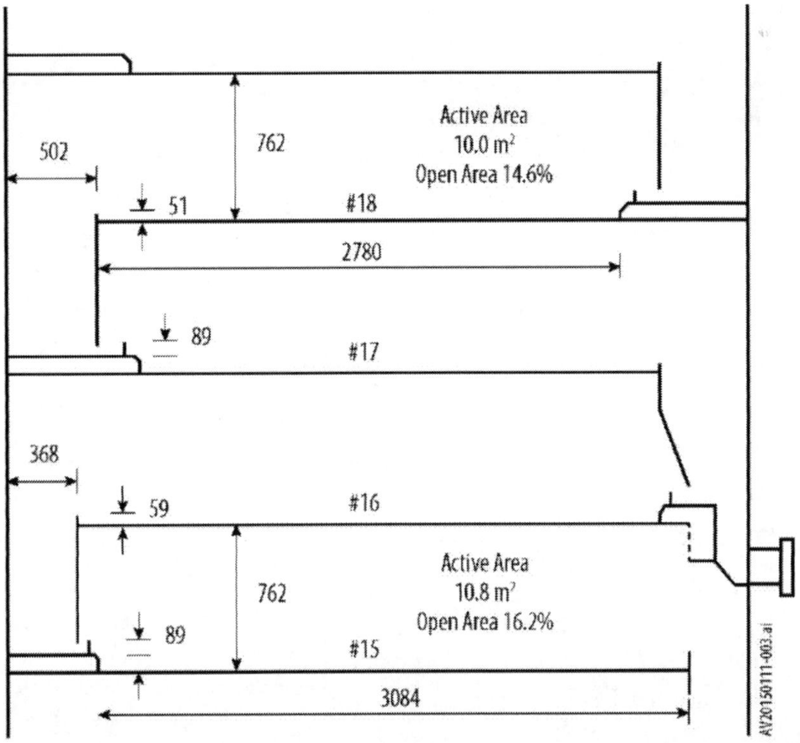

Figure 3. Dimensions of Trays 11 - 20

liquid flow, can reduce the hydraulic gradient and alleviate VCFC[10] as has been demonstrated by one published case study[11] and some others experiences by the authors. The Nye trays used in this tower entered about 10-15% of the hole area in a horizontal direction, providing considerable forward push. This forward push should have been sufficient to counter the hydraulic gradient.

It therefore became necessary to seek further insight into the froth pattern on trays 11 to 20. Channeling is difficult, often impossible, to diagnose with conventional troubleshooting techniques such as vendor software, ΔP measurement, and conventional single-chord qualitative gamma scans. Judicious multi-chordal gamma scans with quantitative analysis is the best tool for diagnosing maldistribution on trays. The principles and practice of this technique were recently described in detail elsewhere[2]. This technique was applied here to give a concise definition of the trays hydraulics and to diagnose the root cause of the premature flood.

TABLE 1. INTERNAL FLOWS AND HYDRAULIC EVALUATION, TRAYS 11-20

Operation	Before operational debottleneck		After operational debottleneck	
Crude charge rate, t/d	5220 (38,700 BPSD)		6620 (49,100 BPSD)	
Kero drawn?	No		Yes	
Tower top pressure, bar g	0.85		1.23	
Tray number	16	18	16	18
Moving valve type	venturi	small-valve	venturi	small-valve
Temperature, °C	213	203	238	203
Vapor rate, t/h	232	225	245	261
Liquid rate, t/h	172	171	134	188
C-factor**, based on active area, m/s	0.091	0.097	0.092	0.107
Weir load, m^3/h m outlet weir	116	100	89	110
% Jet flood, modified FRI	78	78	74	86
% Jet flood, Glitsch (eq.13)*	82	86	78	95
% DC backup	40	45	37	51
DC inlet velocity, m/s	0.128	0.082	0.098	0.088
Dry pressure drop, mm liquid	44	51	45	63

*Not accounting for capacity enhancement by Nye units

** $C = U_A \sqrt{\frac{\rho_G}{\rho_L - \rho_G}}$ where U_A is the vapor velocity based on the active area, m/s, and ρ_G and ρ_L,are vapor and liquid densities in consistent units.

Multichordal Gamma Scans with Quantitative Analysis

Four gamma scan chords were shot through the active areas of trays 9 to 22. The chords were parallel to the downcomers, at 300 mm and 1070 mm either side of the centerline. In addition, two gamma scan chords were shot perpendicular to the downcomers at 1070 mm either side of the centerline. These perpendicular chords were intended to pass through the active areas only and to miss the downcomers. All chords were shot when the tower was just below flood at 37,500 BPSD charge rate ("incipient flood") and repeated when the tower was flooded at 44,300 BPSD charge rate. During the scans, column operation was kept extremely stable and steady. This was vital, as for the scans to be consistent with one another no change could be allowed to the process for the considerable time required to complete all 4 chordal scans.At only one time instability was experienced for a short time, and the data obtained during this instability were discarded.

Additional chords were shot through the downcomers to explore for any possible downcomer bottleneck. The downcomer scans showed no downcomer bottlenecks. They were validated by downcomer neutron backscatter measurements that showed the same. It was concluded that the problem is in the active areas, not in the downcomers.

Verification of data validity and consistency is a must for multichordal gamma scanning[2], as one set of bad data is sufficient to lead to an incorrect diagnostic. Several consistency checks were conducted. Some chords were repeated and were shown to produce near-perfect repeatability.

Over the three months period preceding the tests, there were several scans conducted, at both flooded and unflooded conditions, along the same or similar scan chords to those used in the tests. Those served as invaluable consistency checks. The agreement was nothing short of amazing, commending the excellent work of the scanners. It also showed us that in this tower the flood state appeared to be uniform and stable.

Among the earlier scans, there were scans shot at completely unflooded conditions, not even incipient flood. These were shot at a crude charge rate of 35,000 BPSD. These were also analyzed, and provided invaluable insight to the investigation, in particular to the clear liquid heights and hydraulic gradient evaluation.

The two scans perpendicular to the downcomers were near-identical to each other at incipient flood, demonstrating absence of channeling in this direction, and giving one more consistency check.

Gamma Scan Results

Scan peaks obtained in a quantitative gamma scan were analyzed to provide froth heights, froth densities, clear liquid heights and an entrainment index. The procedures used were discussed in great detail elsewhere[2, 9, 12].

Multichordal quantitative analysis produces a considerable volume of data that are difficult to keep track of. These data need to be put together with the details of the tray geometry in a way that provides a clear picture for problem diagnosis. We use the "Kistergrams" developed by Kister et. al.[2] for representation of the results of quantitative analysis. These are elevation diagrams of the column and trays drawn to scale, with superimposed quantitative information derived from gamma scan analysis. The information plotted includes froth (or spray) heights (Figure 4), entrainment (Figure 5), froth densities (Figure 6), and clear liquid heights (Figure 7). The Kistergrams in this work use the following conventions:

1. All scales are linear.

2. The x-axis shows the exact position of the scan chord on the trays.

3. Froth height data are plotted at the same scale as the diagram. If the determined froth height equals or exceeds the tray spacing, it is plotted right on the tray above. Zero froth height is plotted as a point on the tray floor.

4. The froth density scale shows a froth density of 0.5 as a point on the tray above (the full tray spacing).

5. Clear liquid height is plotted so that a height of half the tray spacing is plotted as a point on the tray above. Zero height is plotted as a point on the tray floor.

6. An Entrainment Index of 2 (flood) is plotted as a point right at the tray above. An Entrainment Index of zero is plotted as a point on the tray floor.

Froth heights (Figure 4). Flood is indicated when a froth height point is plotted on the top of the tray spacing, i.e., at the floor of the tray above. Figure 4 shows the following:

- Flooding took place only on trays 16 and up. Trays 11 to 15 were not flooded.

- Froth heights tend to be low at tray inlets. In both the incipient flood and the flooded scans, the inlet froth heights are all below about 500-550 mm, the only exception being tray 16 where it is about 700 mm. There is no flooding in any inlet regions.

- At incipient flood, flood was approached at the middle of the trays, but not at the tray outlets. On trays 16 through 19, the froth heights reached the tray spacing in the middle, but the froth heights remained about 100 mm below the tray spacing at the outlets. Only tray 20 had a froth reaching the tray spacing at the outlet, but note the different shape of its inlet downcomer.

- Flood occurs in the middle of the trays.

Entrainment (Figure 5) The entrainment index EI is defined as the ratio of the gamma ray transmission count (as picked by the detector) at clear vapor, divided by the maximum transmission of gamma rays at the vapor space above the tray, less 1.0. If there is no entrainment, the gamma ray transmission at the vapor space above the tray would reach the transmission count at clear vapor, the ratio would be unity, and the entrainment index would be zero. In our experience, an entrainment index of 0.5 signifies light entrainment, 1.0 moderate entrainment, 1.5 heavy entrainment, and 2.0 or greater flood.

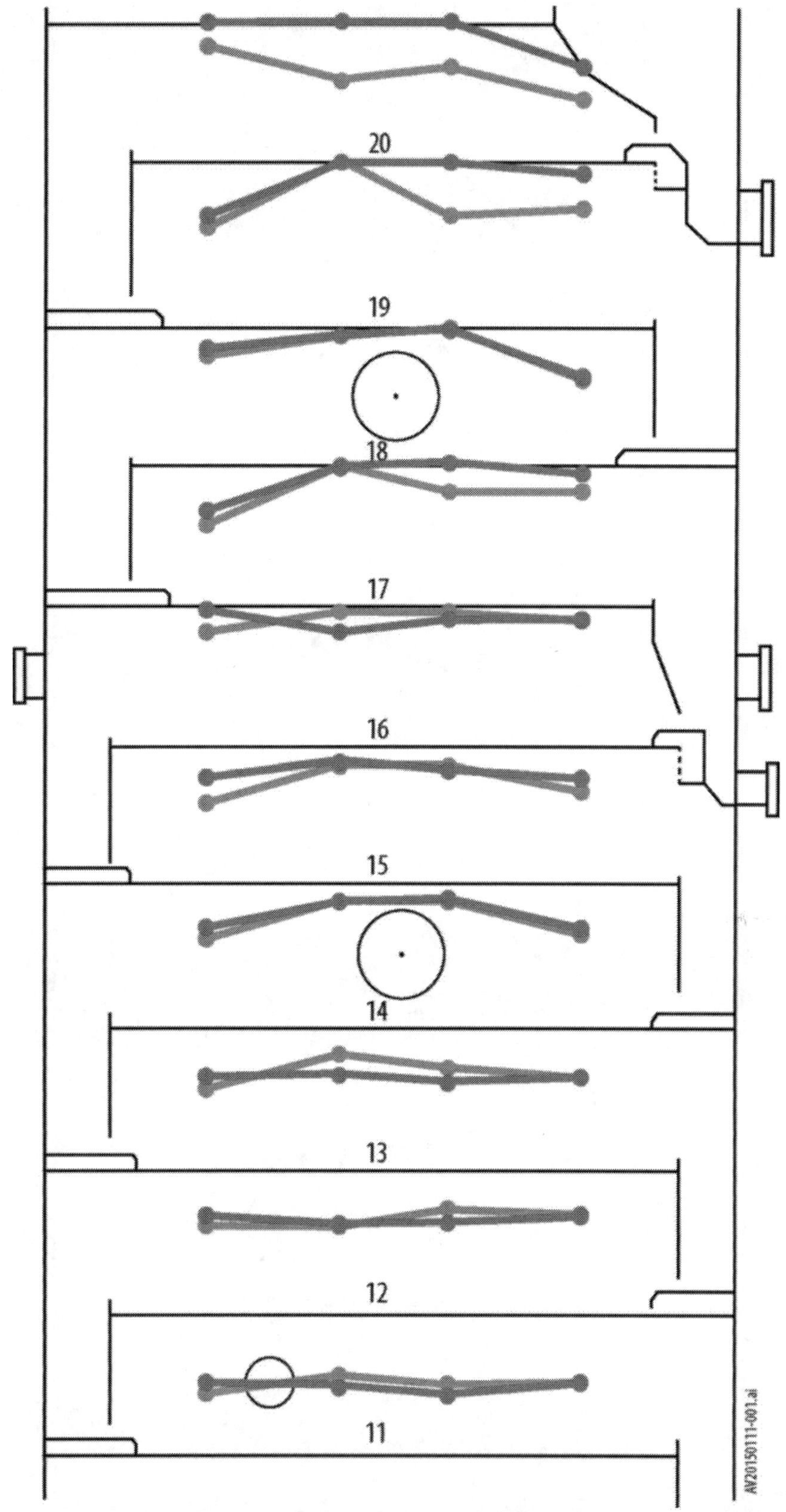

Figure 4. Froth Heights, Tray 11 – 20
Blue, incipient flood; Red, flood

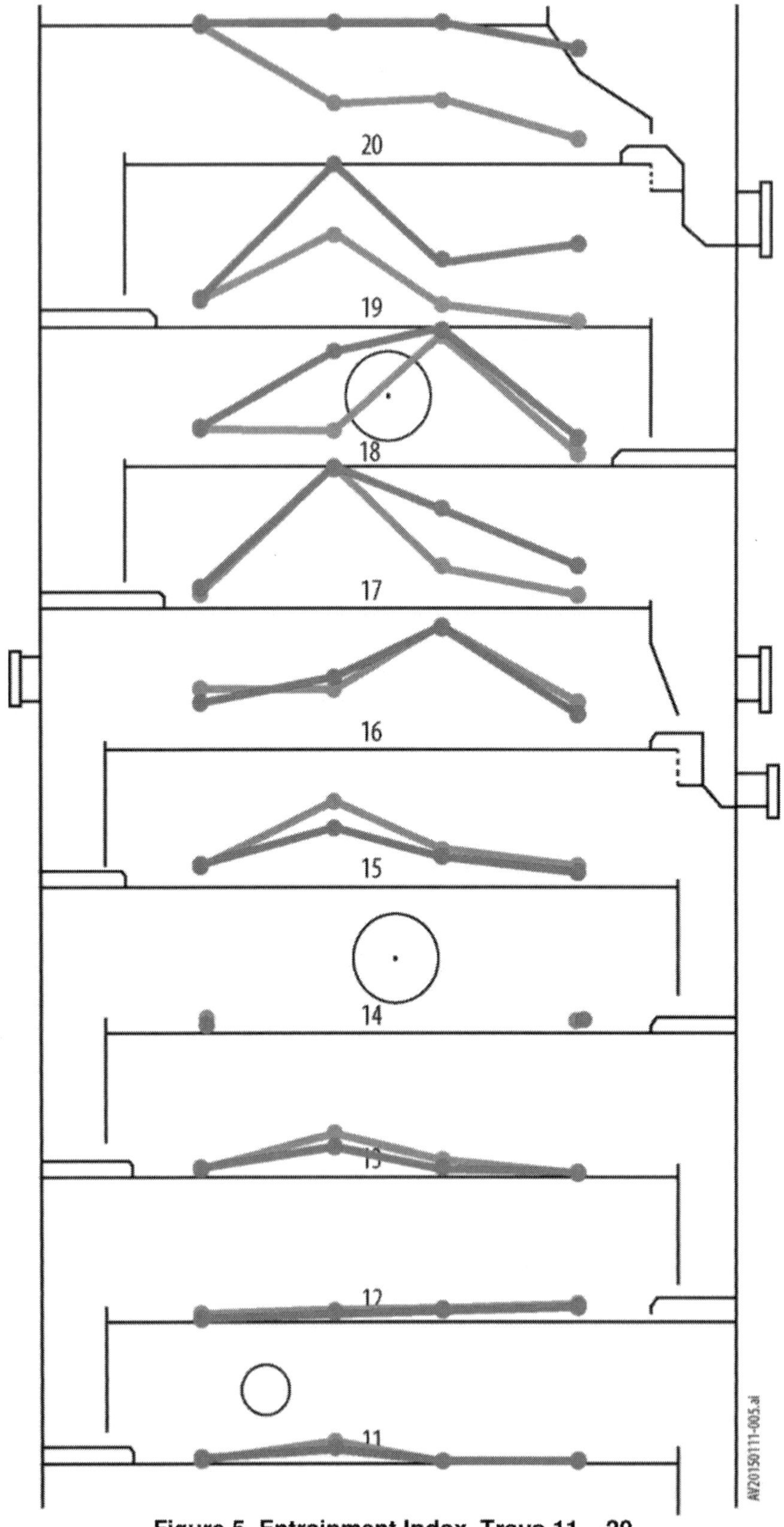

Figure 5. Entrainment Index, Trays 11 – 20
Blue, incipient flood; Red, flood

Flood is indicated when an entrainment index point is plotted on the top of the tray spacing, i.e., at the floor of the tray above. This point corresponds to an entrainment index of 2. Figure 5 shows the following:

- Flooding, recognized by an entrainment index near 2, took place only on trays 16 and up. Trays 11 to 15 were not flooded, and show low entrainment indices.

- Flooding took place only in the second quadrants of trays 16 to 19, and in all but the inlet quadrant of tray 20. In the incipient flood scans, entrainment from other quadrants was light to moderate (entrainment indices 0.1 to 0.8), indicating that these quadrants were not even close to flood. This provides strong evidence to support flood initiation in the second quadrant of the trays.

- Entrainment was light at tray inlets. In both the incipient flood and the flooded scans, the inlet entrainment indices were all below 0.5, the only exception being tray 20 at the flood where it was about 1.6. There was no flooding in any inlet regions.

Froth densities (Figure 6). A froth density of 50% liquid is plotted on the top of the tray spacing, i.e., at the floor of the tray above. A point half way between the tray below and that above indicates a froth density of 25% liquid. Figure 6 shows the following:

- Froth densities are low at the inlets and outlets, both in the flooded and the incipient flood scans, all in the range of 10% to 25%. The only exception is tray 20 where the outlet froth densities reached 30% in both the flooded and incipient flood scans.

- Froth densities in the middle of the trays at incipient flood (16 to 19) were higher, ranging from 22 to 30%.

- Froth densities in the middle of the flooded trays 16 to 19 rose from 22-30% in the incipient flood scans to about 26-38% in the flooded scans.

Clear liquid heights (Figure 7). With clear liquid heights, the unflooded scans provided invaluable insight so they were added to the diagram along with the flooded and incipient flood scans. A clear liquid height of 380 mm is plotted on the top of the tray spacing, i.e., at the floor of the tray above. A point half way between the tray below and that above indicates a clear liquid height of 190 mm. Figure 7 shows the following:

- In the unflooded scans, the inlet clear liquid heights on trays 13 to 20 were 15 to 30 mm lower than the clear liquid heights at the tray middle and outlets. The clear liquid heights in the middle and outlets of trays 13 to 20 were between 80 and 110 mm, except on tray 16 where they were about 10-20 mm higher. The differences between the middle and outlet clear liquid heights on each of these trays were within 10-15 mm. Going from non-flooded to incipient flood, the upward gradient from inlet towards the middle of the trays became magnified. There was a large rise in clear liquid height in the middle of the trays. On trays 17 to 19, the rise was about 100 mm, while on trays 13 to 16 it was 50 to 80 mm. This is an increase by a factor of 1.5 to 2. The rise was smaller, about 20 to 50 mm, at the tray 16- 19 outlets, and was zero for trays 13 to 15 outlets. The inlet clear liquid

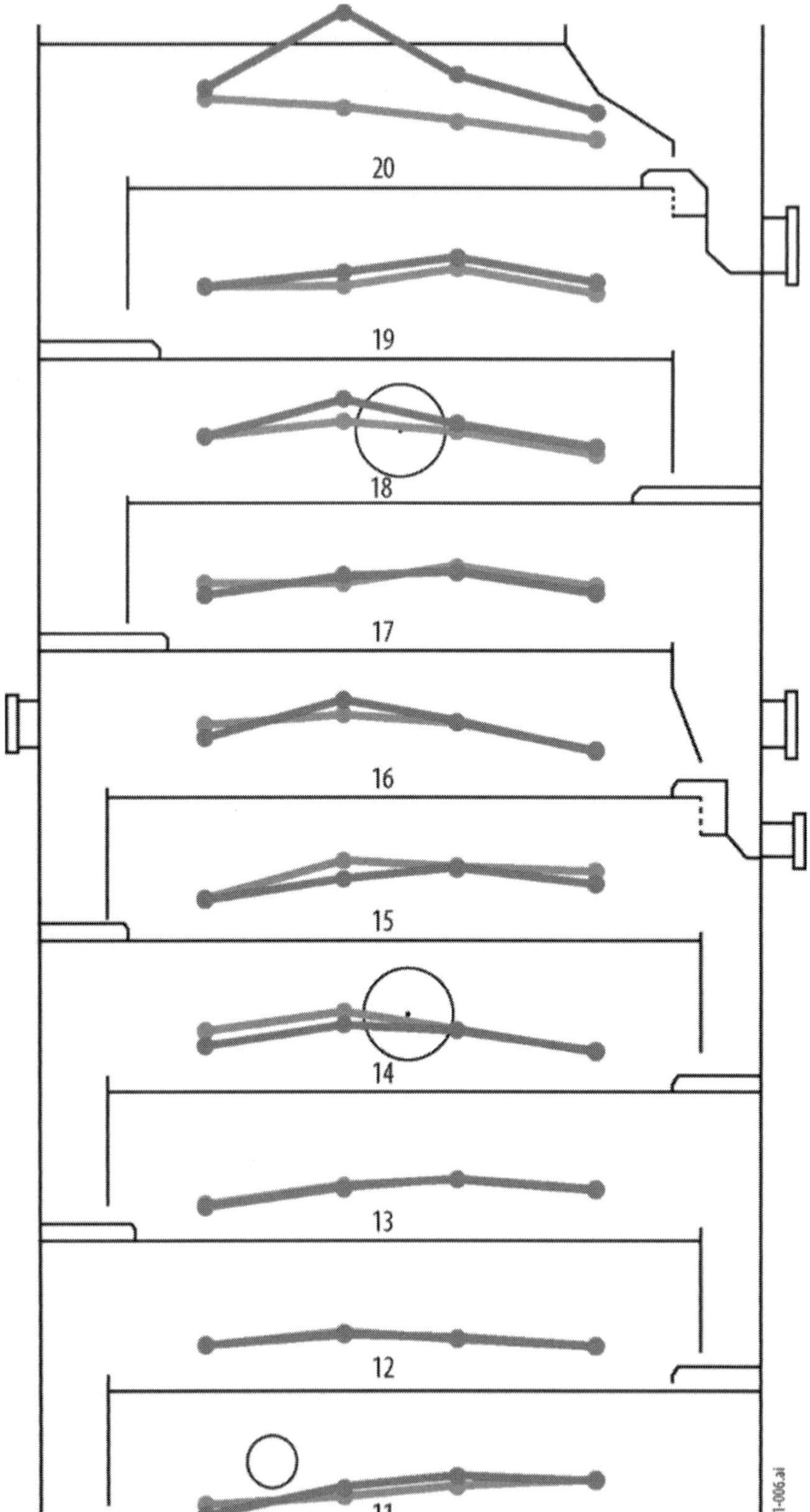

Figure 6. Froth Densities
Blue, incipient flood; Red, flood

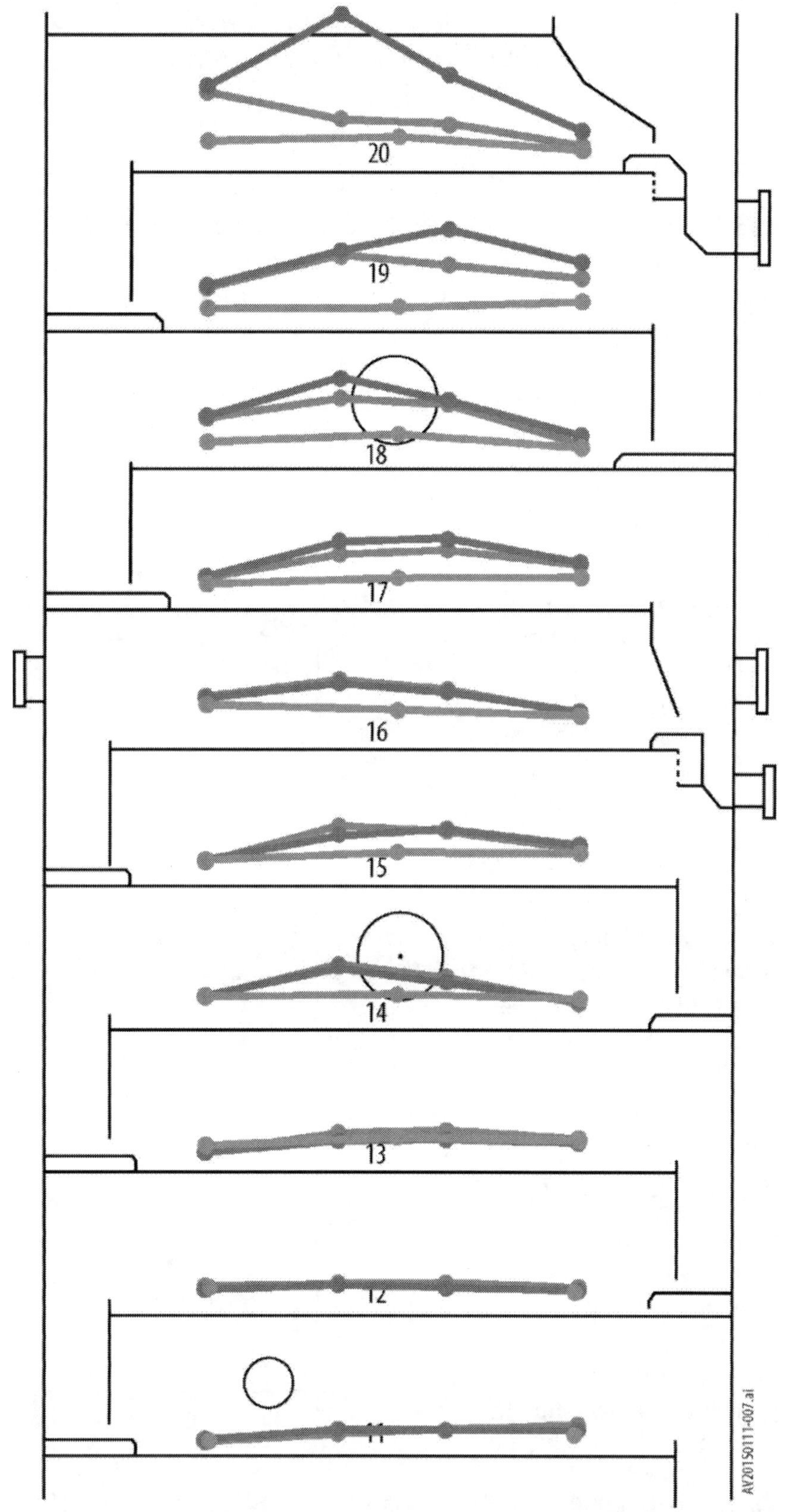

Figure 7. Clear Liquid Heights, Trays 11 – 20
Green, no flood; Blue, incipient flood; Red, flood

- height rise from non-flooded to incipient flood was of the order of 20-30 mm in trays 17 to 19, and none on trays 13 to 16.

- At incipient flood, clear liquid heights in the middle of trays 16 to 19 reached 180 to 230 mm, and a little lower, 150 to 180 mm, on trays 14-15.

- Going from incipient flood to full flood, there was little change in the clear liquid heights on trays 13 to 16. On trays 17 to 19, there were no major changes at the inlet, outlet and second quadrant, but the clear liquid heights on the third quadrant of trays 17 to 19 and at the outlet of tray 19 jumped by 80 – 100 mm.

Reverse Vapor Cross Flow Channeling (RVCFC)

Figure 8 shows how the hydraulic gradient on trays 11 through 20 change from non-flooded to flooded conditions. The hydraulic gradient plotted is the maximum clear liquid height on the tray less the minimum clear liquid height on the tray. We term this a "reverse hydraulic gradient". In all the scans, the minimum clear liquid height was always at the tray inlet. For the incipient flood and the flooded scans, the maximum on trays 12 through 19 was always in the tray middle. For tray 20 at incipient flood, as well as for many trays in the non-flooded scan, the maximum was at the tray outlet. So for the incipient flood and flood conditions on most trays the reverse hydraulic gradient is the difference between the clear liquid heights at the middle and inlet.

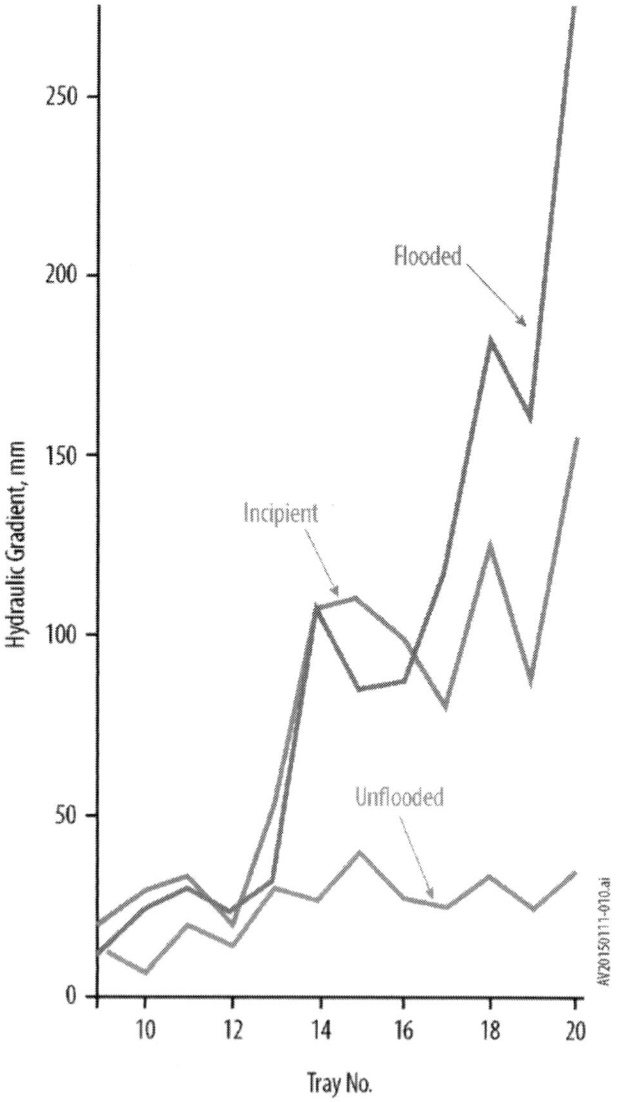

Figure 8. Reverse Hydraulic Gradients, Trays 11 - 20

Figure 8 shows that for non-flooded conditions, the hydraulic gradients are of the order of 20 – 30 mm. This is well in-line with values reported from measurements on conventional trays[9], except that on normally operating conventional trays the gradient is from inlet to outlet, ie clear liquid is highest at the inlet and lowest at the outlet. Here the gradient is reversed, ie, the clear liquid is highest in the middle of the tray, and lowest at the inlet, so the hydraulic gradient is from the middle to the inlet. At incipient flooding, there was a sudden buildup of this reverse hydraulic gradient on trays 14 to 20. This means a lot of liquid channeled towards the middle of the tray, raising the reverse hydraulic gradients to 80 – 120 mm. This is a huge hydraulic gradient. Going from incipient flood to full flood, there was no further change on the initiating tray 16 and the two below. However, as liquid accumulated on trays 17 to 20 above, the reverse hydraulic gradient increased by another 50 - 100 mm, indicating intensification of the channeled pattern. The resulting channeling pattern had vapor breaking through the inlet regions, with liquid building up in the middle of the trays. We term this type of channeling Reverse Vapor Cross Flow Channeling (RVCFC).

Figure 9 assembles this information to construct the likely hydraulic pattern on the trays. The Nye unit at the inlet aerates the incoming liquid and shoots a lot of it forward, generating a low liquid head region near the tray inlet. The liquid from this region is blown to the middle of the tray, generating a high-liquid head region there. At low rates, this is likely to lead to excessive weep from the middle of the tray, which may negatively impact tray efficiency and turndown. The efficiency loss with RVCFC is smaller than in vapor cross flow channeling (VCFC), because in VCFC the weep occurs from the tray inlet, which is much more detrimental to tray efficiency that the weep from the middle of the tray that occurs with RVCFC.

At higher rates, more vapor channels through the tray inlet, generating a high-velocity region at the tray inlet, with liquid blown towards the tray above and the middle of the tray. This shows up as increases in entrainment and froth height in the middle of the tray. Some of the blown liquid falls to the tray floor at the middle of the tray, generating higher liquid heads there. This channels more vapor to the tray inlet. This starts a self-accelerating mechanism which ends up with low froth heights, low liquid heads and low clear liquid heights at the tray inlet, and high froth densities, froth heights, entrainment and clear liquid heights at the tray middle. The result is premature flooding, as experienced in this crude tower.

Similar to VCFC, per the mechanism proposed by Davies[13] , and later discussed for sieve and valve trays[5,6] , the channeling tendency is countered by the dry pressure drop of the tray. When the dry tray pressure drop is high relative to the hydraulic gradient, this pressure drop will distribute the vapor evenly across the tray and prevent channeling from setting in. In the current case, trays 11 to 16 had venturi valve trays, which have rounded orifices that halve the valve pressure drop compared to those with sharp orifices. Also, these valve units had long legs, leading to excessive open slot areas, which also lowered dry pressure drop.

The channeling criterion proposed by Davies[13] is that the hydraulic gradient needs to be kept below 40% of the dry pressure drop to avoid channeling. Table 1 shows that tray 16 has 44 mm of liquid dry pressure drop, so as long as the hydraulic gradient is less than about 20 mm, there will not be significant channeling, as seen in the unflooded scan. However, should the forward push action generate a hydraulic gradient exceeding 40 mm, it is likely that the dry pressure drop will not be large enough to counter the channeling which will then initiates the self-accelerating mechanism. The best prevention is to boost the dry pressure drop.

The small valves on trays 17 to 20 had sharp orifices, and less open areas, so they had higher dry pressure drop (Table 1) and were more resistant to channeling. However, these are also lower pressure-drop valves and are also prone to channeling, albeit to a lesser extent.

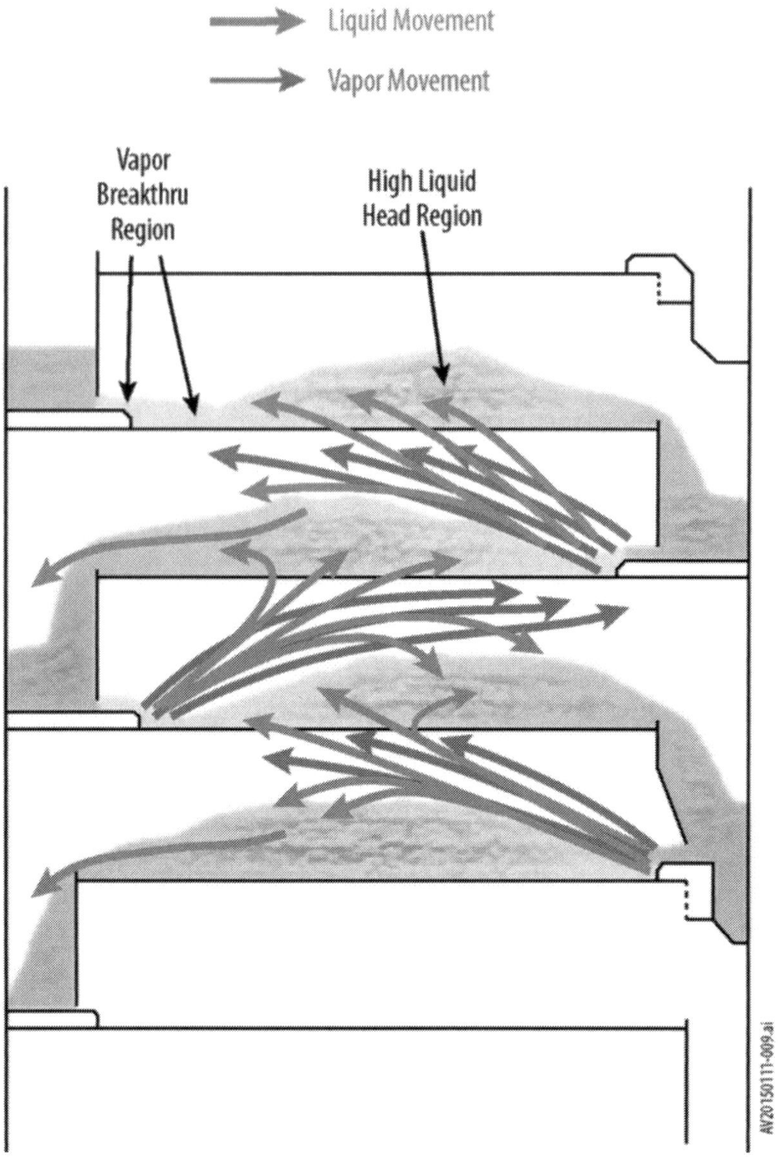

Figure 9. Reverse Vapor Cross-Flow Channeling

Like in VCFC, long flow path to tray spacing ratios generate the geometry shown in Figure 9 (these are drawn to scale), and promote RVCFC. Short flow path compared to tray spacing will not permit development of the RVCFC pattern. Finally, like in VCFC, high weir loads are likely to promote this type of channeling.

The RVCFC pattern requires a source of forward push to generate this pattern. So while push valves have been effective in countering VCFC, with some geometries the forward push can generate its own channeling pattern. We have seen other cases of RVCFC. In one case, RVCFC in a refinery fractionator pumparound caused excessive weeping, generating cavitation at the pump. In a recent case, RVCFC appeared to have combined with multi-pass trays maldistribution to give RVCFC in the middle panels of 4-pass trays[8].

The key, therefore to preventing RVCFC from forming is following the good design practices that prevent VCFC[5,6]. The key is avoiding excessive open area, venturi valves, and avoiding large ratios of flow path length to tray spacing with tray geometries that may be prone to channeling.

Operational Debottleneck

The solution to the excursion problem[1] in Part 1 of our paper also provided an operational debottleneck for the current flood problem. That solution was completing the kerosene side-draw system and operating in the kerosene mode. Following the tests, the kerosene mode became the normal mode of operation. There was little economic penalty to drawing a relatively small amount of kerosene product stream, as it could have been blended with the diesel when not desired as a product.

Operating at the kerosene mode circumvented not only the diesel draw bottleneck, which caused the excursion problem, but also the flooding bottleneck. With both bottlenecks overcome, the tower charge rates could be raised from 38,700 BPSD (5,200 t/d) to 49,100 BPSD (6,600 t/d), much closer to the design 52,500 BPSD (7,070 t/d). At 6,600 t/d, the flood limit re-occurred but the excursion did not. At 6,600 t/d charge rate, the capacity loss was tolerable and an imminent shutdown was avoided.

The red print in Figure 1 shows the tower material balance following the operational debottleneck. The diagram permits comparison to the material balance before the debottleneck, shown in black print.

One puzzle is how drawing kerosene debottlenecked the flood in the tower. This can be addressed by reviewing the values listed in Table 1. As expected, drawing kerosene greatly reduced the liquid rate and weir load on tray 16, where the capacity bottleneck was. However, channeling phenomena such as VCFC and RVCFC produce vapor floods. Table 1 shows that even though the tower charge rate rose by more than 25% in the operational debottleneck, the internal vapor rate on tray 16 only rose by 6%. At the same time, the tower pressure for the operational debottleneck was higher, reducing internal vapor velocities. Table 1 shows that the C-factor for tray 16 at the operational debottleneck was only 1% higher than before. Considering the lower liquid load, the trays hydraulic loads diminished from 78% of flood to 74% of flood despite the rise in tower charge rate, giving it a larger margin from the premature flood.

The next question is why the internal vapor flow rate only increased by 6% when the tower charge rate increased by over 25% with the same crude. Our simulation (Table 1) shows that before the kerosene draw was operated, tray 16 operated at 213° C. Drawing the kerosene made trays 11 to 16 much hotter, with tray 16 operating at 238° C. This 25° C temperature increase carries a lot of heat up the tower. With the tray 18 temperature remaining unchanged, this means a much smaller internal vapor rate was used to maintain the heat balance.

While this helped debottleneck tray 16, trays 17 to 19 did not enjoy these large temperature increases, and the internal vapor rate through them rose. On tray 18, the internal vapor flow rate rose by 16% following the operational debottleneck, and the C-factor increased by 10% (Table 1). With this tray being above the draw, it also saw a 10% increase in weir load.

Following the operational debottleneck, the tower was again gamma scanned using the same technique of multichordal gamma scans with quantitative analysis. Figure 10 (equivalent to Figure 4) shows the froth heights for the operational debottleneck conditions at incipient flood (47,000 BPSD) and at flood (51,000 BPSD). Figure 10 shows no flooding below tray 17. The flood initiated on tray 18.

The flood experienced was still premature, and occurred at 86% of the normal jet flood. This suggests that the RVCFC still occurred, this time in the small valve section. It occurred at higher hydraulic loads than it did prior to the operational debottleneck, and at higher C-factors than previously. The reason is that the small valves with the sharp orifices and lower open area were significantly less prone to channeling than the venturi valves with the larger open area. This is supported by the higher dry pressure drop shown in Table 1 for these valve trays. The hydraulic gradients shown on Figure 11 for the operational debottleneck scans confirm that RVCFC is still in action, this time on tray 18.

Permanent Fix and Takeaway

Based on this diagnosis, for a permanent fix, the RVCFC needed to be eliminated. The tower was retrayed with the same type of tray, Nye trays, but with a design that gave significantly larger friction pressure drop. The new trays also had two passes which halved the ratio of flowpath length to tray spacing and lowered the weir loads. The modifications completely eliminated the bottleneck and the RVCFC, permitting higher-than-design capacities to be achieved.

As previously pointed out[1], modifications were also made to fully eliminate the excursion, with and without the kerosene side draw. To achieve, the diesel draw from tray 11 was replaced by a seal-welded chimney tray. The diesel draw and diesel rundown line were increased to 16 inches diameter.

Following these modifications, the tower continuously operated well at a feed capacity of 7,070 t/d (52,500 BPSD), with or without the kerosene side draw. The tower was tested and shown to operate well at a feed capacity as high as 7,750 t/d (57,500 BPSD). The high capacity Nye trays (which were the same type, but different design, compared to the originals) well-exceeded their original design capacity. Excursions and flooding no longer occurred, soon to become a faded memory.

The investigation unveiled a previously unknown phenomenon which we termed "Reverse Vapor Cross Flow Channeling (RVCFC)". This phenomenon is closely related to vapor cross flow channeling (VCFC), but is induced by the forward push on high capacity trays rather than by the normal hydraulic gradient on the trays. This forward push generates reverse hydraulic gradients, with the clear liquid heights highest at the tray center and outlet and lowest at the tray inlets. Like VCFC, this phenomenon sets in when there is a significant channeling initiator, while the friction pressure drop of the trays is too low to counter the channeling, and when the ratio of flowpath length to tray spacing and the weir loads are high.

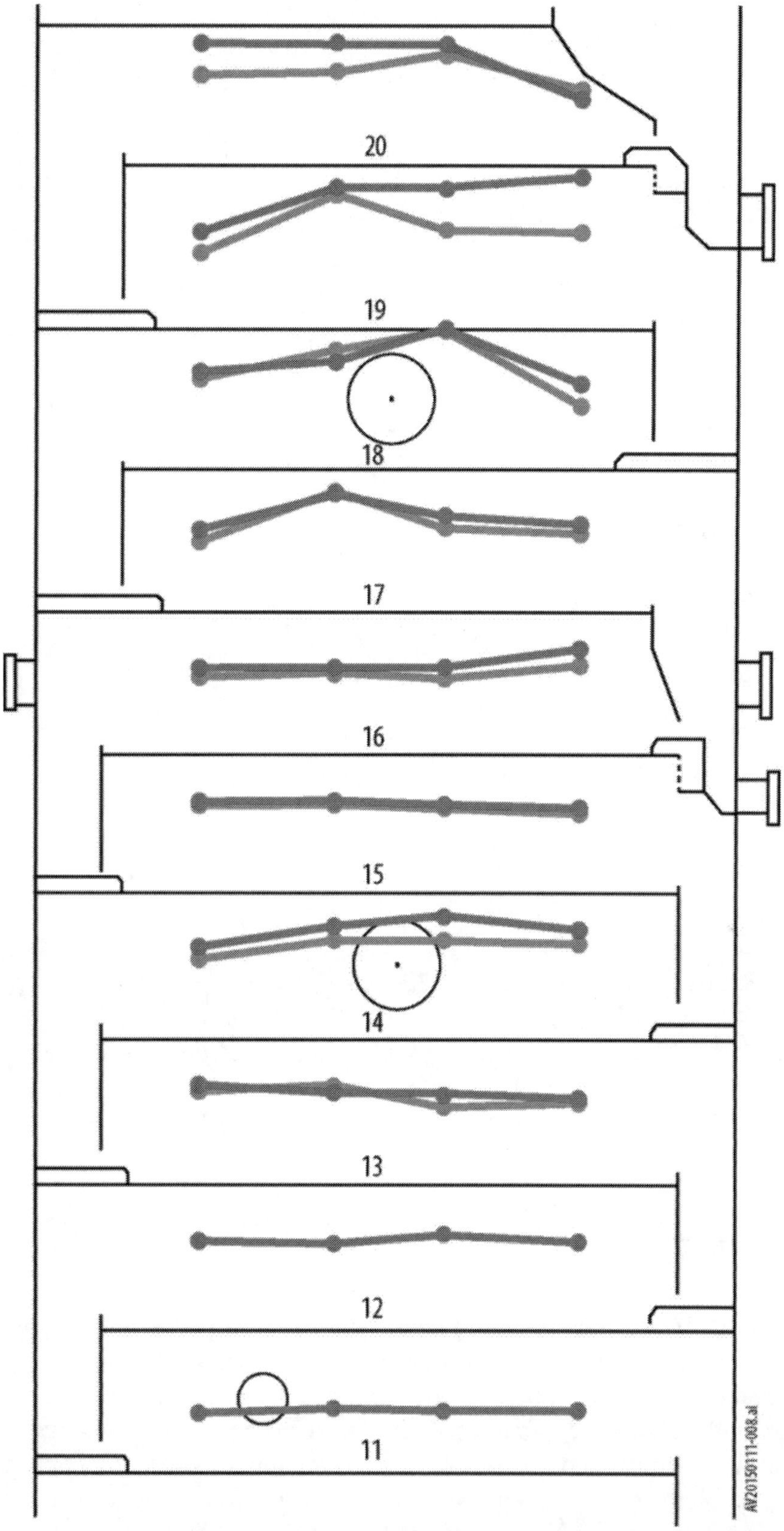

Figure 10. Froth Heights, Following Operational Debottleneck Blue, incipient flood; Red, flood

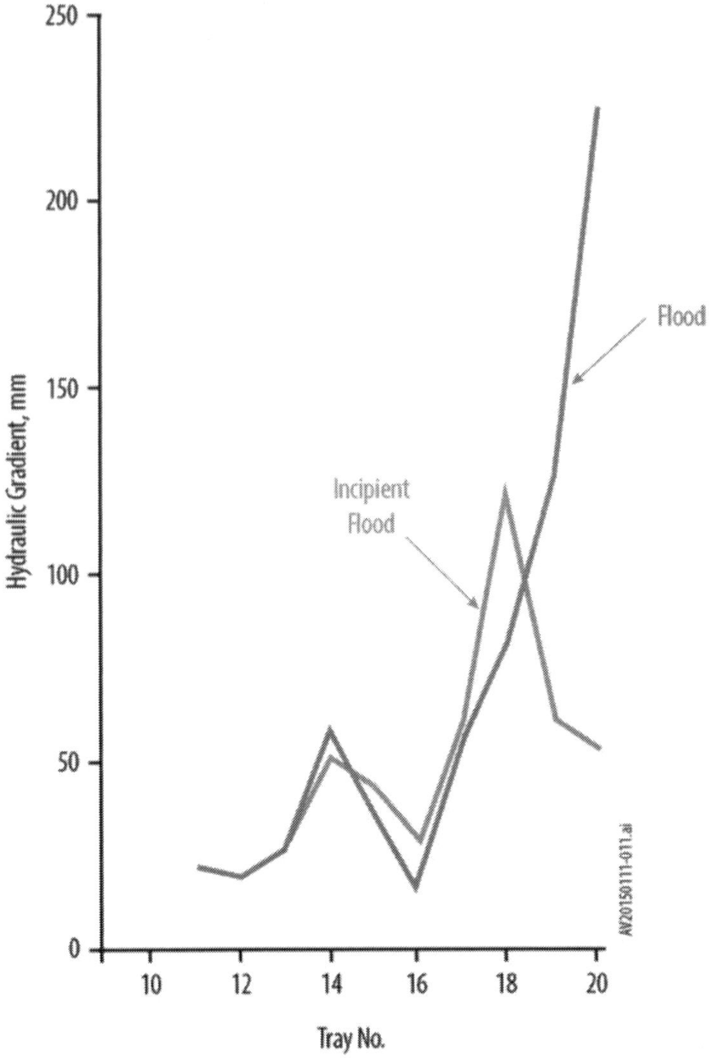

Figure 11. Reverse Gradients, Following Operational Debottleneck Blue, incipient flood; Red, flood

Acknowledgement

As stated, the quantitative gamma scanning requires high-quality gamma scan data. With this in mind, the authors wish to specially acknowledge two gamma scanning maestro, Dave Fradgley from Tracerco, and his Tracerco team, whose excellent work provided us with outstanding-quality data we could bank on, data that led us to a correct diagnosis and solution. Our multitude of validation checks demonstrated that these were among the very best quality scans we have ever seen. Many problems along the way were solved thanks to their ideas and expertise. Without their expertise, we would have been unable to unveil RVCFC.

References

1. Kister, H. Z., N. O'Shea, and D. Cronin, "Systematic Field Tests Turn Loss into Gain in High-Capacity Trays", Proceedings of the Distillation Topical Conference of the AIChE National Spring Meeting, Houston, TX, April 2007.

2. Kister, H. Z., "Use Quantitative Gamma Scans to Troubleshoot Maldistribution on Trays", Chem. Eng. Progr., February 2013, p.33.

3. Lieberman, N. P., "Process Engineering for a Small Planet", John Wiley & Sons, NJ, 2010.

4. Anon., "Tray Converts Dead Area to Vapor/Liquid Contact Use", Chem. Proc., p.44, December 1993.

5. Kister, H. Z., K. F. Larson and P. E. Madsen "Vapor Cross Flow Channeling on Sieve Trays: Fact or Myth?" Chem. Eng. Prog., p.86, November 1992.

6. Kister, H. Z. "Can Vapor Cross Flow Channeling Occur on Valve Trays?", Chem. Engnr. (London), June 24, 1993.

7. Kister, H. Z., K. F. Larson, J. M. Burke, R. J. Callejas, and F. Dunbar "Troubleshooting a Water Quench Tower", Proc. of the 7th Annual Ethylene Producers Conference, Houston, Texas, March 1995.

8. Kister, H. Z., B. Clancy-Jundt, and R. Miller, "Troubleshooting a C_3 Splitter Tower, Part 2: Root Cause and Solution", PTQ, p. 39, Q1 2015.

9. Kister, H. Z., "Is the Hydraulic Gradient on Sieve and Valve Trays Negligible?" Paper presented at the Topical Conference on Distillation, the AIChE Meeting, Houston, TX, April 2012.

10. Resetarits, M. R., and N. Pappademos, "Factors Influencing Vapor Crossflow Channeling", Paper Presented at the AIChE Fall Annual Meeting, Reno, Nevada, November 2001.

11. Hartman, E. L., "New Millennium, Old Problems: Vapor Cross Flow Channeling on Valve Trays", in "Distillation 2001: Frontiers in a New Millennium", Proceedings of Topical Conference, p. 108, AIChE Spring National Meeting, Houston, Texas, April 22-26, 2001.

12. Harrison, M. E., "Gamma Scan Evaluation for Distillation Column Debottlenecking", Chem. Eng. Prog. 86 (3), p.37-44, March 1990.

13. Davies, J. A., "Bubble Trays – Design and Layout", Pet. Ref. 29(8), p.93, 1950, and 29(9), p.121, 1950.

Should You Consider Moving to Fixed Valves?

Scott Hebert, Koch-Glitsch, LP
Neil Sandford, Koch-Glitsch, LP

Paper presented at the Distillation Topical Conference
AIChE Spring Meeting
Austin, Texas
April 2015

Abstract

In recent years we have seen a general shift from using floating valves towards using fixed valves on distillation trays. This is particularly the case in the refining industry where the loss of some operational flexibility is often considered an acceptable tradeoff for the benefits that fixed valves bring, including better reliability, fouling resistance and more robust designs that can better withstand operational upsets. This paper will present a review of the different types of fixed devices and their benefits including R&D and CFD development work and industrial case studies.

Article

Introduction

Refiners depend on distillation trays in numerous services to make required separations. Of critical importance to any refiner's separation is the type of mass transfer device used on distillation trays. For conventional cross-flow trays, refiners generally have two choices for mass transfer devices: moving valves or fixed valves.

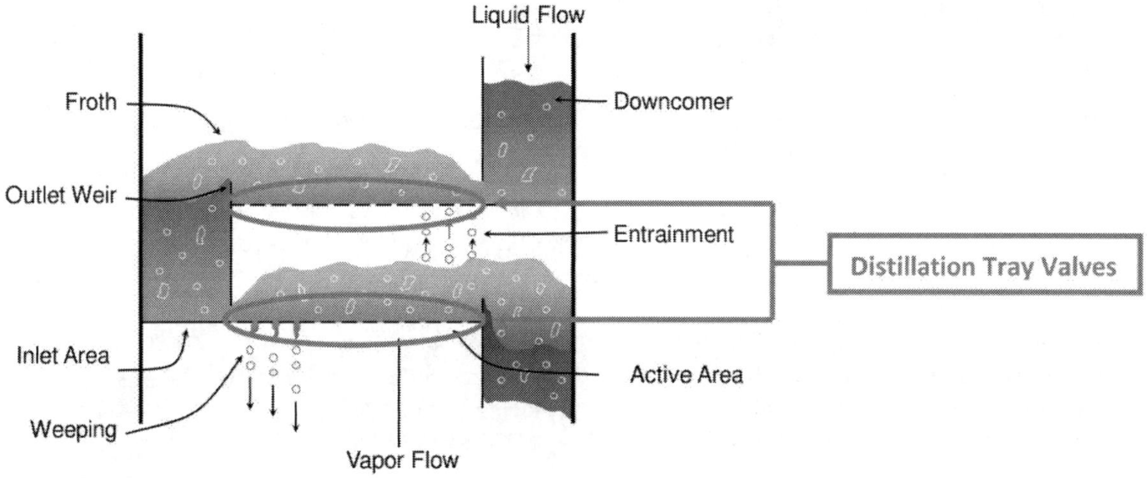

Figure 1. Simplified view of a conventional cross-flow tray

Conventional distillation tray devices have developed according to the simplified timeline shown below:

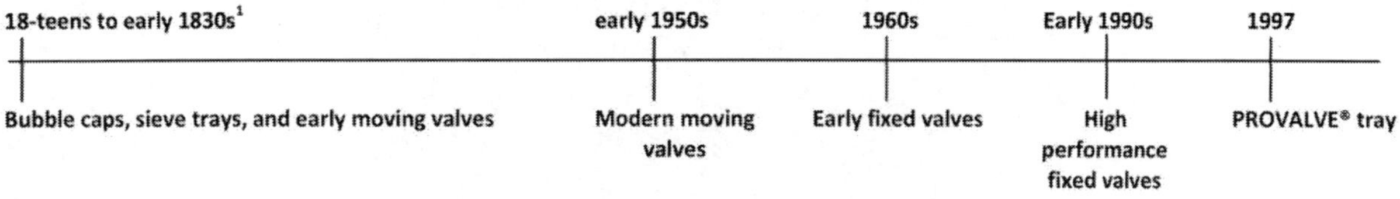

Figure 2. Simplified history of distillation mass transfer devices

Bubble cap and Sieve trays

Bubble cap trays, which are complex in design, have been in use since the early 1800s. Due to the intensive labor and materials required to produce these trays, they are significantly more expensive (by approximately a factor of 2-3) than extruded fixed valves. Very common in older refinery designs, bubble-cap trays have fallen out of favor for general distillation use due to expense. Bubble-cap trays do have their place in some low-liquid rate niche applications and can operate efficiently with widely varying vapor load. The weeping tendency of bubble-cap trays is a function of the mechanical design and installation, not pressure drop. A well-sealed bubble-cap tray can have almost infinite vapor turndown (if tray activity can be ignored). For discussion of limits of vapor turndown caused by tray activity, see Moving Valve Trays section.

Figure 3. Bubble cap tray

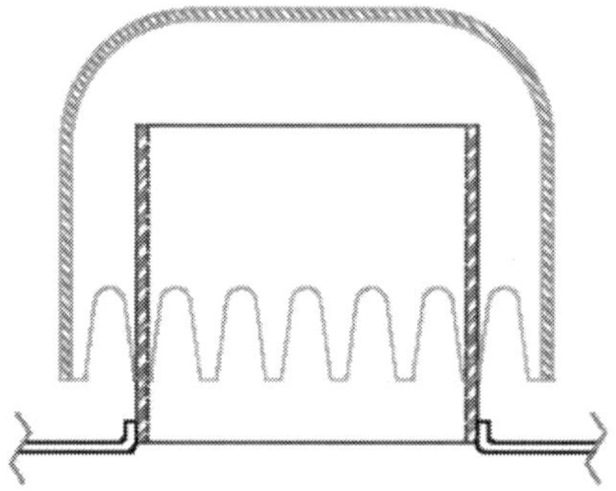

Figure 4. Bubble cap cross-section

Sieve trays were developed very close in time to bubble cap trays. In contrast to bubble-cap trays, sieve trays are relatively easy to produce, as production involves making round holes in the tray deck. Sieve trays have an advantage of low production cost; however they are generally outperformed by other designs. Why? The general answer is, "because of capacity and fouling", although sieve trays may be useful in certain applications.

Sieve trays, which contain only round holes cut or punched through the tray deck, have a lower capacity than moving or fixed valves. Vapor exiting a sieve hole travels vertically, while vapor exiting a valve travels in a more relative horizontal direction. The result of this vapor "deflection" in fixed or moving valves is increased hydraulic capacity. Sieve trays begin to experience entrainment at lower rates than fixed or moving valves because the vertically directed exiting vapor results in higher froth heights.

Figure 5. FLEXITRAY™ sieve tray

The vapor "deflection" is also the reason that valve trays offer more fouling resistance than small sieve holes, as increased vapor velocity on the tray deck around fixed or moving valves helps prevent zones of stagnation where solids can deposit, polymerization can begin, or corrosion can occur. The vapor deflection helps keep the tray deck cleaner.

Moving valves

Following its introduction in the early 1950s, the modern moveable (or floating) valve became a distillation standard for the refinery industry. It has the chief advantage of allowing a wide operating range due to its variable movement. The valve's movement is defined by either the length of the valve legs that protrude through the tray deck orifice or by a cage assembly that restricts vertical movement. The resulting variable vertical movement of the valve (refer to Figures 6-10 for examples) enables the valve to function in a relatively wide operating window.

Figure 6. FLEXITRAY® A-valve

Figure 7. FLEXITRAY® T-valve

Figure 8. MV-1 type MINIVALVE® tray

Figure 9. FLEXITRAY® A-valve tray

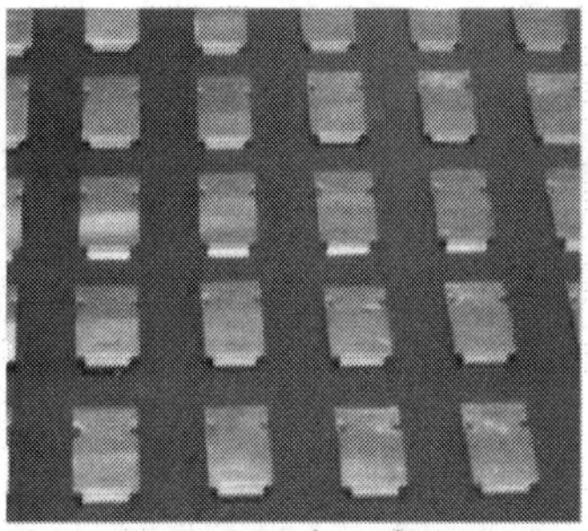

Figure 10. Rectangular valve tray

Hence, the refiner is able to turn up or turn down rates within the operating window; a nominal turndown range of 4:1 is generally achievable with moving valves. Turndown is primarily a function of available pressure drop. As the vapor rate decreases so does the tray pressure drop. When the tray pressure drop becomes too low, the tray begins to weep, and efficiency begins to decline. With moving valves, however, as the vapor rate decreases, some of the valves will close, reducing the effective open area of the tray and limiting the tendency to weep.

An additional factor to consider is tray activity – at reduced vapor rates, not all of the valve devices on the tray deck will be actively bubbling vapor. This is particularly important for multi-pass trays where a higher vapor rate will typically be needed to ensure all passes are actively bubbling vapor. If one pass becomes inactive, tray efficiency will decline sharply. The more flow passes a tray has, the more difficult it is to ensure complete tray activity. One way to improve the turndown performance of moving valve trays is to install the valves in rows of alternating metal thickness (for example 14 gauge/16 gauge). This results in a dual-weighted arrangement, which gives the designer control over the location of the valves that close first (as vapor flow rate decreases); this practice can extend the operating range even further. Turndowns of 10:1 with moving valves are possible with proper design (provided available pressure drop and tray spacing are reasonable and the system is non-foaming).

While the standard moveable valves provide a high turndown and are more efficient than sieve trays, they do have their disadvantages:

- Moveable valves are more prone to fouling than fixed valves, as there tends to be more surface area available for contaminant deposition. This is a huge concern for refiners, as refining services tend to be much more severe than most chemical applications, with much higher potential for particulate deposition caused by dirt and debris, various forms of corrosion (e.g., from acids), dewpoint salt formation, and in some cases, polymerization.
- The valve legs may touch the edges of the orifice as the valves open and close, resulting in erosion and enhancing corrosion. Valves do not typically spin freely in operation, but in some cases anti-spin tabs are added to the orifices to prevent rotation.
- The valves can stick to the deck as a result of fouling or polymer deposition. Most valves include "dimples" that prevent complete contact between the valve cap and the tray deck and provide a small escape area between the deck and the valve. This eliminates "sticking"; however, it can result in increased liquid weeping at low vapor rates.
- If subjected to an unexpected upset, moving valves can "pop" free from their orifices, resulting in the creation of de-facto sieve holes, which have less capacity and efficiency than valves.
- Moveable valves are more expensive than conventional fixed valves, as more labor and material is required to manufacture the valves. A general rule of thumb is that moving valves are 10-15% more expensive than conventional fixed valves.

Fixed Valves

The simplest form of fixed valves are those that are extruded directly from the tray deck, such that the tray deck and valves are all formed from the same piece of metal. Large metal punching (or stamping) machines are used to impart a tremendous amount of energy into a relatively small area of metal sheet, which results in the extrusion of a valve from the tray deck. As a result of this process, the metal is actually stretched from its initial state. (See below for examples).

Figure 11. VG-0 type MINIVALVE® valve

Figure 12. FLEXITRAY® VG-10 valve

Figure 13. VG-0 type MINIVALVE® valve tray

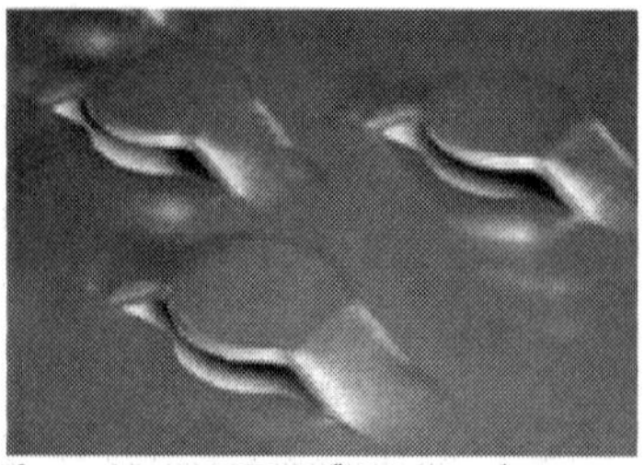

Figure 14. FLEXITRAY® VG-10 valve tray

As a result of this manufacturing process, these fixed valve trays have a rugged and durable construction and have no moving parts. This eliminates the problems associated with moving valves (valve sticking, erosion/corrosion, and "popping"). Extruded fixed valves are very resistant to fouling, given the relatively large opening size between the valve and the deck. The valve net rise (the vertical distance from tray deck to the valve apex underside) can be manufactured with different heights, and can be increased, if needed, to provide more resistance to fouling.

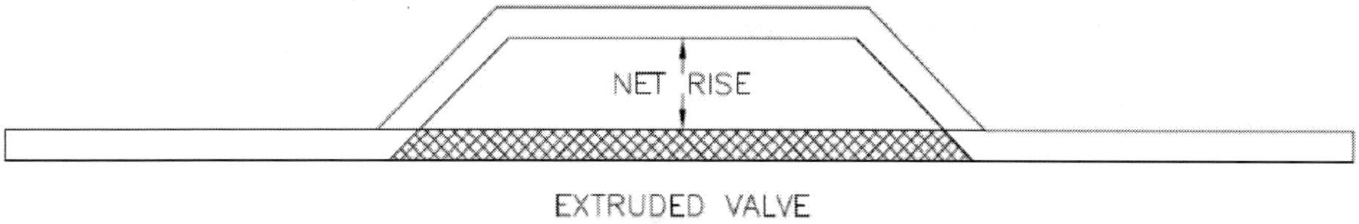

Figure 15. Extruded fixed valve net rise

In addition, as extruded valve opening size decreases, hydraulic capacity increases. Also, as the net rise decreases, less entrainment can be expected. As a result, the smaller (in size and net rise) VG-0 type MINIVALVE® trays are often used in heavily loaded tower sections, such as crude unit pumparounds.

One other benefit of fixed extruded valves is the ability to withstand process upsets. This is especially the case when combined with FLEXILOCK® tray construction (an interlocking tray joint that reduces installation time and strengthens joint and uplift tolerance). Due to their durable construction, the mechanical reliability of fixed valves makes them good choices for towers where uplift rating is required to guard against damage. Possible causes include introduction of water to hot oil, leading to high volume expansion of water to steam; loss of bottoms pumps; loss of circulating pumps; high liquid bottoms level that covers reboiler vapor returns, etc.

An Enhanced Fixed Valve

In the late 1990s, Koch-Glitsch was in the process of developing extruded fixed valve trays from thicker and more exotic metals than typical low-alloy stainless steels. During this process, it was determined that as the metal thickness increased and/or the elongation properties of the metal decreased, the extrusion process became more limited. In other words, the net rise achievable for extruded valves from these materials was decreased due to the inability of thicker and more exotic materials to successfully be stretched further. Since then, Koch-Glitsch has developed different manufacturing methods and know-how to successfully extrude from these materials, although certain manufacturing limitations still exist.

Initially, Koch-Glitsch engineers developed a solution to the thicker and more exotic problem by creating a fixed valve that consisted of a large orifice in the deck combined with a separate cover piece. This cover piece provides overlap around the orifice in the deck and has a unique tapered design. This shape cannot be fabricated with the extrusion process. The valve tray that resulted from this design process is known as the PROVALVE® tray.

Figure 16. PROVALVE® valve

Figure 17. PROVALVE® tray

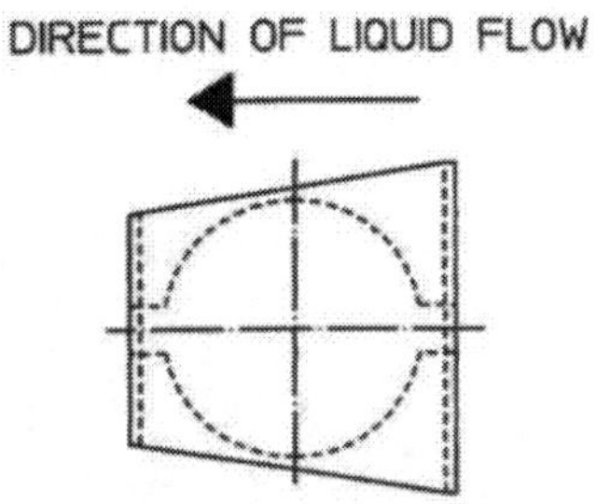

Figure 18. PROVALVE® valve is directional

The PROVALVE® tray valves provide excellent anti-fouling properties, far superior to moving valves and conventional extruded fixed valves, due to the following:

- Large orifice diameter
    - Similar in size to moving valve orifices, but sheltered by the cover
    - Larger than conventional, high-performance fixed valves
- Unique tapered cover design
    - The tapered shape imparts a push to liquid flowing across the tray deck, which helps prevent areas of liquid stagnation on the tray deck. This feature also suppresses weeping, as the orifice is well-protected from the momentum of liquid crossing the tray deck.
    - The cover redirects the vapor horizontally from the valve cover, which promotes an even and low froth height, promoting vapor/liquid contacting and suppressing jet flood.

Koch-Glitsch conducted testing during the development of the PROVALVE® tray to try and dislodge the cover; the testing showed that the valve construction was very durable and not prone to the problem of dislodging that sometimes occurs with traditional moving valves – in most situations the tray structure would fail long before the valves could be dislodged. Hence, like the extruded fixed valves, the PROVALVE® tray construction is reliably durable.

Koch-Glitsch also conducted computational fluid dynamic (CFD) modeling of conventional extruded fixed valves and the PROVALVE® valves. The modeling showed a much flatter vapor profile from the PROVALVE® valve, which allows for enhanced mass transfer and aids in preventing fouling material from settling on the tray or around the orifices. The CFD study results are presented below.

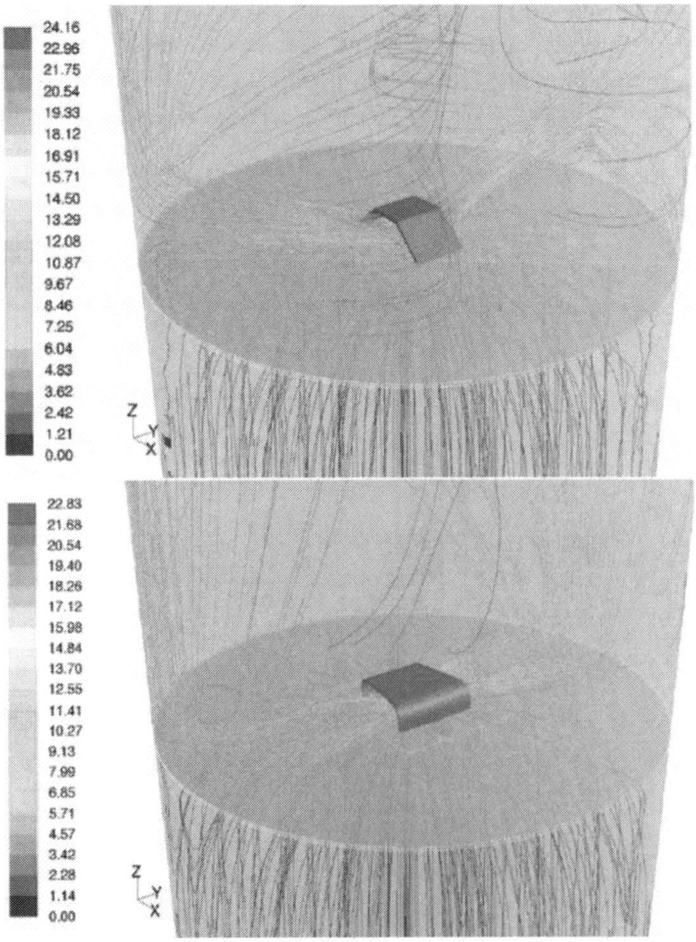

Figure 19. CFD study results – conventional extruded fixed valve (top) and PROVALVE® valve (bottom)

As evident in the CFD results, the PROVALVE® valve provides a much more uniform and flatter distribution of vapor into the liquid on the tray deck compared with a conventional extruded fixed valve. As noted earlier, a general rule is that smaller valves tend to have higher capacity, and lower net rise valves tend to have less entrainment. While the PROVALVE® valve is a larger valve, it achieves high performance, with capacity approaching that of a smaller valve.

Due to the construction process with a separate cover piece, the cost of the PROVALVE® tray is slightly more than extruded fixed valves (comparable to the cost of traditional moving valves); however, the benefits of the PROVALVE® tray typically justify the incremental cost.

Valve selection

Two questions that a refiner should ask when selecting distillation tray valves are,

1. "Do I *need* a large operating range for my process? and
2. "Do I have a severely fouling service?"

To address the former, typically, when refining margins are strong, the tendency is to run operating units at full rates (or beyond design rates, if possible), and not much variance is needed in the tray hydraulic capacity. While it may be desirable to run some operating units at minimum rates during periods of low economic margins or during maintenance on related units, one could make the point that the percentage of unit operation at a rate less than 50% of maximum rate is low. With this point in mind, more refiners are shifting towards fixed valves for distillation trays, since a rule of thumb for achievable fixed valve turndown range is 2:1 (although this turndown range can be pushed higher with proper design). This turndown on distillation tray hydraulic capacity is probably in good agreement with hydraulic turndown in the rest of the operating unit (charge pump minimum flow, heater minimum flows, control valve sizing, etc.).

To address the latter question of fouling: some refinery services are severe, as discussed above, and operating units' run times are often limited by fouling that occurs on the distillation trays, which can eventually render some types of valve trays inoperable. When a unit is down for cleanout maintenance, the refiner experiences a lost opportunity cost that can never be recovered with delayed processing; anything a refiner can do to extend run length is paramount with regard to profitability. In a fouling service, moving valves will result in downtime more often. Two specific examples of problematic fouling occur in the refinery's crude tower:

- Refiners that process high-acid "opportunity" crudes may be exposed to corrosion in the crude tower that can result in deposition of fouling corrosion products on trays.
- The very recent substantial increase of light domestic crude oil production in the United States has created readily available alternative feeds for refineries; however, some of these lighter crude oils are either inherently incompatible with themselves, or become incompatible when blended with other crude oils[2]. This incompatibility can result in asphaltene precipitation[2] and subsequent deposition of fouling products in crude distillation towers.

Refer to the spectrum of distillation deck device choices with regard to fouling resistance below:

PROVALVE® trays

Large extruded fixed valve trays

Large-holed (3/4 inch [19 mm] and up) sieve trays

Small extruded fixed valve trays

Caged moving valve trays

Single piece moving valve trays

Small-hole sieve trays

Figure 20. Some distillation valve choices and relative fouling resistance

There is a loss of operating flexibility associated with the decision to use fixed valves instead of moving valves. But is this loss of flexibility (a narrower turndown range that precludes the ability to run at significantly lower rates) a reasonable price to pay to avoid the following problems associated with moving valves?

- Increased fouling of moving valves compared to fixed valves (more places for dirt and debris to collect)
- The potential "sticking" of moving valve covers to the deck, leading to reduced capacity and tray efficiency
- Loss of tray performance due to moving valves "popping" free from the deck, leading to potential increased entrainment through unrestricted orifices
- Increased capital expenditure to install more expensive moving valves

Many refiners have performed a cost/benefit analysis and determined that installation of fixed valves with a smaller operating window is adequate for their process and helps avoid some of the disadvantages inherent with moving valves.

Case Studies

Sour Water Stripper[3]

Sour water strippers are often difficult to design and operate for the following reasons: they must process feed that often varies significantly in composition, and they often experience severe foaming as well as fouling. Refiner A had two identical sour water strippers in its West Coast USA refinery. Each tower processed the same sour feed and had sieve trays with ½"orifices. These towers suffered chronic fouling problems. Since the fouling of the sieve trays was so severe, each tower required several shutdowns each year for maintenance. The parallel feed arrangement of the two identical sour water stripper towers created a unique opportunity to test the performance of new trays under the same conditions as the existing sieve trays.

In the first test, two Koch-Glitsch PROVALVE® trays were installed in one stripper at a location in which the fouling was most severe. After a typical run length, this first stripper was shut down for maintenance, and the trays were examined. The two PROVALVE® trays showed negligible visual evidence of fouling, while the vertically-adjacent sieve trays showed considerable fouling with the flow area reduced by about 90% (Figures 21 and 22).

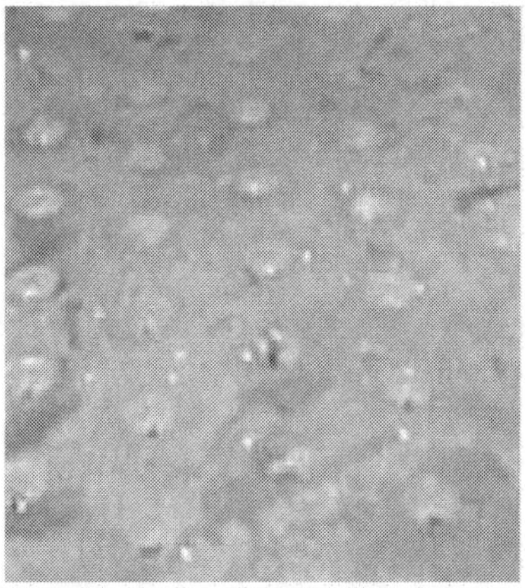

Figure 21. Fouling of sieve orifices (top view of tray deck)

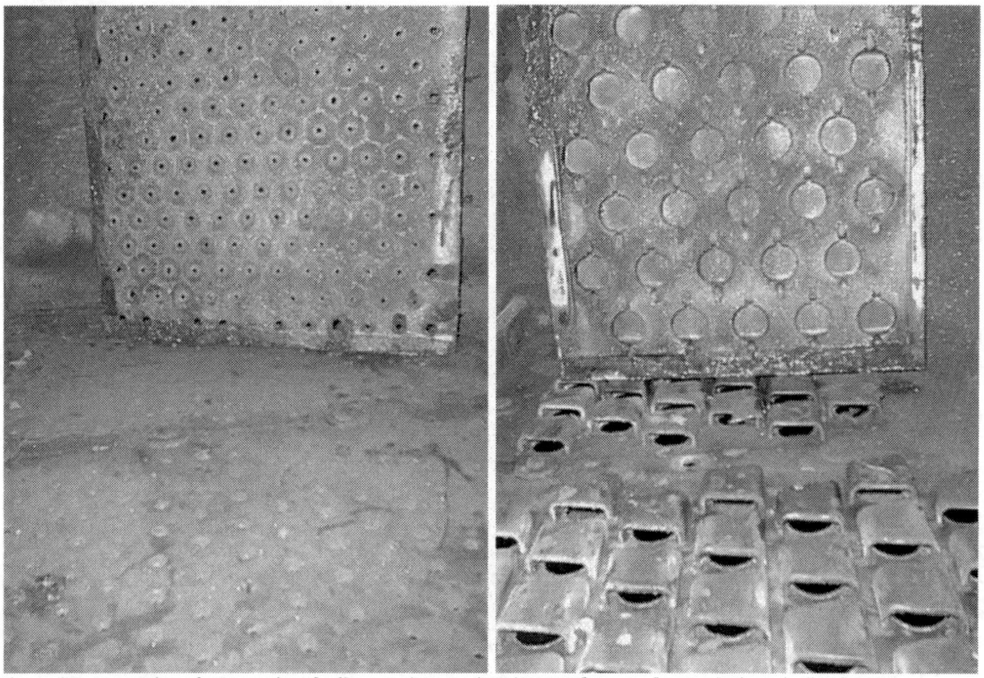

Figure 22. Inspection photographs of adjacent sieve and PROVALVE® trays after typical run in first tower

Based on the success of the first tower test, the entire second stripper was re-trayed with PROVALVE® trays and then restarted. Approximately two weeks later, the first tower was cleaned and then restarted, with the existing sieve trays still in place. After a typical run length (130 days), the first tower with the sieve trays had to be shut down for cleanout. After the same run time, the second sour water stripper with the PROVALVE® trays was processing *more* sour water feed, yet with no significant increase in pressure drop (Table 1).

Table 1. Comparison of Sour Water Strippers

	Tray type	Sour Water Feed (Start)	Sour Water Feed (130 Days)	Column Pressure Drop (130 Days)
First tower	½" sieve trays	480 gpm	200 gpm	Flooded
Second tower	PROVALVE® trays	600 gpm	550-600 gpm	No increase

A process upset provided a 3rd opportunity for testing, after the sour water feed was contaminated with caustic, which greatly increased the tendency for fouling in the towers. At the time of the upset, the sieve tray tower had been online for 42 days since the last cleanout, while the PROVALVE® trayed tower had been online for 191 days. On Day 8 following the upset, the sieve tray tower pressure drop was so excessive that rate was cut in half.

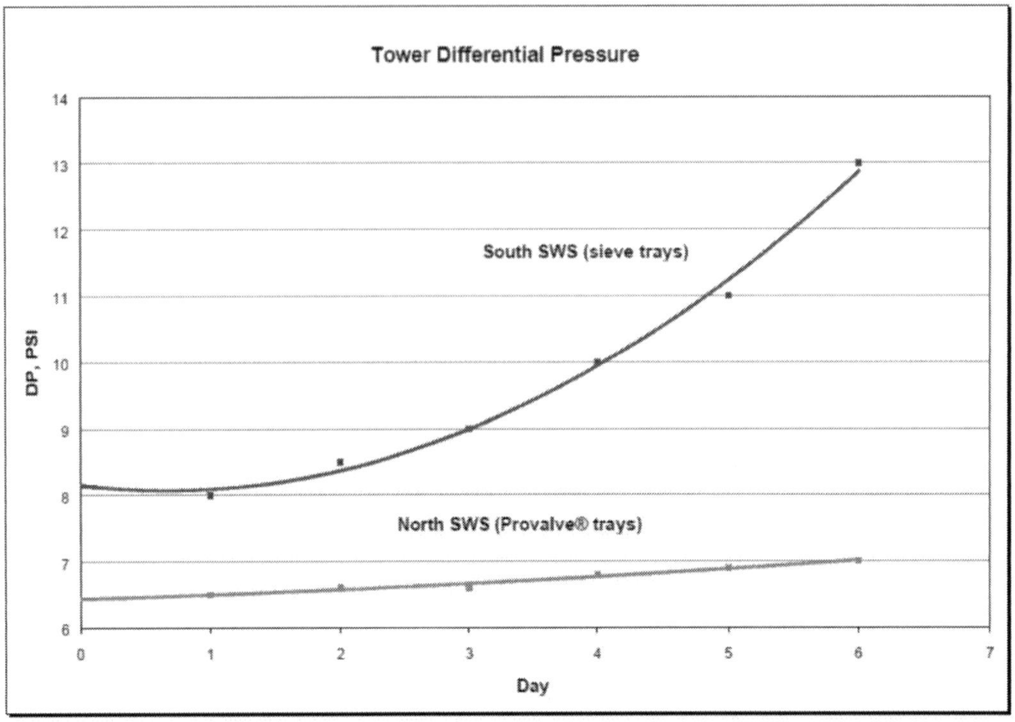

Figure 23. Sour water stripper pressure profiles after caustic contamination

30 days after the upset, the sieve tray tower was taken offline for cleanout. Meanwhile, the other tower with PROVALVE® trays stayed online, although rate was limited to 350-450 gpm. 350 days after the upset, the PROVALVE® trayed tower was still online and processing 350 gpm. Following these events, the sieve tray tower was revamped with PROVALVE® trays. The test results clearly demonstrated the improved performance and fouling resistance of the PROVALVE® fixed valve trays in sour water service.

Solvent Recovery Dehydration Column

Another case study focuses on a Solvent Recovery Dehydration Column, which experienced run-length limiting severe fouling of existing sieve trays. Shutdowns were required for cleanout when tower pressure drop became excessive, after approximately 120 days.

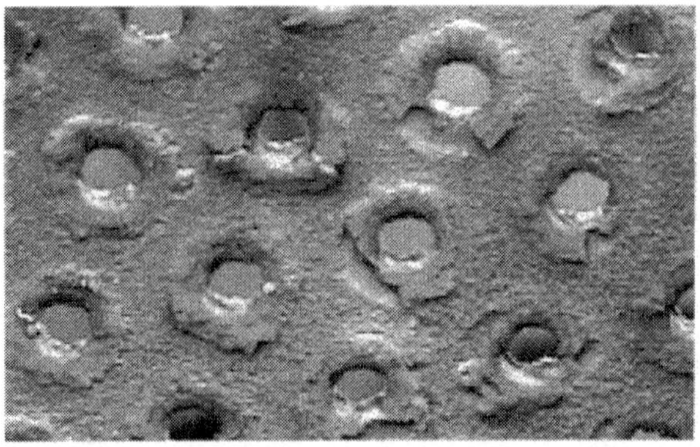

Figure 24. Fouling in Solvent Recovery Dehydration Column

The fouling material was a black flaky polymer. To correct the problem, the customer installed extruded fixed valves; unfortunately, these fixed valves were too small and actually had less open area than the prior sieve trays. As a result, the tower soon became completely fouled. (Remember – valve selection is critical!) The customer noted that the most severe fouling was toward the bottom of the tower.

Due to the problems with the fixed valves, the customer initially considered reverting back to all sieve trays; however, PROVALVE® trays were provided for the bottom section of the tower, with the same downcomer area, same weir height, and same downcomer clearance as the previous trays. The tower was restarted with sieve trays in the top and PROVALVE® trays in the bottom. The customer achieved a record run length of 197 days (an increase of 65%) and shut down the column for inspection purposes only. While the sieve trays in the top section showed considerable fouling, the PROVALVE® tray orifices showed no signs of fouling, with minimal polymer coating on the tray decks.

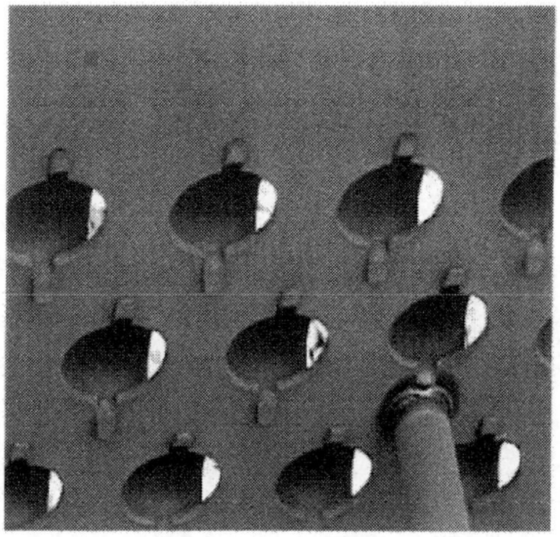

Figure 25. Inspection of PROVALVE® tray in Solvent Recovery Dehydration Column after run

The tower pressure drop at this point in the run was attributed to the top section of sieve trays. With this in mind, the operator took great care to carefully clean the top section and then began another process run. At 90 days into the run, the tower pressure drop was lower than previous with an all-sieve tray arrangement. As a result of these tests, the operator installed PROVALVE® trays for three identical towers.

Heavy Oil Service (Coker Fractionator)

For coker fractionators that process dirty feed in harsh conditions, fouling is a major design consideration. While not a case study per sé, the following discussion illustrates the benefit of fixed valves in this service. Consider the following pictures, taken in a coker fractionator that had been in service with moving caged valves.

Figure 26. Inspection of moving caged valves in coker fractionator

Figure 27. Inspection of another tray with moving caged valves in coker fractionator

Figure 28. Inspection of underside of tray with moving caged valves in coker fractionator

Fixed valves are strongly recommended for coker fractionator service. PROVALVE® tray technology has been installed in over 25 coker fractionators worldwide, in towers ranging from 6.5 ft to 25 ft diameter. The following pictures show PROVALVE® trays that were inspected after many years of service in a coker fractionator; notice that while there is some slight accumulation of fouling material against the valve cover, the orifice and the area around the orifice are virtually clean.

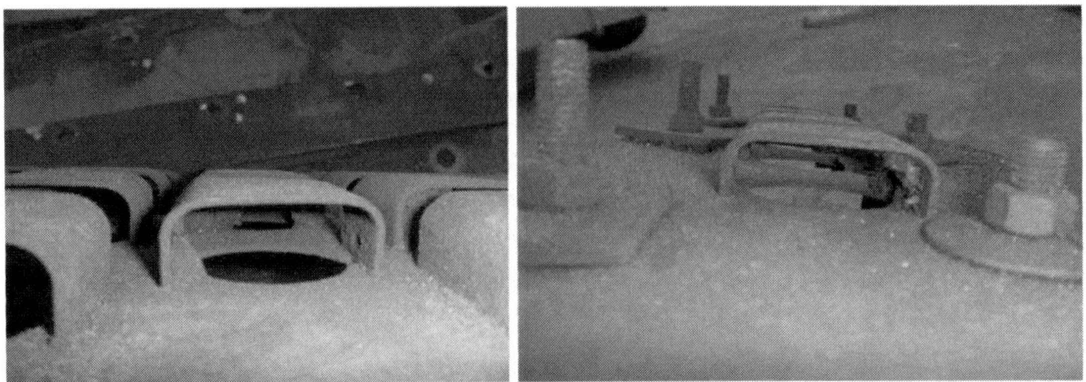

Figures 29 and 30. Inspection of PROVALVE® trays after several years service in coker fractionator

Fixed valve trays, and especially PROVALVE® trays, have been successful choices for refinery services that can be considered severe, where high temperature, fouling, and corrosion are significant concerns. A partial list of these services is shown below in Table 2:

Table 2. Examples of Severe Refining Services

Tower/Location	Problem(s)
Crude tower – top trays	Dewpoint salt formation; acid
Crude tower – pumparounds; sidestrippers	High traffic
Crude tower – gas oil sections	High corrosion potential due to concentration of naphthenic acids
Crude tower – stripping sections	High temperature; high fouling potential
Crude vacuum tower – stripping sections	High temperature; high fouling potential; high coking and polymerization potential
Coker fractionators	High temperature; high fouling potential; high coking and polymerization potential
FCC main fractionators	High temperature; high fouling potential; salt formation
Sour water strippers	High fouling potential

Conclusion

In today's refining world, supply of economic tower feeds must continuously be evaluated, and choosing the correct distillation valve is critical. While moving valves still have their place in refining distillation applications, moving valves should generally be reserved for non-fouling services where high turndowns are absolutely necessary. Consequently, there is a shift in the refining industry from moving valves to fixed valves. Fixed valves offer several key advantages, including higher fouling resistance, durable construction, uplift resistance, higher capacity (in some cases), reasonable turndown ranges, and lower capital cost. The PROVALVE® tray offers superior fouling resistance and capacity boosts (in moderate to high liquid loads) compared to conventional extruded fixed valves and is well suited for severe services.

REFERENCES

1. Fair, J.R., "Historical Development of Distillation Equipment," AIChE Symposium Series No. 235, Vol. 80, 1983.
2. "Soluble Solutions," "Crude Oil Compatibility Testing," Wiehe, I., www.solublesolutions.com/?page_id=14 (accessed April 10, 2015)
3. Hauser, R., and Kirkey, T., "Refinery Tests Demonstrate Fixed Valve Trays Improve Performance In Sour Water Stripper", Koch-Glitsch, LP, AIChE 2003 Spring National Meeting, New Orleans, Louisiana, March 31, 2003.

Optimal Design and Synthesis of Ternary Distillation Systems involving Complex Column Configurations

Chafika Adiche[1], Belaid Ait Aissa [2]

[1] Technical university Darmstadt, Chair of Process Engineering, Otto-Berndt-Straße 2, D-64287 Darmstadt, Germany. [2] Technical University Clausthal, Institute for Thermal Process Engineering, Leibnizstrasse 15, D-38678 Clausthal-Zellerfeld, Germany, present address: Evonik Industries, Germany.

Abstract

The synthesis of the optimal distillation column sequence / design for the separation of multicomponent mixtures is until today a challenging task particularly if complex column configurations are involved. Indeed, the available short-cut methods used for the screening of different complex distillation column alternatives including single-feed columns with a side stream above or below the feed section, columns with side stripper / rectifier, and side stream columns with prefractionator are approximate and have many restrictions. In this paper, distillation sequences including the above mentioned complex column configurations for the separation of ternary mixtures were evaluated with respect to both the total vapor flow rate provided by the reboiler (s) for a given distillation system, and the minimum total number of theoretical stages of the respective column (s). So that the relevant screening criteria required to estimating both operating and capital costs may be considered. For this purpose, new short-cut methods developed recently were used for the determination of optimal design parameters for each investigated complex column configuration. These short-cut methods are based on (i) the assumptions of (1) mixtures with constant relative volatilities and (2) constant molar flow rates within each mass transfer section, (ii) the characterization of each mass transfer section by its Eigenfunction and (iii) the identification of all possible pairs of constant composition zones in a column including both those around the feed sections and those around the side-stream sections. Furthermore, the minimum total number of theoretical stages for each column configuration was established with taking into account the corresponding separation geometry as well as the mass balance around each feed section. The design tools mentioned previously were successfully implemented for the assessment of optimal distillation systems for the separation of ideal behaving ternary mixtures as well as ternary azeotropic mixtures approximated by pseudo-ideal mixtures.

1. Introduction

The Available design methods like the empirical correlations of Fenske-Gilliland-Kirkbride [1] and the boundary value method [2] based on geometrical and graphical approaches are approximate design methods. Their incorporation within a systematic procedure to screening different distillation column designs is not reliable, particularly if complex configurations are involved. In this work, we show that the new short-cut methods developed recently [3-5] for different complex column configurations [Figure 1] are well suitable for the synthesis of optimal distillation systems for the separation of both ideal behaving and azeotropic ternary mixtures.

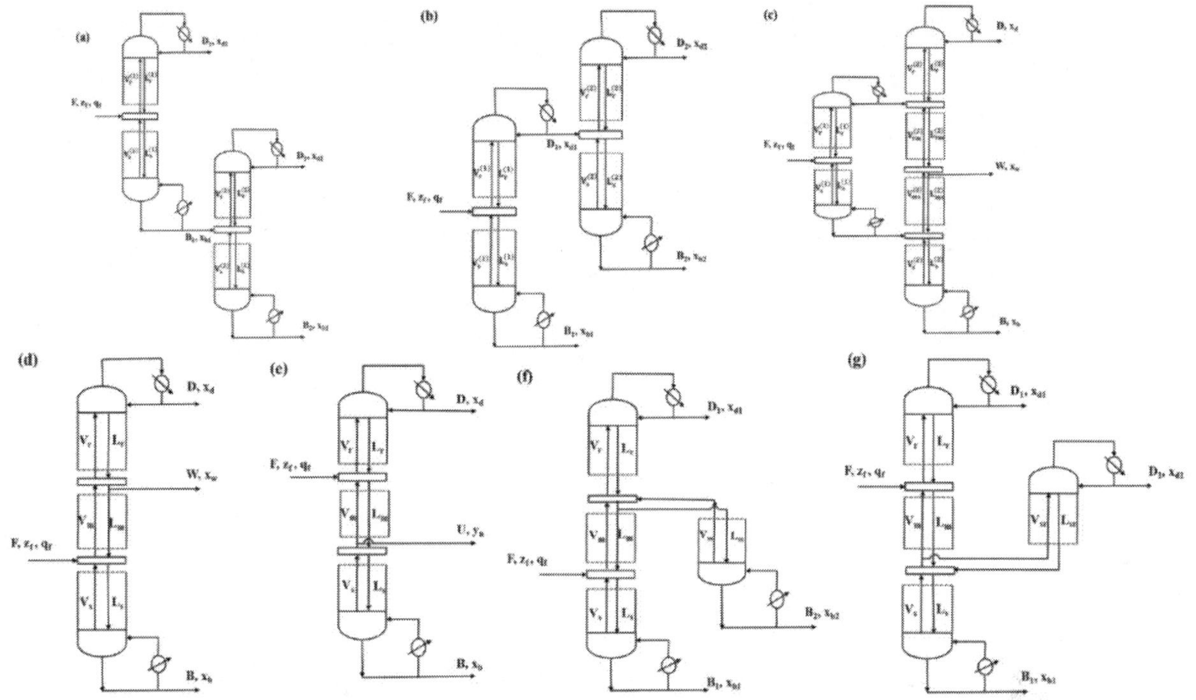

Figure 1. Basic distillation column designs for the separation of a ternary mixture

2. Theoretical background

We assume constant relative volatilities α and constant molar flow rates within each mass transfer section. Furthermore, each mass transfer section k in a column, is defined by its characteristic equation also called Eigenfunction [3] given by

$$\sum_1^3 \frac{\alpha_i . a_{k,i}}{\alpha_i - r_k . E_k} = 1, \tag{1}$$

where $r_k = L_k / V_k$ is the internal reflux ratio in section k and $a_{k,i}$ is the composition of the net flow for component i in section k.

The three Eigenvalues E_k of section k are determined by solving eq. (1); allowing thereby the calculation of the composition of the respective pinch points delimiting the distillation space [3] of section k according to

$$M^L_{k,i,j} = \frac{a_{k,i}\, E_{k,j}}{\alpha_i - r_k\, E_{k,j}}, \ i = 1,..,3, \ j = 1,..,3 \tag{2}$$

and $M^V_{k,i,j} = r_k M^L_{k,i,j} + a_{k,i}$, i =1,.., 3, j =1,.., 3 (3)

for the liquid and vapor phase respectively.

The nature of each pinch point i.e. stable node, unstable node or saddle in a section k, may be established graphically by tracing the field of separation driving force vectors $y_i^* - y_i$ in section k [4]. Figure 2 shows the field of separation driving force vectors for the rectifying section in a simple column. All vectors $y_i^* - y_i$ close to the left node are pointing away from this node, the vectors near the top node are pointing partially towards and partially away from this node, whereas all the vectors near the right node are pointing towards the node. Accordingly, the former pinch point is an unstable node, the intermediate one is a saddle and the last one is a stable node.

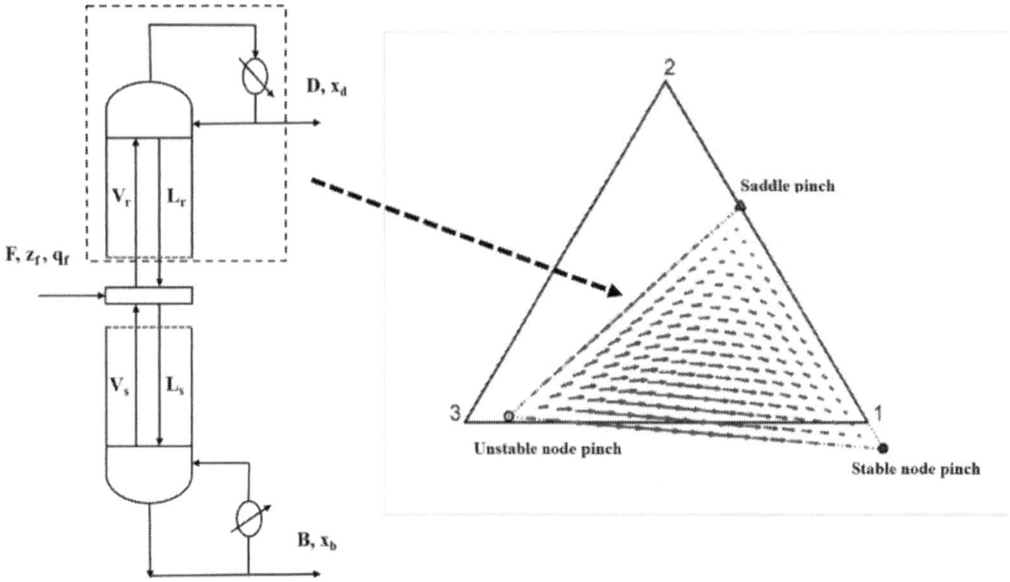

Figure 2. Field of separation driving force vectors for a rectifying section, $\alpha = [6.4\ 2.7\ 1]$, $q_f = 1$, $D/F = 0.28$, $V_s/F = 0.8$, $z_f = [0.3\ 0.4\ 0.3]$, , $x_d = [0.95\ 0.0499\ 10^{-4}]$.

For a given initial point (stream) in section k with a composition x_0, the distillation line starting at x_0 and constituting the composition profile in section k is given by

$$\frac{X_{k,2}}{X_{k,1}} = C_k \cdot \left(\frac{X_{k,3}}{X_{k,1}}\right)^{e_k}, \quad C_k = \frac{X_{0,k,2}/X_{0,k,1}}{\left(X_{0,k,2}/X_{0,k,1}\right)^{e_k}}, \quad e_k = \frac{\ln E_{k,1} - \ln E_{k,2}}{\ln E_{k,1} - \ln E_{k,3}}, \quad \sum_{i=1}^{3} X_{k,i} = 1,$$

with $x = M_k^L X$, $y = M_k^V Y$, and $X_{0,k} = (M_r^L)^{-1} x_0$. (4)

Let us consider a split lkc / hkc taking place in a mass transfer section k, where lkc is the light component and hkc is the heavy key component. It results that the number of theoretical stages at operating conditions in this section writes

$$N_k = \frac{\ln\left[\left(\frac{X_{k,lkc}}{X_{k,hkc}}\right)\cdot\left(\frac{X_{k,hkc,0}}{X_{k,lkc,0}}\right)\right]}{\ln\left(\frac{E_{k,lkc}}{E_{k,hkc}}\right)}. \quad (5)$$

3. Shortcut methods

The short-cut methods used in this work for the conceptual design of all investigated distillation column designs [Figure 1] are based on three main procedures [3]: (i) initialization procedure for the calculation of the design parameters at pinch conditions along with the corresponding separation geometry, and the establishment of the initial location of the feed section (s), (ii) optimization procedure for the calculation of the minimum number of total theoretical stages in the whole column sequence / design by adjusting the appropriate design parameters (Eq. (5)), and finally (iii) the calculation of the composition profiles in a column sequence / design by means of the analytical formulation of the distillation line (Eq. (4)) for each mass transfer section.
The design parameters at pinch conditions were determined by means of a new generalized design approach [4]. The latter is based on the analytical characterization of each mass transfer section (Eq. (1)), as well as on the identification of all possible pairs of constant composition zones in a column including both those around the feed sections and those around the side-stream sections. Accordingly, no conceptual splitting [6-7] of a column with side stripper / rectifier in a sequence of simple columns is

required. Furthermore, the analysis of the geometry of sharp separations for a column with a side stripper / rectifier provides the optimal composition of the column side stream (feed of the side section) at operating conditions [5].

4. Results and discussion

4.1. Ideal behaving mixtures

The short-cut methods presented previously, were used for the calculation of optimal design parameters for all investigated column configurations in this work. For this purpose, operating reflux / boil-up flow rates taken as 1.2 times their minimum values were set for the separation of the ideal behaving mixture of benzene(1), toluene(2) and O-xylene(3). The composition profiles at optimal design and operating conditions for the studied complex distillation columns are depicted in Figures 3 (a) to (e).

The comparison between the design parameters obtained for all column configurations [Table 1] – including direct and indirect sequences (configurations a and b respectively)- shows that, for almost pure component products, the column with a side rectifier constitutes the optimal configuration in terms of both operating and capital costs. In the case however, if the specification on the intermediate component is not very strict (i.e. a pure intermediate product is not required), the single-feed column with a side stream above the feed section will be the most economical.

The design parameters calculated for each column design [Figure 1], were successfully validated by comparing them with rigorous simulation results using RadFrac (Aspen plus) for the column model and Wilson equations for the calculation of thermodynamic properties. Indeed, for each column design, the calculated design parameters were introduced into the corresponding elaborated distillation model with RadFrac (Aspen plus). The accordingly obtained product compositions show a very good agreement with those obtained by the short-cut methods for all investigated distillation column configurations, as illustrated in Table 2.

4.2. Azeotropic mixtures

The design approach presented in this work was extended to the synthesis of optimal distillation systems separating ternary azeotropic mixtures. For this purpose, azeotropic mixtures were approximated by appropriate pseudo-ideal mixtures. This approximation approach requires (i) the knowledge of the thermodynamic structure of the azeotropic mixture along with the corresponding residue curve map, and (ii) the analytical characterization of the relevant distillation region where the separation is occurring. Accordingly, an appropriate set of relative volatilities of the singular points (pure component (s) and azeotrope (s)) forming the corresponding pseudo-ideal mixture may be determined in a transformed system of compositions by means a rigorous thermodynamic model calculated at feed composition [3]. In this work, separation examples of the highly non-ideal azeotropic ternary mixture of methanol, isopropanol and water were investigated. The activity coefficients and the vapor pressures were calculated using Wilson model and Antoine equation, respectively. The corresponding data were taken from the DECHEMA liquid-vapor equilibrium data collection [8]. This mixture (methanol, isopropanol and water) exhibits a minimum boiling temperature azeotrope constituted of isopropanol and water; giving rise to two distillation regions separated by a simple distillation boundary line [Fig. 4 (a)]. For all considered column configurations, the separation of the mixture of methanol, isopropanol and water (3) is taking place inside the distillation region delimited by methanol, azeotrope and water as illustrated in Fig. 4 (b).

As compared to ideal behaving distillation systems, the design parameters obtained for the azeotropic distillation systems tested in this work, do not present very good agreement with those obtained by rigorous simulation [Tables 3 and 4]. However, they provide good initial values for the assessment of optimal column design using rigorous simulation. In this case, we found that the single-feed column with a side stream below the feed section (configuration e) is the optimal column configuration in terms of both capital and operating costs. Furthermore, as compared to a single-feed with a side stream above the feed section (configuration d), the composition of the side stream for this configuration (e) is closer to that of the azeotrope; allowing thereby a direct recycling in the feed.

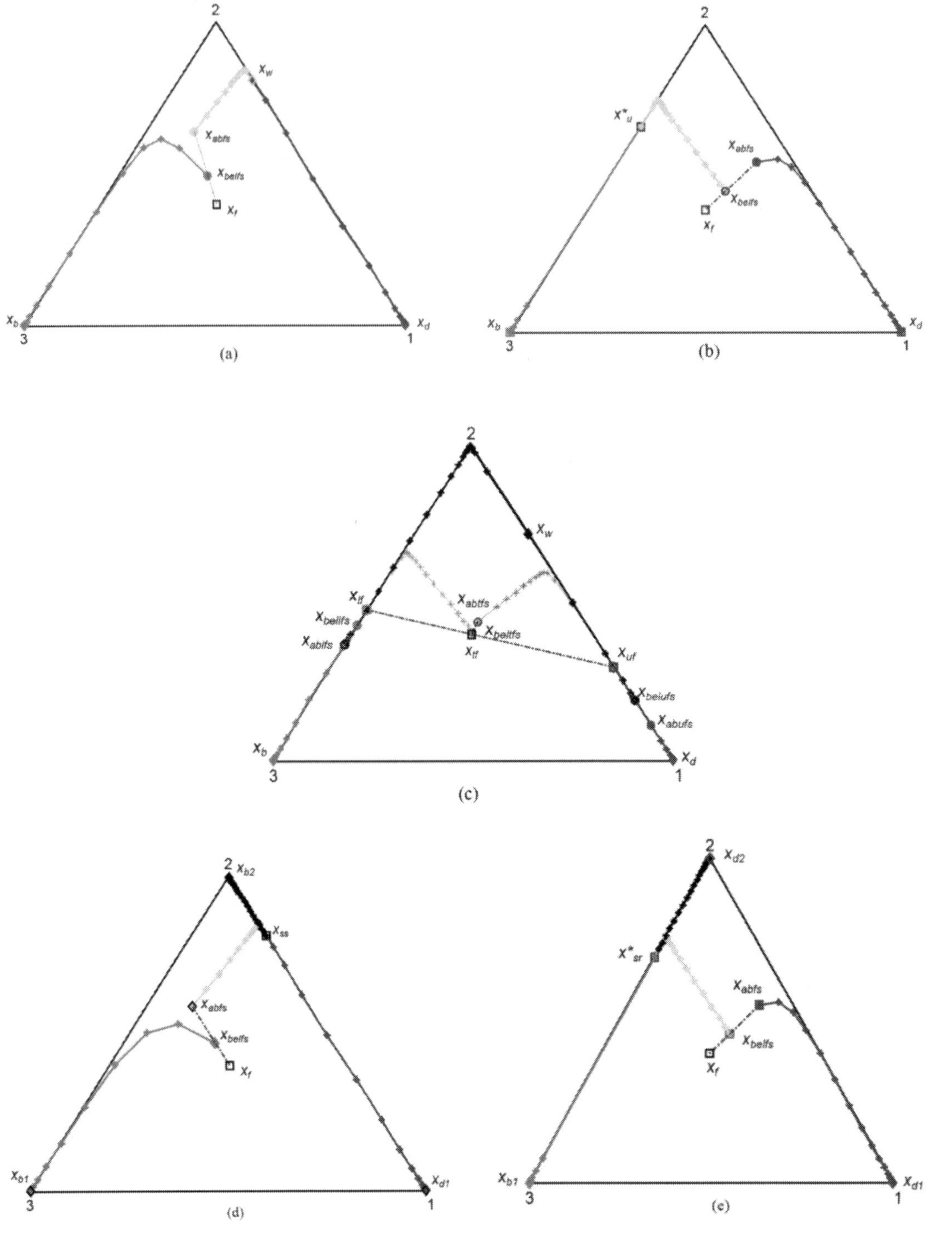

Figure 3. Composition profiles at optimal design and operating conditions in (a) a single-feed distillation column with a side stream above the feed section, (b) a single-feed distillation column with a side stream below the feed section, (c) a side stream column with a prefractionator, (d) a distillation column with a side stripper and (e) a distillation column with a side rectifier, ideal behaving ternary mixture of benzene (1), toluene (2) and o-xylene (3), $\alpha = [6.4\ 2.7\ 1]$, $z_f = [0.3\ 0.4\ 0.3]$, $q_f = 1$.

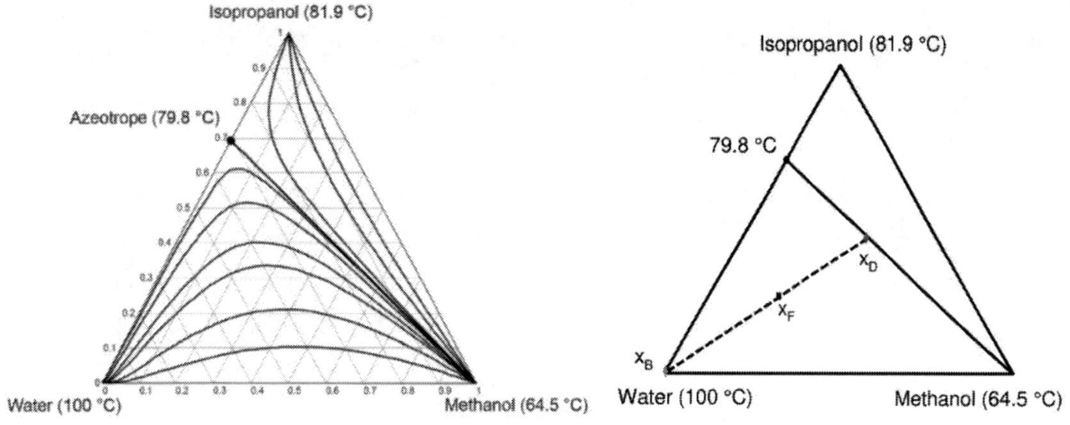

(a) (b)

Figure 4. (a) Residue curve map of the azeotropic mixture of methanol, isopropanol and water, (b) distillation region for $x_f = [0.20\ 0.25\ 0.55]$.

Table 1. Design parameters for different column configurations for an ideal behaving mixture

	a	b	c	d	e	f	g
$V_s^{(1)}$	0.96	1.35	0.65	1.35	1.43	1.35	1.67
$V_s^{(2)}$	0.90	0.91	0.73				
W / U			0.56	0.49	0.47		
V_{ss} (L_{ss})						0.49 (0.89)	
L_{sr} (V_{sr})							0.37 (0.77)
$N_s^{(1)}$	26	12	16	13	10	12	11
$N_r^{(1)}$	15	21	12	12	17	14	17
$N_s^{(2)}$	14	18	11				
$N_r^{(2)}$	17	17	7				
$N_m^{(1)}$				13	22	18	19
$N_{ms}^{(2)}$			24				
$N_{mr}^{(2)}$			13				
N_{ss} / N_{sr}						31	21
ΣV_s	1.86	2.26	1.38	1.35	1.43	1.84	1.67
N_t	72	68	83	38	49	67	68

Table 2. Product compositions for different column configurations for an ideal behaving mixture

		Product 1		Product 2		Product 3	
Configuration	Component	This work	Radfrac	This work	Radfrac	This work	Radfrac
a	Benzene	0.999	0.999	0.0001	trace	0.0009	10 PPM
	Toluene	0.0009	0.001	0.0009	0.001	0.999	0.998
	O-xylene	0.0001	Trace	0.999	0.999	0.0001	0.002
b	Benzene	0.0001	33 PPB	0.999	1.000	0.0009	242 PPM
	Toluene	0.0009	596 PPM	0.0009	322 PPM	0.999	0.999
	O-xylene	0.999	0.999	0.0001	Trace	0.0001	447 PPM
c	Benzene	0.999	0.999	0.0001	Trace	0.2802	0.2780
	Toluene	0.0009	0.0001	0.0009	0.004	0.7193	0.7117
	O-xylene	0.0001	Trace	0.999	0.996	0.0005	0.0102
d	Benzene	0.999	0.999	0.0001	10 PPB	0.1891	0.184
	Toluene	0.0009	819 PPM	0.0009	0.001	0.8104	0.815
	O-xylene	0.0001	Trace	0.999	0.999	0.0006	633 PPM
e	Benzene	0.999	1.000	0.0001	Trace	0.0001	67 PPM
	Toluene	0.0009	105 PPM	0.0009	0.003	0.8462	0.850
	O-xylene	0.0001	Trace	0.999	0.997	0.1537	0.150
f	Benzene	0.999	1.000	0.0001	122 PPB	0.0009	0.005
	Toluene	0.0009	192 PPM	0.0009	0.007	0.999	0.995
	O-xylene	0.0001	Trace	0.999	0.993	0.0001	27 PPM
g	Benzene	0.999	1.000	0.0001	Trace	0.0009	316 PPM
	Toluene	0.0009	421 PPM	0.0009	0.002	0.999	0.998
	O-xylene	0.0001	Trace	0.999	0.998	0.0001	0.001

Table 3. Design parameters of different column configurations for a ternary azeotropic mixture

	a	b	d	e	f	g
$V_s^{(1)}$	1.17	1.53	1.54	1.65	1.54	2.01
$V_s^{(2)}$	1.20	1.03				
W / U			0.42	0.48		
V_{ss} (L_{ss})					0.68 (1.04)	
L_{sr} (V_{sr})						0.58 (0.94)
$N_s^{(1)}$	31	18	21	14	17	16
$N_r^{(1)}$	21	26	18	22	19	22
$N_s^{(2)}$	20	26				
$N_r^{(2)}$	19	23				
$N_m^{(1)}$			21	33	23	28
N_{ss} / N_{sr}					42	24
ΣV_s	2.37	2.56	1.54	1.65	2.22	2.01
N_t	91	93	60	69	93	90

Table 4. Product compositions for different column configurations for a ternary azeotropic mixture

		Product 1		Product 2		Product 3	
Configuration	Component	This work	Radfrac	This work	Radfrac	This work	Radfrac
a	Methanol	0.999	0.897	0.0001	trace	0.0009	0.055
	Isopropanol	0.0009	0.090	0.0009	trace	0.6911	0.642
	Water	0.0001	0.013	0.999	1.000	0.3080	0.303
b	Methanol	0.0001	trace	0.999	0.999	0.0009	408 PPM
	Isopropanol	0.0009	trace	0.0009	735 PPM	0.6911	0.692
	Water	0.999	1.000	0.0001	383 PPB	0.3080	0.307
d	Methanol	0.9990	0.983	0.0001	0.028	0.1593	0.119
	Isopropanol	0.0006	0.016	0.0006	0.011	0.5815	0.579
	Water	0.0004	590 PPM	0.9993	0.962	0.2592	0.302
e	Methanol	0.9990	0.994	0.0001	trace	0.0001	0.064
	Isopropanol	0.0006	0.006	0.0006	trace	0.5177	0.519
	Water	0.0004	55 PPM	0.9993	1.000	0.4822	0.417
f	Methanol	0.9990	0.998	0.0001	trace	0.0009	969 PPM
	Isopropanol	0.0006	0.002	0.0006	trace	0.6911	0.693
	Water	0.0004	7 PPM	0.9993	1.000	0.3080	0.306
g	Methanol	0.9990	0.999	0.0001	64 PPM	0.0009	0.113
	Isopropanol	0.0006	0.001	0.0006	0.062	0.6911	0.618
	Water	0.0004	3 PPM	0.9993	0.937	0.3080	0.268

5. Conclusion

In this work, we investigated the synthesis of optimal distillation systems involving complex column configurations for the separation of ternary mixtures. For all considered column designs including complex configurations, the new synthesis approach requires a minimum number of specifications, is based on the distillation theory of ideal mixtures, and on the analysis of the separation geometry for each column design. This approach was tested successfully for the assessment of optimal distillation system separating both ideal behaving and azeotropic ternary mixtures.

Literature

(1) Henley, E.J. and Seader J.D., 1981. Equilibrium-stage Separation Operations in Chemical Engineering, John Wiley & Sons, Inc.

(2) Rooks RE, Malone MF, Doherty MF. A Geometric Design Method for Side-Stream Distillation Columns. Ind. Eng. Chem. Res. 1996, 35, 3653-3664.

(3) Adiche C and Vogelpohl A. Short-cut Methods for the Optimal Design of Simple and Complex Distillation columns. Chem. Eng. Res. Des. Part A. 2011, 89, 1321-1332.

(4) Adiche C. and B. Ait Aissa, A Generalized Approach for the Conceptual Design of Distillation Columns with Complex Configurations, (accepted, Chem. Eng. Res. Des. Part A., 2015).

(5) Adiche C. and B. Ait Aissa, A New short-cut Method for the optimal design of Distillation columns with Side Sections (submitted, 2015).

(6) Glinos, K.N. and Malone, M.F., 1985. Minimum Vapor Flows in a Distillation Column with a Sidestream Stripper, Ind. Eng. Chem. Process Des. Dev. 24, 4, 1087-1090.

(7) Carlberg NA and Westerberg AW. Temperature-Heat Diagrams for Complex Columns. 2. Underwood's Method for Side Strippers and Enrichers. Ind. Eng. Chem. Res., 1989, 28, 1379-1386.

(8) Gmehling, J.; Onken, V.; Arlt, W., since 1977. Vapour-liquid equilibrium data collection, Dechema Chemistry Data Series, Frankfurt/M.

CH2MHILL.

Crude Unit Heat Integration – Best Practices

Andrew W. Sloley – CH2M HILL, Bellingham, WA

Prepared for Presentation at the AIChE 2015 Spring Meeting, Austin, TX, 26-30 April, 2015

Best Practices and Economics

2

Background

- Working and publishing on
  - Fundamentals
  - Heat integration types
  - Overall crude distillation sequence optimization
  - Revamps
  - Preflash drums
  - Preflash towers
  - Complex configurations
  - Crude types
  - Condensate splitters
- Always struggling with
  - What is the best base configuration for the comparison?

CH2MHILL.

Why Is the 'Best' Configuration Important?

- Examining changes
- Compare the 'Best option A' to 'Best option B'
- Helps understand fundamentals of crude units
- Poor comparisons give poor solutions

CH2MHILL.

Comparisons Tried

- Constant yields
- Constant separations
  - D86 or D2892 gaps-overlaps
- Constant heat recovery
- Always try to compare 'best' versus 'best'
  - Basis changes results
  - How to justify a basis?
- What is the tradeoff?
  - Energy recovery versus fractionation
  - Investment differences

CH2MHILL.

The Approach

Take a Different Approach

- Define problem boundary
- Best separation
- Best energy recovery
- Understand what moves the optimum between the boundaries
- Basic technical tradeoff defined by Packie in 1941
- Technical solution must support economic solution

- Packie does not include
  - Explicit energy and yield
  - Equipment Limits
- No differential value to heat at varied temperatures
- Maximum boundary – max heat recovery at every level
  - Limited by liquid rate inside tower
- Minimum boundary – zero heat recovery
  - All duty to overhead

Packie, J. W. Distillation equipment in the oil refining industry. *Trans AIChE*, 37 (1), Feb 25, 1941: 51-78.

CH2MHILL

Heat Recovery Boundary – Upper Limit

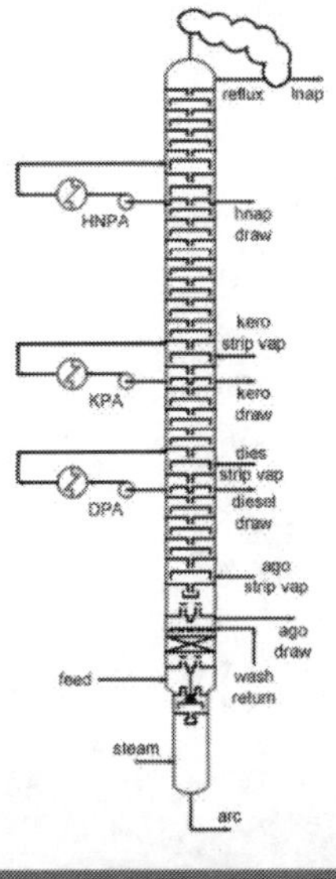

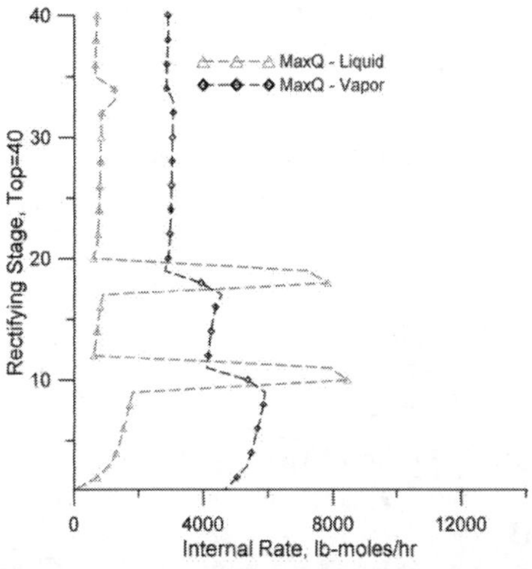

CH2MHILL

Heat Recovery Boundary – Lower Limit (Zero)

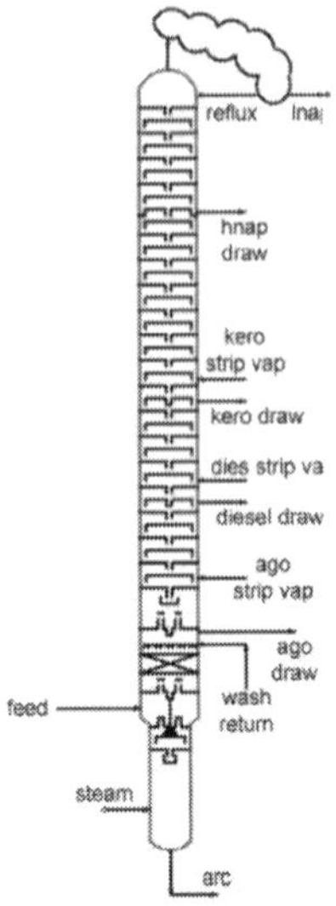

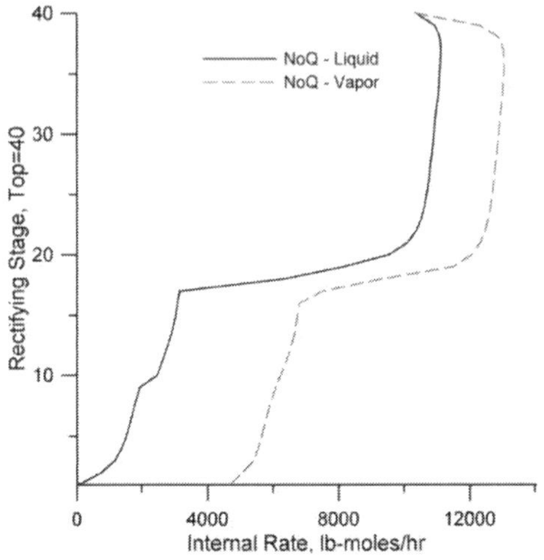

Heat Recovery versus Fractionation

- Maximum heat recovery =
- Minimum fractionation

- Minimum heat recovery =
- Maximum fractionation

- What is heat recovery worth?
- What is fractionation worth?
- What are investment differences?

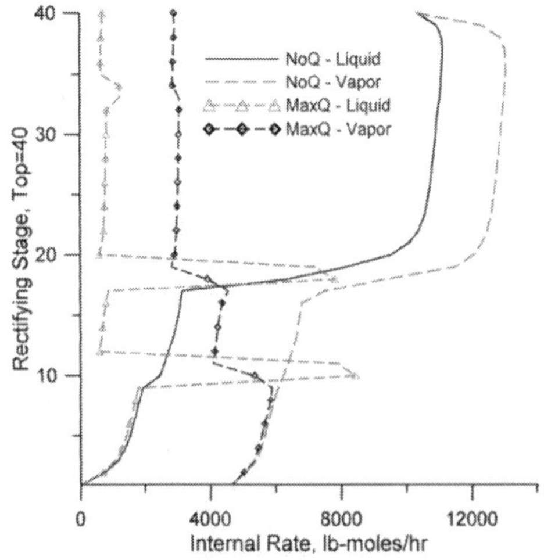

Problem Approaches

- Many documented approaches
- Either too general
  - Usually ended up with no clear answer
  - But perhaps this is the reality?
- Or too specific
  - Many case studies
  - But how often are special cases repeated?

Problem and Solution

- Problem

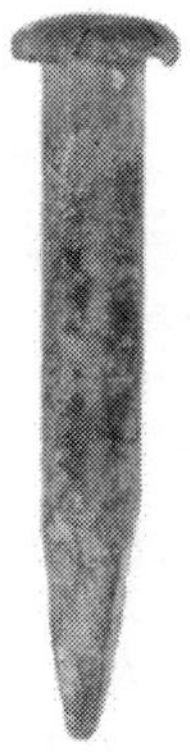

- Solution

Solution

- Yields
- Heat balance
- Sizing

- Sizing
- Local optimization

- Economics
- Analysis

Simulation (and Some Optimization)

- ~620,000 cases
- Define heat balance
  - Minimum liquid rates constrain upper heat recovery
- Determine yields
- Equipment
  - Size tower – sets minimum liquid rates – back to heat balance
  - Size exchanger – preliminary
  - Size pumps

Optimization

- Local optimization
- Equation solving to
  - Minimize exchanger area
  - Limit number of exchanger shells
  - Balance metallurgy costs of circuits
  - Balance fouling at different rates
- Recalculate heat recovered

Economics

- Probabilistic analysis
  - Product values
  - Heat values
- Started with current values
- Examined historical values
- Uses discounted cash flow (DCF) to select 'best'
- Net present value (NPV) may give different results
- DCF preferred when alternate investments are available

Starting Point

- Tia Juana Light crude
- 31.6° API (0.8676 SG)
- Close to average US crude
- 100,000 bpd (6,290 cmpd)
- Close to average US refinery size
- 4 $/Mbtu (3.79 $ GJ) gas price

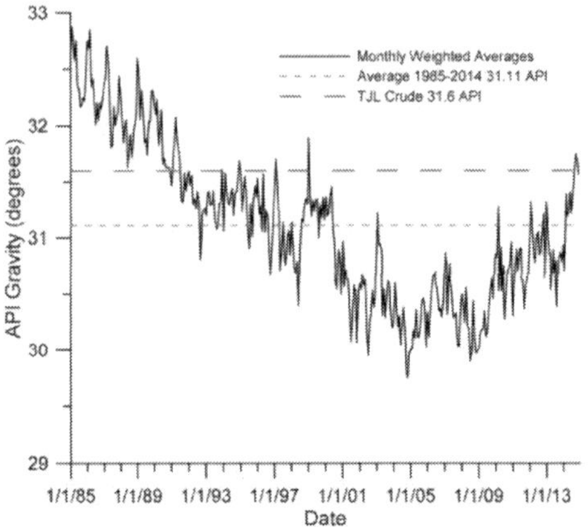

CH2MHILL.

Product Values

- Energy Information Agency data on prices
  - Gasoline (proxy for naphtha)
  - Kerosene (jet) (proxy for kerosene)
  - Diesel (proxy for diesel)
- Simplified model to estimate
  - Gas oil
  - Atmospheric residue values
  - Based on 10-process unit linear program for an 'average' refinery

CH2MHILL.

EIA Based Margins

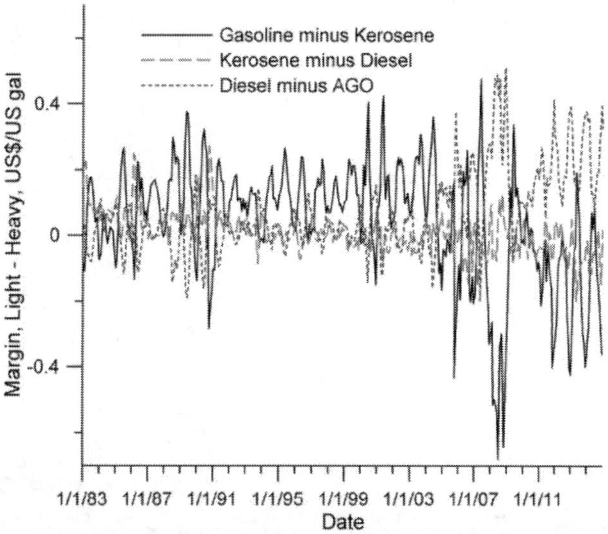

CH2MHILL.

19

Current Results – Work Ongoing

20

Fractionation Stages versus Reflux

- Preliminary work showed very strong driver towards maximum trays
  - Maximum trays = minimum reflux
  - Minimum reflux = minimum energy downgrade
  - True even with low energy price level of $4 per Mbtu gas price (3.79 $ GJ)
  - More expensive energy makes this driver even more powerful
- ~15 trays is economic limit
- Same result as logic leading to ASTM D2892 (TBP)

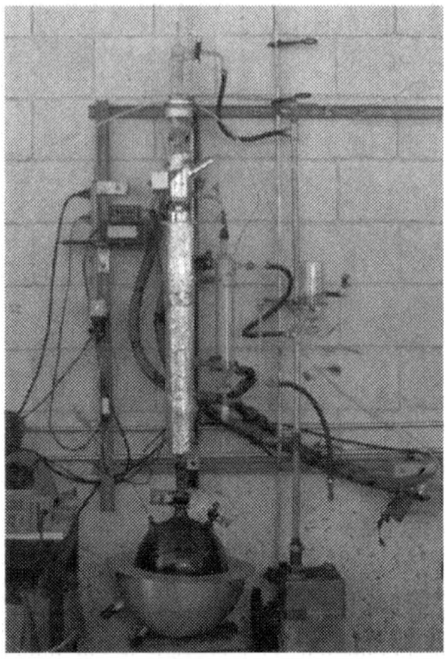

Splits

- Naphtha versus heavy naphtha fractionation
  - Nearly any differential value for split drives toward 15 trays
- Heavy naphtha versus kerosene split
  - Nearly any differential value for split drives toward 15 trays
- Kerosene versus diesel split
  - Overlap between products makes economics of more trays less powerful
  - Higher energy prices still drive units to 15 trays
- Diesel versus atmospheric gas oil split
  - Liquid wetting limits potential performance of trays
  - Packed sections become attractive at high kerosene-diesel splits

Heat Removal Creates?

- Product at that heat removal level
- Reflux for fractionation below

- Heat shifts result from
  - Reflux changes
  - Product rate (specification) changes

Define Limits

		Min Q	Max Q				
Start	$/Mbtu						
End	$/Mbtu						
HNPA	Mbtu/hr	0	5				
KPA	Mbtu/hr	0	78				
DPA	Mbtu/hr	0	78				

- Heavy naphtha pumparound duty low – no heavy naphtha product

Zero Differential Product Values

		Min Q	Max Q	Zero Val			
Start	$/Mbtu						
End	$/Mbtu						
HNPA	Mbtu/hr	0	5	33			
KPA	Mbtu/hr	0	78	46			
DPA	Mbtu/hr	0	78	77			

- Duty shift from KPA to HNPA
- Shift reflects reducing stripping steam in kerosene sidestripper versus differential value of heat based on temperature

Heat Shifts on Tower

+Light product value

+Heavy product value

Energy value

Economic Operating Zones – Duty Shift Direction

	Kero > Diesel	Kero = Diesel	Kero < Diesel
Nap > Kero	?	?	?
Nap = Kero	?	-	↓
Nap < Kero	?	↓	↓

- Higher values on heavy products reinforce energy values

Zero Differential Product Value = Base Case

		Min Q	Max Q	Zero Val			
Start	$bbl/Mbtu						
End	$bbl/Mbtu						
HNPA	Mbtu/hr	0	5	**33**			
KPA	Mbtu/hr	0	78	**46**			
DPA	Mbtu/hr	0	78	**77**			

Increase Kerosene Value

- Duty quickly shifts from diesel pumparound to kerosene pumparound
- Shift starts at 0.70 $ bbl / Mbtu (0.02 $ gal / $ Mbtu)
- Shift finished at 1.05 $ bbl / Mbtu (0.03 $ gal / $ Mbtu)
- Remember energy value is difference between DPA and KPA
- Controlled by overlap on kerosene-diesel products
- For 1/83-1/15, differential value of 1.05 $ bbl / Mbtu exceeded 20% of the time
  - Current trends favors diesel
- An extra shift of 2.10 $ bbl / Mbtu in favor of kerosene would create a situation exceeded 80% of the time (0.05 $ gal / $ Mbtu)

CH2MHILL.

Kerosene Value Plateau

		Min Q	Max Q	Zero Val	High K		
Start	$bbl/Mbtu				0.70		
End	$bbl/Mbtu				1.05		
HNPA	Mbtu/hr	0	5	33	29		
KPA	Mbtu/hr	0	78	46	124		
DPA	Mbtu/hr	0	78	77	0		

CH2MHILL.

Increase Naphtha Value

- Relatively high differential value required to shift duty levels
- Occurs in two plateaus
- First plateau starts at 8.40 $ bbl / Mbtu (0.20 $ gal / $ Mbtu)
- Plateau finished at 10.15 $ bbl / Mbtu (0.24 $ gal / $ Mbtu)
- Controlled by blending on naphtha-kerosene split
- Shifts KPA duty to HPA
- For 1/83-1/15, this value was exceeded 3% of the time
- Not credible flexibility case
- Second plateau starts at 13.65 $ bbl / Mbtu (0.32 $ gal / $ Mbtu)
- Shifts KPA duty to overhead
- Not credible flexibility case

CH2MHILL.

Flexibility Range

		Min Q	Max Q	Zero Val	High K	Plat 1	Plat 2
Start	$bbl/Mbtu				0.70	8.40	13.65
End	$bbl/Mbtu				1.05	10.15	15.40
HNPA	Mbtu/hr	0	5	**33**	**29**	48	48
KPA	Mbtu/hr	0	78	**46**	**124**	102	94
DPA	Mbtu/hr	0	78	**77**	**0**	0	0
				Bound 1	**Bound 2**		

- Current economics strongly favor zero value
- Some flexibility to shift DPA to KPA might be useful

CH2MHILL.

Loading Comparison

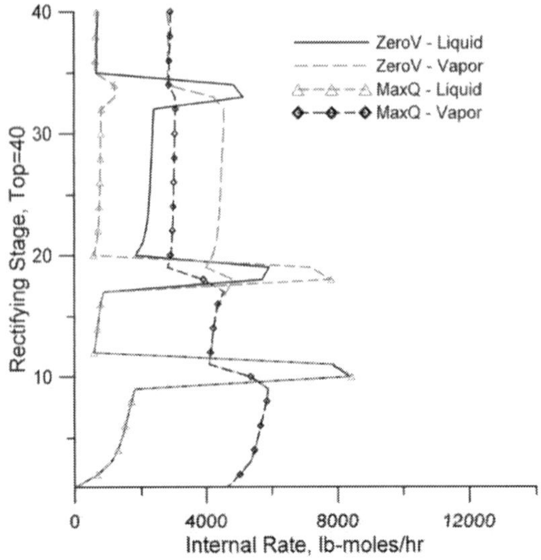

CH2MHILL.

Implications of Loading Comparison

- Reasonable liquid rates between heavy naphtha and kerosene – trays work fine
- Low liquid rate just above diesel pumparound – tray wetting limits heat recovery

CH2MHILL.

Sensitivity Analysis

				Zero Val		Error	Error
						Mbtu/hr	%
HNPA	Mbtu/hr			**33**		2 – 5	<15
KPA	Mbtu/hr			**46**		2 – 5	<11
DPA	Mbtu/hr			**77**		2 – 5	<6

CH2MHILL.

Strategy

- Maximize recovered energy to minimum liquid loads
- Check kerosene steam versus HNPA duty economics
- Check KPA versus DPA heat recovery at high kerosene values

- Depends upon local economics
- Revamps often plant specific

CH2MHILL.

What Moves Economics

- Higher energy values strongly shift energy recovery to lower down the tower
- Freeze-cloud specifications shift energy recovery slightly to more up the tower
- Lighter crudes shift energy recovery to lower down the tower
  - More lighter products made
  - More low level heat available
  - Heat recovery pinches against LMTD at low temperatures
  - Increases value of high temperature heat

CH2MHILL.

How Economics Move

- Moves in plateaus
- Relatively short ranges of values cause heat to shift
- Then heat stays close to constant for a large change
- Plateaus are roughly 4 times wider than shift ranges
- If economics are in a plateau, the exact values of products do not change answers
- If economics are in the shifting regions, small economics value changes create large changes in optimum heat allocation
- Example, at 4 $US / Mbtu (3.79 $US / GJ), looking at naphtha more valuable than kerosene
  - Economics flat for energy recovery for naphtha up to 0.24 US$ /gal
  - Economics shift rapidly from 0.24 to 0.29 US$/gal n-k change
  - Economics flat for energy recovery for naphtha from 0.30 to 0.44 US$ /gal
  - Economics shift somewhat from 0.44 to 0.49 US$/gal

CH2MHILL.

Open Questions – Future Work?

1. Three naphtha-kerosene split options
   - More main column reflux versus imported steam to sidestripper (used here)
   - Reboiled sidestripper versus steam sidestripper
   - Imported steam versus steam geneator inside the crude unit
2. Economies of scale
   - Very small units – modular plants for light crudes
   - Very large units – 250,000 bpd and larger
3. No kerosene products
   - Petrochemical naphtha product
   - Regional markets

- Some preliminary work has been done on #2. Will be completed at some point
- Further investigation of #1 and #3 not planned

CH2MHILL.

Questions ?

Revisions

- 2015-04-22: Rev 0
- 2015-04-30: Rev 4 – for handouts
  - Many thanks for the audience at the AIChE presentation of this paper on 2015-04-30 for their many comments.
  - Clarifications for questions asked during the meeting have been added.

CH2MHILL.

Fluidized packed beds (FPB) – an alternative for process gas treatment in "dirty" systems?

Dr. J. F. Maćkowiak[1], *Prof. Dr. J. Maćkowiak*[1], *Dr. J. Szust*[1]

[1]*ENVIMAC Engineering GmbH, 46149 Oberhausen, Germany*

Abstract

Fluidized packed beds might be an alternative for gas-liquid separation processes, when columns with trays or packings fail due to the high solid content of the phases. Literature on fluidized packed bed is scarce thus the aim of this work is to present the results of extensive experimental investigation of fluidized beds. Hydraulic behavior and mass transfer of a fluidized bed have been investigated in a two-tray test column with a diameter of $d_C = 0.6$ m. The hydraulics of fluidized beds with a packing made of polypropylene have been investigated using the air/water system under ambient conditions in the gas load range up to the flooding point and at specific liquid loads between 0 and 80 $m^3m^{-2}h^{-1}$. Beside deriving model equation for prediction of fluidization point and pressure drop in the operating range, the separation efficiency and effective mass transfer area have been determined using the standard absorption test system CO_2/air-1n NaOH and gas controlled mass transfer has been determined using NH_3/air-1 n H_2SO_4 system. The results have been compared with random 50 mm (2") Pall rings. It has been shown, that mass transfer performance of the fluidized packing in the studied range and for the studied systems is in a similar order of magnitude as that of 50 mm Pall rings, but the specific pressure drop of the fluidized packed bed is significantly higher. The design data developed in this work has successfully been used for the design of an industrial plant.

Keywords: Fluidized bed, absorption, fluid dynamics, separation efficiency, column design, dirty systems

1. Introduction

Columns with fixed internals, such as trays, random or structured packings, are well known and widely used in fluid separation processes such as absorption, stripping and distillation. In gas cleaning processes, packed columns are widely used. Especially random packings cover a large part of such applications in industrial practice, especially when solid impurities are present in the liquid or in the gas phase. In these cases random packings have advantages compared to structured packings. However, there are applications where the immobilized packings fail or cause serious operational problems due to their susceptibility to clogging during the treatment of solid loaded media, even using special random packings with large open area, so called lattice type packings, which have been developed for gas-liquid applications with solid impurities [1], see Figure 1.

Figure 1: Fouling in a packed column, even modern lattice type random packings have might have problems in handling very dirty systems.

Fluidized packed beds are discussed as an alternative for gas-liquid operations in systems with a high solid content. Continuous movement of the packings allows self-cleaning of the packing and thus promises stable operation even in systems with very high solid content. Unless the possible advantage in such dirty systems, fluidized packed beds are far less common compared to fixed packed beds, this is partly due to the lack of reliable design data for this type of packed column.

Experimental data of hydraulics and mass transfer characteristics of fluidized beds is scarce in literature and limited to specific packing types and a narrow operating range [2]. In order to provide reliable data for design tasks, the results of an extensive experimental investigation of hydraulics and mass transfer on fluidized beds in a technical scale test plant is presented.

2. Experimental set-up

The ENVIMAC test plant for standardized absorption measurements has been adapted to perform the experimental investigation of the hydraulics and mass transfer characteristics of the fluidized packing. Two trays with a tray spacing of 1 m have been installed inside the column with a diameter of d_C = 600 mm. The static packing height on one tray was H_{stat} = 0,45 m, thus a total static height of 0,9 m was achieved, see Figure 2.

As a fluidized bed packing (FBP), ellipsoidal packings made from polypropylene have been used, the characteristic size of one packing is d_P = 35 mm (FBP 35). The fluidized bed packings have a closed surface and are hollow from the inside. In the test column, various liquid distributors can be tested, such as spray nozzles and pipe distributor whereas gas is flowing from the bottom of the column upwards through the packed bed.

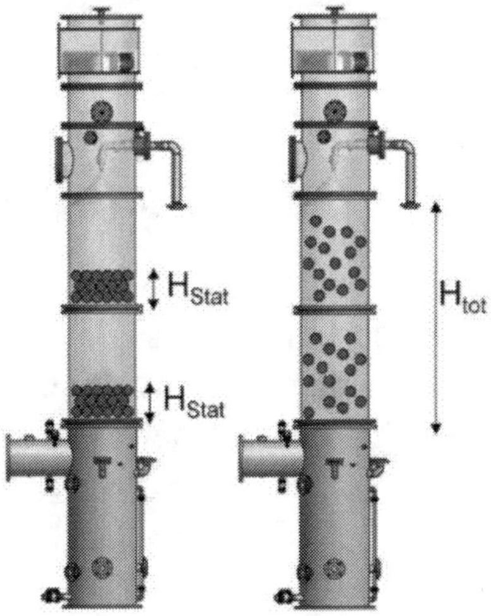

Figure 2: Layout of the test plant with static bed (left side) and fluidized bed (right side).

The column walls of the test plant are made of transparent PVC, allowing the determination of the expanded bed height and the observation of the bed movement during operation. The test column is equipped with state-of the art measurement devices and control technology, minimizing measurement errors and allowing electronic data storage and evaluation.

The characteristic parameters of the studied fluidized bed packing have been summarized in Table 1. The fluidized bed packing (FBP) has an experimentally determined surface area of the static bed of a_{stat} = 120 m^2m^{-3} which is comparable to the widely known 50 mm Pall ring. In order to allow a better comparison of the achieved results, the characteristic data of a 50 mm Pall ring studied in the same test column have been added to Table 1.

Table 1: Parameters of the investigated ellipsoid fluidized bed packing (FBP 35) referred to static bed in comparison to a 50 mm Pall ring.

Parameter	Symbol	Unit	FBP 35	Pall ring 50mm M
Packing size	d_P	[mm]		50
Void fraction	ε_{stat}	[m^3 m^{-3}]	0,44	0,952
Packing density	G_{stat}	[kg m^{-3}]	185	380
Spec. area (stat.)	a_{stat}	[m^2 m^{-3}]	120	110
Packing density	N_{stat}	[m^{-3}]	30.484	6.100

The hydraulics of fluidized beds have been investigated using the air/water system under ambient conditions in the gas load range up to the flooding point and at specific liquid loads between 0 and 80 $m^3m^{-2}h^{-1}$. The separation efficiency of the fluidized bed packing has been studied in the same test plant using the standard test system

CO_2/air-1n NaOH under atmospheric conditions. Based on these results, the effective mass transfer area has been determined. Additionally, the test system NH_3-H_2SO_4/water has been used to determine the mass transfer performance in a gas side controlled system. All these data have been used in a comparison with a 50 mm Pall ring.

3. Hydraulics of fluidized beds

3.1. Results of experimental investigation

The fluidized bed can be operated in both the static mode or in fluidized mode, whereas the latter is the true operating mode. In the static range, gas flows upwards through the bed of packings and liquid flows counter currently not affecting the packing which lays statically on the tray. While increasing the gas load, the packed bed is lifted up after exceeding a specific gas load leading to a highly mixed three phase fluidized zone. The lifting point is called the fluidization point, marking the lower point of the operating range of the fluidized bed. The upper limit or flooding point of the operating range of the fluidized bed is characterized by a gas load where the packing in being thrown out of the column or pressed against the upper tray.

Figure 3 shows the results of the hydraulics investigation of the fluidized packing in the studied range. As can be seen in Figure 3, the pressure drop rises quickly with increasing gas load factor F_V until the fluidization point.

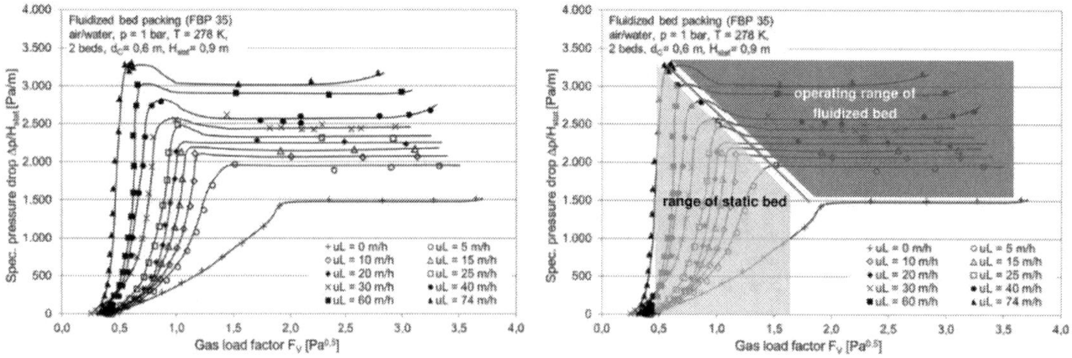

Figure 3: Pressure drop characteristics of the studied fluidized bed packing in dependency of the gas load factor F_V for various liquid loads for 0 to 80 $m^3m^{-2}h^{-1}$ (left side). The diagram on the left side shows the operating range of the static (blue) and fluidized bed (red), the red line dividing the ranges is the line of fluidization points.

Once the fluidization has started and the packing starts to move inside the bed finally reaching its expanded bed height H_{exp}, the pressure drop remains almost constant, even when gas load is further increased. The pressure drop increases again, when the flooding point is reached.

According to Figure 3, the pressure drop increases for higher liquid loads because liquid hold-up h_L rises, leaving less free volume for the gas to pass upwards. In comparison to immobilized packings, the pressure drop of a fluidized bed is almost constant for a given liquid load throughout the entire operation range.

3.2. Modeling of the hydraulics of fluidized beds

For design purposes, a prediction of the fluidization begin and the pressure drop at that point are crucial, as these two parameters determine the most important points of the operation range of a fluidized bed packing. A model equation, based on the SBD model [1], has been derived and presented elsewhere [4], allowing a prediction of the fluidization point based on the assumption, that the fluidization point of the fluidized bed is equal to the flooding point of the static bed. The model equation previously developed for droplet flow in packed columns [1] has been adopted and adjusted to the fluidized bed characteristic data, see eq. (1).

The presented equation allows a prediction of the gas velocity at the fluidization point $u_{V,Fl}$ with an average deviation of 5,9%.

$$u_{V,Fl} = 0,693 \cdot \psi_{Fl}^{-1/6} \cdot \varepsilon^{6/5} \cdot \left[\frac{d_h}{d_T}\right]^{1/4} \cdot \left[\frac{d_T \cdot \Delta\rho \cdot g}{\rho_V}\right]^{1/2} \cdot \left(1 - h_{L,Fl}^0\right)^{7/2} \cdot K_{\rho_V} \quad \left[ms^{-1}\right] \qquad (1)$$

With liquid hold-up at the flooding point $h_{L,Fl}^0$ acc. to eq. (2) for different phase flow ratios λ_0 [1], d_h representing the hydraulic diameter of the packed bed, d_T the droplet diameter acc. to Sauter and $K_{\rho v}$ being a correction factor for gas density [1].

$$h_{L,Fl}^0 = \frac{\sqrt{\lambda_0^2 \cdot (m+2)^2 + 4 \cdot \lambda_0 \cdot (m+1) \cdot (1-\lambda_0)} - (m+2) \cdot \lambda_0}{2 \cdot (m+1) \cdot (1-\lambda_0)} \quad \left[m^3 m^{-3}\right]$$

$$m \text{ acc. to } [1],\ \lambda_0 = \left(u_L / u_{V,Fl}\right) \qquad (2)$$

$$\psi_{Fl_m} \cong 1,66 \text{ (from experiments)}$$

Following the SBD-model, the specific pressure drop can be calculated based on the modeling concept developed for packed columns [1], eq. (3) shows the adjusted relation for the studied fluidized bed packings:

$$\frac{\Delta p}{H_{stat}} = 3,8 \cdot \mu \cdot \frac{1-\varepsilon}{\varepsilon^3} \cdot \frac{F_V^2}{d_P \cdot K} \cdot \left(1 + \frac{h_{L,Fl}}{1-\varepsilon}\right) \cdot \left(1 - \frac{h_{L,Fl}}{\varepsilon}\right)^{-3} \quad \left[Pa\,m^{-1}\right] \qquad (3)$$

With K being the wall factor correction and d_P the particle diameter [1].

Eq. (3) for the determination of the specific pressure drop related to the static packing height allows the prediction of the pressure drop at the flooding point of the static bed and thus in the operating range of the fluidized bed up to ca. 85% of the flooding point with an average accuracy of 2,6%.

Compared to standard random packings, like the 50 mm Pall ring, the specific pressure drop of the fluidized bed packing is approx. 10-20 times higher under the same operating conditions.

4. Mass transfer and effective separation area of FBP 35

As many of the mass transfer applications in gas cleaning technology are limited by the liquid side, the absorption of carbon dioxide (CO_2) in caustic soda under ambient conditions has been studied, allowing the determination of the effective separation area acc. to the procedure presented in literature [3].

Figure 4 shows the dependency of the volumetric mass transfer coefficient ($k_{OV}a_e$) of the fluidized packing determined from the experiments with the standard test system CO_2/air-1n NaOH evaluated based on the HTU-NTU-model as a function of the specific liquid load u_L, see eq. (4).

$$H = HTU_{OV} \cdot NTU_{OV} = \frac{u_V}{k_{OV} \cdot a_e} \cdot NTU_{OV} \quad [m] \tag{4}$$

Based on the mass transfer measurements, the effective mass transfer area can be determined according to the following equation, presented by Duss et al. [3]:

$$a_e = \left(\frac{1}{\beta_V} + \frac{H_{pc,CO_2}}{RT \cdot \sqrt{D_{L,CO_2} \cdot k_2 \cdot C_{NaOH}}} \cdot \frac{u_V}{H} \cdot \ln\left(\frac{y_{CO_2,in}}{y_{CO_2,out}} \right) \right) \quad [m^2/m^3] \tag{5}$$

The following Figure 4 shows the calculated effective mass transfer area according to eq. 5. It shows that effective mass transfer area is increasing with increasing liquid load. In the studied range it seems to be higher than the geometrical surface area symbolized by the straight line. It has to be pointed out, that the results shown in Figure 4 have been related to the static height of the packing, which of course is lower than the expanded height in fluidized mode. As the expanded height is not constant, the static height is a good point of reference.

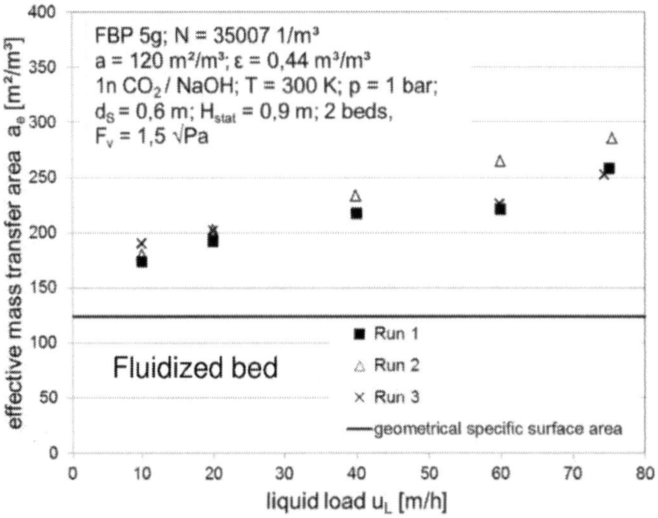

Figure 4: Effective separation area of the fluidized bed packing referred to static height, determined with the standard test system CO_2/air-1n NaOH under ambient conditions.

The gas side controlled mass transfer has been determined with the standard test system NH_3/air-1 n H_2SO_4, where NH_3 is absorbed into 1 n sulfuric acid solution, the results are shown in Figure 5.

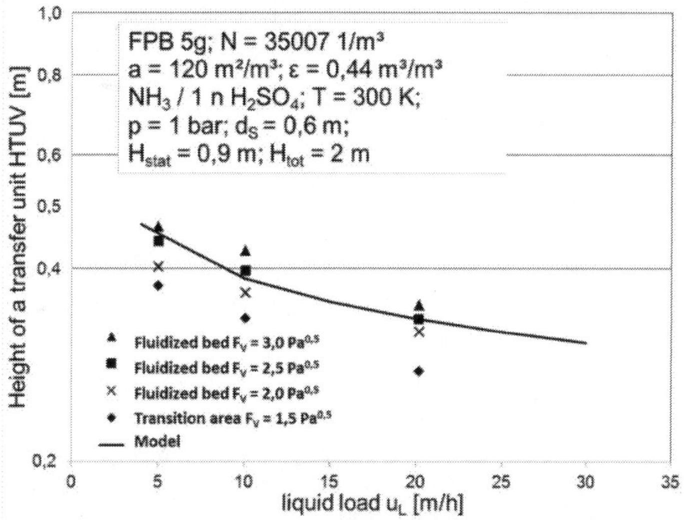

Figure 5: Height of a transfer unit (HTU_V) of the fluidized bed packing referred to static height, determined with the standard test system NH_3/air-1n H_2SO_4 under ambient conditions.

Based on the results shown in Figure 5, an empirical dimensionless model has been derived according to eq. (6).

$$\frac{HTU_V}{d_h} = 65 \times Re_L^{-0,21} \times Sc_V^{2/3} \qquad for\ 10 < Re_L < 200 \tag{6}$$

The 50 mm (2") Pall ring is one of the standard random packings often used as a benchmark for comparison of mass transfer performance. As the geometrical specific surface area of a 50 mm Pall ring is similar to the geometrical specific surface area of the studied FBP 35, it has been chosen as a benchmark packing in this work as well. In order to compare the performance of the FBP 35 with the 50 mm Pall ring, the NH_3/air-1 n H_2SO_4 has been used.

Figure 6 shows the volumetric mass transfer coefficient $k_{OV}a$ in dependency of the liquid load for the static FBP 35 bed, the FBP 25 fluidized bed referred to the total height and the 50 mm Pall ring determined at a packing height same as the total height in the fluidized bed. The diagram shows, that the mass transfer efficiency of the FBP 35 is in the same range as that of the 50 mm Pall ring.

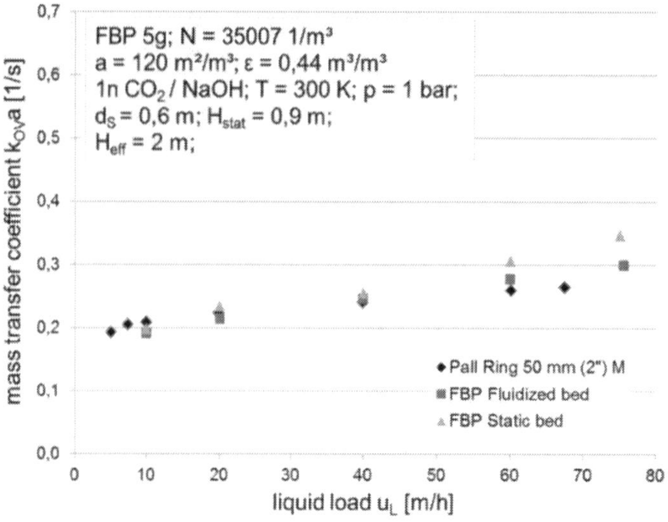

Figure 6: Mass transfer coefficient of both the static and fluidized bed packing referred to total height, determined with the standard test system NH_3/air-1n H_2SO_4 under ambient conditions, compared to the Pall ring 50 mm mass transfer coefficient, determined in the same column at the same total height.

5. Conclusions

5.1 In this work, hydraulics and mass transfer of a fluidized packed bed (FPB) have been investigated in a technical scale test column (d_C = 0,6 m). Based on the experiments, new model for fluidized packed beds describing the fluidization point and the pressure drop has been presented. Compared to 50 mm Pall rings, the pressure drop is approx. 10-20 times larger, although it remains constant in the fluidized operating range. Separation efficiency and effective mass transfer area of the FPB have been determined and compared with 50 mm Pall ring. The separation efficiency of the FPB 35 is in the same range as that of the 50 mm Pall rings.

5.2 The design rules developed in this work have been used to design an industrial scale column for a high pressure gas cleaning application with a diameter of d_C = 1,4 m. The plant is in continuous operation for more than 2 years now, all design parameters have been met, see exemplarily Figure 7.

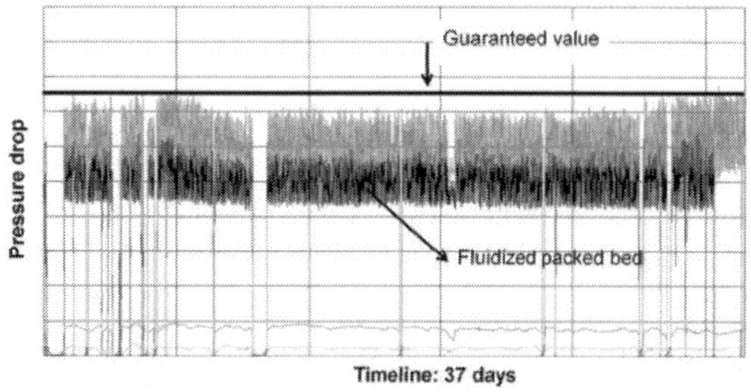

Figure 7: Pressure drop curve for over one month of industrial operation always showing values lower than guaranteed values.

5.3 Fluidized beds thus can be recommended as a solution for gas-liquid applications with a very high solid load. Compared to random packings, a significantly higher pressure drop is the "price" for the trouble-free handling and stable operation of dirty systems.

6. Literature

[1] MAĆKOWIAK, J., Fluid Dynamics of Packed Columns. Springer, Berlin, 2010.

[2] BILLET, R., Packed Towers. VCH, Weinheim, 1995.

[3] DUSS, M., MAIERHOFER, H., NUTTER, D.E., Effective interfacial area and liquid holdup of Nutter rings at high liquid loads. Chem. Eng. Technol. 24 (2001), p. 716-724.

[4] MAĆKOWIAK, J., MAĆKOWIAK, J.F. SZUST, J.: Hydraulics and mass transfer characteristics of fluidized packed beds for gas-liquid systems. In: Book of full papers pp. 600, 10th International Conference on DISTILLATION & ABSORPTION 2014, Friedrichshafen, Germany

Investigation of fluid dynamic parameters in packed distillation columns using X-ray computed tomography

S. Schug and W. Arlt

Institute of Separation Science and Technology, Friedrich-Alexander-University Erlangen-Nuremberg, Erlangen, Germany

Abstract

The flow behavior of one or more liquid phases in packed distillation columns seemed to be understood (film theory). Our investigations do not cope these findings.
In this work a X-ray customized computed tomography (CT) is used for measurements of flows in vertical objects. The detector and the X-ray source are rotating around the column. Fluid dynamic parameters in distillation columns (like interfacial area, liquid hold-up and liquid film thickness on the packing or between two liquid phases) can be directly estimated from the recorded CT images of a packing cross section at different height positions along the column during operation. For the evaluation several steps of image treatment were applied to finally obtain segmented images with specific gray values for each component.
In a former study we showed that the packing is not evenly wetted and the liquid film thickness is not constant over the entire packing. Hence we developed a method to determine the liquid distribution and the flow morphology on a packing cross section and the packing height.
Due to the fact that the packing surface is not evenly wetted we determined the interfacial area between the liquid on the packing and the surrounding to compare these values to the geometrical packing surface area. To evaluate the effect of the maldistribution of the liquid along the column height on the hold-up and interfacial area, the packing cross sections were assigned in radial sections and evaluated over the entire packing height. The flow morphology was analyzed using three dimensional visualizations on several height positions.
The final aim of this work is to develop a model for the design and process optimization of packed distillation columns, using reliable local data for the flow behavior of one or more liquid phases on a packing.

Keywords: computed tomography, fluid dynamics, distillation, packed column

1 Introduction

For the modelling of distillation processes it is necessary to determine reliable experimental data for the flow morphology and the fluid dynamics inside a packed distillation column. Due to the fact that the flow must not be influenced by any measuring device, a non-invasive measuring technique like computed tomography (CT) is required. [1] Another non-invasive measuring technique is magnetic resonance tomography but with a lower spatial resolution. Furthermore the experimental set-up has to be free of metallic components due to the strong induced magnetic field. [2]
First investigations with tomography were performed from Mewes [3] to investigate the fluid dynamic properties of two-phase flows in irregular packings.
The research group of Marchot and Toye [4-8] investigated the hold-up and interfacial area distribution of structured packings using computed tomography. The authors used a comparison between dry and wetted packing for the evaluation of the recorded CT-images. For this method it is necessary to perform two scans at exactly the same height position to avoid a shift of the packing in the images what is not fulfilled in their experimental set-up. The results show a deviation of the locally determined values from the integral values over the entire packing.

Eldridge et al. [9-12] used computed tomography to visualize the flow pattern in a packed bed of a vapor-liquid contactor. The image evaluation was also performed with a comparison of dry and wetted images at the same height positions.

2 Experimental set-up

The aim of this work is to obtain reliable local experimental data for the liquid hold-up and the wetted packing surface area. The flow morphology is evaluated using three dimensional visualizations of short packing sections. For the evaluation of the recorded CT-images a new method was established to determine all values without a comparison to dry packing images.

For the experiments a custom-designed computer tomograph (figure 1) was used which was developed and manufactured from the Fraunhofer Development Center for X-ray Technology in Fuerth. The X-ray source has an acceleration voltage of 140 kV and enables an investigation of objects with a wall thickness of an equivalent up to 40 mm aluminum. The spatial resolution is about 80 μm and the temporal resolution is 1000 projections per second, leading to one cross-sectional image every second. The great advantage of this computer tomograph is the possibility to scan upright objects without moving the object itself. During the measurement the detector and X-ray source are rotating around the column with one rotation per second and record 1000 projections. The obtained CT-images are 8-bit images and therefore they have a range of gray values between 0 and 255. [13]

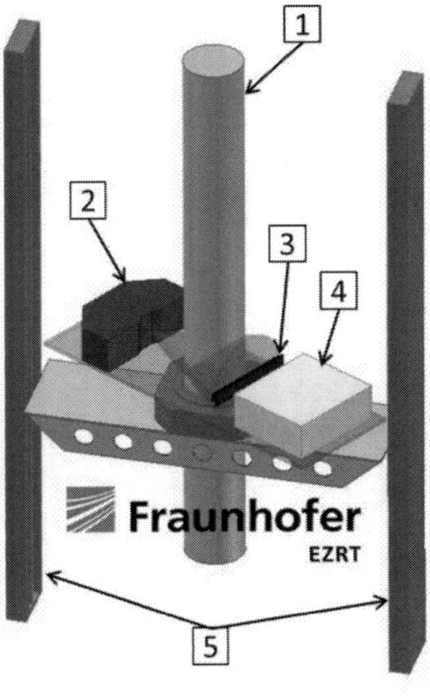

figure 1: Custom-designed computer tomograph at Institute of Separation Science and Technology, University of Erlangen-Nuremberg; 1: measuring object, 2: X-ray source, 3: detector, 4: computing device, 5: stroke axis (source: Fraunhofer EZRT, Fuerth)

The experimental set-up is shown in figure 2. For all measurements a glass column with an inner diameter of 100 mm was used. The packing height was 1000 mm consisting of five packing segments. The experiments were performed with five special designs of SULZER Mellapak 500.Y which are basically three main types: a smooth, fluted and gridded texture with holes. For the smooth and gridded ones are additional designs without holes available. Water is fed from a storage tank on the packing by a specially designed distributor and the liquid load is varied between 7 and 50 $m^3\ m^{-2}\ h^{-1}$. The CT-scanner takes images at several height positions of the packing and parameters like the liquid hold-up, interfacial area and liquid film thickness can be determined directly from the images.

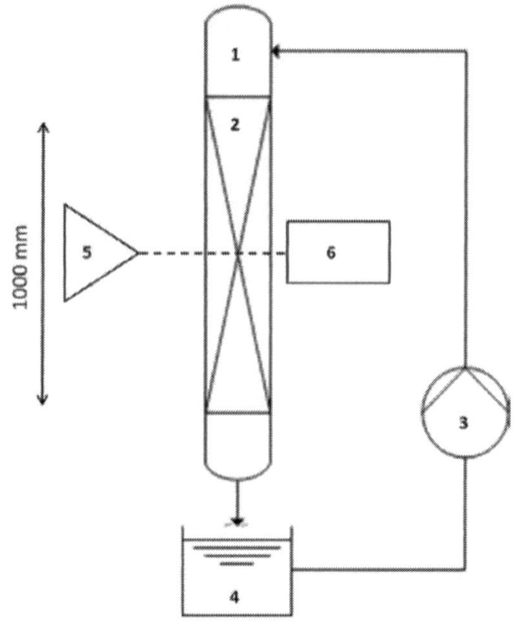

figure 2: Experimental set-up; 1: glass column, 2: packing, 3: pump, 4: storage tank, 5: x-ray source, 6: detector

The integral hold-up values were compared to predictive models from the literature. For the calculations of the hold-up the correlations from Suess and Spiegel [14] (1) and Olujic et al. [15] (2) were used.

$$h_L = c a_P^{0.83} B^x \left(\frac{\eta}{\eta_0}\right)^{0.25} \tag{1}$$

with c = 0.0169 and x = 0.37 for B < 40 $m^3\ m^{-2}\ h^{-1}$ and c = 0.0075 and x = 0.59 for B > 40 $m^3\ m^{-2}\ h^{-1}$

$$h_L = a_P \sqrt[3]{\frac{3B\eta}{a_P \rho g \sin(\alpha)}} \tag{2}$$

For the determination of the wetted packing surface area all pixel edge lengths with contact to the surrounding were counted. Pixels with two contact edges would overestimate the real shape of the contact area because of the rectangular shape of the pixels. This procedure was used by Toye [7] and therefore an empirical factor of 1.65 was considered instead of two pixel edge lengths for two contact edges.

The evaluation of the obtained CT-images was performed with the software ImageJ and MATLAB.

3 Results

On the left side in figure 3 a raw CT-image of a wetted packing is shown. The outer circle represents the glass column and the smaller inner circle is the wall wiper around the packing. The corrugated structure is the packing and the gray areas are the liquid on the packing. In this image the phase boundaries are hardly determinable, so that first of all three steps of image treatment are applied. In a first step the image is cut on the inner side of the glass column to reduce the image size. After that a mathematical filter is applied which averages every pixel at the same position over seven successively images to reduce the noise. The last step is to set an upper and lower threshold for an enhancement of the contrast and to sharp the edges of the occurring phases. The result is shown in the middle of figure 3.

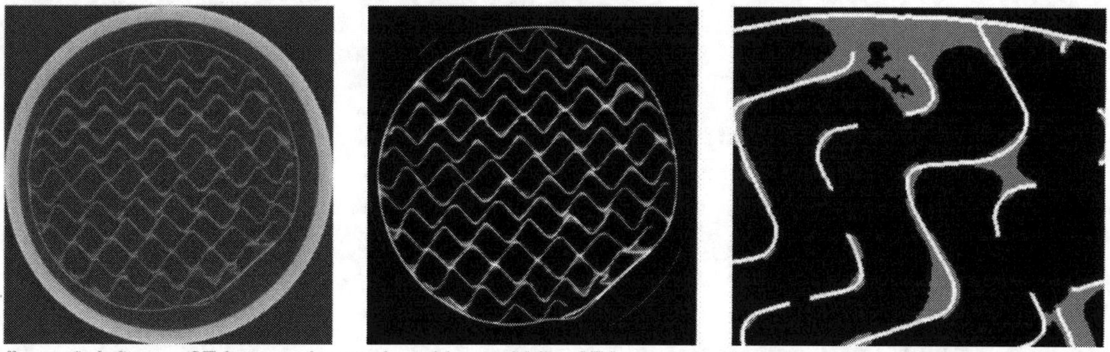

figure 3: left: raw CT-image of wetted packing, middle: CT-image after image treatment, right: enlargement of segmented CT-image

For a clearly classification of the occurring phases a segmentation algorithm was developed. It assigns the surrounding the gray value 0, the liquid 128 and the packing 255. This is represented on the right side of figure 3. The evaluation of the CT-image is now easily possible and all relevant parameters can be read out directly from the recorded images.

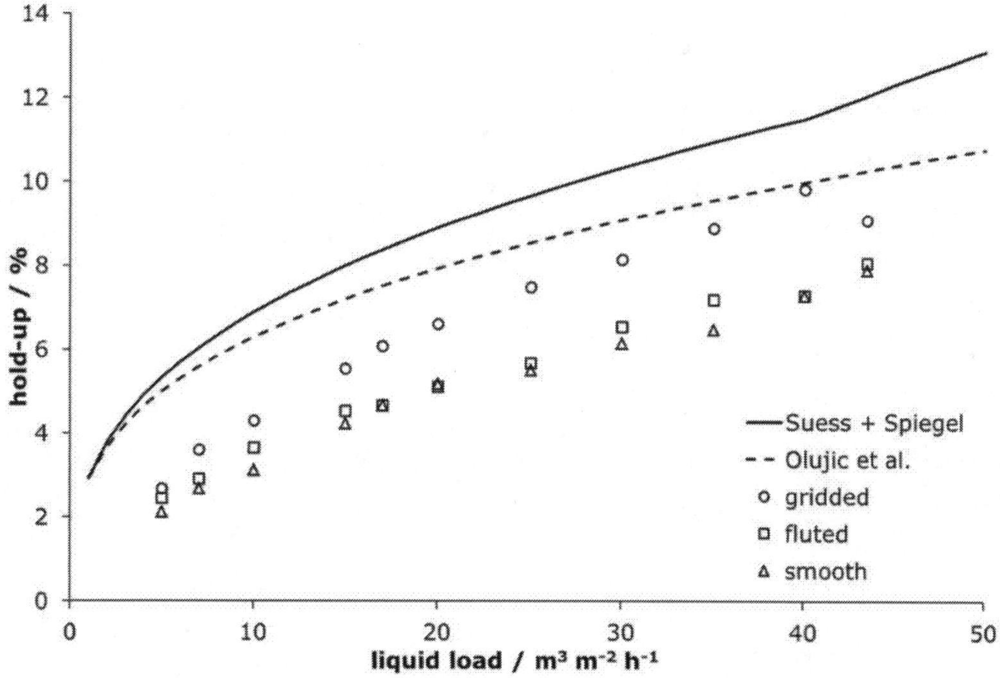

figure 4: Experimental hold-up values as function of the liquid load compared to the correlations from Olujic et al. and Suess and Spiegel

The hold-up for the three packings with holes as function of the liquid load is plotted in figure 4. The progress is as expected, with increasing liquid load the hold-up values are increasing. The standard Mellapak packing shows the highest hold-up values however the experimental data are lower than predicted with the correlations from the literature. One of several reasons for this effect is that the hold-up models are made for big columns and maldistribution and boundary effects are dominant in the smaller laboratory column which is used in this study.

For an evaluation of the local liquid distribution a hold-up height profile for the standard packing type with holes is shown in figure 5 for three different liquid loads.

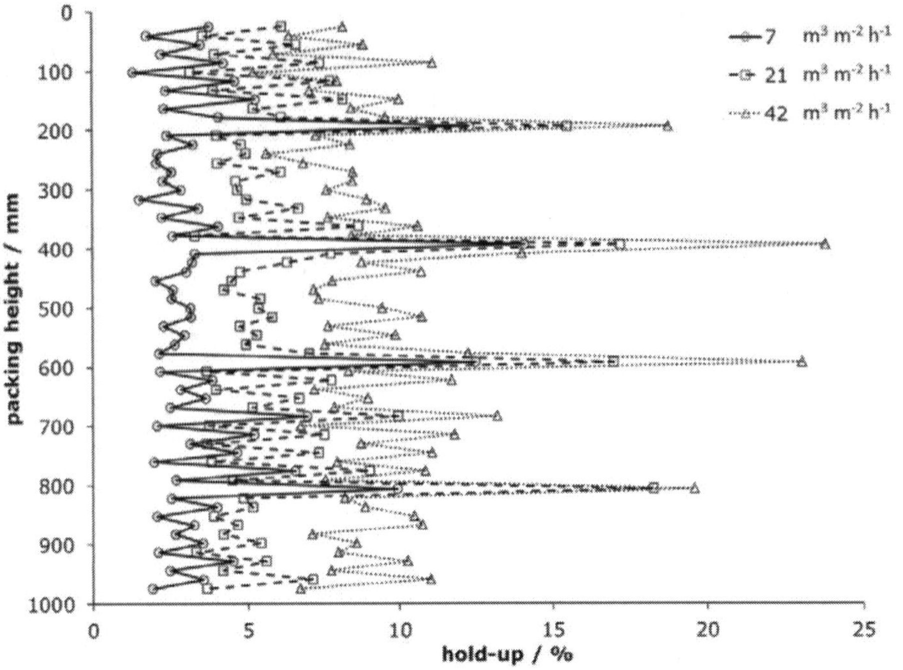

figure 5: height profile of the hold-up for three liquid loads for the standard packing type with holes

It is observable that the hold-up fluctuates in a high range along the packing height, but the minimum and maximum values always occur at the same position. With increasing liquid load the local hold-up values are increasing. The fluctuation arises from the uneven packing surface area due to the holes in the packing and the wall wiper. At measuring levels with holes in the packing and no wall wiper around low hold-up values are obtained. In contrast to positions with the full packing material and a wall wiper those yield in high hold-up values. At the intersections between two packing segments the hold-up arises to very high values. This effect can be explained with the surface tension of the liquid and the decreasing flow velocity due to the flow inversion of the liquid at the end of the packing segment. It is shown that most of the liquid is located at these intersections.

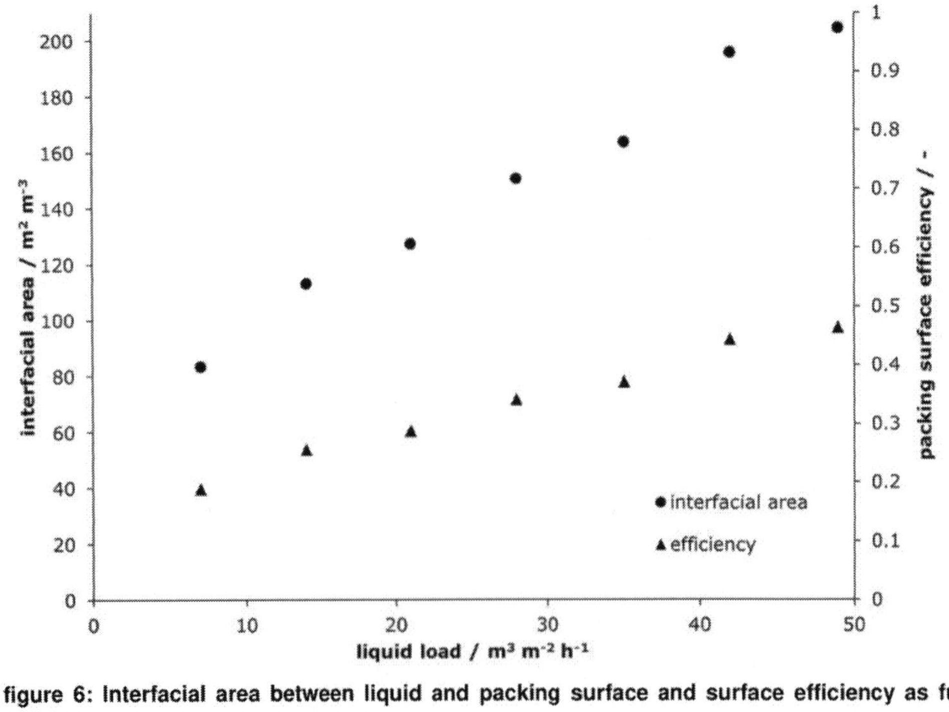

figure 6: Interfacial area between liquid and packing surface and surface efficiency as function of the liquid load on the standard packing

For an evaluation of the distribution of the liquid on the packing the packing surface efficiency was determined as ratio between wetted packing and geometrical packing surface. The result is shown in figure 6 as function of the liquid load on the Mellapak standard packing. It is observable that even at high liquid loads the wetted packing surface area is less than half of the geometrical packing surface area. The amount of liquid on the packing is in the scale as assumed due to models from the literature, so that this leads to the conclusion that most of the liquid forms not an even film on the packing instead there have to be other flow patterns. For the investigation of the flow morphology on the packing three dimensional visualizations have been generated. For a column with an inner diameter of 50 mm two visualizations are shown in figure 6 for water and n-heptane at almost the same liquid load.

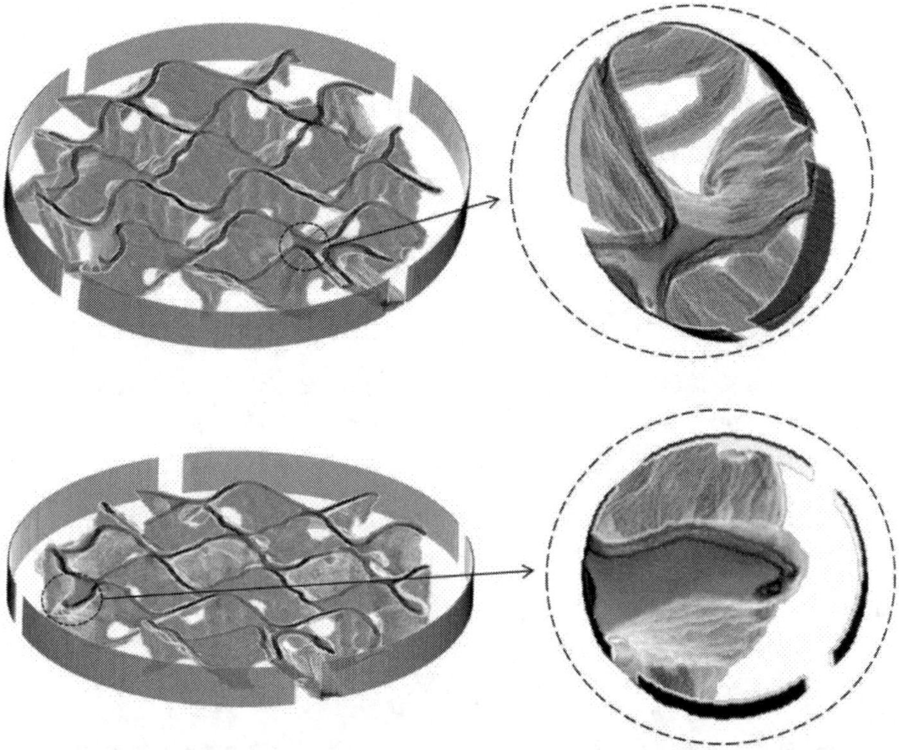

figure 7: Three dimensional visualization of a short packing section for water (above) and n-heptane (below) at 30 $m^3\ m^{-2}\ h^{-1}$

The liquid is represented in yellow / orange and the gray areas are the packing surface. It is observable that almost all liquid is flowing as rivulets in windings of the packing or between two packing plates. For n-heptane some areas with film flow can be found, but even for this organic compound with low surface tension the amount of rivulets is proportional greater than the film flow.

4 Conclusions

This study shows the successful application of X-ray computed tomography for the investigation of fluid dynamic parameters in packed distillation columns. For the interpretation of the recorded CT-images an evaluation method was developed. After the image treatment a segmentation algorithm is applied for the clearly determination of the relevant parameters out of the images. The liquid hold-up on different packing types show lower values than calculations according to correlations from the literature. Along the packing height high fluctuations of the hold-up are observable due to the uneven packing surface. The wetted packing surfaced is less than half of geometrical surface area even at high liquid loads. The three dimensional visualizations showed that most of the liquid flows as rivulet and not as film on the packing. These results are a contrast to existing assumptions in the literature.

5 Notation

a_P	packing surface area
B	liquid load
c	empiric parameter
g	gravity constant
h_L	hold-up
x	empiric parameter
α	packing inclination angle
η	dynamic viscosity
η_0	dynamic viscosity of water at 20 °C
ρ	density

6 References

[1] KALENDER W. A., (2006), *Computertomographie. Grundlagen, Gerätetechnologie, Bildqualität, Anwendungen*, Publicis Corp. Publ.

[2] LAIBLIN T., ARLT W., KAISER R., KIRSCH R., (2007), *Möglichkeit der Messung von Geschwindigkeiten in einer präparativen HPLC-Säule mit Hilfe der Magnet-Resonanz-Tomographie*, Chem. Ing. Tech., 79 1213-1216

[3] REINECKE N., MEWES D., (1997), *Investigation of the two-phase flow in trickle-bed reactors using capacitance tomography*, Chem. Ing. Tech., 52 2111-2127

[4] MARCHOT P., TOYE D., PELSSER A.-M., CRINE M., L'HOMME G., OLUJIC Z., (2001), *Liquid distribution images on structured packing by X-ray computed tomography*, AIChE J., 47 1471-1476

[5] TOYE D., CRINE M., MARCHOT P., (2005), *Imaging of liquid distribution in reactive distillation packings with a new high-energy x-ray tomograph*, Meas. Sci. Technol., 16 2213-2220

[6] AFERKA S., CRINE M., SAROHA A. K., TOYE D., Marchot P., (2007), *In situ measurements of the static liquid holdup in Katapak-SP12TMSP12TM packed column using X-ray tomography*, Chem. Eng. Sci., 62 6076-6080

[7] AFERKA S., MARCHOT P., CRINE M., TOYE D., (2010), *Interfacial area measurement in a catalytic distillation packing using high energy X-ray CT*, Chem. Eng. Sci., 65 511-516

[8] JANZEN A., STEUBE J., AFERKA S., KENIG E. Y., CRINE M., MARCHOT P., TOYE D., (2013), *Investigation of liquid flow morphology inside a structured packing using X-ray tomography*, Chem. Eng. Sci., 102 451-460

[9] SCHMIT C. E., CARTMEL D., ELDRIDGE R. B., (2000), *Process Tomography: An Option for the Enhancement of Packed Vapor–Liquid Contactor Model Development*, Ind. Eng. Chem. Res., 39 1546-1553

[10] SCHMIT C. E., CARTMEL D. B., ELDRIDGE R. B., (2001), *The experimental application of X-ray tomography to a vapor–liquid contactor*, Chem. Eng. Sci., 56 3431-3441

[11] SCHMIT C. E., ELDRIDGE R. B., (2004), *Investigation of X-ray imaging of vapor–liquid contactors. 1. Studies involving stationary objects and a simple flow system,* Chem. Eng. Sci., 59 1255-1266

[12] GREEN C. W., FARONE J., BRILEY J. K., ELDRIDGE R. B., KETCHAM R. A., NIGHTINGALE B., *Novel Application of X-ray Computed Tomography: Determination of Gas/Liquid Contact Area and Liquid Holdup in Structured Packing,* Ind. Eng. Chem. Res., 46 5734-5753

[13] SCHUG S., ARLT W., (2015), *In-situ-Stoffübergangsmessungen mit Computertomographie: Auswahl des Testsystems,* Chem. Ing. Tech., 87 656-659

[14] SUESS P., SPIEGEL L., (1992), *Hold-up of Mellapak structured packings,* Chem. Eng. Proc., 31 119-124

[15] OLUJIC Z., BEHRENS M., SPIEGEL L., (2007), *Experimental Characterization and Modeling of the Performance of a large-specific-area high-capacity Structured Packing,* Ind. Eng. Chem. Res., 46 883-893

ANALOGY OF MASS TRANSFER IN ABSORPTION AND DISTILLATION COLUMNS: WETTED WALL COLUMN STUDY

Jan Haidl[a], Frantisek J. Rejl[a], Lukas Valenz[a], Vaclav Linek[a] and Michael Schultes[b]
[a] *University of Chemistry and Technology, Prague, Czech Republic*
[b] *RASCHIG GmbH, Ludwigshafen am Rhein, Germany*

Introduction

The absorption and distillation performed in packed columns can be treated as analogical processes. In both processes the liquid flows down on the packing where it is in contact with the gas/vapor and the mass transfer occurs between the phases. The volumetric mass-transfer coefficients, k_La and k_Ga, are usually used as the mass-transfer rate parameters which relate the component flux through the gas/liquid interface and the mass-transfer driving force – difference in component concentration at the interface and in the phase bulk. The values of the mass-transfer coefficients depend on the system properties and process conditions. If the assumption of the process analogy is valid it should be possible to build identical relations for the mass-transfer coefficients calculation which would be applicable for both processes.

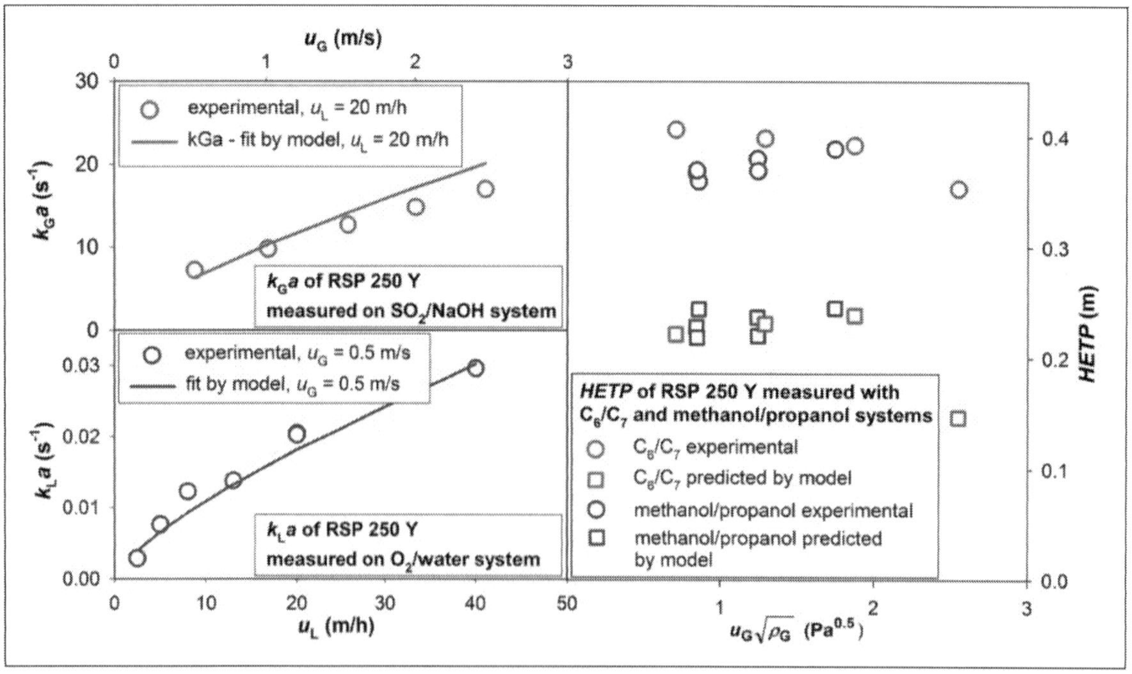

Fig. 1. Comparison of measured *HETP* values of RSP 250-Y packing with values calculated with model of Billet and Schultes[1] adjusted to the results of absorption experiments

According to the principle of the universal model the mass-transfer characteristics measured under absorption conditions can be re-calculated to the distillation ones providing reliable estimation of the packing separation performance (usually in form of *HETP*). However as it is shown in Fig. 1, when the Billet and

Schultes[1] model* is adjusted to fit well the absorption mass-transfer coefficients, then it seriously underestimates the *HETP* values for distillation systems. This discrepancy lays the questions: is it really possible to calculate the distillation performance out of the absorption data and vice versa and if so then how to do it.

This study is focused on the possibility of transfer the mass-transfer coefficients between absorption and distillation. The absorption and distillation experiments will be performed in the wetted wall column which provides identical interfacial area for both processes eliminating its possible influence on the mass-transfer coefficients.

Furthermore, being aware of the fact that the physical properties of standard absorption systems usually differ considerably from properties of standard distillation systems we chose to use such absorption systems and conditions which are similar to the distillation ones. For the scope of this paper the process conditions and system properties are mostly characterized by dimensionless Reynolds and Schmidt numbers of the phases. As the mixtures of primary alcohols are considered as systems suitable for distillation experiments, those alcohols were also used for the k_L measurements. The properties of the gas phase for the k_G measurements were varied using three gases differing considerably in density.

In this paper the mass-transfer characteristics of the wetted wall column measured under absorption conditions utilizing systems with properties close to those of distillation systems are presented. In further work these characteristics will be used for the prediction of the column distillation performance in terms of *HETP* and mass-transfer coefficients and compared to the experimentally determined values.

Experimental setup and procedures

The experiments were performed in two glass columns of 25 mm inner diameter (geometrical area a_G = 160 m^2/m^3) and lengths of 1.4 m (k_L measurements) and 2.1 m (k_G and a_e measurements). The columns are equipped with 5 (7) pairs of holes which enable placing the liquid and gas sampling devices along the column. The liquid is fed to the column by means of the needle distributor which is capable to create a stable film for the flow rates higher than 110 ml/min for pure water and 80 ml/min for NaOH solution and organic solvents. Schematic drawing of the setup is shown in Figs. 2 and 3.

k_L measurements

The k_L was measured by desorption of O_2 from the liquid into the stream of nitrogen. The absorption liquid (water, methanol, ethanol, n-propanol, n-heptane and equimolar mixture of methanol and *n*-propanol) was saturated with pure oxygen at desired temperature in the tempered and well stirred vessel and fed to the column. An additional glass exchanger was installed right in front of the liquid distributor to ensure the liquid entering the column has the desired temperature. Flow rates from 60 up to 600 ml/min and temperatures from 20 to 75 °C were utilized. The liquid samples were analyzed by means of the optical probe (distance between the sampling points, h = 0.6 m). The k_L was calculated according to the Eq. 1.

$$k_L = \frac{Q_L}{\pi \cdot d \cdot h} \ln\left(\frac{p_{O_2,top}}{p_{O_2,bottom}}\right) \qquad (1)$$

---

* The model of Billet and Schultes (1999) is chosen as it is derived as an universal mass-transfer model for absorption and distillation and enables to fit its parameters to the experimental data

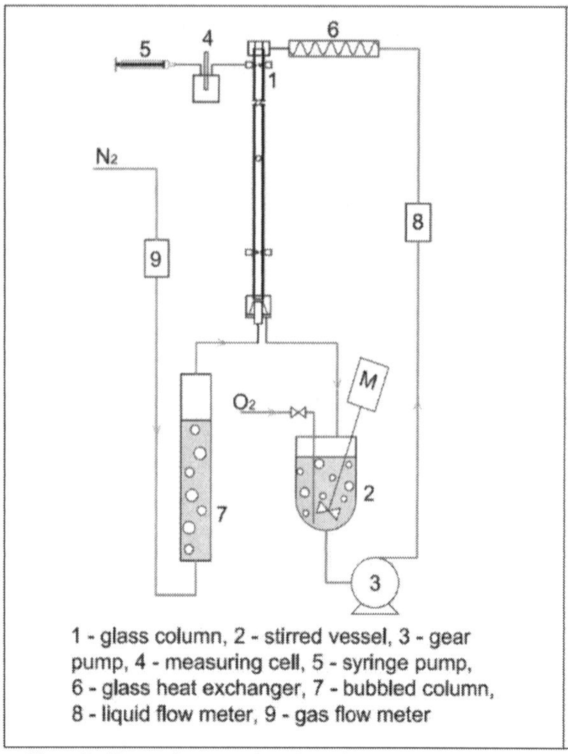

Fig. 2. Experimental setup for the absorption k_L measurements

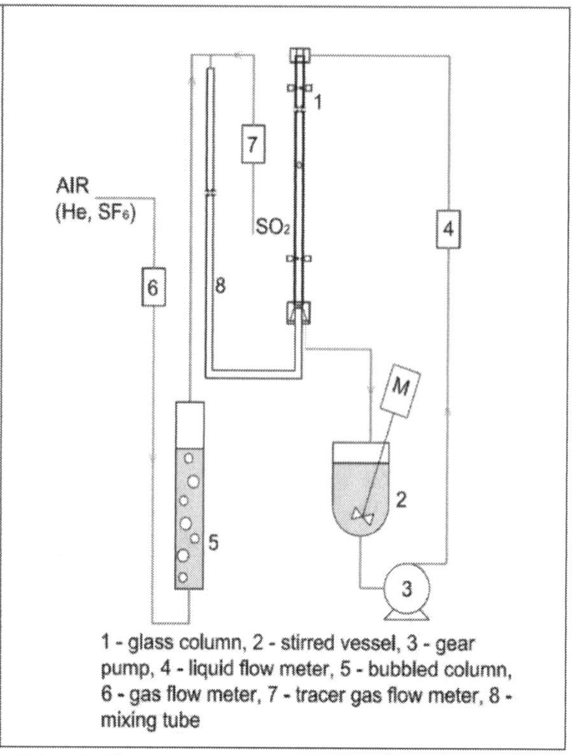

Fig. 3. Experimental setup for the absorption k_G and a_e measurements

k_G measurements

The k_G measurements were measured using the standard SO_2/NaOH system. Three carrier gases (air, He, SF_6) saturated with the vapors of water were utilized providing wide range of the gas phase physical properties (Sc_G = 0.45 – 2.75). SO_2 was injected into the carrier gas in a 2.5 m long mixing tube before entering the column. The gas samples collected from the column (sample points distance, h = 1.8 m) were analyzed by IR spectrometer. Eq. 2 was used for the k_G calculation.

$$k_G = \frac{u_G \cdot d}{4 \cdot h} \ln\left(\frac{c_{SO_2,bottom}}{c_{SO_2,top}}\right) \tag{2}$$

a_e measurements

The effective interfacial area, a_e, was measured using the air/CO_2/NaOH system. The same column setup as for the k_G measurement was used, the gas samples were analyzed by the IR spectrometer. The concentration of the NaOH (1 kmol/m^3) was determined by titration with HCl solution. The effective interfacial area was calculated by the Eq. 3. The parameter of K_a = 0.00243 m/s related to 1 kmol/m^3 NaOH solution was determined independently by still level experiments.

$$a_e = \frac{u_G}{h \cdot K_a} \ln\left(\frac{c_{CO_2,bottom}}{c_{CO_2,top}}\right) \tag{3}$$

The conditions utilized for the measurements were chosen with respect to the conditions expected in the distillation column: Re_L = 100 – 400, $Sc_L \approx 130$, Re_G = 5000 – 20000, Sc_G = 0.5 – 1.

Results

Effective Interfacial area

Effective interfacial area of the column, a_{eff}, measured with CO_2/NaOH system has been found 158 -161 m^2/m^3 for several gas and liquid velocities which is in excellent agreement with the geometrical area of the column (a_G = 160 m^2/m^3).

Mass transfer in the liquid phase

The k_L data acquired for all organic solvents at all temperatures for $Re_L > 80$ have been correlated in dimensionless form by Eq. 4

$$Sh_L = 0.33 Re_L^{-0.12} Sc_L^{0.47} \quad (4)$$

with RSD of 6 %. The Eq. 4 is valid for Re_L = 80 – 600 and Sc_L = 66 – 1200 and covers the range of conditions expected in distillation experiments.

As there is no reliable source for the diffusivities of oxygen in organic solvents at various temperatures, the diffusivities have been estimated by Tyn-Calus method as recommended by Reid et al.[2]

Our Sh_L data has been compared with those of Hikita et al.[3] and Won and Mills[4] which were measured with non-aqueous systems. The data of Hikita were found to fit well to our correlation, whereas the data of Won and Mills were found to be approx. 20 % lower than predicted from our correlation.

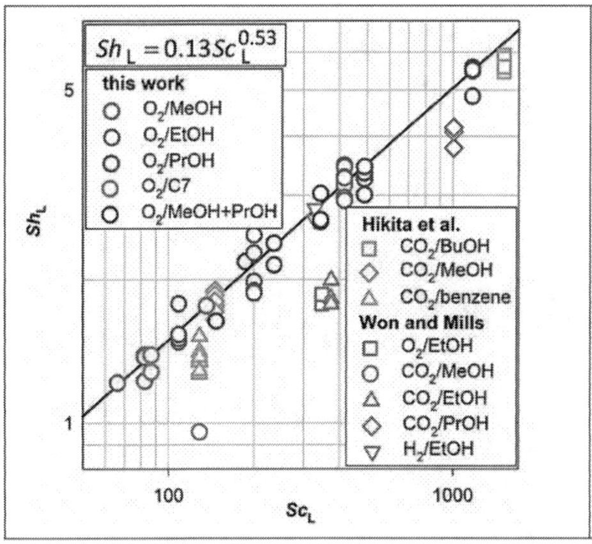

Fig. 4. The experimental Sh_L data of organic systems as a function of Sc_L. Comparison to data of Hikita et al.[3] and Won and Mills[4] obtained with non-aqueous systems

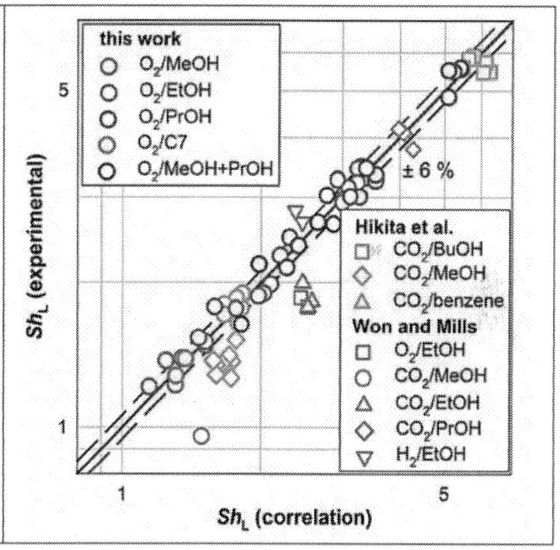

Fig. 5. The experimental Sh_L values correlated by Eq. 4 as a function of Re_L and Sc_L. Comparison to data of Hikita et al.[3] and Won and Mills[4] obtained with non-aqueous systems

The shape of the correlation determines $k_L \propto D^{0.53}$ close to the prediction of the Higgbie's penetration theory. The mass-transfer data in terms of Sh_L obtained for different solvents, flows and temperatures are fitted well by the correlation and mutually overlap, supporting the rationality of their description in the terms of chosen dimensionless numbers Re_L and Sc_L. General success of the correlation suggests that in the range of Reynolds and Schmidt numbers used it shall be valid for similar organic solvents. Agreement with the Hikita's data measured with carbon dioxide suggests its applicability also for different solutes.

Mass transfer in the gas phase

The mass transfer in the gas phase of the wetted-wall column has been studied by chemisorption of SO_2 from the air, He, and SF_6 into the aqueous solution of sodium hydroxide. The k_G data measured on all three systems were correlated together in dimensionless form resulting in Eq. 5 which is valid for Re_{LG} = 1200 - 12000 and Sc_G in the range 0.45-2.75 with RSD of 4% (see Fig. 7).

$$Sh_G = 0.012 Re_{LG}^{0.90} Sc_G^{0.61} \qquad (5)$$

In the range of its validity the correlation agrees well with the results of Gilliland and Sherwood[5] and within 20 % also with those obtained by Crause and Nieuwoudt[6] and Johnstone and Pigford[7] (see Fig. 7).

The power of Reynolds criteria 0.9 falls in between 0.68-1.05 found by the other investigators. The correlation provides 0.61 power of the Schmidt number, which is higher than that found by Gilliland and Sherwood[5] and also higher than 1/3 utilized in Chilton-Colburn analogy. The value is close, however, to the 0.566 value found by Crause and Nieuwoudt[6].

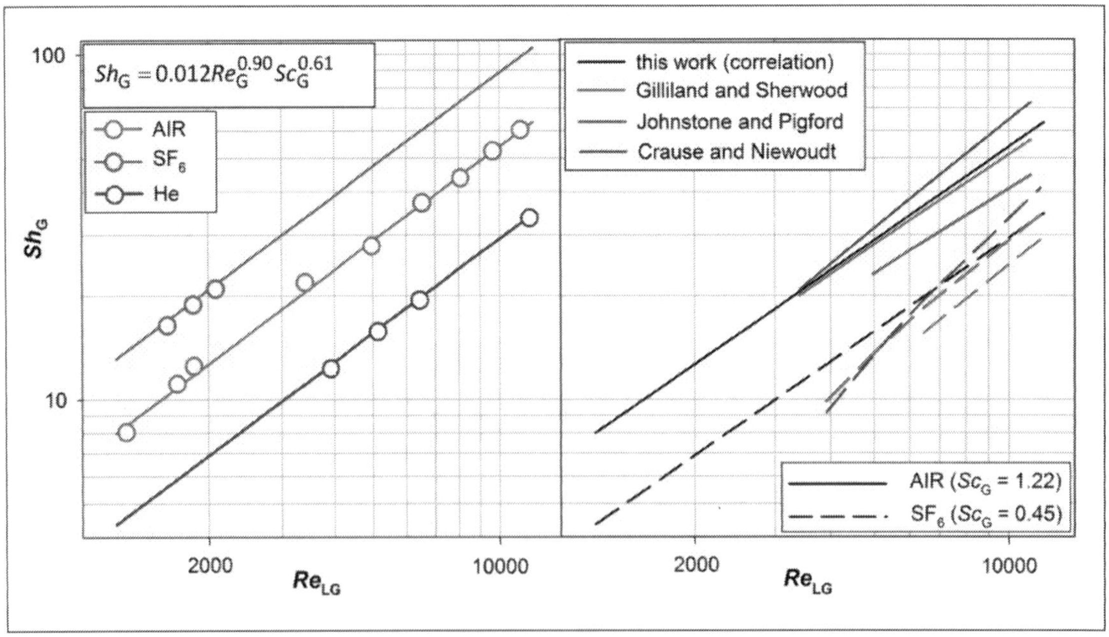

Fig. 7. The Sh_G data measured with SO_2/air, SF_6 and He systems correlated by Eq. 5. Comparison with correlations of Gilliland and Sherwood[5], Johnstone and Pigford[7] and Crause and Nieuwoudt[6]

Conclusions

This paper brings mass-transfer data of the organic liquid phase and of the several carrier gasses in a wetted wall column for absorption/desorption conditions. Such data have been only rarely provided in the literature.

There was shown that obtained correlations of the mass-transfer data in terms of the dimensionless *Sh*, *Re* and *Sc* numbers fit the mass-transfer data for any of the studied system and therefore seem to be universal. Both correlations seem to agree well with available literature data, the highest difference between experimental and literature data (30 %) was found for the k_L data measured by Won and Mills[4] with O_2/ethanol and CO_2/ethanol systems.

The aim of the work has been to obtain absorption data under conditions approaching those found for atmospheric distillation of alcoholic systems in the same column. Comparison of the data will be used in the broader scope of study of the transfer of the data between absorption and distillation. The revealed correlations predict relative resistance to the mass-transfer in the liquid up to 45 % (methanol-*n*-propanol system) which shall be sufficient for its more precise determination.

Acknowledgement

Authors are grateful to Raschig GmbH for financial support of the project, to the Czech Science Foundation for supporting the project through Grant 13-01251S, and to the financial support from specific university research (MSMT No 20/2015).

References

1. Billet, R. and Schultes, M. (1999), „Prediction of Mass Transfer Columns with Dumped and Arranged Packings. Updates Summary of the Calculation Method of Billet and Schultes.", in *Trans IChemE*, 77 (Part A), p. 498-504.

2. Reid, R.C., Prausnitz, J.M., Poling, B.E. (1987), „The Properties of Gases and Liquids, 4.th ed.", *MacGraw Hill, New York.*

3. Hikita, H., Nakaniski, K., Kataoka, T. (1959), „Liquid-phase mass transfer in wetted-wall columns.", in *Kagaku Kogaku*, 23, p. 459-466.

4. Won, Y. S. and Mills, A. F. (1982), „Correlation of the effects of viscosity and surface tension on gas absorption rates into freely falling turbulent liquid films.", in *International Journal of Heat and Mass Transfer*, 25(2), p. 223-229.

5. Gilliland, E.R. and Sherwood, T.K. (1934), „Diffusion of Vapors into Air Streams", in *Industrial and engineering chemistry*, 26(5), p. 516-523.

6. Crause, J.C. and Nieuwoudt, I. (1999), „Mass Transfer in a Short Wetted-Wall Column. 1. Pure Components", in *Ind. Eng. Chem. Res.*, 38, p. 4928-4932.

7. Johnstone, H.F. and Pigford, R.L. (1942), „Distillation in a wetted-wall column", in *Transactions of American Institute of Chemical Engineers*, 38, p. 25-51.

Symbols

a_e	effective interfacial area of the column, $m^2 \cdot m^{-3}$
a_G	geometrical area of the column, $a_G = 4/d$, $m^2 \cdot m^{-3}$
c	molar concentration, $mol \cdot m^{-3}$
D	diffusivity, $m^2 \cdot s^{-1}$
d	column diameter, m

g	gravity acceleration, 9.82 m·s^{-2}
h	height of the column, distance between sampling points, m
K_a	reactive-diffusion parameter in Eq. 3.3, K_a = 0.00243 m/s
k_G	gas phase mass-transfer coefficient, m/s
$k_G a$	volumetric gas phase mass-transfer coefficient, s^{-1}
k_L	liquid phase mass-transfer coefficient, m/s
$k_L a$	volumetric liquid phase mass-transfer coefficient, s^{-1}
p	pressure, Pa
Q	volumetric flow rate, m^3·s^{-1}
u_G	gas velocity, m/s
u_L	liquid film velocity according to the Nusselt's theory, $u_L = \left(\frac{Re_L^2 \mu_L g}{48\rho_L}\right)^{1/3}$, m/s
u_{LG}	gas veloctity relative to the liquid suface, $u_{LG} = u_L + u_G$, m/s

Greek symbols

μ	viscosity, Pa·s
ρ	density, kg·m^{-3}

Subscripts

bottom	bottom position
G	gas phase
L	liquid phase
top	top position

Dimensionless numbers

$Re_L = \frac{4 \cdot Q_L \cdot \rho_L}{\pi \cdot d \cdot \mu_L}$	liquid phase Reynolds number
$Re_{LG} = \frac{d \cdot u_{LG} \cdot \rho_G}{\mu_G}$	gas phase Reynolds number relative to the liquid surface
$Sc = \frac{\mu}{\rho \cdot D_{12}}$	Schmidt number
$Sh_G = \frac{k_G d}{D_{12,G}}$	gas phase Sherwood number
$Sh_L = \frac{k_L}{D_{12,L}}\left(\frac{\mu_L^2}{\rho_l^2 \cdot g}\right)^{1/3}$	liquid phase Sherwood number

AUTHOR INDEX